Southeast
East
Asia
EUROPE
SOUTH AMERICA
SOUTHWEST ASIA
& NORTH AFRICA
EUROPE
SAHARAN
AUSTRALIA
THE EURASIAN
REPUBLICS
South
Asia

Southeast Asia
North America
East Asia
South America
Southwest Asia & North Africa
Europe
Russia & the Eurasian Republics
Southeast Asia
Central America & the Caribbean
Sub-Saharan Africa
Australia, the Pacific Realm & Antarctica
South Asia
Southeast Asia
Central America & the Caribbean
East Asia
North America
Southwest Asia & North Africa
Sub-Saharan Africa
South America
North America

NATIONAL GEOGRAPHIC

World Cultures and Geography

WESTERN HEMISPHERE with EUROPE

Go interactive with **myNGconnect.com**

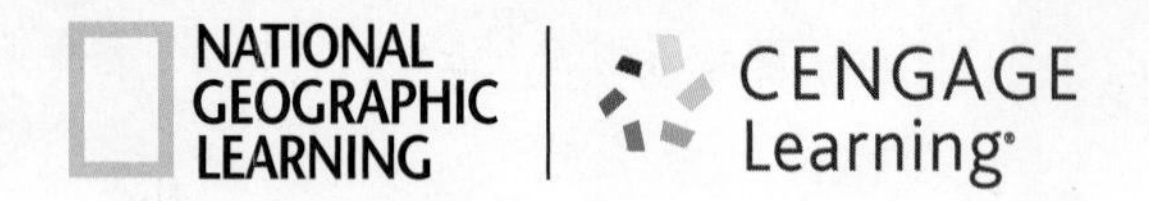

Acknowledgments

Grateful acknowledgment is given to the authors, artists, photographers, museums, publishers, and agents for permission to reprint copyrighted material. Every effort has been made to secure the appropriate permission. If any omissions have been made or if corrections are required, please contact the Publisher.

Photographic Credits

Front Cover: © Roger Ressmeyer/Corbis
Back Cover: © Neale Clark/Robert Harding World Imagery/Getty Images

Acknowledgments and credits continued on page R97.

Visit National Geographic Learning online at www.NGSP.com
Visit our corporate website at www.cengage.com

Printed in the USA
RR Donnelley

ISBN: 978-07362-8999-3

14 15 16 17 18 19 20

10 9 8 7 6 5

Andrew J. Milson

Andrew Milson is a professor of social science education and geography at the University of Texas at Arlington. He taught middle school history and geography near Dallas, Texas. Andy conducts research on geographic education and the use of geospatial technologies in education. He has published more than 30 articles and is an elected member of the Executive Board of the National Council for Geographic Education. He serves as an associate editor of the *Journal of Geography.*

Peggy Altoff

Peggy Altoff's experience includes teaching middle school and high school students, supervising teachers, and serving as adjunct university staff. Peggy served as a state social studies specialist in Baltimore and as a K–12 facilitator in Colorado Springs. She was president of the National Council for the Social Studies (NCSS) in 2006–2007 and was on the task force for the new NCSS National Curriculum Standards.

Mark H. Bockenhauer

Mark Bockenhauer is a professor of geography at St. Norbert College and a former geographer-in-residence at the National Geographic Society. Mark has extensive experience in teacher professional development. He co-wrote *Our Fifty States* and the *World Atlas for Young Explorers, 3rd edition*—both for National Geographic. Mark is coordinator of the Wisconsin Geographic Alliance, and he served as president of the National Council for Geographic Education in 2007.

Janet Smith

Jan Smith is an associate professor of geography at Shippensburg University. Jan began her teaching career as a high school teacher in Virginia where she served as a teacher consultant for the Virginia Geographic Alliance for many years. Her primary research interest focuses on how children develop their spatial thinking skills. Jan served as president of the National Council for Geographic Education in 2008, and she is currently the coordinator for the Pennsylvania Geographic Alliance.

Michael W. Smith

Michael Smith is a professor in the Department of Curriculum, Instruction, and Technology in Education at Temple University. He became a college teacher after 11 years of teaching high school English. His research focuses on how experienced readers read and talk about texts, as well as what motivates adolescents' reading and writing. Michael has written many books and monographs, including the award-winning *"Reading Don't Fix No Chevys": Literacy in the Lives of Young Men.*

David W. Moore

David Moore is a professor of education at Arizona State University. He taught high school social studies and reading before entering college teaching. He currently teaches teacher preparation courses and conducts research in adolescent literacy. David has published numerous professional articles, book chapters, and books, including *Developing Readers and Writers in the Content Areas* and *Principled Practices for Adolescent Literacy.*

Teacher Reviewers

Kayce Forbes
Deerpark Middle School
Austin, Texas

Michael Koren
Maple Dale School
Fox Point, Wisconsin

Patricia Lewis
Humble Middle School
Humble, Texas

Julie Mitchell
Lake Forest Middle School
Cleveland, Tennessee

Linda O'Connor
Northeast Independent School District
San Antonio, Texas

Leah Perry
Exploris Middle School
Raleigh, North Carolina

Robert Poirier
North Andover Middle School
North Andover, Massachusetts

Heather Rountree
Bedford Heights Elementary
Bedford, Texas

Erin Stevens
Quabbin Regional Middle/High School
Barre, Massachusetts

Beth Tipper
Crofton Middle School
Crofton, Maryland

Mary Trichel
Atascocita Middle School
Humble, Texas

Andrea Wallenbeck
Exploris Middle School
Raleigh, North Carolina

Reviewers of Religious Content

The following individuals reviewed the treatment of religious content in selected pages of the text.

Charles Haynes
First Amendment Center
Washington, D.C.

Shabbir Mansuri
Institute on Religion and Civic Values
Fountain Valley, California

Susan Mogull
Institute for Curriculum Reform
San Francisco, California

Raka Ray
Chair, Center for South Asia Studies
University of California
Berkeley, California

National Geographic Society

The National Geographic Society contributed significantly to *World Cultures and Geography.* Our collaboration with each of the following has been a pleasure and a privilege: National Geographic Maps, National Geographic Education Programs, National Geographic Missions Programs, National Geographic Digital Motion, National Geographic Digital Studio, and National Geographic Weekend. We thank the Society for its guidance and support.

Katey Walter Anthony
Aquatic Ecologist and Biogeochemist
National Geographic Emerging Explorer

Katey Walter Anthony explores ways to use a greenhouse gas for energy.

Ken Banks
Mobile Technology Innovator
National Geographic Emerging Explorer

Ken Banks develops mobile technology to connect groups in remote areas.

Katy Croff Bell
Archaeological Oceanographer
National Geographic Emerging Explorer

Katy Croff Bell uses deep sea technology to explore the depths of the ocean and the Black Sea.

Christina Conlee
Archaeologist
National Geographic Grantee

Christina Conlee researches geoglyphs and Nasca lines etched into the earth in South America.

Sylvia Earle
Oceanographer
National Geographic Explorer-in-Residence

Sylvia Earle's research focuses on exploring and preserving marine ecosystems.

SYLVIA EARLE
Oceanographer

Fredrik Hiebert
Archaeologist
National Geographic Fellow
Fredrik Hiebert uncovers mysteries of the past and has traced ancient trade routes, including the Silk Road.

Sam Meacham
Cave Diver
National Geographic Grantee
Sam Meacham explores, preserves, and maps aquifers, caves, and other groundwater sources in Mexico.

Johan Reinhard
Anthropologist
National Geographic Explorer-in-Residence
Johan Reinhard conducts field research in South America and investigates the cultural practices of Andes people.

Enric Sala
Marine Ecologist
National Geographic Explorer-in-Residence
Enric Sala dedicates his life to finding ways to reverse the damage humans have caused to the seas.

Paola Segura
Sustainable Agriculturalist
National Geographic Emerging Explorer
Paola Segura teaches sustainable farming techniques to farmers in Brazil, conserving the rain forest.

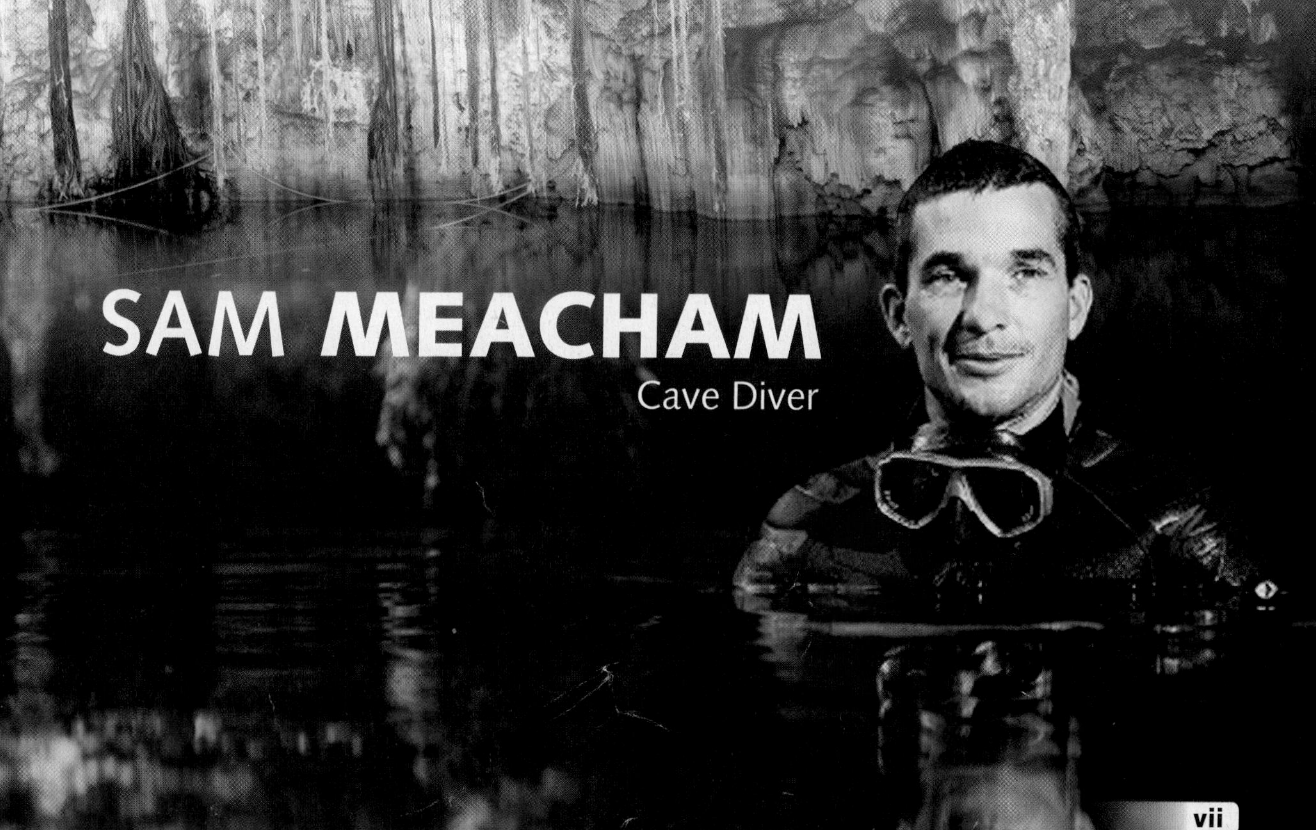

Beth Shapiro
Molecular Biologist
National Geographic Emerging Explorer

Beth Shapiro studies samples of ancient DNA to find out how species change over time.

José Urteaga
Marine Biologist and Conservationist
National Geographic Emerging Explorer

José Urteaga uses creative ideas to help save sea turtles from extinction.

Cid Simoes
Sustainable Agriculturalist
National Geographic Emerging Explorer

Cid Simoes teaches farmers in Brazil how to farm in economically efficient and ecologically sound ways.

Spencer Wells
Population Geneticist
National Geographic Explorer-in-Residence

Spencer Wells studies human migration patterns. He is the director of National Geographic's Genographic Project.

JOSÉ URTEAGA
Marine Biologist and Conservationist

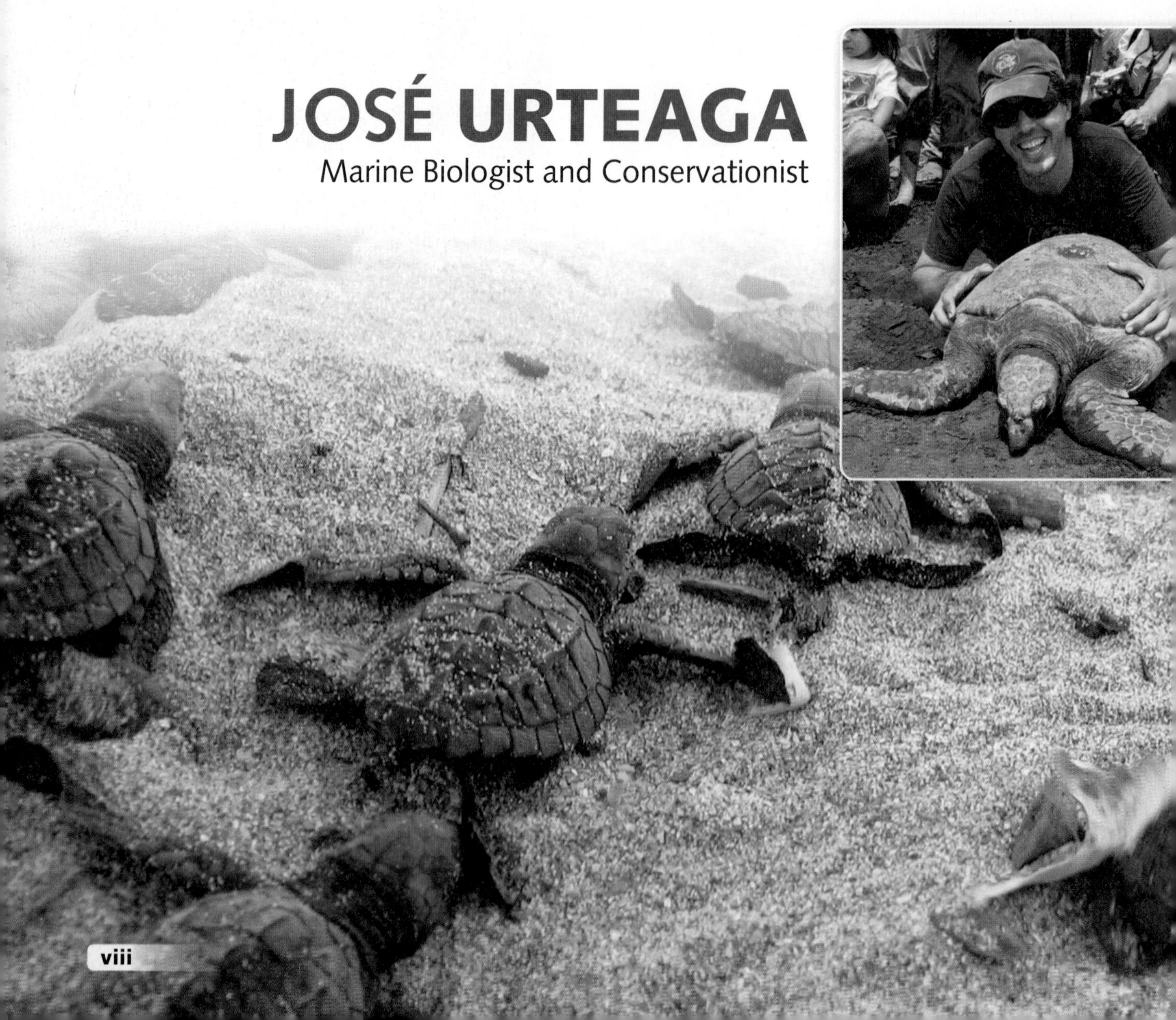

THE ESSENTIALS OF GEOGRAPHY

TECHTREK

myNGconnect.com

Digital Library
Unit 1 GeoVideo
Introduce the Essentials of Geography

Explorer Video Clip
Sylvia Earle, Oceanographer
National Geographic Explorer-in-Residence

NATIONAL GEOGRAPHIC PHOTO GALLERY
Photos of Earth's physical features and world cultures

Maps and Graphs
Interactive Map Tool

Interactive Whiteboard GeoActivities
- Draw the Stages of an Earthquake
- Build a Climograph
- Map the Spread of Buddhism

Magazine Maker
Create your own presentations

Archaeologists at work

UNIT 2

NORTH AMERICA

CHAPTER 3
North America Geography & History

CHAPTER 4

TECHTREK

myNGconnect.com

Digital Library
Unit 2 GeoVideo
Introduce North America

Explorer Video Clip
Sam Meacham, Cave Diver
National Geographic Grantee

NATIONAL GEOGRAPHIC **PHOTO GALLERY**

Regional photos, including the Grand Canyon and Great Plains, Mexico City, Vancouver, and New York City

Maps and Graphs
Interactive Map Tool

Interactive Whiteboard GeoActivities
- Compare Climates
- Illustrate the Rain Shadow Effect
- Create a Sketch Map of Tenochtitlán

Magazine Maker
Create your own presentations

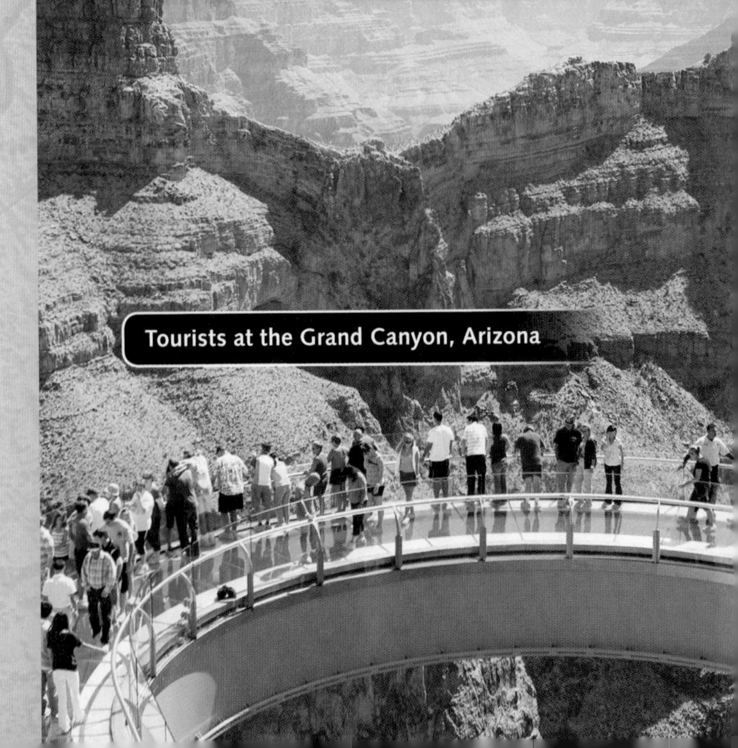

Tourists at the Grand Canyon, Arizona

UNIT 3 Central America & the Caribbean

TECHTREK 

myNGconnect.com

Mayan girl

Digital Library
Unit 3 GeoVideo
Introduce Central America & the Caribbean

Explorer Video Clip
José Urteaga, Marine Biologist and Conservationist
National Geographic Emerging Explorer

NATIONAL GEOGRAPHIC PHOTO GALLERY
Regional photos, including the rain forest, the Andes Mountains, and Montserrat

Music Clips
Audio clips of music from the region

Maps and Graphs
Outline Maps

Interactive Whiteboard GeoActivities
- Research Rain Forest Species
- Build a Time Line of Colonial Rule
- Analyze Push-Pull Factors

Connect to NG
Research links and current events in Central America and the Caribbean

UNIT 4 SOUTH AMERICA

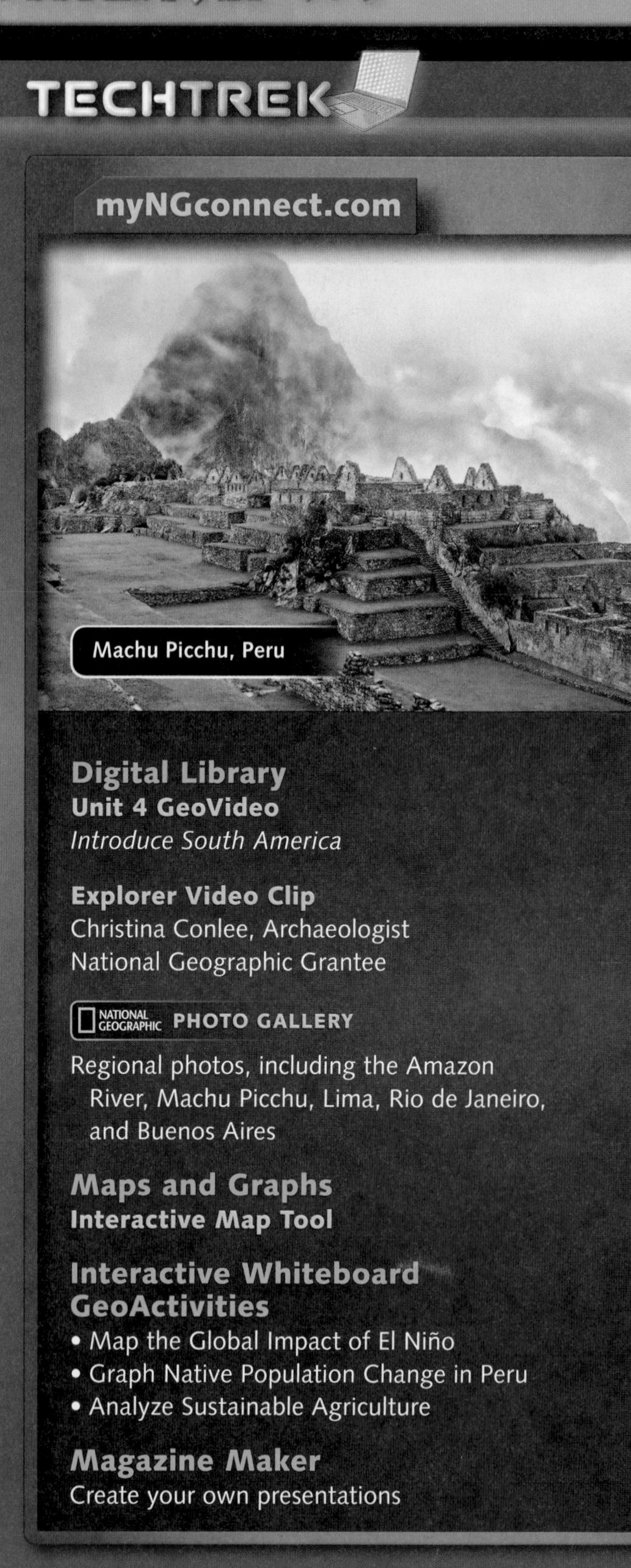

UNIT 5 EUROPE

CHAPTER 9

CHAPTER 10

TECHTREK

myNGconnect.com

Digital Library

Unit 5 GeoVideo
Introduce Europe

Explorer Video Clip
Enric Sala, Marine Ecologist
National Geographic Explorer-in-Residence

NATIONAL GEOGRAPHIC PHOTO GALLERY

Regional photos, including Florence, Paris, Amsterdam, and Budapest

Maps and Graphs

Interactive Map Tool

Interactive Whiteboard GeoActivities

- Map Europe's Land Regions
- Compare Greek and Roman Governments
- Analyze Causes and Effects of World War I

Connect to NG

Research links and current events in Europe

Louvre Museum, Paris, France

UNIT 6

Russia & THE EURASIAN REPUBLICS

CHAPTER 11

CHAPTER 12

REFERENCE

TECHTREK

myNGconnect.com

Digital Library
Unit 6 GeoVideo
Introduce Russia & the Eurasian Republics

Explorer Video Clip
Katey Walter Anthony, Aquatic Ecologist
National Geographic Emerging Explorer

NATIONAL GEOGRAPHIC PHOTO GALLERY

Regional photos, including St. Basil's Cathedral in Moscow, the Ural Mountains, Siberian tundra, and St. Petersburg

Maps and Graphs
Interactive Map Tool

Interactive Whiteboard GeoActivities
- Analyze Central Asian Economies
- Graph Napoleon's March Through Russia
- Explore the Trans-Siberian Railroad

Magazine Maker
Create your own presentations

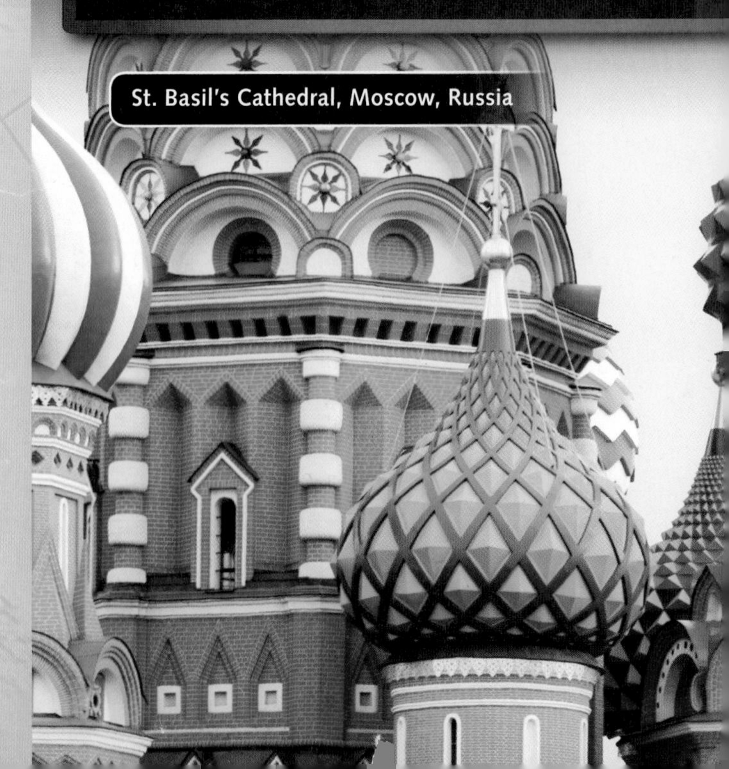

St. Basil's Cathedral, Moscow, Russia

SPECIAL FEATURES

National Geographic Explorers

Graphs

Charts and Tables

Infographics

Maps

Unit 1

UNIT 2

UNIT 3

UNIT 4

UNIT 5

UNIT 6

NATIONAL GEOGRAPHIC ATLAS

World Physical

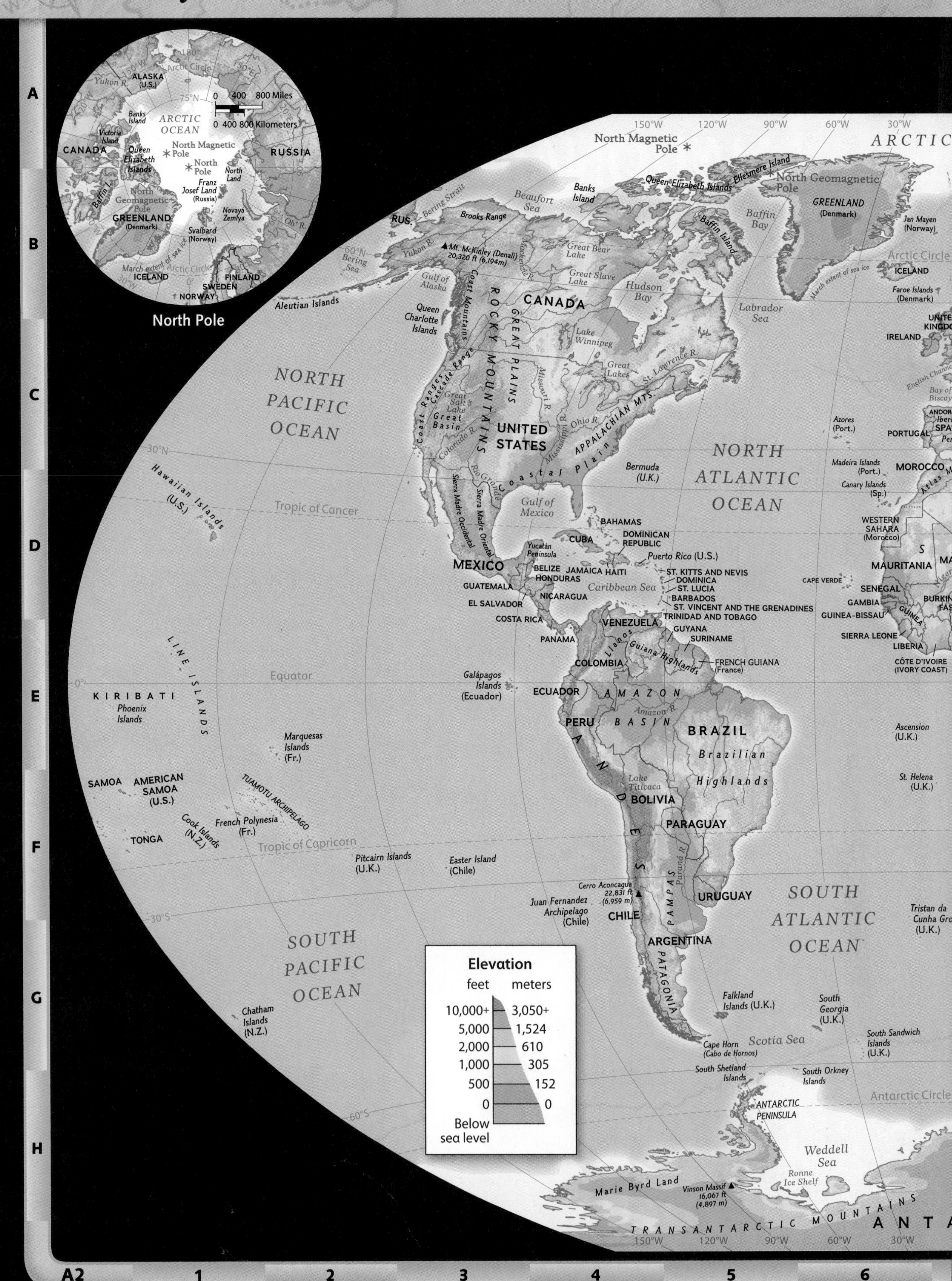

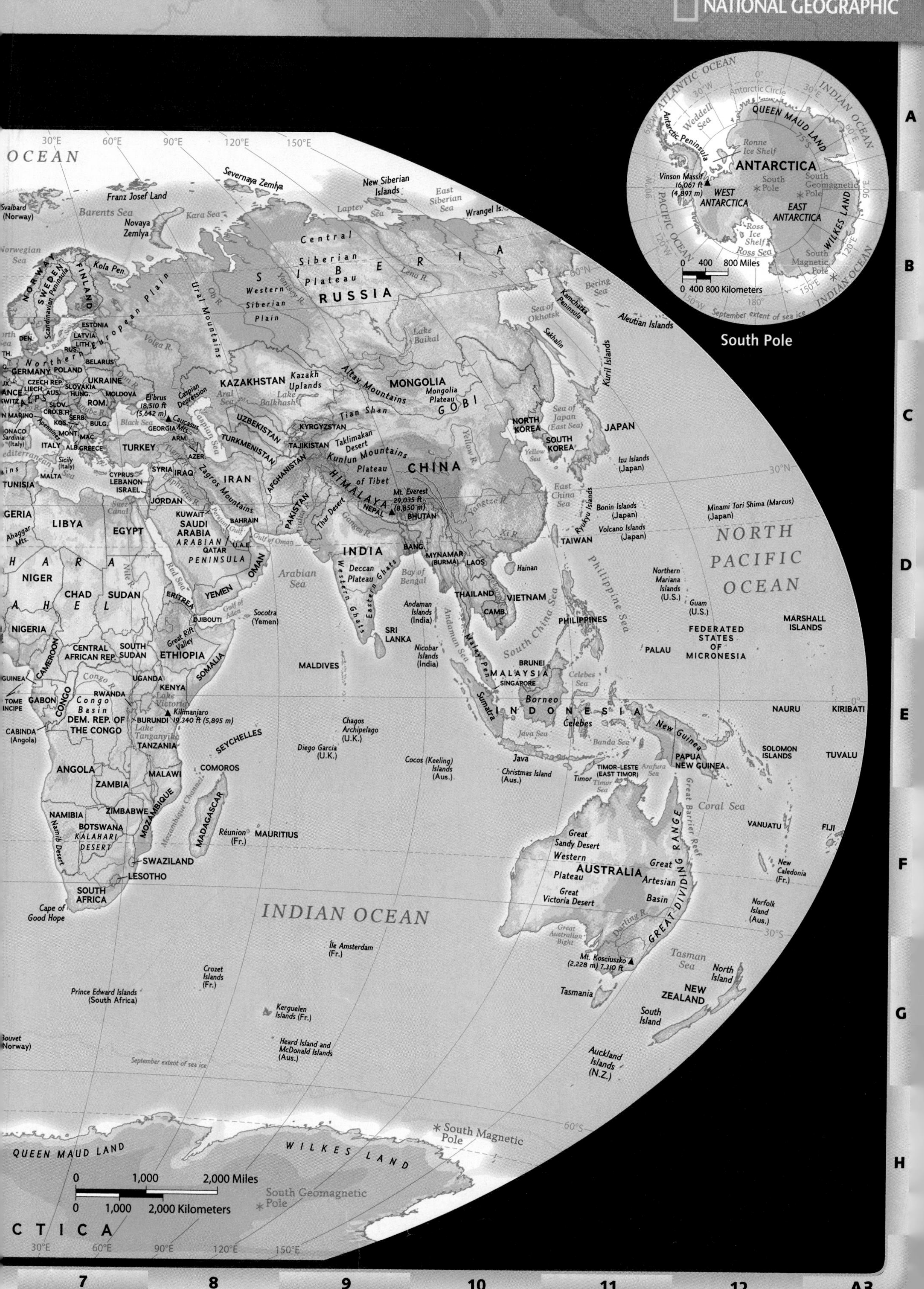
ATLANTIC OCEAN
INDIAN OCEAN
Antarctic Circle
Weddell Sea
QUEEN MAUD LAND
Antarctic Peninsula
Ronne Ice Shelf
ANTARCTICA
Vinson Massif 16,067 ft (4,897 m)
WEST ANTARCTICA
South Pole
South Geomagnetic Pole
EAST ANTARCTICA
PACIFIC OCEAN
Ross Ice Shelf
Ross Sea
WILKES LAND
South Magnetic Pole
INDIAN OCEAN
0 400 800 Miles
0 400 800 Kilometers
September extent of sea ice
South Pole
OCEAN
Svalbard (Norway)
Franz Josef Land
Barents Sea
Novaya Zemlya
Kara Sea
Severnaya Zemlya
New Siberian Islands
Laptev Sea
East Siberian Sea
Wrangel Is.
Norwegian Sea
NORWAY
SWEDEN
FINLAND
Scandinavian Peninsula
Kola Pen.
Central Siberian Plateau
SIBERIA
RUSSIA
Western Siberian Plain
Ural Mountains
Ob R.
Yenisey R.
Lena R.
Bering Sea
Kamchatka Peninsula
Sea of Okhotsk
Aleutian Islands
Sakhalin
Lake Baikal
Kuril Islands
ESTONIA
LATVIA
LITH.
RUS.
DEN.
BELARUS
GERMANY
POLAND
CZECH REP.
SLOVAKIA
UKRAINE
AUS.
HUNG.
MOLDOVA
SLOV.
CRO.
B.H.
ROM.
SERB.
KOS.
MONT.
BULG.
MAC.
ALB.
GREECE
ITALY
Sardinia (Italy)
Sicily (Italy)
MALTA
TUNISIA
Northern European Plain
Volga R.
Elbrus 18,510 ft (5,642 m)
Caucasus Mts.
GEORGIA
ARM.
AZER.
Black Sea
Caspian Depression
Caspian Sea
KAZAKHSTAN
Kazakh Uplands
Aral Sea
Lake Balkhash
Altay Mountains
MONGOLIA
Mongolia Plateau
GOBI
Tian Shan
UZBEKISTAN
TURKMENISTAN
KYRGYZSTAN
TAJIKISTAN
Taklimakan Desert
Kunlun Mountains
NORTH KOREA
SOUTH KOREA
Sea of Japan (East Sea)
JAPAN
Yellow Sea
TURKEY
CYPRUS
LEBANON
ISRAEL
SYRIA
IRAQ
IRAN
Zagros Mountains
AFGHANISTAN
Plateau of Tibet
CHINA
HIMALAYA
Mt. Everest 29,035 ft (8,850 m)
NEPAL
BHUTAN
PAKISTAN
Thar Desert
Indus R.
Ganges R.
Izu Islands (Japan)
East China Sea
Ryukyu Islands
Bonin Islands (Japan)
Volcano Islands (Japan)
Minami Tori Shima (Marcus) (Japan)
TAIWAN
JORDAN
KUWAIT
BAHRAIN
SAUDI ARABIA
ARABIAN PENINSULA
QATAR
U.A.E.
OMAN
Persian Gulf
Gulf of Oman
LIBYA
EGYPT
Suez Canal
Ahaggar Mts.
SAHARA
SAHEL
NIGER
CHAD
SUDAN
Nile R.
Red Sea
ERITREA
YEMEN
Gulf of Aden
DJIBOUTI
Socotra (Yemen)
Arabian Sea
INDIA
BANG.
Deccan Plateau
Western Ghats
Eastern Ghats
Bay of Bengal
MYANMAR (BURMA)
LAOS
Hainan
NORTH PACIFIC OCEAN
Philippine Sea
Northern Mariana Islands (U.S.)
Guam (U.S.)
THAILAND
VIETNAM
CAMB.
Andaman Islands (India)
Andaman Sea
SRI LANKA
Nicobar Islands (India)
PHILIPPINES
South China Sea
FEDERATED STATES OF MICRONESIA
PALAU
MARSHALL ISLANDS
NIGERIA
CAMEROON
CENTRAL AFRICAN REP.
SOUTH SUDAN
ETHIOPIA
Great Rift Valley
SOMALIA
MALDIVES
Malay Pen.
BRUNEI
MALAYSIA
SINGAPORE
Celebes Sea
Borneo
Sumatra
INDONESIA
Celebes
EQUATORIAL GUINEA
GABON
CONGO
SAO TOME AND PRINCIPE
Congo R.
Congo Basin
RWANDA
UGANDA
KENYA
Lake Victoria
Kilimanjaro 19,340 ft (5,895 m)
BURUNDI
DEM. REP. OF THE CONGO
Lake Tanganyika
TANZANIA
SEYCHELLES
CABINDA (Angola)
Chagos Archipelago (U.K.)
Diego Garcia (U.K.)
Java Sea
Banda Sea
New Guinea
NAURU
KIRIBATI
SOLOMON ISLANDS
TUVALU
PAPUA NEW GUINEA
Java
Cocos (Keeling) Islands (Aus.)
Christmas Island (Aus.)
Timor
TIMOR-LESTE (EAST TIMOR)
Arafura Sea
Timor Sea
ANGOLA
COMOROS
ZAMBIA
MALAWI
MOZAMBIQUE
Mozambique Channel
MADAGASCAR
NAMIBIA
ZIMBABWE
BOTSWANA
KALAHARI DESERT
Namib Desert
Réunion (Fr.)
MAURITIUS
SWAZILAND
LESOTHO
SOUTH AFRICA
Cape of Good Hope
Great Barrier Reef
Coral Sea
VANUATU
FIJI
Great Sandy Desert
Western Plateau
AUSTRALIA
Great Artesian Basin
GREAT DIVIDING RANGE
New Caledonia (Fr.)
Great Victoria Desert
Darling R.
Norfolk Island (Aus.)
INDIAN OCEAN
Île Amsterdam (Fr.)
Great Australian Bight
Mt. Kosciuszko (2,228 m) 7,310 ft
Tasman Sea
North Island
NEW ZEALAND
Crozet Islands (Fr.)
Prince Edward Islands (South Africa)
Tasmania
South Island
Kerguelen Islands (Fr.)
Bouvet (Norway)
Heard Island and McDonald Islands (Aus.)
Auckland Islands (N.Z.)
September extent of sea ice
South Magnetic Pole
QUEEN MAUD LAND
WILKES LAND
0 1,000 2,000 Miles
0 1,000 2,000 Kilometers
South Geomagnetic Pole
ANTARCTICA
30°E
60°E
90°E
120°E
150°E
60°N
30°N
0°
30°S
60°S
A
B
C
D
E
F
G
H
7
8
9
10
11
12

World Political

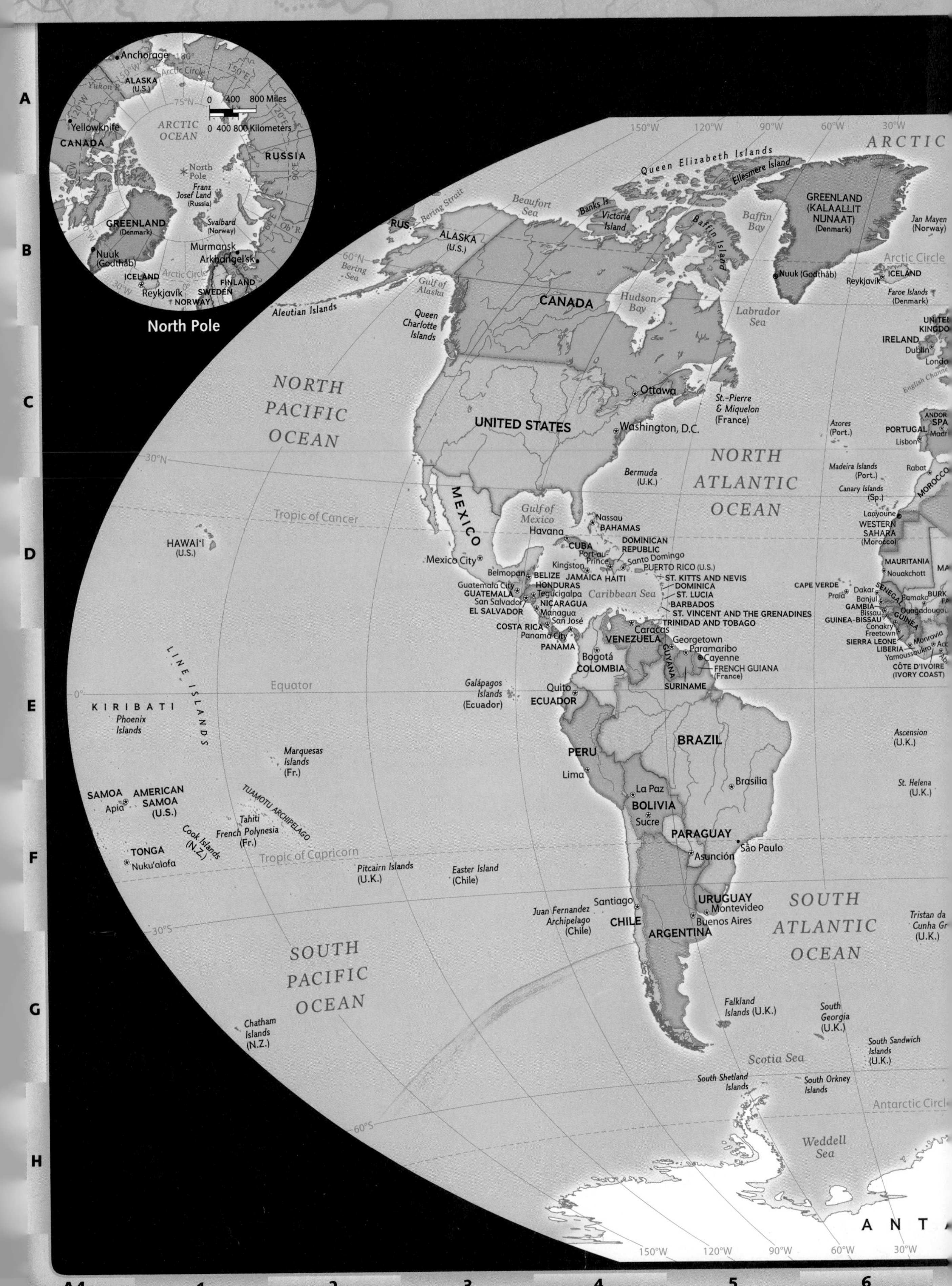

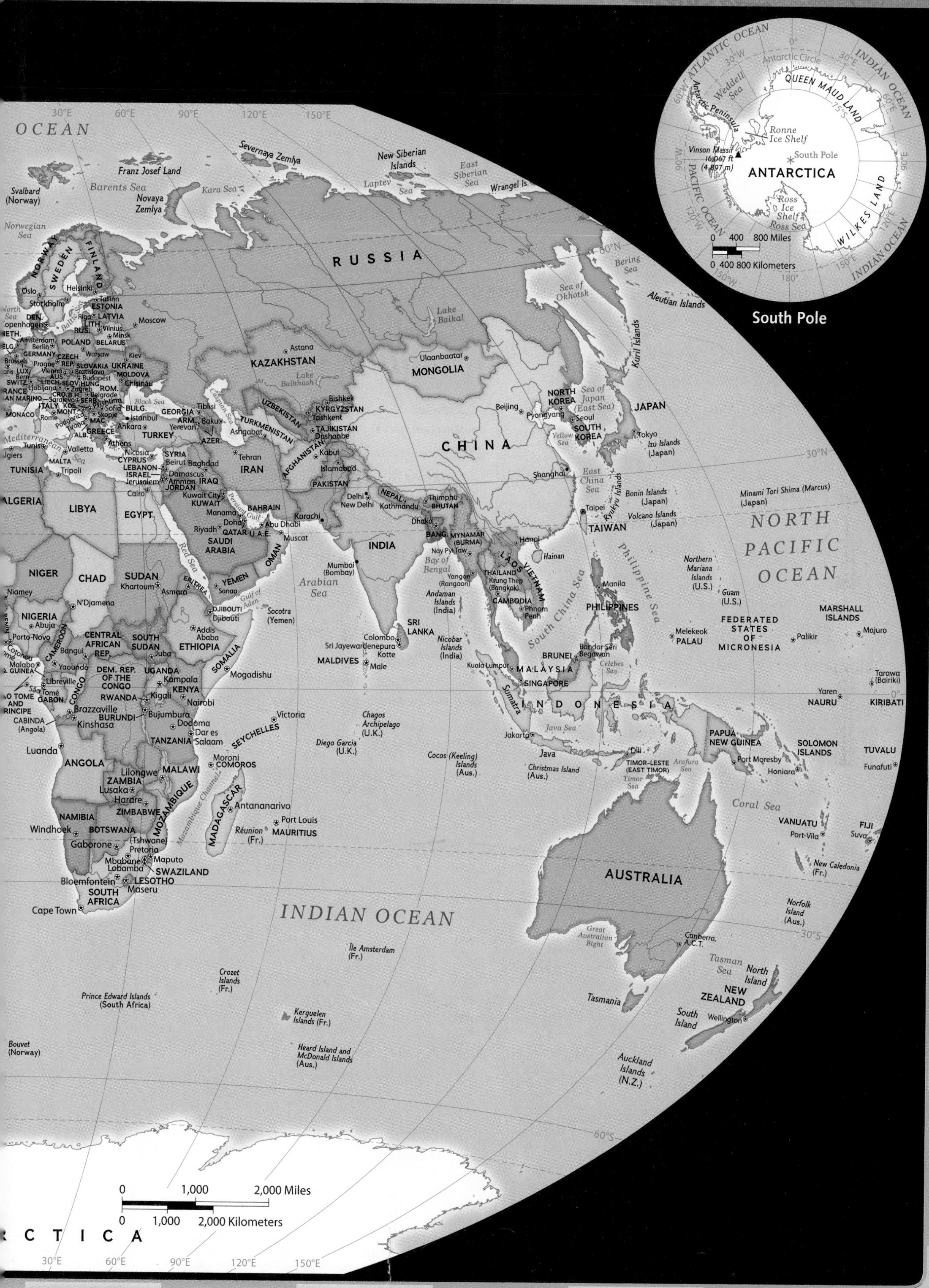

South Pole

ATLANTIC OCEAN
INDIAN OCEAN
Antarctic Circle
Weddell Sea
QUEEN MAUD LAND
Antarctic Peninsula
Ronne Ice Shelf
Vinson Massif 16,067 ft (4,897 m)
South Pole
ANTARCTICA
PACIFIC OCEAN
Ross Ice Shelf
Ross Sea
WILKES LAND
INDIAN OCEAN
0 400 800 Miles
0 400 800 Kilometers
0°
30°W
30°E
60°W
60°E
75°S
90°W
90°E
120°W
120°E
150°W
150°E
180°

OCEAN
30°E 60°E 90°E 120°E 150°E
Svalbard (Norway)
Franz Josef Land
Severnaya Zemlya
New Siberian Islands
East Siberian Sea
Laptev Sea
Wrangel Is.
Barents Sea
Kara Sea
Novaya Zemlya
Norwegian Sea
NORWAY
SWEDEN
FINLAND
RUSSIA
60°N
Bering Sea
Sea of Okhotsk
Aleutian Islands
Lake Baikal
Kuril Islands
Oslo
Stockholm
Helsinki
Tallinn
ESTONIA
LATVIA
Riga
LITH.
Vilnius
RUS.
Minsk
BELARUS
North Sea
DEN.
Copenhagen
NETH.
Amsterdam
Berlin
POLAND
Warsaw
GERMANY
CZECH REP.
Prague
Brussels
LUX.
Paris
Vienna
AUS.
Bern
SWITZ.
LIECH.
Ljubljana
SLOV.
HUNG.
Bratislava
Budapest
SLOVAKIA
UKRAINE
Kiev
MOLDOVA
Chisinau
ROM.
CRO.
B.H.
Zagreb
Belgrade
Sarajevo
SERB.
MONT.
KOS.
SAN MARINO
ITALY
MONACO
Rome
Podgorica
Tirana
Skopje
Sofia
BULG.
MAC.
ALB.
GREECE
Athens
Moscow
Black Sea
Istanbul
Ankara
TURKEY
GEORGIA
Tbilisi
ARM.
Yerevan
Baku
AZER.
Caspian Sea
Astana
KAZAKHSTAN
Lake Balkhash
UZBEKISTAN
TURKMENISTAN
Ashgabat
Tashkent
Bishkek
KYRGYZSTAN
TAJIKISTAN
Dushanbe
Ulaanbaatar
MONGOLIA
Beijing
NORTH KOREA
Pyongyang
Seoul
SOUTH KOREA
Sea of Japan (East Sea)
JAPAN
Tokyo
Izu Islands (Japan)
Yellow Sea
CHINA
30°N
Mediterranean Sea
Tunis
Valletta
MALTA
Algiers
TUNISIA
Tripoli
Nicosia
CYPRUS
LEBANON
Beirut
SYRIA
Damascus
ISRAEL
Jerusalem
Amman
JORDAN
Baghdad
IRAQ
Tehran
IRAN
AFGHANISTAN
Kabul
Islamabad
PAKISTAN
Shanghai
East China Sea
Bonin Islands (Japan)
Minami Tori Shima (Marcus) (Japan)
ALGERIA
LIBYA
Cairo
EGYPT
Kuwait City
KUWAIT
Persian Gulf
BAHRAIN
Manama
Doha
QATAR
Riyadh
U.A.E.
Abu Dhabi
SAUDI ARABIA
Muscat
OMAN
Karachi
Delhi
New Delhi
NEPAL
Kathmandu
Thimphu
BHUTAN
Dhaka
BANG.
INDIA
MYANMAR (BURMA)
Nay Pyi Taw
Hanoi
LAOS
Taipei
TAIWAN
Ryukyu Islands
Volcano Islands (Japan)
NORTH PACIFIC OCEAN
Red Sea
NIGER
CHAD
SUDAN
Khartoum
Asmara
ERITREA
YEMEN
Sanaa
Gulf of Aden
Mumbai (Bombay)
Arabian Sea
Bay of Bengal
Yangon (Rangoon)
THAILAND
Krung Thep (Bangkok)
VIETNAM
Hainan
Philippine Sea
Northern Mariana Islands (U.S.)
Niamey
N'Djamena
DJIBOUTI
Djibouti
Socotra (Yemen)
Andaman Islands (India)
CAMBODIA
Phnom Penh
South China Sea
Manila
PHILIPPINES
Guam (U.S.)
MARSHALL ISLANDS
NIGERIA
Abuja
Porto-Novo
CENTRAL AFRICAN REP.
Bangui
SOUTH SUDAN
Juba
Addis Ababa
ETHIOPIA
SOMALIA
Mogadishu
Colombo
SRI LANKA
Sri Jayewardenepura Kotte
Nicobar Islands (India)
Melekeok
PALAU
FEDERATED STATES OF MICRONESIA
Palikir
Majuro
Lomé
Cotonou
Malabo
EQ. GUINEA
CAMEROON
Yaoundé
Libreville
GABON
SAO TOME AND PRINCIPE
São Tomé
CONGO
DEM. REP. OF THE CONGO
UGANDA
Kampala
KENYA
RWANDA
Kigali
Nairobi
MALDIVES
Male
BRUNEI
Bandar Seri Begawan
Celebes Sea
Kuala Lumpur
MALAYSIA
SINGAPORE
Sumatra
Tarawa (Bairiki)
Yaren
NAURU
KIRIBATI
0°
Brazzaville
BURUNDI
Bujumbura
Kinshasa
CABINDA (Angola)
Dodoma
Dar es Salaam
TANZANIA
SEYCHELLES
Victoria
Chagos Archipelago (U.K.)
INDONESIA
Java Sea
Jakarta
Java
PAPUA NEW GUINEA
SOLOMON ISLANDS
Luanda
ANGOLA
Moroni
COMOROS
Diego Garcia (U.K.)
Cocos (Keeling) Islands (Aus.)
Christmas Island (Aus.)
Dili
TIMOR-LESTE (EAST TIMOR)
Arafura Sea
Timor Sea
Port Moresby
Honiara
TUVALU
Funafuti
Lilongwe
MALAWI
ZAMBIA
Lusaka
Harare
ZIMBABWE
MOZAMBIQUE
Mozambique Channel
MADAGASCAR
Antananarivo
Coral Sea
NAMIBIA
Windhoek
BOTSWANA
Gaborone
Pretoria (Tshwane)
Maputo
Mbabane
Lobamba
SWAZILAND
Port Louis
Réunion (Fr.)
MAURITIUS
VANUATU
Port-Vila
FIJI
Suva
New Caledonia (Fr.)
Bloemfontein
LESOTHO
Maseru
SOUTH AFRICA
Cape Town
AUSTRALIA
INDIAN OCEAN
Norfolk Island (Aus.)
Great Australian Bight
Canberra, A.C.T.
30°S
Île Amsterdam (Fr.)
Tasman Sea
North Island
NEW ZEALAND
Crozet Islands (Fr.)
Prince Edward Islands (South Africa)
Tasmania
South Island
Wellington
Kerguelen Islands (Fr.)
Bouvet (Norway)
Heard Island and McDonald Islands (Aus.)
Auckland Islands (N.Z.)
60°S

0 1,000 2,000 Miles
0 1,000 2,000 Kilometers
ANTARCTICA
30°E 60°E 90°E 120°E 150°E

A B C D E F G H

CALIFORNIA CONDOR
Condors are threatened by pesticides and habitat loss. A captive breeding program has raised numbers in the wild from 22 in 1987 to more than 160 today.

It's About Habitat

Wild animals need space to live. That means room to roam, room to find food, and room to hide. Quality habitat also means clean water and clean air. However, as the human population grows, the places where people live tend to spread into quality habitat. As the habitat is lost the number of animals decreases. If too much of a species' habitat is lost, the animal is at risk of becoming extinct. The International Union for Conservation of Nature (IUCN) works to save habitat. By identifying the level of endangerment with simple categories, the IUCN helps target the habitats that need to be conserved and the species to be protected.

NORTH AMERICA
NORTH ATLANTIC OCEAN
PACIFIC OCEAN
SOUTH AMERICA

CONSERVATION STATUS
- Critically Endangered
- Endangered
- Vulnerable

Asian elephant	Hawksbill turtle
Black rhino	Mountain gorilla
Blue whale	Polar bear
California condor	Snow leopard
Giant armadillo	Tiger
Giant panda	Whooping crane

BLUE WHALE
Commercial whaling is no longer a danger, but climate change threatens the blue whale's food source, krill. Fewer than 5,000 individuals exist in the wild.

HAWKSBILL TURTLE
The Hawksbill's habitat is severely threatened, but a thriving trade in turtle products poses a greater danger.

POLAR BEAR
Climate change is the polar bear's biggest threat. They hunt on Arctic and sub-Arctic ice flows. As the ice melts, they lose access to their main food source, seals.

TIGER

With a territory that once ranged across Asia, tigers remain in only a few pockets of the continent, primarily Southeast Asia. There may be as few as 3,200 left in the wild.

GIANT PANDA

The giant panda faces an uncertain future. Though fiercely protected, its habitat is threatened by the roads, railroads, and pollution that are part of China's expanding economy.

BLACK RHINOCEROS

Black rhinos suffered a huge decline between 1970 and 1992 because of poaching. Although in severe danger, their numbers have been slowly rising.

THE WHOOPING CRANE *Success Story*

The whooping crane was near extinction, with only about 20 birds left in the wild. Thanks to intensive recovery efforts, the species now boasts a population of over 600, with more growth expected. The species is still endangered, but scientists believe the recovery is sustainable.

North America Physical

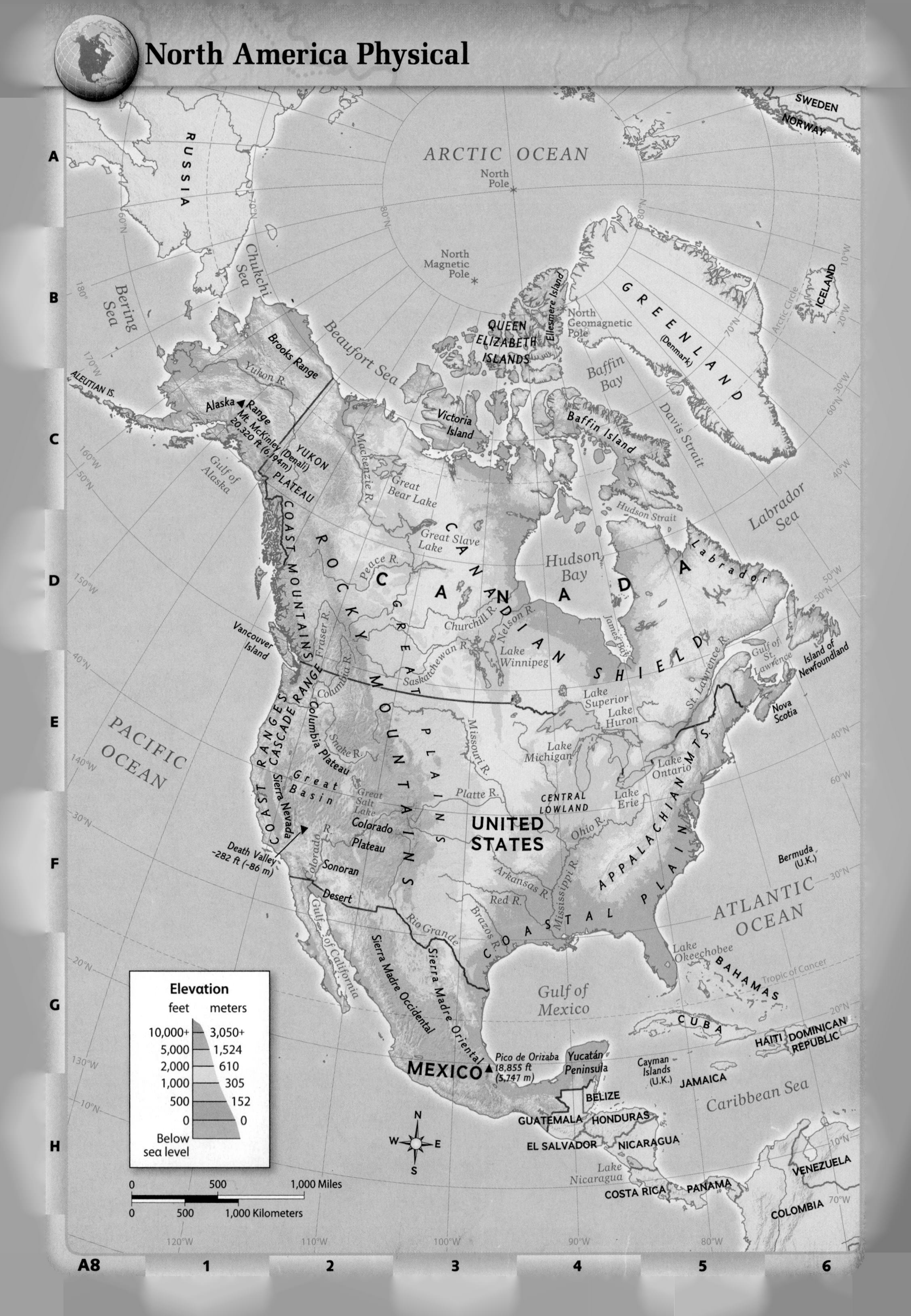

North America Political
NATIONAL GEOGRAPHIC
ARCTIC OCEAN
North Pole
RUSSIA
Chukchi Sea
Bering Sea
Beaufort Sea
ALASKA (U.S.)
Yukon R.
Anchorage
Gulf of Alaska
GREENLAND (KALAALLIT NUNAAT) (Denmark)
ICELAND
SWEDEN
NORWAY
Arctic Circle
Baffin Bay
Davis Strait
Hudson Strait
Labrador Sea
Mackenzie R.
Great Bear Lake
Great Slave Lake
Peace R.
Hudson Bay
James Bay
CANADA
Edmonton
Calgary
Churchill R.
Nelson R.
Saskatchewan R.
Lake Winnipeg
Winnipeg
Fraser R.
Columbia R.
Vancouver
Seattle
Lake Superior
Lake Huron
Lake Michigan
Lake Ontario
Lake Erie
St. Lawrence R.
Gulf of St. Lawrence
St.-Pierre & Miquelon (France)
Québec
Montréal
Ottawa
Toronto
Halifax
Boston
New York
Philadelphia
Washington, D.C.
Detroit
Chicago
Minneapolis
Missouri R.
Snake R.
Great Salt Lake
Platte R.
San Francisco
Denver
UNITED STATES
Ohio R.
Colorado R.
Los Angeles
Phoenix
Memphis
Arkansas R.
Mississippi R.
Atlanta
Red R.
Brazos R.
El Paso
Rio Grande
Houston
Miami
Bermuda (U.K.)
ATLANTIC OCEAN
BAHAMAS
Tropic of Cancer
Gulf of California
Chihuahua
Monterrey
Gulf of Mexico
MEXICO
CUBA
HAITI
DOMINICAN REPUBLIC
Cayman Islands (U.K.)
JAMAICA
PACIFIC OCEAN
Mexico City
Veracruz
Acapulco
BELIZE
GUATEMALA
HONDURAS
EL SALVADOR
NICARAGUA
Lake Nicaragua
COSTA RICA
PANAMA
Caribbean Sea
VENEZUELA
COLOMBIA
N
W
E
S
0
500
1,000 Miles
0
500
1,000 Kilometers
A
B
C
D
E
F
G
H
1
2
3
4
5
6
A9

United States Political
C A N
MEXICO
PACIFIC OCEAN
WASHINGTON
Olympia
Seattle
Portland
Salem
Columbia R.
OREGON
IDAHO
Boise
Snake R.
MONTANA
Helena
Bozeman
Missouri R.
Yellowstone R.
NORTH DAKOTA
Bismarck
James R.
SOUTH DAKOTA
Pierre
WYOMING
Cheyenne
N. Platte R.
NEBRASKA
Platte R.
S. Platte R.
NEVADA
Carson City
Las Vegas
Lake Mead
CALIFORNIA
Sacramento
San Francisco
Los Angeles
San Diego
Salton Sea
UTAH
Great Salt Lake
Salt Lake City
Green R.
Lake Powell
Colorado R.
COLORADO
Denver
Colorado Springs
Arkansas R.
KANSAS
Wichita
ARIZONA
Phoenix
Gila R.
NEW MEXICO
Santa Fe
Albuquerque
Rio Grande
El Paso
OKLAHOMA
Oklahoma City
Canadian R.
Red R.
TEXAS
Brazos R.
Pecos R.
Austin
San Antonio
Laredo
RUSSIA
ARCTIC OCEAN
Barrow
Chukchi Sea
Beaufort Sea
Arctic Circle
ALASKA
CANADA
Anchorage
Juneau
Bering Sea
Gulf of Alaska
Aleutian Islands
0 300 Miles
0 300 Kilometers
50°N
40°N
30°N
70°N
60°N
120°W
110°W
100°W
180°
170°W
160°W
150°W
140°W
130°W
A B C D E F G H
1 2 3 4 5 6

NATIONAL GEOGRAPHIC
A D A
Lake Superior
Lake Michigan
Lake Huron
Lake Ontario
Lake Erie
St. Lawrence R.
Red River of the North
MINNESOTA
Duluth
Minneapolis
St. Paul
WISCONSIN
Madison
Milwaukee
MICHIGAN
Lansing
Detroit
Grand R.
IOWA
Des Moines
Des Moines R.
Missouri R.
Mississippi R.
Lincoln
ILLINOIS
Chicago
Springfield
INDIANA
Indianapolis
Wabash R.
OHIO
Columbus
Cincinnati
Ohio R.
WEST VIRGINIA
Charleston
NEW YORK
Buffalo
Albany
Hudson R.
New York
Long Island
VERMONT
Montpelier
NEW HAMPSHIRE
Concord
MAINE
Augusta
MASSACHUSETTS
Boston
Providence
RHODE ISLAND
Hartford
CONNECTICUT
PENNSYLVANIA
Harrisburg
Philadelphia
Trenton
NEW JERSEY
Dover
DELAWARE
Baltimore
Annapolis
Washington, D.C.
MARYLAND
VIRGINIA
Richmond
James R.
Virginia Beach
Roanoke R.
opeka
Jefferson City
St. Louis
MISSOURI
Louisville
Frankfurt
KENTUCKY
Ohio R.
Cumberland R.
NORTH CAROLINA
Raleigh
Nashville
TENNESSEE
Memphis
Arkansas R.
Little Rock
ARKANSAS
Tennessee R.
SOUTH CAROLINA
Columbia
Atlanta
Savannah R.
ALABAMA
Alabama R.
Montgomery
GEORGIA
Jackson
MISSISSIPPI
LOUISIANA
Baton Rouge
New Orleans
Tallahassee
St. Johns R.
FLORIDA
Lake Okeechobee
Miami
Florida Keys
ouston
Gulf of Mexico
ATLANTIC OCEAN
BAHAMAS
Tropic of Cancer
N
E
S
W
0 150 300 Miles
0 150 300 Kilometers
INSET MAP
HAWAI'I
Honolulu
Hilo
0 100 Miles
0 100 Kilometers
22°N
20°N
160°W 158°W 156°W
90°W
80°W
70°W
50°N
40°N
30°N
A
B
C
D
E
F
G
H
7
8
9
10
11
12

Central America & the Caribbean Physical

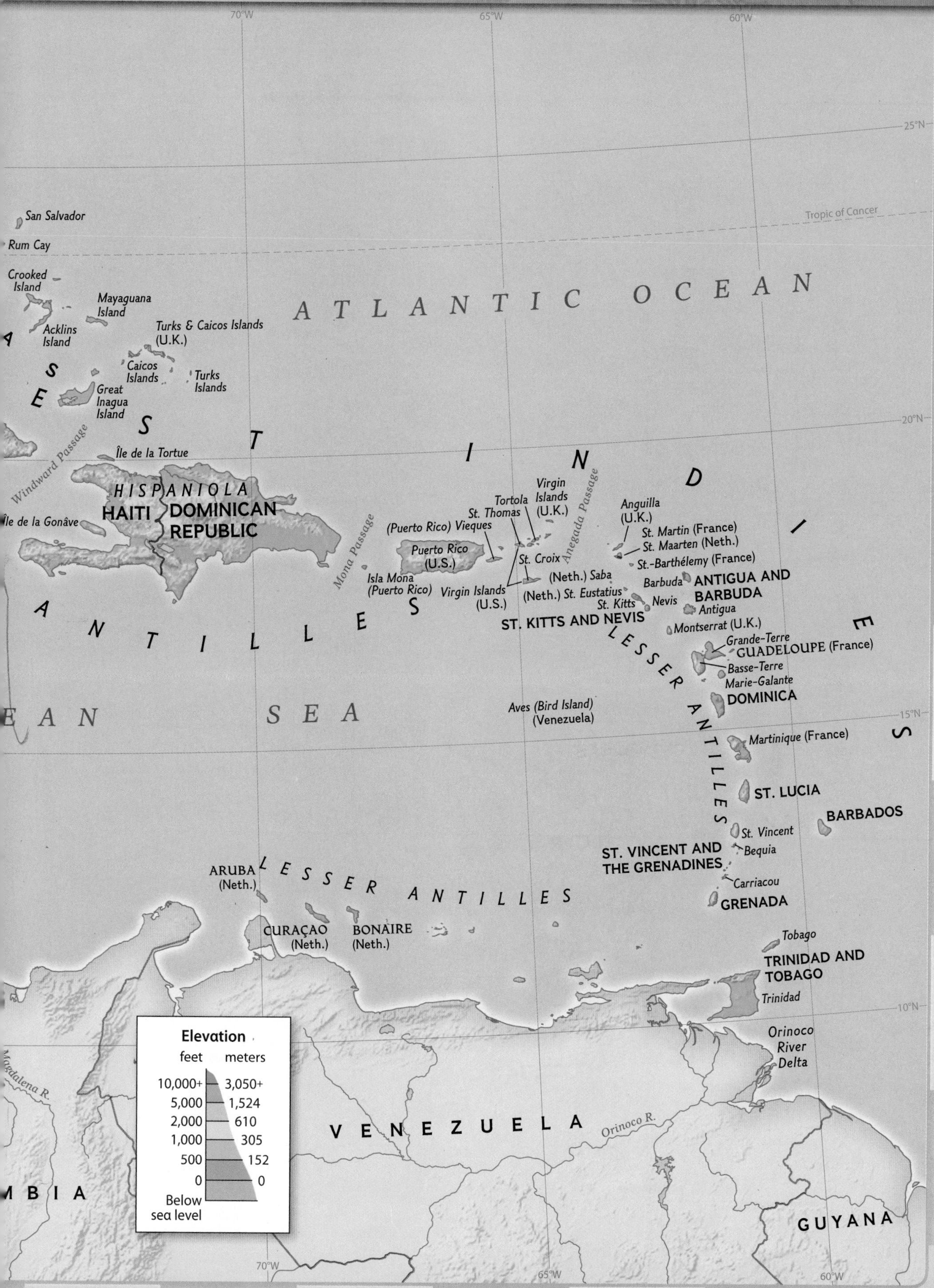

70°W
65°W
60°W
A
25°N
B
Tropic of Cancer
San Salvador
Rum Cay
Crooked Island
Mayaguana Island
Acklins Island
Turks & Caicos Islands (U.K.)
Caicos Islands
Turks Islands
Great Inagua Island
ATLANTIC OCEAN
C
20°N
Windward Passage
Île de la Tortue
W E S T I N D I E S
HISPANIOLA
HAITI
DOMINICAN REPUBLIC
Île de la Gonâve
Mona Passage
Puerto Rico (U.S.)
(Puerto Rico) Vieques
Isla Mona (Puerto Rico)
St. Thomas
Tortola
Virgin Islands (U.K.)
Anegada Passage
St. Croix
Virgin Islands (U.S.)
Anguilla (U.K.)
St. Martin (France)
St. Maarten (Neth.)
St.-Barthélemy (France)
(Neth.) Saba
(Neth.) St. Eustatius
St. Kitts
Nevis
Barbuda
ANTIGUA AND BARBUDA
Antigua
ST. KITTS AND NEVIS
Montserrat (U.K.)
D
GREATER ANTILLES
Grande-Terre
GUADELOUPE (France)
Basse-Terre
Marie-Galante
DOMINICA
LESSER ANTILLES
E
CARIBBEAN SEA
Aves (Bird Island) (Venezuela)
15°N
Martinique (France)
ST. LUCIA
BARBADOS
St. Vincent
Bequia
ST. VINCENT AND THE GRENADINES
Carriacou
GRENADA
F
ARUBA (Neth.)
LESSER ANTILLES
CURAÇAO (Neth.)
BONAIRE (Neth.)
Tobago
TRINIDAD AND TOBAGO
Trinidad
10°N
G
Orinoco River Delta
Magdalena R.
VENEZUELA
Orinoco R.
COLOMBIA
GUYANA
H

Elevation

feet	meters
10,000+	3,050+
5,000	1,524
2,000	610
1,000	305
500	152
0	0
Below sea level	

70°W
65°W
60°W

Central America & the Caribbean Political

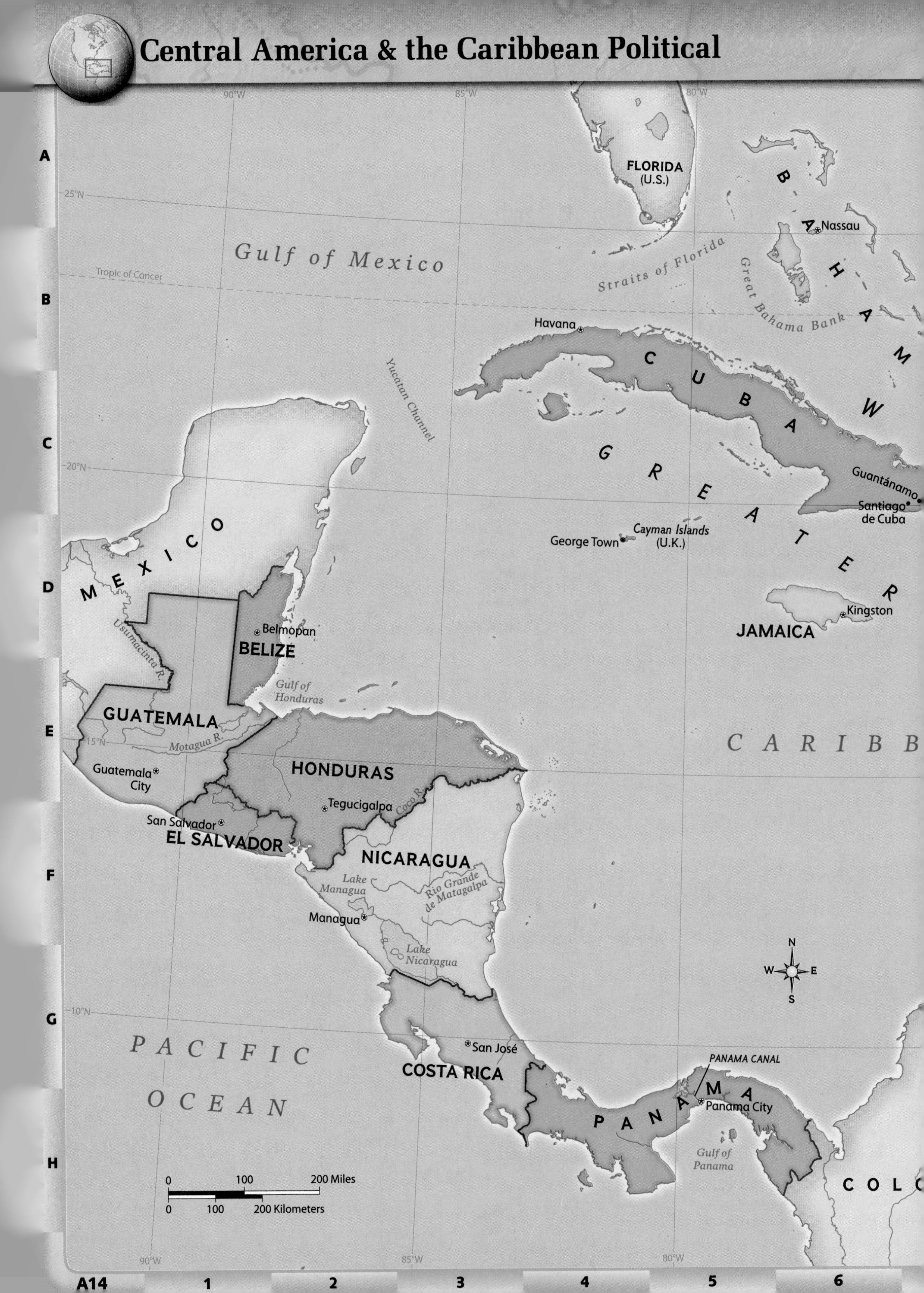

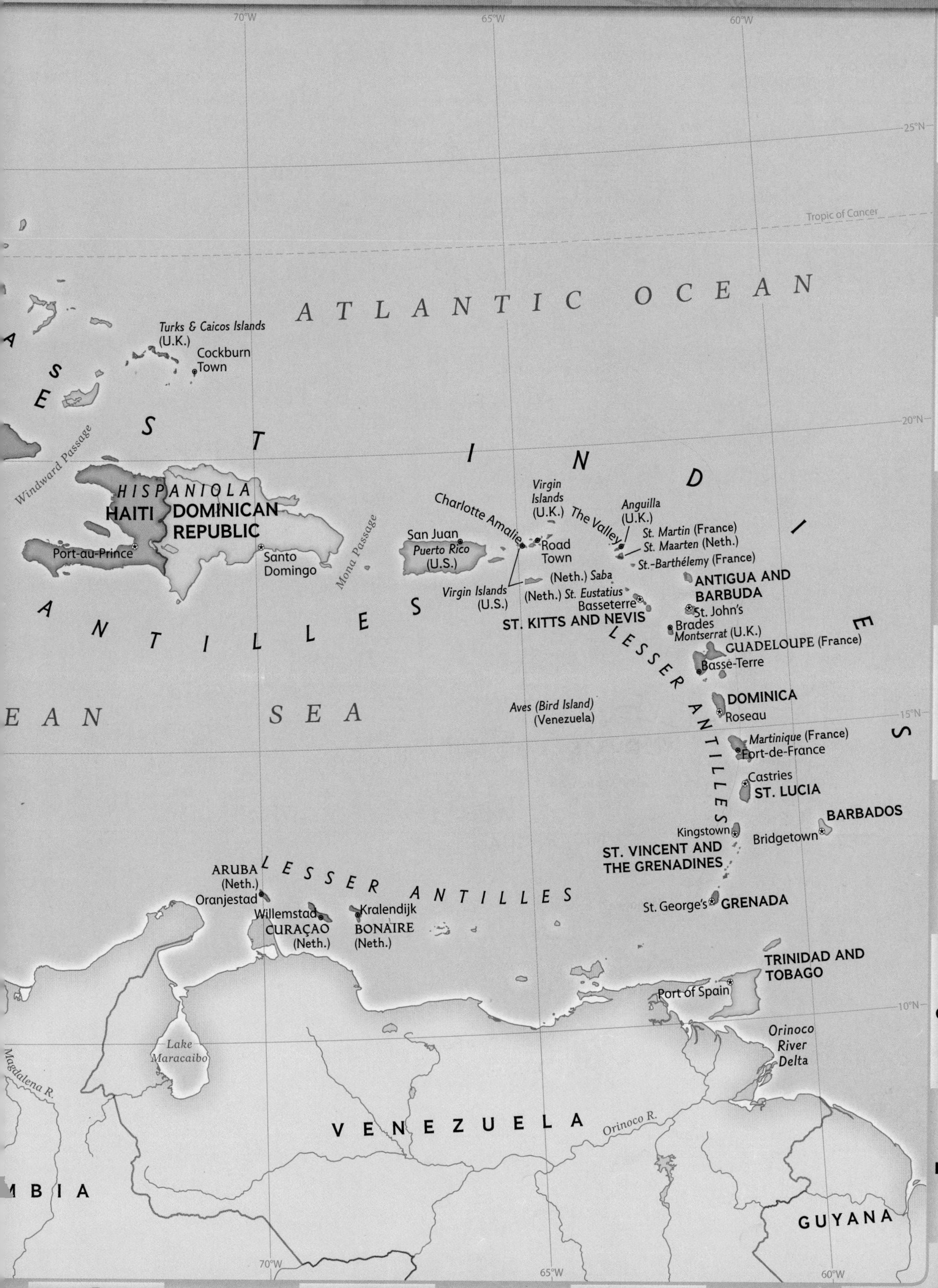

70°W
65°W
60°W
A
25°N
Tropic of Cancer
B
ATLANTIC OCEAN
Turks & Caicos Islands (U.K.)
Cockburn Town
C
20°N
WEST INDIES
Windward Passage
HISPANIOLA
HAITI
DOMINICAN REPUBLIC
Port-au-Prince
Santo Domingo
Mona Passage
San Juan
Puerto Rico (U.S.)
Charlotte Amalie
Virgin Islands (U.K.)
Road Town
The Valley
Anguilla (U.K.)
St. Martin (France)
St. Maarten (Neth.)
St.-Barthélemy (France)
D
Virgin Islands (U.S.)
(Neth.) Saba
(Neth.) St. Eustatius
Basseterre
ST. KITTS AND NEVIS
ANTIGUA AND BARBUDA
St. John's
Brades
Montserrat (U.K.)
GUADELOUPE (France)
Basse-Terre
GREATER ANTILLES
LESSER ANTILLES
CARIBBEAN SEA
Aves (Bird Island) (Venezuela)
DOMINICA
Roseau
E
15°N
Martinique (France)
Fort-de-France
Castries
ST. LUCIA
BARBADOS
Kingstown
Bridgetown
ST. VINCENT AND THE GRENADINES
F
ARUBA (Neth.)
Oranjestad
LESSER ANTILLES
Willemstad
CURAÇAO (Neth.)
Kralendijk
BONAIRE (Neth.)
St. George's
GRENADA
TRINIDAD AND TOBAGO
Port of Spain
10°N
G
Orinoco River Delta
Lake Maracaibo
Magdalena R.
VENEZUELA
Orinoco R.
H
COLOMBIA
GUYANA
70°W
65°W
60°W

South America Physical

South America Political

Europe Physical

20°E
30°E
40°E
50°E
60°E
70°E
80°E
70°N
60°N
50°N
40°N
30°N
70°E
60°E
50°E
40°E
30°E
20°E

A
B
C
D
E
F
G
H

Barents Sea
KOLA PENINSULA
White Sea
FINLAND
Gulf of Bothnia
Lake Onega
Lake Ladoga
Gulf of Finland
ESTONIA
LATVIA
LITHUANIA
BELARUS
Sea
N
AND
Vistula R.
Pechora R.
Ob R.
Irtysh R.
Northern Dvina R.
RUSSIA
EUROPEAN PLAIN
Volga R.
Kama R.
Ural R.
KAZAKHSTAN
N
E
W
S
Aral Sea
UZBEKISTAN
Don R.
Volga R.
UKRAINE
Dnieper R.
Dniester R.
CARPATHIAN MOUNTAINS
AKIA
ARY
MOLDOVA
ROMANIA
Sea of Azov
CRIMEA
Caspian Sea
TURKMENISTAN
GEORGIA
AZERBAIJAN
ARMENIA
AZERB.
Black Sea
Danube R.
SERBIA
Balkan Mountains
BULGARIA
KOSOVO
MACEDONIA
ALBANIA
Bosporus
Sea of Marmara
Dardanelles
TURKEY
GREECE
Aegean Sea
Crete (Greece)
Rhodes (Greece)
CYPRUS
LEBANON
SYRIA
Euphrates R.
Tigris R.
IRAQ
IRAN

Europe Political

Barents Sea
White Sea
FINLAND
Gulf of Bothnia
Lake Onega
Lake Ladoga
Helsinki
Gulf of Finland
Tallinn
ESTONIA
Riga
LATVIA
LITHUANIA
Kaliningrad
Vilnius
Minsk
BELARUS
Warsaw
Kraków
RUSSIA
Pechora R.
Ob R.
Irtysh R.
Northern Dvina R.
Volga R.
Kama R.
Ural R.
Don R.
Dnieper R.
KAZAKHSTAN
Aral Sea
UZBEKISTAN
TURKMENISTAN
Caspian Sea
UKRAINE
Kiev
Kharkiv
Dnipropetrovs'k
MOLDOVA
Chișinău
Sea of Azov
ROMANIA
Bucharest
Danube R.
Belgrade
SERBIA
BULGARIA
Sofia
Prishtina
KOSOVO
Skopje
MACEDONIA
Tirana
LBANIA
GREECE
Athens
Aegean Sea
Crete (Greece)
Rhodes (Greece)
Black Sea
GEORGIA
ARMENIA
AZERBAIJAN
AZERB.
TURKEY
CYPRUS
LEBANON
SYRIA
IRAQ
IRAN
Euphrates R.
Tigris R.
N
S
E
W
A
B
C
D
E
F
G
H
7
8
9
10
11
12
30°E
40°E
50°E
60°E
70°E
80°E
70°N
60°N
50°N
40°N
30°N

Russia & the Eurasian Republics Physical

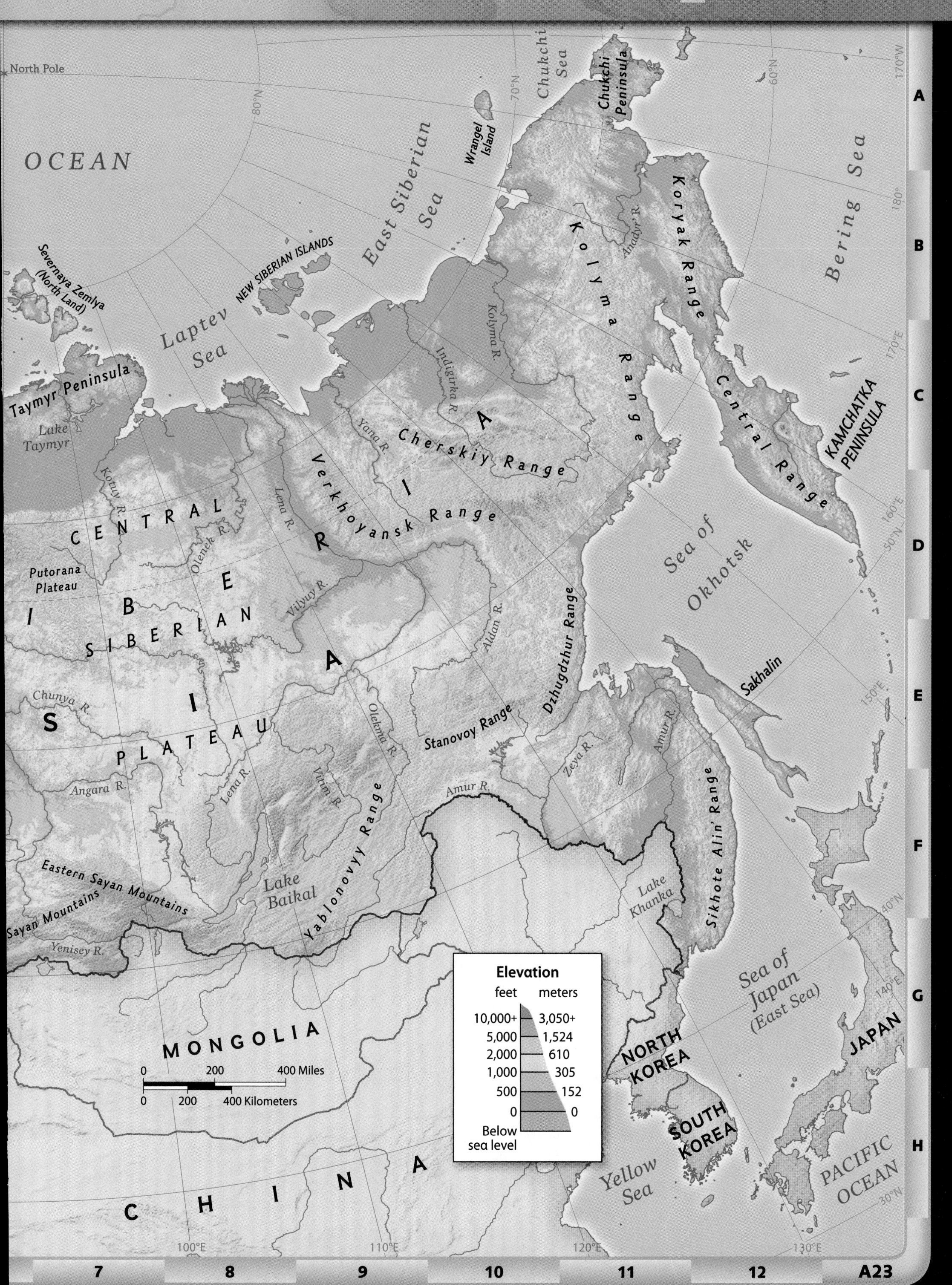

North Pole
OCEAN
Chukchi Sea
Chukchi Peninsula
Wrangel Island
East Siberian Sea
Bering Sea
Koryak Range
Kolyma Range
Anadyr' R.
Severnaya Zemlya (North Land)
NEW SIBERIAN ISLANDS
Laptev Sea
Kolyma R.
Indigirka R.
Taymyr Peninsula
Lake Taymyr
Central Range
KAMCHATKA PENINSULA
Yana R.
Cherskiy Range
Kotuy R.
Lena R.
Verkhoyansk Range
CENTRAL
Olenek R.
Sea of Okhotsk
Putorana Plateau
Vilyuy R.
SIBERIAN
SIBERIA
Aldan R.
Dzhugdzhur Range
Sakhalin
Chunya R.
Olekma R.
Stanovoy Range
PLATEAU
Zeya R.
Amur R.
Angara R.
Lena R.
Vitim R.
Yablonovyy Range
Amur R.
Sikhote Alin' Range
Eastern Sayan Mountains
Sayan Mountains
Lake Baikal
Lake Khanka
Yenisey R.
Elevation
feet meters
10,000+ 3,050+
5,000 1,524
2,000 610
1,000 305
500 152
0 0
Below sea level
Sea of Japan (East Sea)
MONGOLIA
0 200 400 Miles
0 200 400 Kilometers
NORTH KOREA
JAPAN
SOUTH KOREA
PACIFIC OCEAN
Yellow Sea
CHINA
80°N
70°N
60°N
170°W
180°
170°E
160°E
50°N
150°E
140°E
40°N
30°N
100°E
110°E
120°E
130°E
A
B
C
D
E
F
G
H
7
8
9
10
11
12

Russia & the Eurasian Republics Political

North Pole
OCEAN
Chukchi Sea
Wrangel Island
East Siberian Sea
Anadyr'
Anadyr' R.
Bering Sea
Severnaya Zemlya (North Land)
NEW SIBERIAN ISLANDS
Laptev Sea
Kolyma R.
Indigirka R.
Lake Taymyr
Yana R.
Kotuy R.
Lena R.
Magadan
Olenek R.
Sea of Okhotsk
Vilyuy R.
Yakutsk
Aldan R.
Mirny
A
I
S
Sakhalin
Chunya R.
Olekma R.
Amur R.
Zeya R.
Angara R.
Lena R.
Vitim R.
Amur R.
Lake Baikal
Irkutsk
Lake Khanka
Vladivostok
Yenisey R.
Sea of Japan (East Sea)
MONGOLIA
NORTH KOREA
JAPAN
SOUTH KOREA
PACIFIC OCEAN
Yellow Sea
C H I N A
0 200 400 Miles
0 200 400 Kilometers
80°N
70°N
60°N
50°N
40°N
30°N
170°W
180°
170°E
160°E
150°E
140°E
100°E
110°E
120°E
130°E
A
B
C
D
E
F
G
H
7
8
9
10
11
12

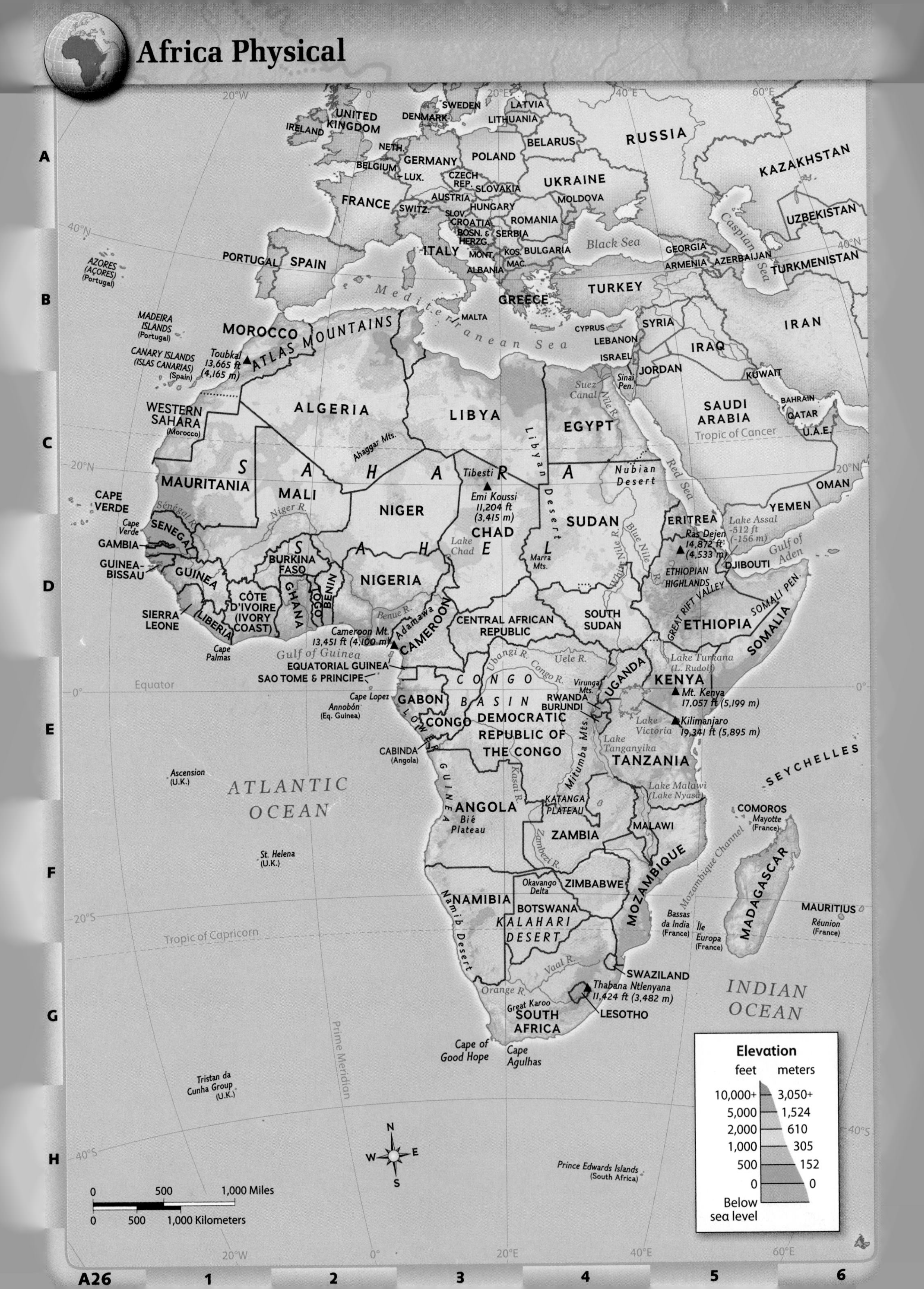
Africa Physical
UNITED KINGDOM
IRELAND
SWEDEN
DENMARK
LATVIA
LITHUANIA
RUSSIA
NETH.
GERMANY
POLAND
BELARUS
BELGIUM
LUX.
CZECH REP.
SLOVAKIA
UKRAINE
KAZAKHSTAN
FRANCE
SWITZ.
AUSTRIA
HUNGARY
MOLDOVA
SLOV.
CROATIA
ROMANIA
BOSN. & HERZG.
SERBIA
UZBEKISTAN
PORTUGAL
SPAIN
ITALY
MONT.
KOS.
BULGARIA
Black Sea
GEORGIA
ALBANIA
MAC.
ARMENIA
AZERBAIJAN
Caspian Sea
TURKMENISTAN
AZORES (AÇORES) (Portugal)
Mediterranean Sea
GREECE
TURKEY
MALTA
MADEIRA ISLANDS (Portugal)
MOROCCO
ATLAS MOUNTAINS
CYPRUS
SYRIA
IRAN
Toubkal 13,665 ft (4,165 m)
CANARY ISLANDS (ISLAS CANARIAS) (Spain)
LEBANON
ISRAEL
IRAQ
JORDAN
Suez Canal
Sinai Pen.
KUWAIT
WESTERN SAHARA (Morocco)
ALGERIA
LIBYA
EGYPT
BAHRAIN
QATAR
SAUDI ARABIA
U.A.E.
Tropic of Cancer
Ahaggar Mts.
Libyan Desert
Nile R.
Tibesti
SAHARA
Nubian Desert
Red Sea
MAURITANIA
MALI
NIGER
Emi Koussi 11,204 ft (3,415 m)
OMAN
YEMEN
CAPE VERDE
Sénégal R.
Niger R.
CHAD
SUDAN
ERITREA
Lake Assal -512 ft (-156 m)
Cape Verde
SENEGAL
Lake Chad
SAHEL
Ras Dejen 14,872 ft (4,533 m)
Gulf of Aden
GAMBIA
BURKINA FASO
Marra Mts.
White Nile R.
Blue Nile R.
DJIBOUTI
GUINEA-BISSAU
GUINEA
NIGERIA
ETHIOPIAN HIGHLANDS
CÔTE D'IVOIRE (IVORY COAST)
GHANA
TOGO
BENIN
SOUTH SUDAN
GREAT RIFT VALLEY
SOMALI PEN.
SIERRA LEONE
LIBERIA
Benue R.
Adamawa
CENTRAL AFRICAN REPUBLIC
ETHIOPIA
SOMALIA
Cameroon Mt. 13,451 ft (4,100 m)
CAMEROON
Cape Palmas
Gulf of Guinea
Ubangi R.
Uele R.
Lake Turkana (L. Rudolf)
EQUATORIAL GUINEA
SAO TOME & PRINCIPE
CONGO BASIN
Congo R.
Virunga Mts.
UGANDA
KENYA
Equator
Cape Lopez
GABON
RWANDA
BURUNDI
Mt. Kenya 17,057 ft (5,199 m)
Annobón (Eq. Guinea)
LOWER GUINEA
CONGO
DEMOCRATIC REPUBLIC OF THE CONGO
Lake Victoria
Kilimanjaro 19,341 ft (5,895 m)
Mitumba Mts.
CABINDA (Angola)
Lake Tanganyika
TANZANIA
SEYCHELLES
Ascension (U.K.)
ATLANTIC OCEAN
Kasai R.
Lake Malawi (Lake Nyasa)
KATANGA PLATEAU
ANGOLA
Bié Plateau
COMOROS
Mayotte (France)
MALAWI
ZAMBIA
Zambezi R.
Mozambique Channel
St. Helena (U.K.)
MOZAMBIQUE
MADAGASCAR
Okavango Delta
ZIMBABWE
NAMIBIA
BOTSWANA
Namib Desert
KALAHARI DESERT
MAURITIUS
Bassas da India (France)
Île Europa (France)
Réunion (France)
Tropic of Capricorn
Vaal R.
SWAZILAND
Orange R.
Thabana Ntlenyana 11,424 ft (3,482 m)
INDIAN OCEAN
Great Karoo
SOUTH AFRICA
LESOTHO
Cape of Good Hope
Cape Agulhas
Prime Meridian
Elevation
feet meters
10,000+ 3,050+
5,000 1,524
2,000 610
1,000 305
500 152
0 0
Below sea level
Tristan da Cunha Group (U.K.)
N
W
E
S
Prince Edwards Islands (South Africa)
0 500 1,000 Miles
0 500 1,000 Kilometers
20°W
0°
20°E
40°E
60°E
40°N
20°N
0°
20°S
40°S
A
B
C
D
E
F
G
H
1
2
3
4
5
6

Africa Political

NATIONAL GEOGRAPHIC

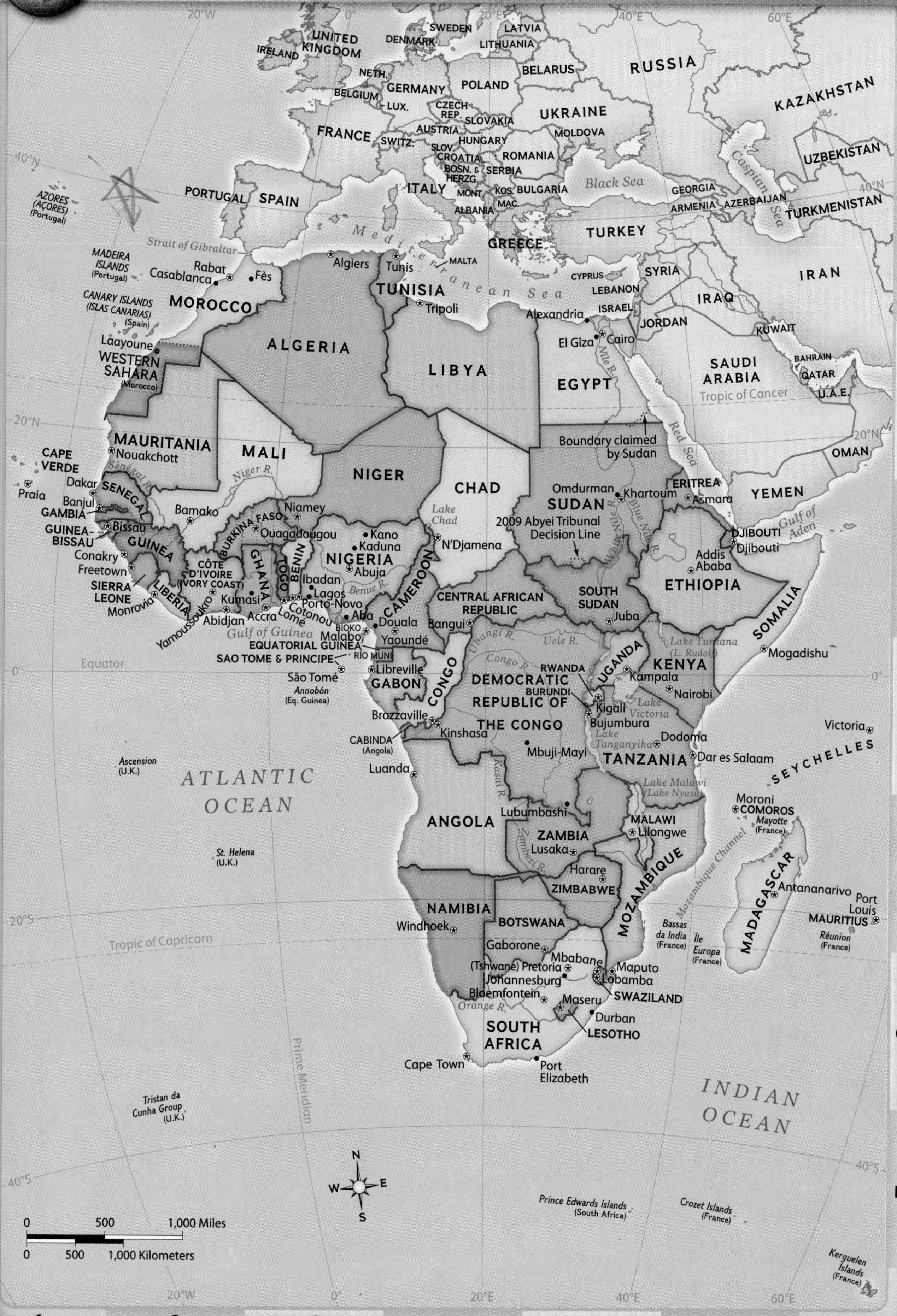

20°W 0° 20°E 40°E 60°E

UNITED KINGDOM
IRELAND
SWEDEN
DENMARK
LATVIA
LITHUANIA
NETH.
BELARUS
RUSSIA
GERMANY
POLAND
BELGIUM
LUX.
CZECH REP.
SLOVAKIA
UKRAINE
KAZAKHSTAN
FRANCE
SWITZ.
AUSTRIA
HUNGARY
MOLDOVA
SLOV.
CROATIA
ROMANIA
BOSN. & HERZG.
SERBIA
UZBEKISTAN
40°N
AZORES (AÇORES) (Portugal)
PORTUGAL
SPAIN
ITALY
MONT.
KOS.
BULGARIA
Black Sea
GEORGIA
Caspian Sea
MAC.
ARMENIA
AZERBAIJAN
TURKMENISTAN
ALBANIA
Mediterranean Sea
TURKEY
Strait of Gibraltar
GREECE
MADEIRA ISLANDS (Portugal)
Rabat
Casablanca
Fès
Algiers
Tunis
MALTA
CYPRUS
SYRIA
IRAN
CANARY ISLANDS (ISLAS CANARIAS) (Spain)
MOROCCO
TUNISIA
Tripoli
LEBANON
ISRAEL
IRAQ
Alexandria
JORDAN
KUWAIT
Laayoune
ALGERIA
El Gîza
Cairo
WESTERN SAHARA (Morocco)
Nile R.
BAHRAIN
LIBYA
SAUDI ARABIA
QATAR
EGYPT
Tropic of Cancer
U.A.E.
20°N
Red Sea
MAURITANIA
Nouakchott
Boundary claimed by Sudan
OMAN
CAPE VERDE
Sénégal R.
MALI
Niger R.
NIGER
Praia
Dakar
SENEGAL
CHAD
Omdurman
Khartoum
ERITREA
Asmara
YEMEN
Banjul
GAMBIA
Bamako
Niamey
Lake Chad
SUDAN
2009 Abyei Tribunal Decision Line
Blue Nile R.
White Nile R.
Gulf of Aden
GUINEA-BISSAU
Bissau
BURKINA FASO
Ouagadougou
Kano
DJIBOUTI
Djibouti
GUINEA
Kaduna
N'Djamena
Conakry
NIGERIA
Abuja
Addis Ababa
Freetown
CÔTE D'IVOIRE (IVORY COAST)
GHANA
TOGO
BENIN
Ibadan
SOUTH SUDAN
ETHIOPIA
SIERRA LEONE
Lagos
Benue R.
CAMEROON
LIBERIA
Monrovia
Yamoussoukro
Kumasi
Porto-Novo
CENTRAL AFRICAN REPUBLIC
Abidjan
Accra
Lomé
Cotonou
Aba
Juba
SOMALIA
Gulf of Guinea
BIOKO
Malabo
Douala
Bangui
Ubangi R.
Uele R.
Lake Turkana (L. Rudolf)
EQUATORIAL GUINEA
Yaoundé
UGANDA
KENYA
Mogadishu
SAO TOME & PRINCIPE
RIO MUNI
Congo R.
RWANDA
Kampala
Equator
0°
Libreville
São Tomé
GABON
CONGO
DEMOCRATIC REPUBLIC OF THE CONGO
BURUNDI
Nairobi
Annobón (Eq. Guinea)
Kigali
Lake Victoria
Brazzaville
Bujumbura
Victoria
Kinshasa
Lake Tanganyika
Dodoma
CABINDA (Angola)
Mbuji-Mayi
SEYCHELLES
Dar es Salaam
Ascension (U.K.)
TANZANIA
Kasai R.
Luanda
ATLANTIC OCEAN
Lake Malawi (Lake Nyasa)
Moroni
COMOROS
Lubumbashi
Mayotte (France)
ANGOLA
MALAWI
Lilongwe
ZAMBIA
Mozambique Channel
St. Helena (U.K.)
Lusaka
Zambezi R.
MOZAMBIQUE
Harare
MADAGASCAR
ZIMBABWE
Antananarivo
Port Louis
NAMIBIA
20°S
Windhoek
BOTSWANA
Bassas da India (France)
Île Europa (France)
MAURITIUS
Réunion (France)
Tropic of Capricorn
Gaborone
Mbabane
(Tshwane) Pretoria
Maputo
Johannesburg
Lobamba
Bloemfontein
SWAZILAND
Maseru
Orange R.
Durban
LESOTHO
SOUTH AFRICA
Cape Town
Port Elizabeth
Prime Meridian
INDIAN OCEAN
Tristan da Cunha Group (U.K.)
40°S
N
W
E
S
Prince Edwards Islands (South Africa)
Crozet Islands (France)
0 500 1,000 Miles
0 500 1,000 Kilometers
Kerguelen Islands (France)

A B C D E F G H
1 2 3 4 5 6

Southwest Asia Physical

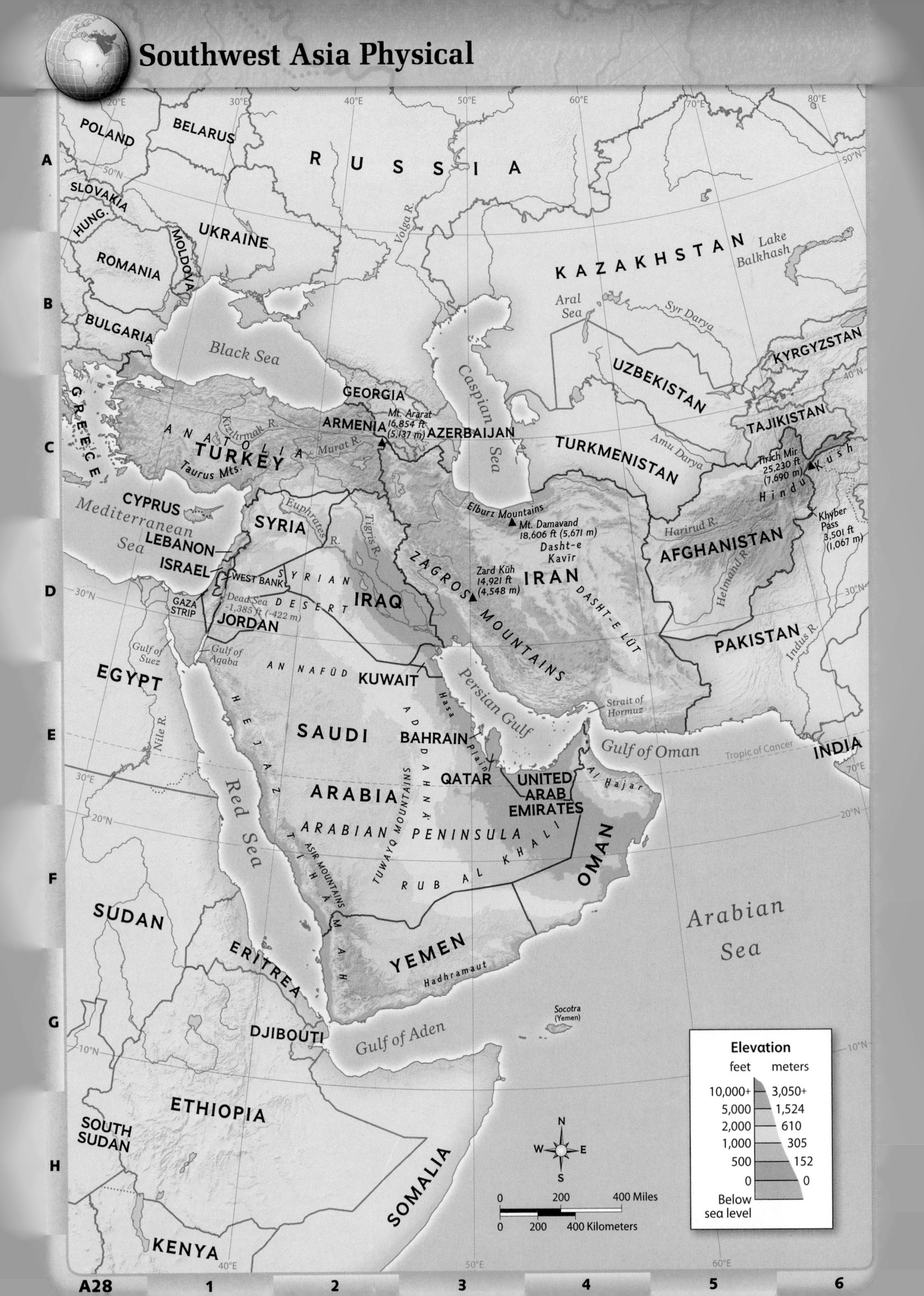

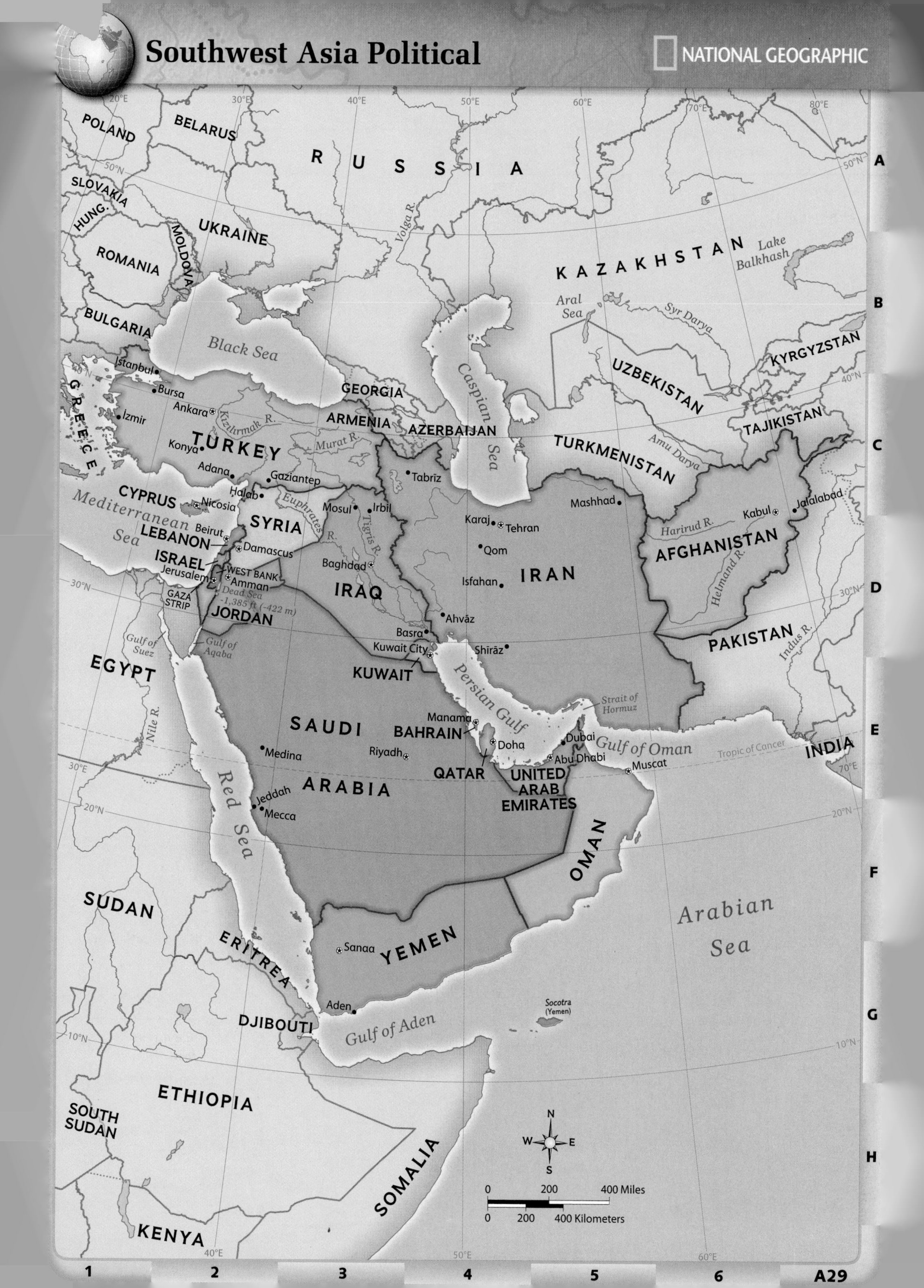

Southwest Asia Political

South Asia Physical

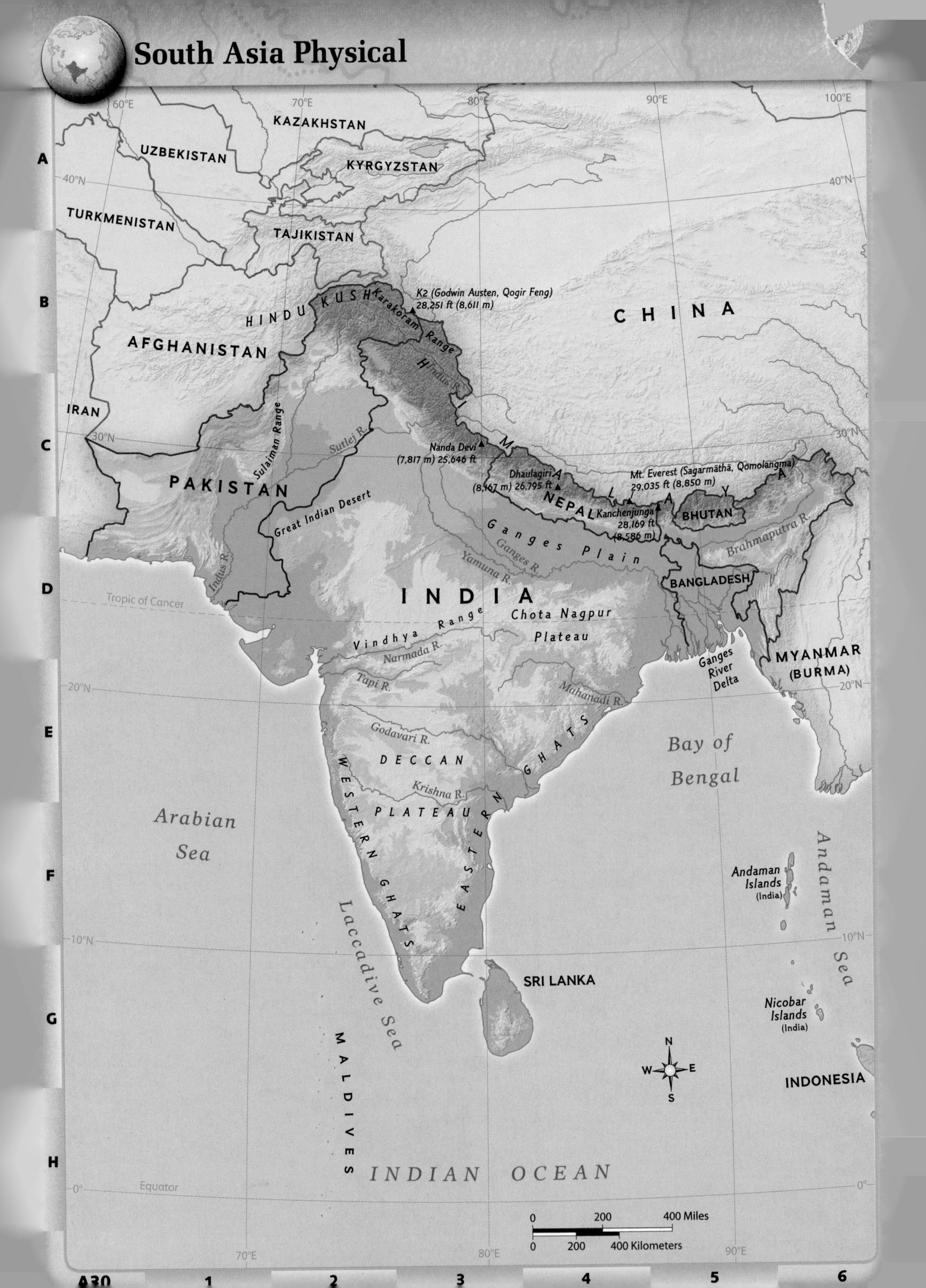

South Asia Political

NATIONAL GEOGRAPHIC

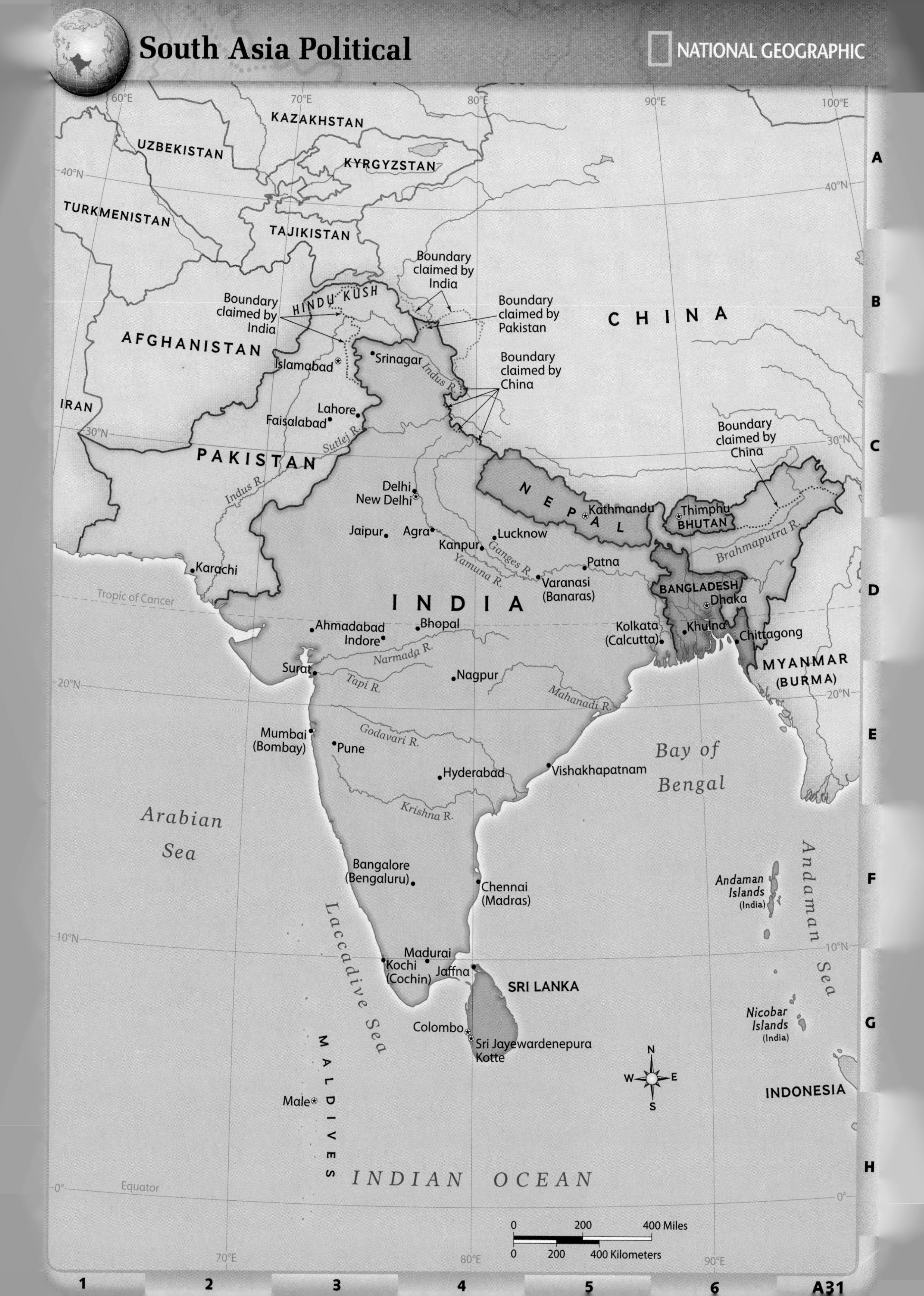

East Asia Physical

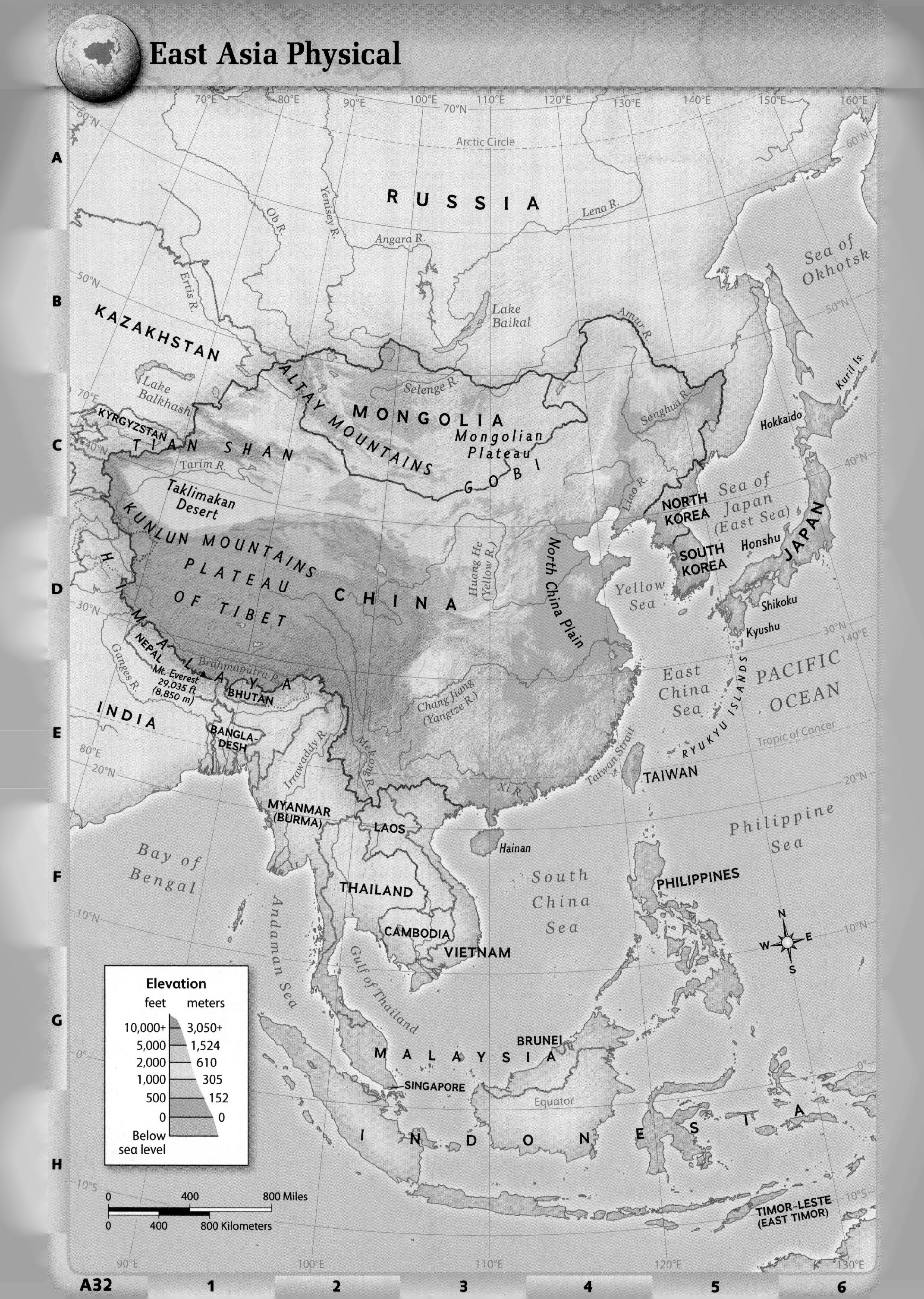

East Asia Political

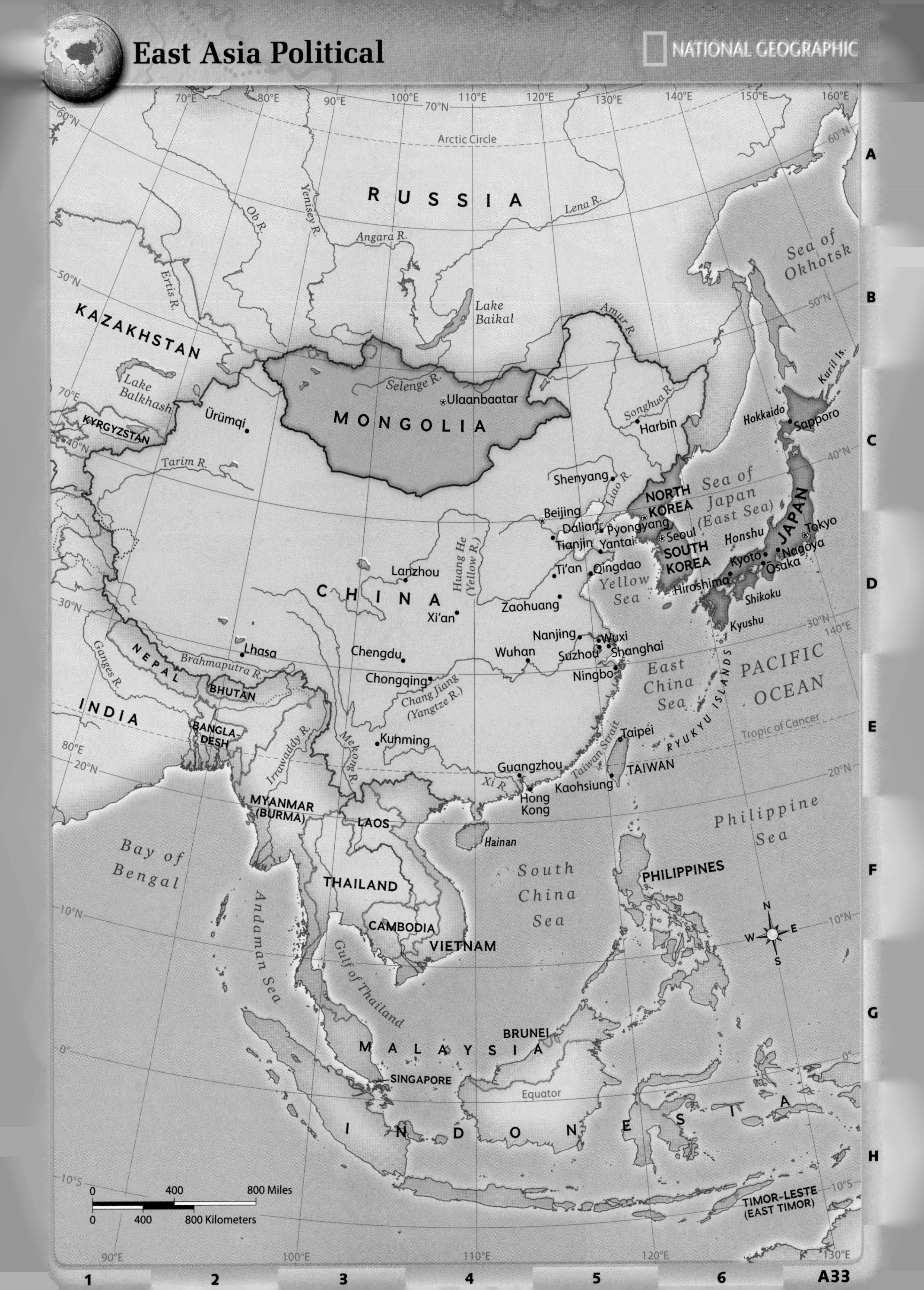

Southeast Asia Physical

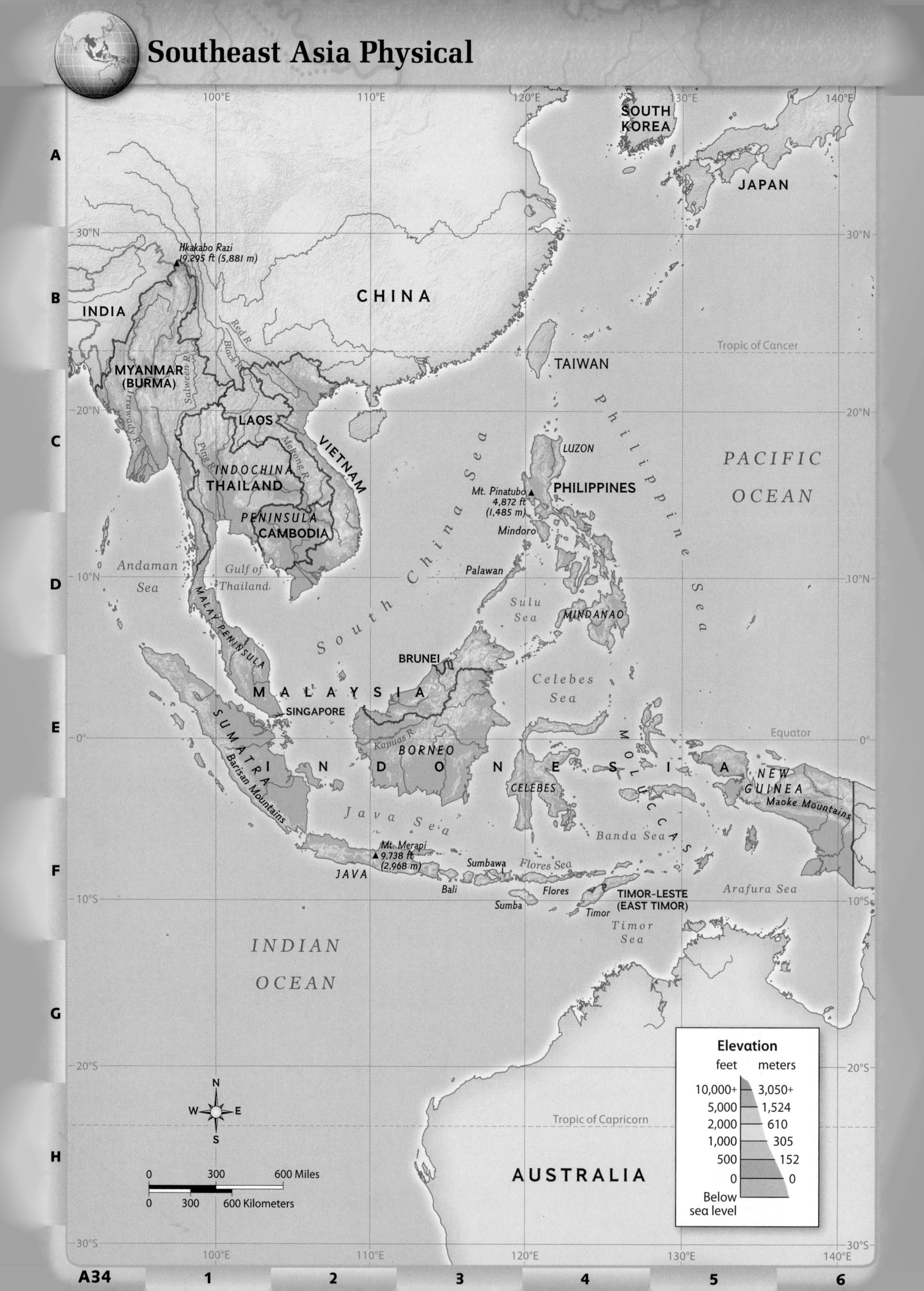

Southeast Asia Political

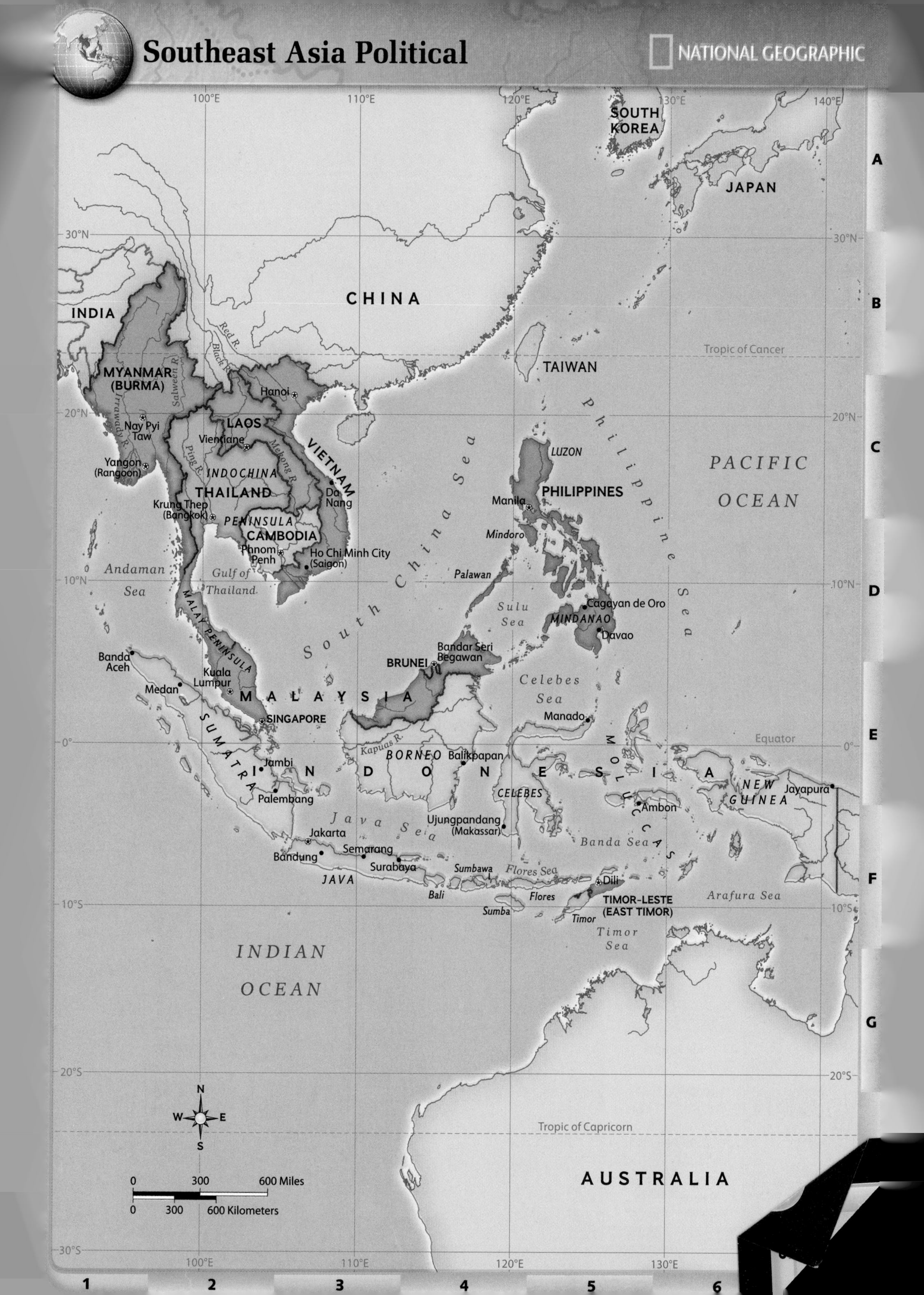

Australia, the Pacific Realm & Antarctica Physical

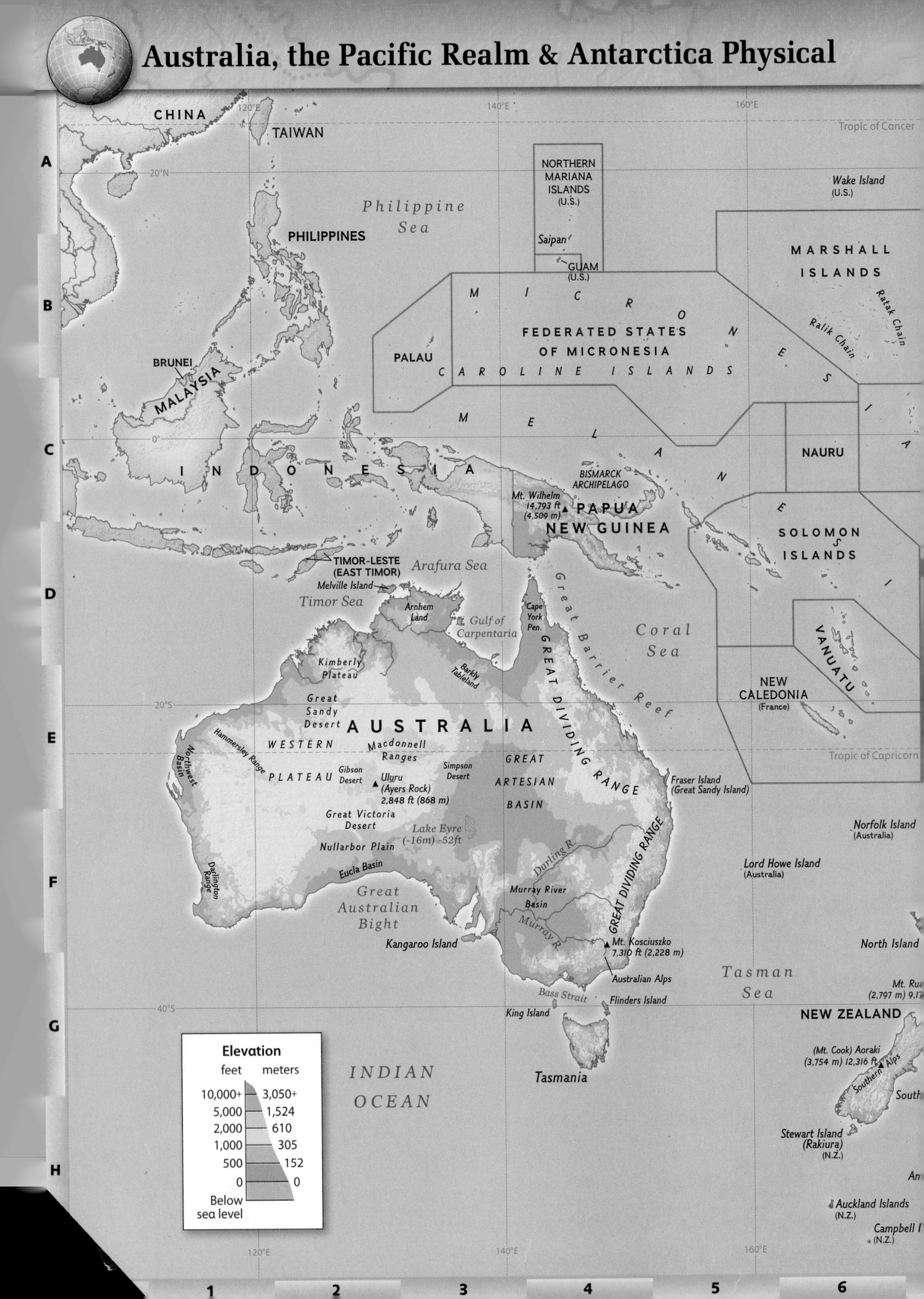

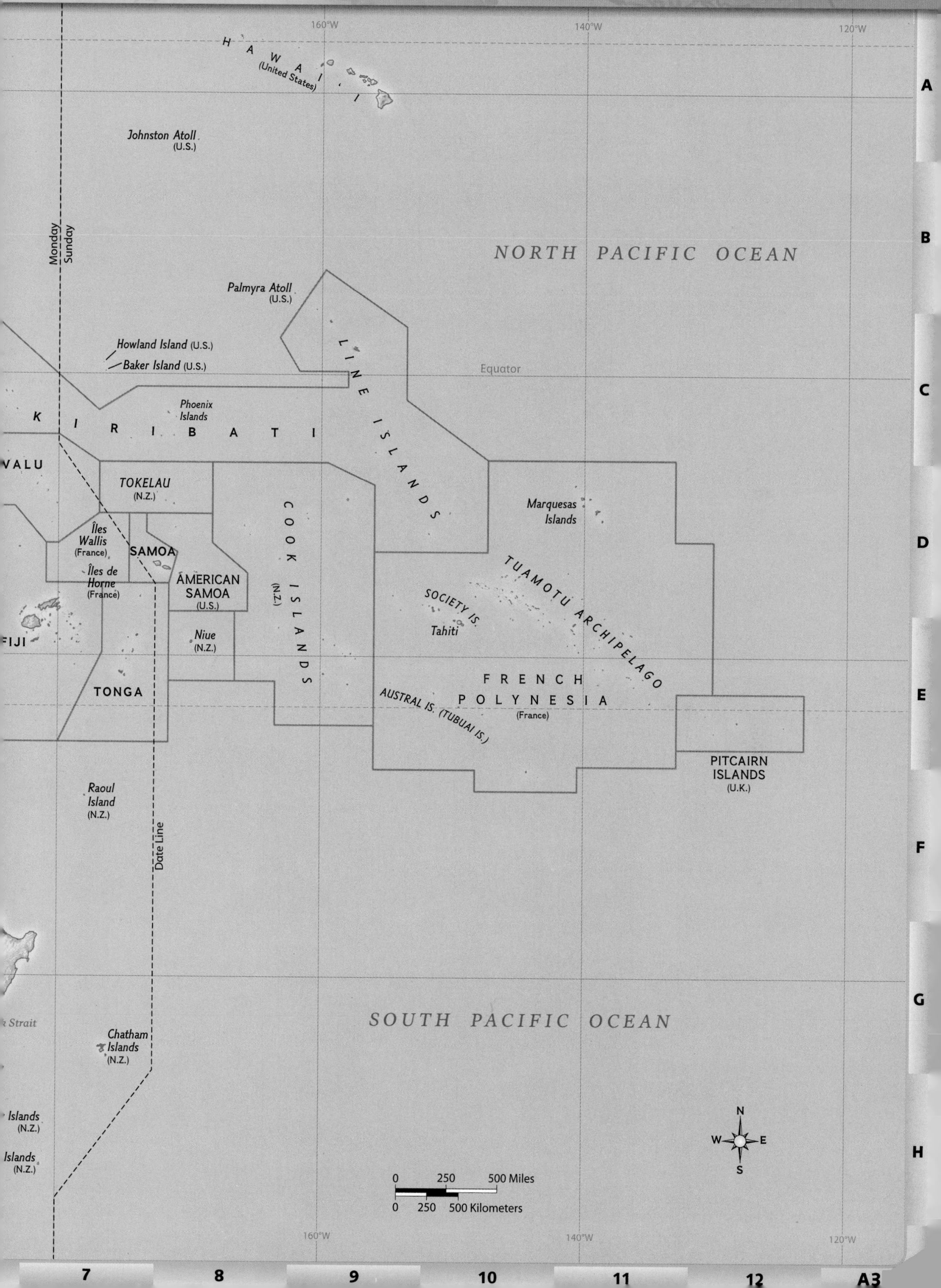

160°W 140°W 120°W

HAWAII (United States)

Johnston Atoll (U.S.)

NORTH PACIFIC OCEAN

Monday
Sunday

Palmyra Atoll (U.S.)

Howland Island (U.S.)
Baker Island (U.S.)

Equator

LINE ISLANDS

Phoenix Islands

KIRIBATI

VALU

TOKELAU (N.Z.)

Îles Wallis (France)

SAMOA

Îles de Horne (France)

AMERICAN SAMOA (U.S.)

COOK ISLANDS (N.Z.)

Marquesas Islands

TUAMOTU ARCHIPELAGO

SOCIETY IS.

Tahiti

FIJI

Niue (N.Z.)

TONGA

FRENCH POLYNESIA (France)

AUSTRAL IS. (TUBUAI IS.)

PITCAIRN ISLANDS (U.K.)

Raoul Island (N.Z.)

Date Line

Strait

SOUTH PACIFIC OCEAN

Chatham Islands (N.Z.)

Islands (N.Z.)

Islands (N.Z.)

N
W E
S

0 250 500 Miles
0 250 500 Kilometers

160°W 140°W 120°W

A B C D E F G H

7 8 9 10 11 12

Australia, the Pacific Realm & Antarctica Political

CHINA
TAIWAN
Tropic of Cancer
Wake Island (U.S.)
NORTHERN MARIANA ISLANDS (U.S.)
Saipan
Capitol Hill
Philippine Sea
PHILIPPINES
(Agana) Hagåtña
GUAM (U.S.)
MARSHALL ISLANDS
FEDERATED STATES OF MICRONESIA
Palikir
Majuro
Melekeiok
PALAU
BRUNEI
MALAYSIA
NAURU
Yaren
INDONESIA
PAPUA NEW GUINEA
Port Moresby
SOLOMON ISLANDS
Honiara
TIMOR-LESTE (EAST TIMOR)
Arafura Sea
Timor Sea
Darwin
Gulf of Carpentaria
Great Barrier Reef
Coral Sea
VANUATU
Port Vila
NEW CALEDONIA (France)
Nouméa
NORTHERN TERRITORY
QUEENSLAND
Tropic of Capricorn
WESTERN AUSTRALIA
AUSTRALIA
Lake Eyre
SOUTH AUSTRALIA
Brisbane
Norfolk Island (Australia)
Lord Howe Island (Australia)
Darling R.
NEW SOUTH WALES
Perth
Great Australian Bight
Sydney
Adelaide
Murray R.
Canberra, AUSTRALIAN CAPITAL TERRITORY
VICTORIA
Melbourne
Tasman Sea
Auckland
Bass Strait
NEW ZEALAND
TASMANIA
Hobart
INDIAN OCEAN
Auckland Islands (N.Z.)
Campbell I. (N.Z.)
120°E
140°E
160°E
20°N
0°
20°S
40°S
A B C D E F G H
1 2 3 4 5 6

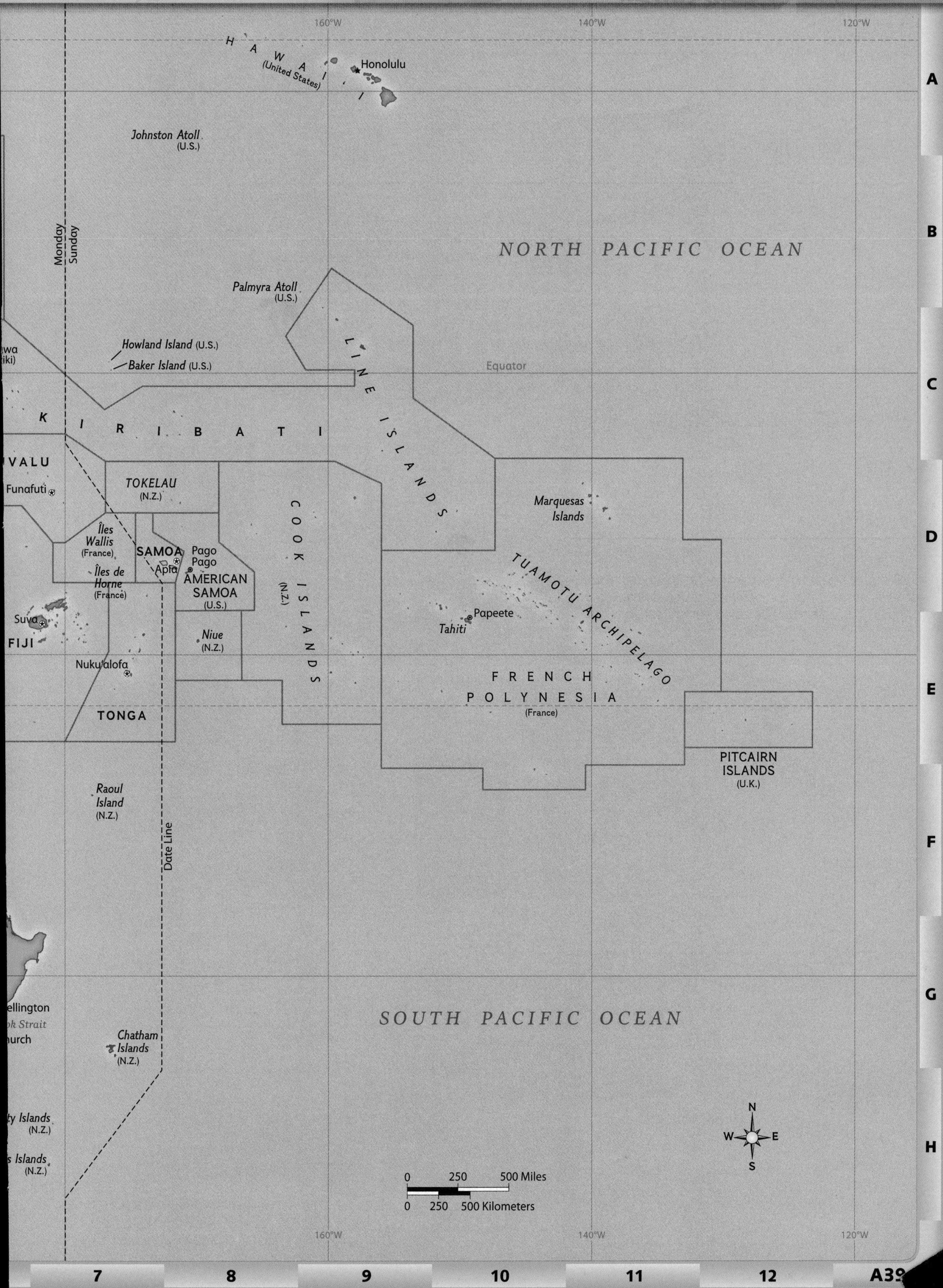

160°W 140°W 120°W

HAWAII (United States)
Honolulu
Johnston Atoll (U.S.)
Monday
Sunday
NORTH PACIFIC OCEAN
Palmyra Atoll (U.S.)
Howland Island (U.S.)
Baker Island (U.S.)
wa
iki)
Equator
LINE ISLANDS
KIRIBATI
VALU
Funafuti
TOKELAU (N.Z.)
COOK ISLANDS (N.Z.)
Marquesas Islands
Îles Wallis (France)
SAMOA
Apia
Pago Pago
AMERICAN SAMOA (U.S.)
Îles de Horne (France)
TUAMOTU ARCHIPELAGO
Papeete
Tahiti
Suva
FIJI
Niue (N.Z.)
Nuku'alofa
FRENCH POLYNESIA (France)
TONGA
PITCAIRN ISLANDS (U.K.)
Raoul Island (N.Z.)
Date Line
ellington
ok Strait
hurch
SOUTH PACIFIC OCEAN
Chatham Islands (N.Z.)
ty Islands (N.Z.)
s Islands (N.Z.)
N
W E
S
0 250 500 Miles
0 250 500 Kilometers

A B C D E F G H

NATIONAL GEOGRAPHIC
Explore the

Inspiring people to care about the planet

-National Geographic Society Mission

For more than 100 years, National Geographic Society has sparked our curiosity about the world. NGS supports a network of explorers whose work in the field is vital to the planet. Some of them are shown below. Through education and exploration, the Society works to protect the physical environment and preserve the world's cultures.

Explorers at WORK

Oceanographer
NG Emerging Explorer Katy Croff Bell explores underwater archaeology.

Research Scientist
NG Emerging Explorer Albert Lin uses technology to study artifacts from Asian civilizations.

Archaeologist
NG Emerging Explorer Beverly Goodman researches the ways in which humans affect nature along coastlines.

Conservationist
NG Emerging Explorer Shafqat Hussain protects endangered snow leopards.

Educator
NG Emerging Explorer Kakenya Ntaiya works to improve girls' education in Kenya.

Filmmaker & Anthropologist
NG Fellow Elizabeth Kapu'uwailani Lindsey strives to preserve the Polynesian culture through documentary film.

Across the world, archaeologists, anthropologists, oceanographers, and linguists represent National Geographic Society in their work. Don't let their long titles confuse you—they are all scientists doing exciting work in the field. New information about Earth's physical features is uncovered every day.

These are some of the explorers at work for National Geographic Society.

From the classroom to the world

Tools for EXPL

National Geographic explorers all depend on tools for exploration. You can use the variety of tools provided with your program at myNGconnect.com to explore the world and its cultures.

Connect to NG
Gateway to research links and current events through National Geographic

my eEdition

Interactive Map Tool
A comprehensive online mapmaker at your fingertips

ORATION

Taj Mahal, India

Digital Library

GeoVideos, Explorer Video Clips, and hundreds of photographs of the world's physical geography and cultures

Interactive Whiteboard GeoActivities

Hands-on activities to learn more about how the world works

Magazine Maker

Tool for creating student magazines using program resources or by uploading your own photos

From the classroom to the world

The knowledge and skills that explorers need are described in the National Geography Standards, shown here. Keep them in mind as you study *World Cultures and Geography*.

HIGH

REPORT

1. How to use maps and other geographic representations, tools, and technologies to acquire, process, and **REPORT** information from a spatial perspective

2. How to use mental maps to organize information about people, places, and environments in a spatial context

3. How to **ANALYZE** the spatial organization of people, places, and environments on Earth's surface

ANALYZE

4. The physical and human characteristics of places

EXPERIENCE

5. How people create regions to **INTERPRET** Earth's complexity

6. How culture and **EXPERIENCE** influence people's perceptions of places and regions

INTERPRET

7. The physical processes that shape the patterns of Earth's surface

8. The characteristics and spatial distribution of ecosystems on Earth's surface

9. The characteristics, distribution, and migration of human population on Earth's surface

ER ORDER THINKING

10. The characteristics, distribution, and complexity of Earth's cultural mosaics

PATTERNS

11. The **PATTERNS** and networks of economic interdependence on Earth's surface

12. The processes, patterns, and functions of human settlement

MODIFY

13. How the forces of cooperation and conflict among people influence the division and control of Earth's surface

14. How human actions **MODIFY** the physical environment

15. How physical systems affect human systems

APPLY

16. The changes that occur in the meaning, use, distribution, and importance of resources

17. How to **APPLY** geography to interpret the past

18. How to **APPLY** geography to interpret the present and plan for the future

THE ESSENTIALS OF GEOGRAPHY

MEET THE EXPLORER

NATIONAL GEOGRAPHIC

Explorer-in-Residence Spencer Wells is director of the Genographic Project. He analyzes DNA samples to trace humankind's ancient migration patterns and learn where we really come from.

INVESTIGATE GEOGRAPHY

In 2002, NASA released its spectacular "blue marble" image of Earth. It is made from satellite observations gathered over several months, and then carefully combined. The image was inspired by a photo taken by astronauts during a 1972 space mission.

CONNECT WITH THE CULTURE

Culture is how people of a certain region live, behave, and think, but culture is not limited by geography. A busy urban area like New York City (shown here) attracts people from around the world, representing a vast variety of cultures. Those cultures can live together in a single place.

THINK LIKE A GEOGRAPHER

Scientists work to uncover the skeleton of an ancient whale in Wadi Al Hitan, Egypt. Fossils, like those shown here, reveal clues about the earth's past.

NATIONAL GEOGRAPHIC CHAPTER 1

THE GEOGRAPHER'S TOOLBOX

PREVIEW THE CHAPTER

Essential Question How do geographers think about the world?

SECTION 1 • GEOGRAPHIC THINKING

KEY VOCABULARY

- spatial thinking
- geographic pattern
- Geographic Information System (GIS)
- absolute location
- Global Positioning System (GPS)
- relative location
- region
- continent
- terrace

ACADEMIC VOCABULARY

significant, categorize

Essential Question How do people use geography?

SECTION 2 • MAPS

KEY VOCABULARY

- globe
- map
- latitude
- equator
- longitude
- prime meridian
- hemisphere
- scale
- cartographer
- elevation
- relief
- projection

ACADEMIC VOCABULARY

distort, theme

TERMS & NAMES

- North Pole
- South Pole
- Northern Hemisphere
- Southern Hemisphere
- Western Hemisphere
- Eastern Hemisphere

Scientists wear breathing aids while collecting gas samples from a spring in the Cueva (Cave) de Villa Luz in Mexico.

SECTION 1 GEOGRAPHIC THINKING

1.1 Thinking Spatially

myNGconnect.com For an online map and photos of cities on water

DENVER, COLORADO

CHICAGO, ILLINOIS

Critical Viewing Denver, Colorado, is located in the foothills of the Rocky Mountains. Chicago, Illinois, is located on the shore of Lake Michigan—one of the five Great Lakes. How might these different locations affect the activities of people who live in these cities?

Main Idea Geographers study the location of places and the people who live there.

Geography is about more than the names of places on a map. It involves **spatial thinking**, or thinking about the space on Earth's surface, including where places are located and why they are there.

Ask Geographic Questions

Geographers use spatial thinking to ask questions, such as "Where is a place located? Why is this location **significant**, or important?"

Look at the photographs of Denver, Colorado, and Chicago, Illinois, above. Denver is near the Rocky Mountains, where gold was once discovered. Chicago is on Lake Michigan, which made it an important shipping center for many years.

Now find New York City on the map. It is on a protected bay. Why is New York's location significant? Like Chicago, it is near water, which is good for trade.

Study Geographic Patterns

By asking and answering many questions, geographers can find patterns. **Geographic patterns** are similarities among places. The location of large cities near water is one example of a geographic pattern.

Many geographers use computer-based **Geographic Information Systems (GIS)**. They create maps and analyze patterns using many layers of data.

Before You Move On

Summarize What do geographers study? How do they study it?

ONGOING ASSESSMENT

MAP LAB

1. **Location** Where is Rio de Janeiro located on the map? Use directional words in your answer.
2. **Make Inferences** Look at the other cities on the map. Where are most of them located? What pattern do you notice?
3. **Summarize** What are geographic patterns, and how do geographers find them?

1.2 Themes and Elements

TECHTREK
myNGconnect.com For photos and Guided Writing

Digital Library

Student Resources

Main Idea Geographers use themes and elements to understand the world.

Geographers ask questions about how people, places, and environments are arranged and connected on Earth's surface. The five themes and six elements will help you **categorize**, or group, information.

The Five Themes of Geography

Geographers use five themes to categorize similar geographic information.

1. **Location** provides a way of locating places. **Absolute location** is the exact point where a place is located. Geographers use a satellite system called the **Global Positioning System (GPS)** to find absolute location. **Relative location** is where a place is in relation to other places. The Great Wall is located near Beijing in northern China.
2. **Place** includes the characteristics of a location. A famous place in the western United States is the Grand Canyon. It has steep rock walls that were carved over centuries by the Colorado River.
3. **Human-Environment Interaction** explains how people affect the environment and how the environment affects people. For example, people build dams to change the flow of rivers.
4. **Movement** explains how people, ideas, and animals move from one place to another. The spread of different religions around the world is an example of movement.
5. **Region** involves a group of places that have common characteristics. North America is a region that includes the United States, Mexico, and Canada.

Critical Viewing Tourists climb the Great Wall of China. What physical land features can you see in the photo?

The Great Wall runs through the Chinese countryside.

Six Essential Elements

Some geographers identify essential elements, or key ideas, to study physical processes and human systems.

1. **The World in Spatial Terms** Geographers use tools such as maps to study places on Earth's surface.
2. **Places and Regions** Geographers study the characteristics of places and regions.
3. **Physical Systems** Geographers examine Earth's physical processes, such as earthquakes and volcanoes.
4. **Human Systems** Geographers study how humans live and what systems they create, such as economic systems.
5. **Environment and Society** Geographers explore how humans change the environment and use resources.
6. **The Uses of Geography** Geographers interpret the past, analyze the present, and plan for the future.

Before You Move On

Make Inferences How do geographers use the themes and elements to better understand the world?

ONGOING ASSESSMENT

WRITING LAB

GeoJournal

1. **Write About Geographic Themes** Write a paragraph in which you use the five themes to describe your community. Explain which theme is the most important in making your community what it is today. Go to **Student Resources** for Guided Writing support.
2. **Categorize** Which theme and which element would you use to categorize information about forms of energy? Explain your answer.
3. **Compare and Contrast** Create a chart like the one below. Use the six essential elements to compare how your town relates to another one that you know about.

ELEMENT	MY TOWN	OTHER TOWN
The World in Spatial Terms	North of the highway	South of the highway
Places and Regions		
Physical Systems		

1.3 World Regions

TECHTREK

myNGconnect.com For an online map and photos of world regions

Maps and Graphs | Digital Library

Main Idea Geographers divide the world into regions. Each region is shaped by shared physical and human processes.

In 1413, a Chinese admiral and explorer, Zheng He, sailed from China to Arabia. When he arrived in Arabia, he saw people dressed in ways he had never seen. Yet, like him, these people wanted to trade.

Regions and Continents

Zheng He saw that regions of Earth have similarities and differences. A **region** is a group of places with common traits. The places within a region are linked by trade, culture, and other human activities. They also share similar physical processes and characteristics, such as climate.

A region often includes an entire continent. A **continent** is a large landmass on Earth's surface. Geographers have identified seven continents: Africa, Asia, Australia, Europe, North America, South America, and Antarctica.

Geographers study the world's regions, but they also take a global perspective when they investigate Earth. They might, for instance, study ocean currents around the globe or how one region affects another. Both ways of looking at the world add to our understanding of it.

Before You Move On

Make Inferences Why do geographers study the world by dividing it into regions?

Visual Vocabulary Chinese field workers view the terraced landscape. A **terrace** is a flat surface that is built into a hillside.

WORLD REGIONS

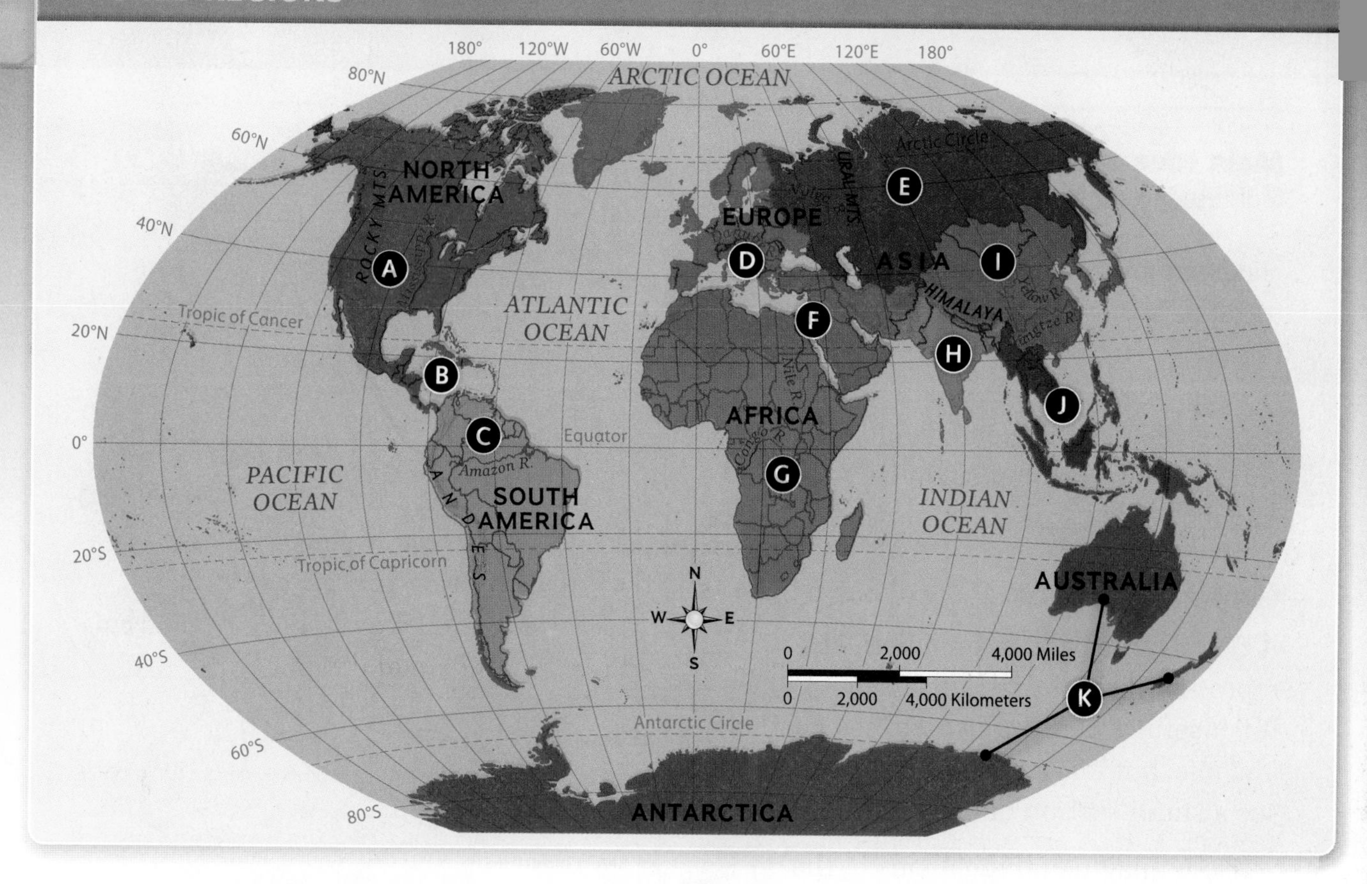

Regions in This Book

A **North America** contains the United States, Canada, and Mexico, along with the Great Lakes—the largest group of freshwater lakes.

B **Central America and the Caribbean** includes the islands of the Caribbean and the countries that connect North and South America.

C **South America** includes Brazil, a growing economic power, and the Amazon Rain Forest, the world's largest tropical rain forest.

D **Europe** includes 29 countries and has nearly 24,000 miles of coastline.

E **Russia and the Eurasian Republics** includes countries that were part of the former Union of Soviet Socialist Republics (U.S.S.R.).

F **Southwest Asia and North Africa** spans two continents—Africa and Asia.

G **Sub-Saharan Africa** includes Africa south of the Sahara, the world's largest desert.

H **South Asia** includes India, one of the world's fastest growing countries.

I **East Asia** includes China, the world's most populous country.

J **Southeast Asia** includes Indonesia, a country made up of over 17,000 islands.

K **Australia, the Pacific Realm, and Antarctica** includes the Pacific island nations north and east of Australia and New Zealand.

ONGOING ASSESSMENT
MAP LAB

1. **Interpret Maps** Which regions span more than one continent?
2. **Pose and Answer Questions** Find your region on the map. Write one geographic question about the region and answer the question.
3. **Region** Identify a physical feature in the region in which you live that makes it different from other regions. How does this feature affect you, or those who live near it?

2.1 Elements of a Map

TECHTREK

myNGconnect.com For online maps of geographic regions and photos of antique maps

 Maps and Graphs

 Digital Library

Main Idea Globes and maps are two different tools used to study places on Earth.

Have you ever needed to figure out how to get to a friend's house? Imagine that the only resource you had was a globe. In order to see enough detail to find your friend's house, the globe would have to be enormous—much too big to carry around in your pocket!

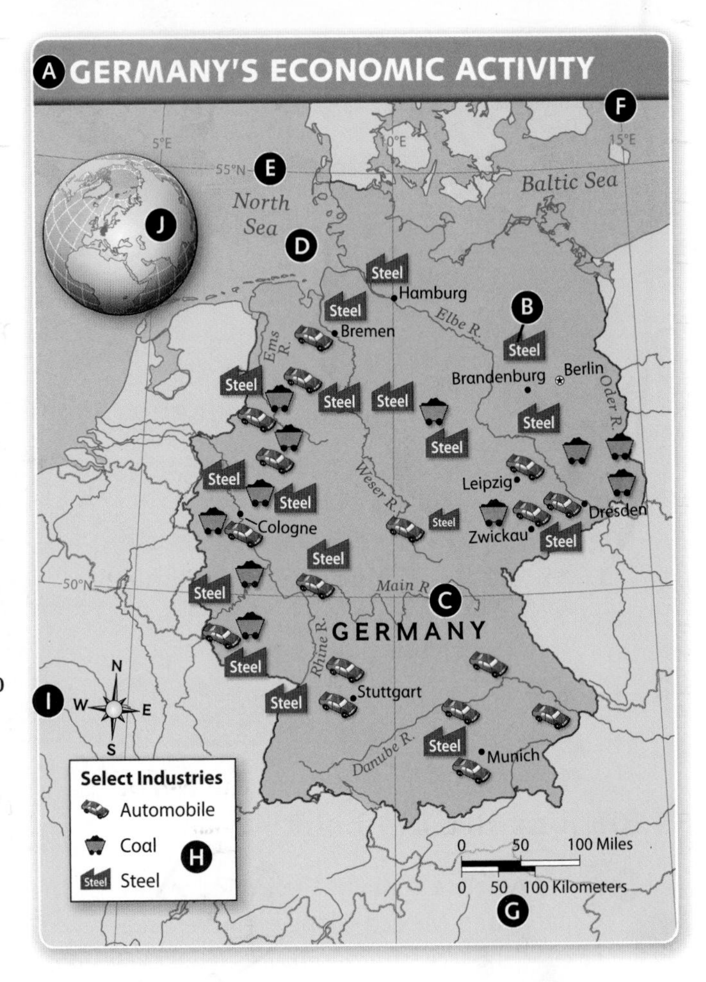

Globes and Maps

A three-dimensional, or spherical, representation of Earth is called a **globe**. It is useful when you need to see Earth as a whole, but it is not helpful if you need to see a small section of Earth.

Now imagine taking a part of the globe and flattening it out. This two-dimensional, or flat, representation of Earth is called a **map**. Maps and globes are different representations of Earth, but they have similar features.

Map and Globe Elements

A A **title** tells the subject of the map or globe.

B **Symbols** represent information such as natural resources and economic activities.

C **Labels** are the names of places, such as cities, countries, rivers, and mountains.

D **Colors** represent different kinds of information. For example, the color blue usually represents water.

E **Lines of latitude** are imaginary horizontal lines that measure the distance north or south of the equator.

F **Lines of longitude** are imaginary vertical lines that measure the distance east or west of the prime meridian.

G A **scale** shows how much distance on Earth is represented by distance on the map or globe. For example, a half inch on the map above represents 100 miles on Earth.

H A **legend**, or key, explains what the symbols and colors on the map or globe represent.

I A **compass rose** shows the directions north (N), south (S), east (E), and west (W).

J A **locator globe** shows the specific area of the world that is shown on a map. The locator globe on the map above shows where Germany is located.

Latitude

Lines of **latitude** are imaginary lines that run east to west, parallel to the equator. The **equator** is the center line of latitude. Distances north and south of the equator are measured in degrees (°). There are 90 degrees north of the equator and 90 degrees south. The equator is 0°. The latitude of Berlin, Germany, is 52° N, meaning that it is 52 degrees north of the equator.

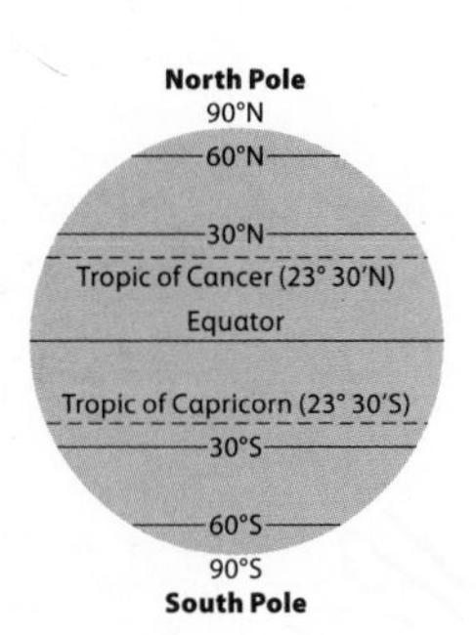

Longitude

Lines of **longitude** are imaginary lines that run north to south from the **North Pole** to the **South Pole**. They measure distance east or west of the prime meridian. The **prime meridian** runs through Greenwich, England. It is 0°. There are 180 degrees east of the prime meridian and 180 degrees west. The longitude of Berlin, Germany, is 13° E, meaning that it is 13 degrees east of the prime meridian.

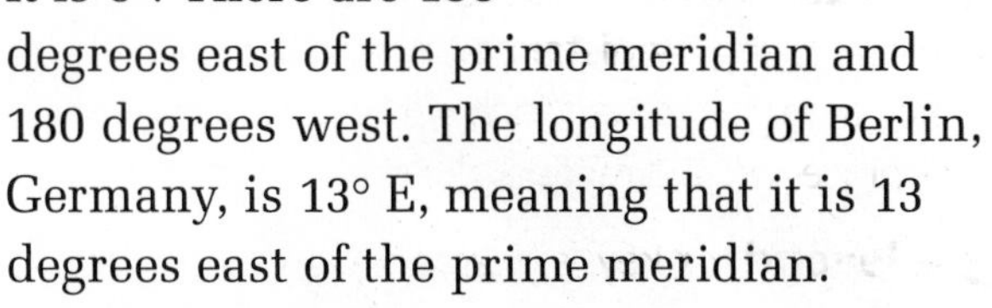

Remember that absolute location is the exact point where a place is located. This point includes a place's latitude and longitude. For example, the absolute location of Berlin, Germany, is 52° N, 13° E. You say this aloud as "fifty-two degrees North, thirteen degrees East."

Hemispheres

A **hemisphere** is half of Earth. The equator divides Earth into the **Northern Hemisphere** and the **Southern Hemisphere**. North America is entirely in the Northern Hemisphere. Most of South America is in the Southern Hemisphere.

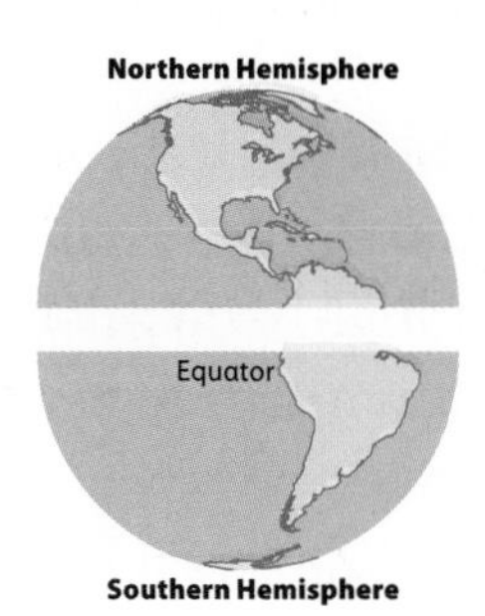

The **Western Hemisphere** is west of the prime meridian. The **Eastern Hemisphere** is east of the prime meridian. South America is in the Western Hemisphere. Most of Africa is in the Eastern Hemisphere.

Before You Move On

Monitor Comprehension How are maps and globes different? How is each one used?

ONGOING ASSESSMENT

MAP LAB

GeoJournal

1. **Interpret Maps** What types of industry are located in Germany? What map elements did you use to find the answers?
2. **Make Inferences** What is the main industry in southern Germany? Why might this industry be located there?
3. **Location** What is the difference between lines of latitude and lines of longitude?

2.2 Map Scale

TECHTREK

myNGconnect.com For online maps and and photos at different scales

Maps and Graphs

Digital Library

Main Idea Maps use different scales for different purposes.

On a walk through a city, such as Charlotte, North Carolina, you might use a highly-detailed map that shows only the downtown area. To drive up the Atlantic coast, however, you would use a map that covers a large area, including several states. These maps have different scales.

Interpreting a Scale

A map's **scale** shows how much distance on Earth is shown on the map. A large-scale map covers a small area but shows many details. A small-scale map covers a large area but includes few details.

A scale is usually shown in both inches and centimeters. One inch or centimeter on the map represents a much larger distance on Earth, such as a number of miles or kilometers.

To use a map scale, mark off the length of the scale several times on the edge of a sheet of paper. Then hold the paper between two points to see how many times the scale falls between them. Add up the distance.

CHARLOTTE, NORTH CAROLINA

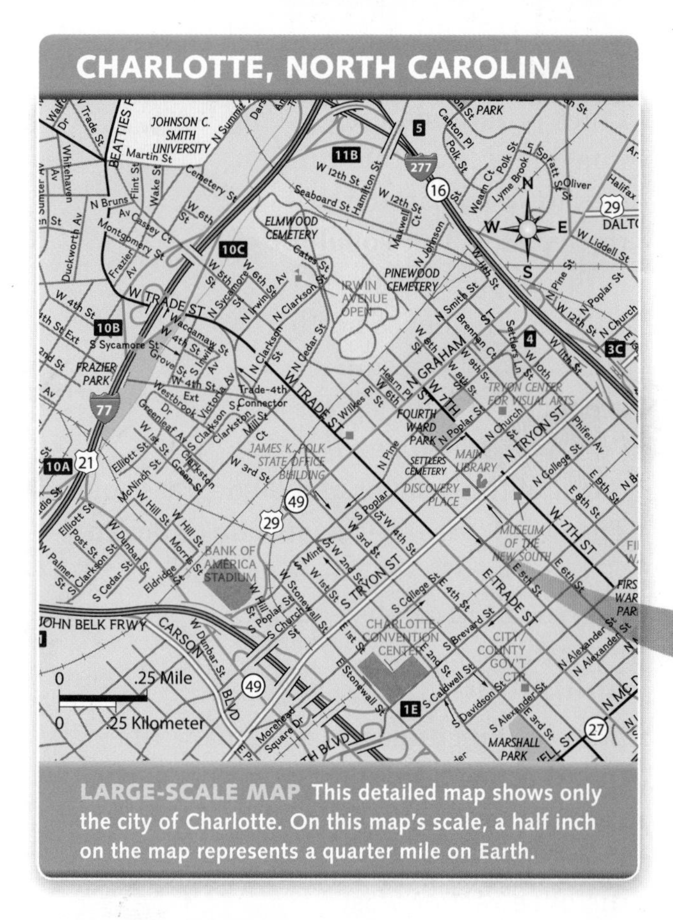

LARGE-SCALE MAP This detailed map shows only the city of Charlotte. On this map's scale, a half inch on the map represents a quarter mile on Earth.

Purposes of a Scale

The scale of a map should be appropriate for its purpose. For example, a tourist map of Washington, D.C., should be large-scale, showing every street name, monument, and museum.

Critical Viewing The Atlantic Ocean rolls in along the dunes on the Outer Banks of North Carolina. Which map labels the location of the Outer Banks?

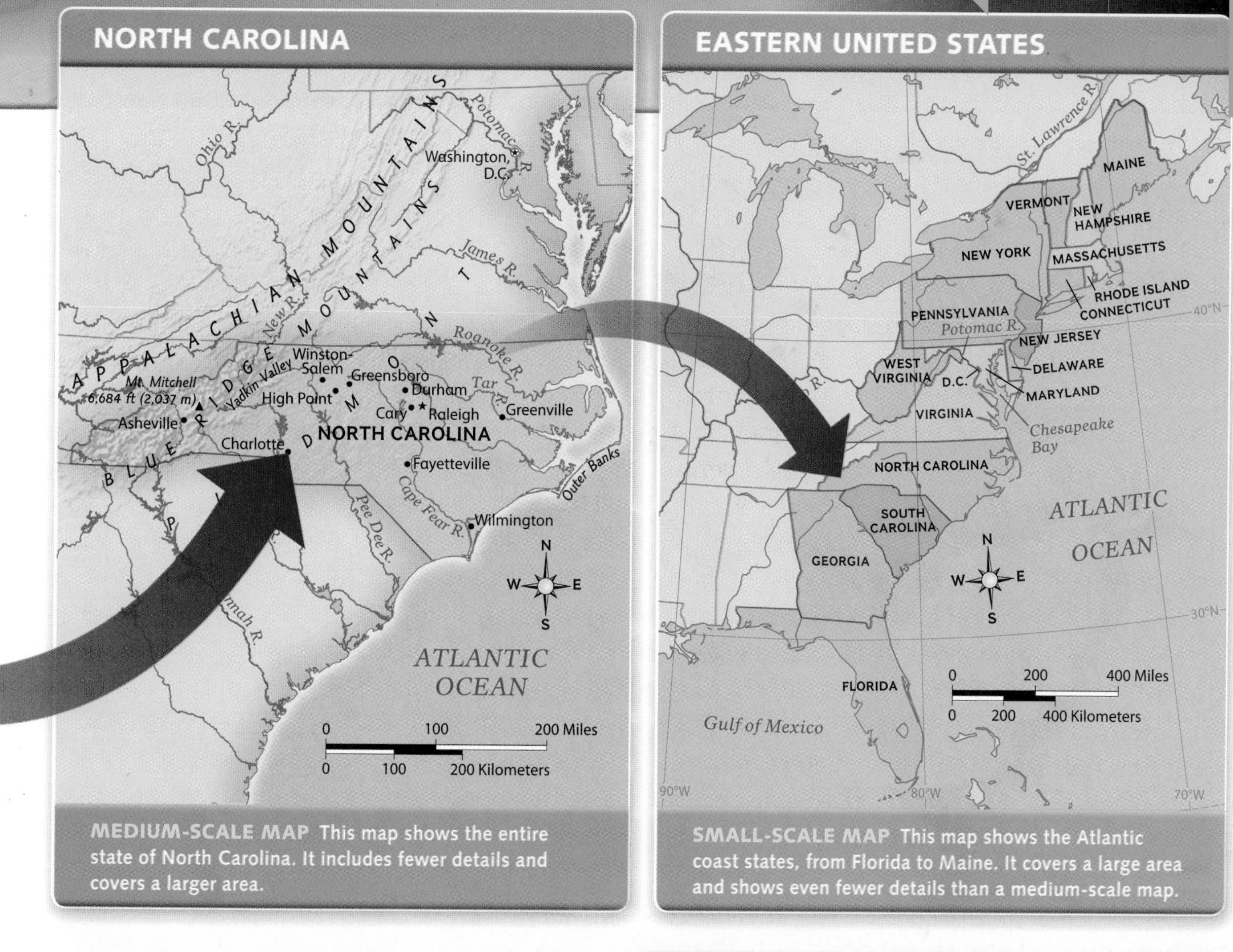

MEDIUM-SCALE MAP This map shows the entire state of North Carolina. It includes fewer details and covers a larger area.

SMALL-SCALE MAP This map shows the Atlantic coast states, from Florida to Maine. It covers a large area and shows even fewer details than a medium-scale map.

Maps of any scale show geographic patterns. The map of Washington, D.C., for instance, would show that many government buildings are in one area.

Before You Move On

Summarize What are the purposes of a small-scale map and a large-scale map?

ONGOING ASSESSMENT

MAP LAB

1. **Interpret Maps** On the map of North Carolina, approximately how many inches represent 200 miles? On the map of the eastern United States, how many miles does one inch represent?
2. **Location** How far is Washington, D.C., from the southwestern border of Maine? Which map did you use to find the distance? Why?
3. **Synthesize** What geographic pattern do you see in the location of cities in North Carolina? How might you explain this pattern?

2.3 Political and Physical Maps

TECHTREK

myNGconnect.com For online political and physical maps of world regions

Maps and Graphs

EAST ASIA POLITICAL

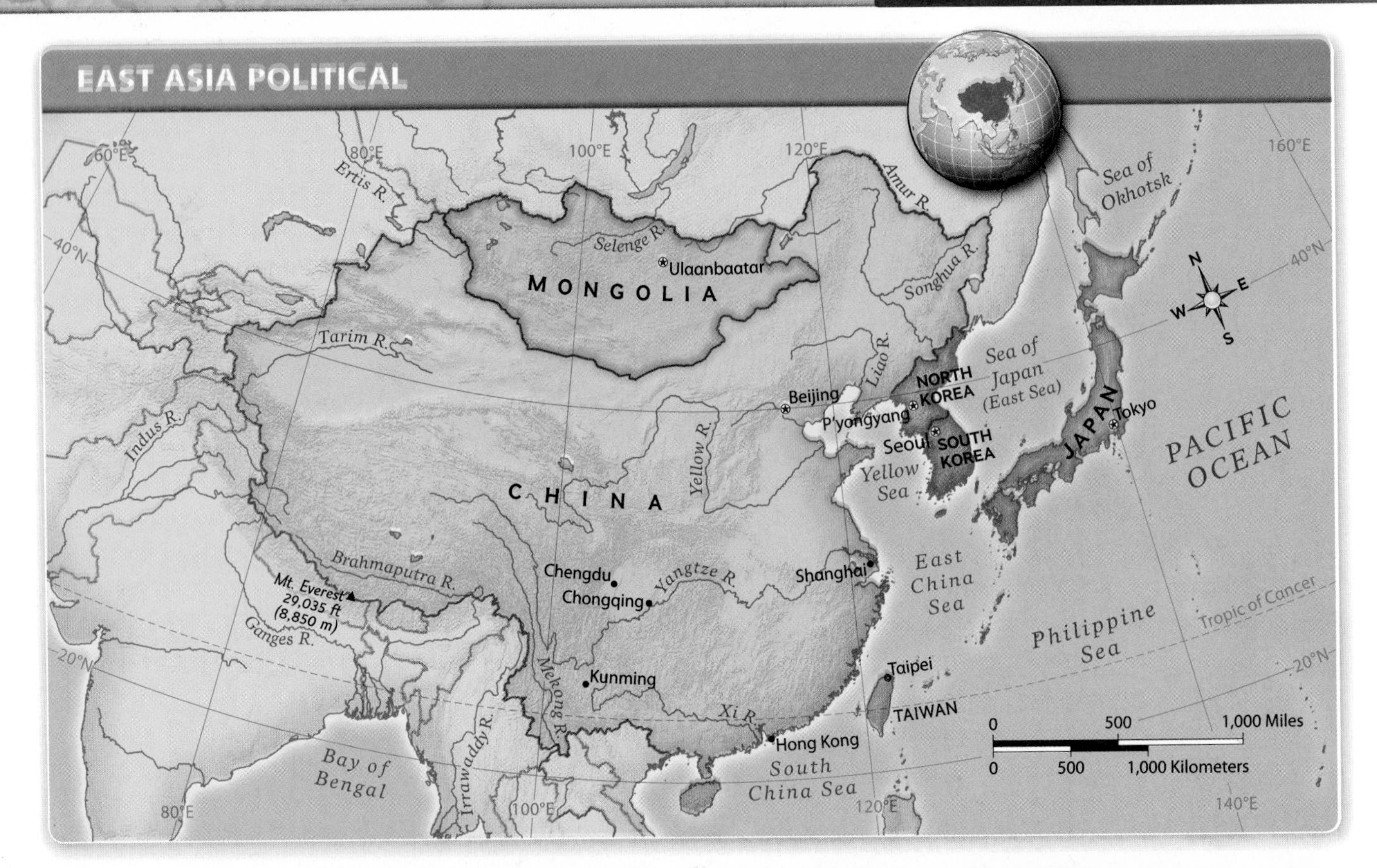

EAST ASIA PHYSICAL

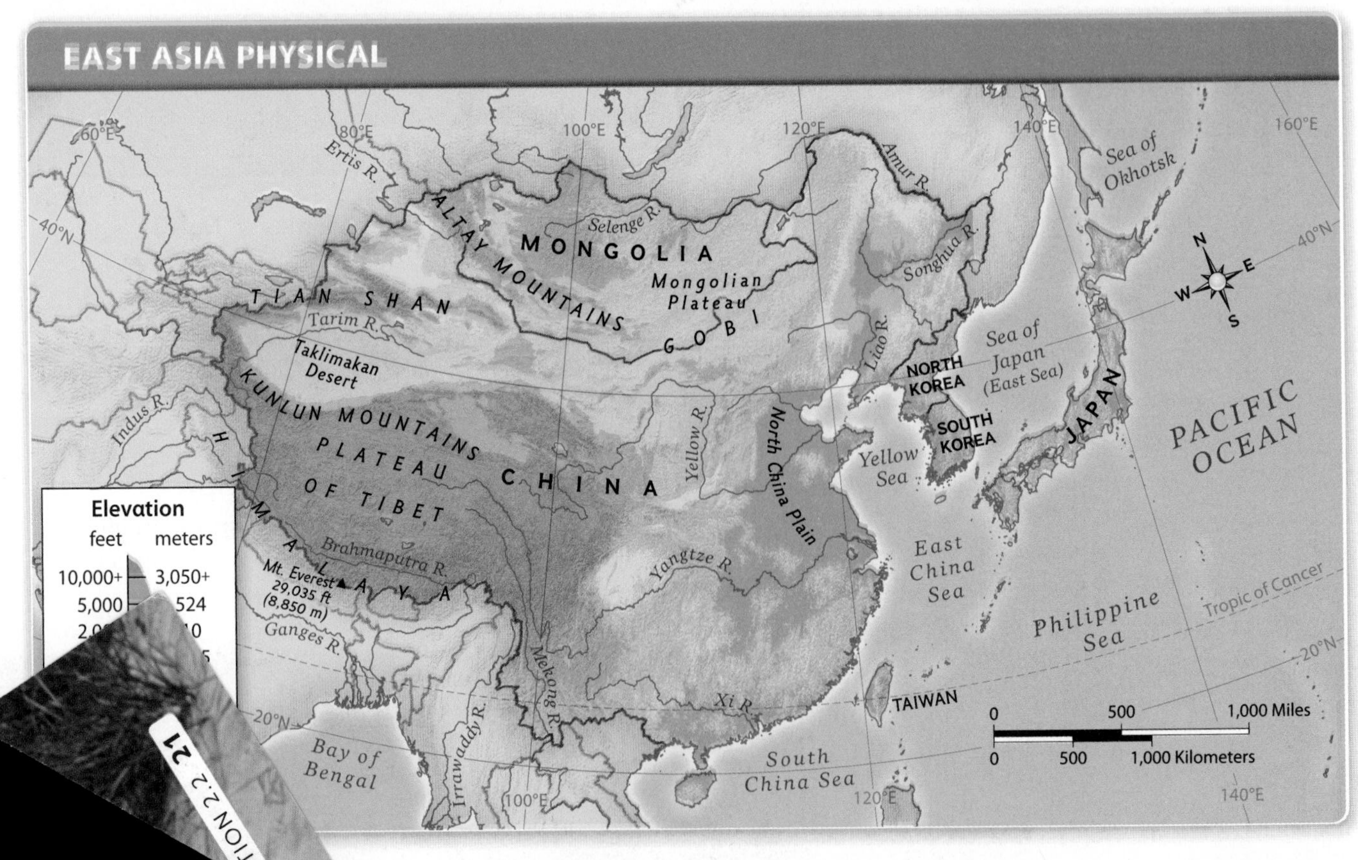

Main Idea Political maps show features that humans have created on Earth's surface. Physical maps show natural features.

The governor of a state needs a map that shows counties and cities. A mountain climber needs a map that shows cliffs, canyons, and ice fields. **Cartographers**, or mapmakers, create different kinds of maps for these different purposes.

Political Maps

A political map shows features that humans have created, such as countries, states, provinces, and cities. These features are labeled, and lines show boundaries, such as those between countries.

Physical Maps

A physical map shows natural features of physical geography. It includes landforms, such as mountains, plains, valleys, and deserts. It also includes oceans, lakes, rivers, and other bodies of water.

A physical map can also show elevation and relief. **Elevation** is the height of a physical feature above sea level. **Relief** is the change in elevation from one place to another. Maps show elevation by using color. The physical map at left uses seven colors for seven ranges of elevation.

Before You Move On

Monitor Comprehension How is a political map different from a physical map?

Critical Viewing The Sobaek Mountains cut diagonally across South Korea. Which map best indicates the location of these mountains?

ONGOING ASSESSMENT

MAP LAB

GeoJournal

1. **Interpret Maps** What is the most mountainous country in East Asia? How did you find the answer?
2. **Human-Environment Interaction** Based on elevations shown on the map, what economic activity would you expect to find on the North China Plain?
3. **Draw Conclusions** What do the locations of Hong Kong, Shanghai, and Tokyo have in common? What conclusion can you draw about the location of cities around the world?

2.4 Map Projections

TECHTREK
myNGconnect.com For additional maps in a variety of projections

Maps and Graphs

Main Idea Cartographers use various projections to show Earth's curved surface on a flat map.

The world is a sphere, but maps are flat. As a result, maps **distort**, or change, shapes, areas, distances, and directions found in the real world. To reduce distortion, mapmakers use **projections**, or ways of showing Earth's curved surface on a flat map. Five common map projections are the azimuthal, Mercator, homolosine, Robinson, and Winkel Tripel. Each projection has strengths and weaknesses—each distorts in a different way.

When cartographers make maps, they need to choose a map projection. The type of projection depends on the map's purpose. Which elements are acceptable to distort? Which are not acceptable to distort? For example, if a cartographer is creating a navigation map, it is important that directions are not distorted. It may not matter, however, if some areas or shapes are distorted.

Before You Move On

Make Inferences How do cartographers decide which projection to use?

AZIMUTHAL PROJECTION

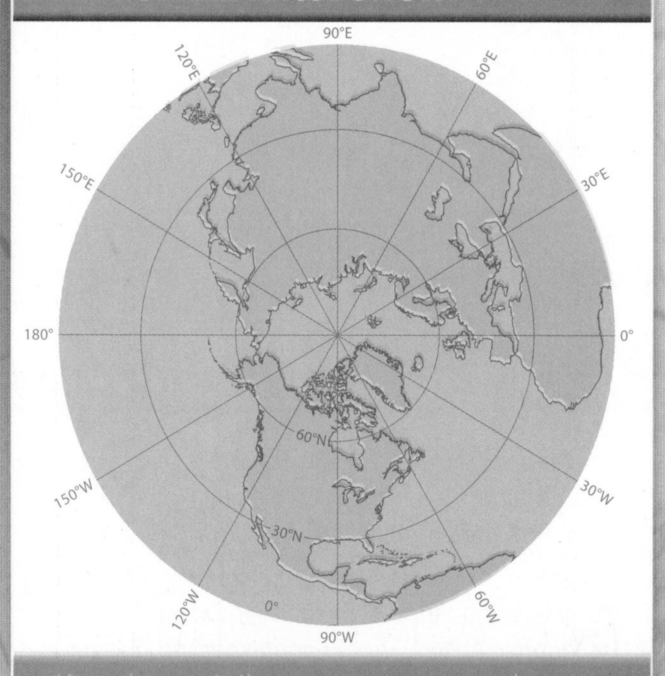

Mapmakers create the azimuthal projection by projecting part of the globe onto a flat surface. The projection shows directions accurately but distorts shapes. It is often used for the polar regions.

MERCATOR PROJECTION

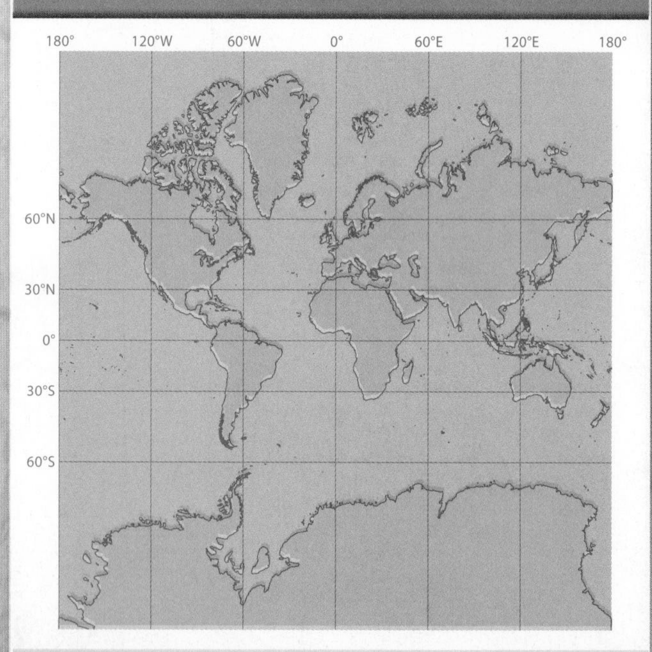

This Mercator projection shows much of Earth accurately, but it distorts the shape and area of land near the North and South Poles. This projection shows direction accurately, so it is good for navigation maps.

HOMOLOSINE PROJECTION

The **homolosine projection** resembles the flattened peel of an orange. It accurately shows the shape and area of landmasses by cutting up the oceans. However, it does not show distances accurately.

ROBINSON PROJECTION

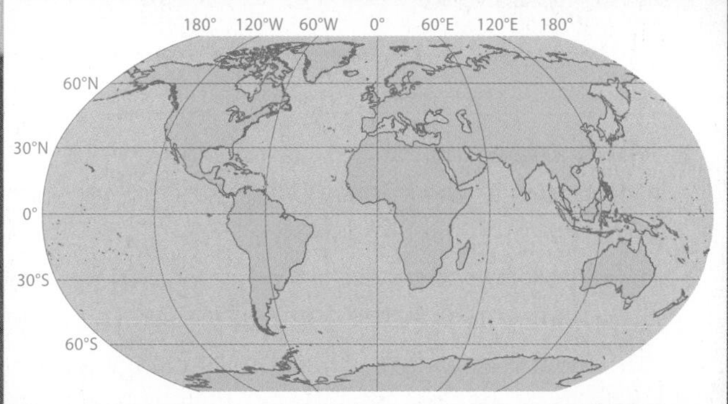

The **Robinson projection** combines the strengths of other projections. It shows the shape and area of the continents and oceans with reasonable accuracy. However, the North and South Poles are distorted.

WINKEL TRIPEL PROJECTION

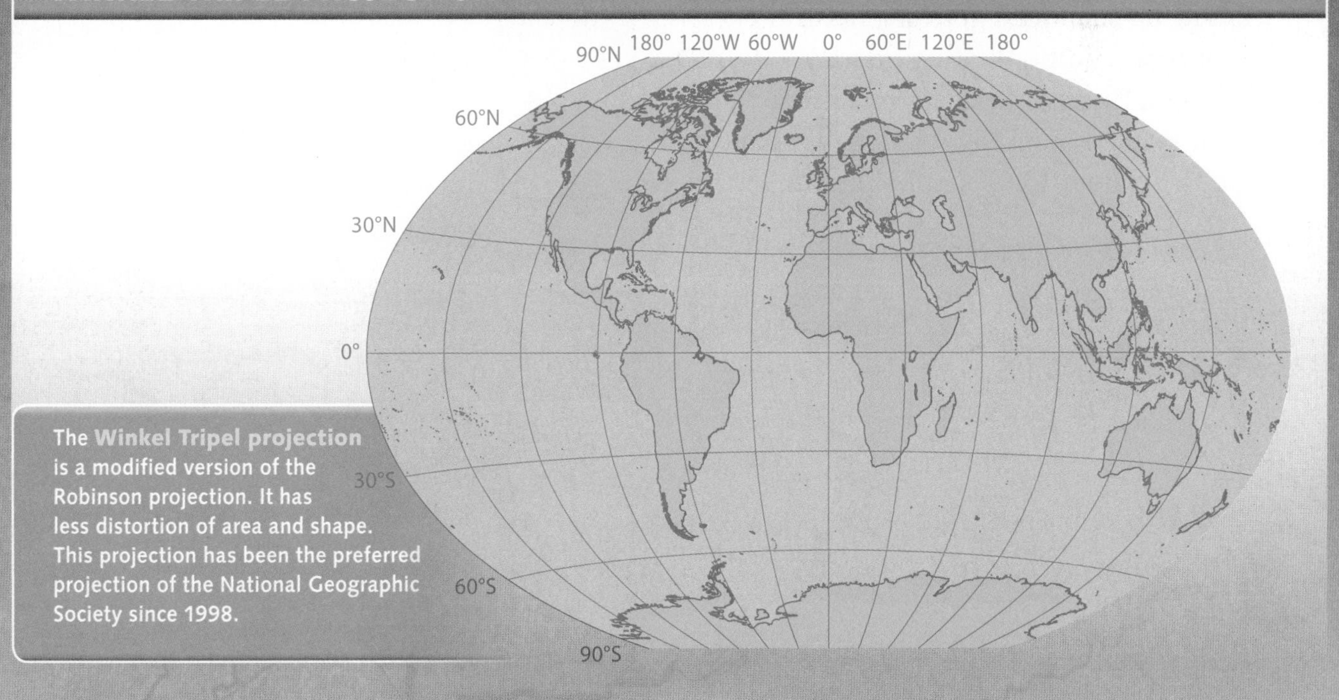

The **Winkel Tripel projection** is a modified version of the Robinson projection. It has less distortion of area and shape. This projection has been the preferred projection of the National Geographic Society since 1998.

ONGOING ASSESSMENT

MAP LAB

1. **Compare and Contrast** Locate Greenland on the Mercator projection and on the Robinson projection. What is similar and different in the two maps? Why?
2. **Location** What does the azimuthal projection show about the relative location of Alaska and Russia?

2.5 Thematic Maps

TECHTREK 

myNGconnect.com For additional examples of thematic maps

Maps and Graphs

Main Idea Thematic maps focus on specific topics, such as the population density or economic activity in a region or country.

Suppose you wanted to create a map showing the location of sports fields in your community. You would create a thematic map, which is a map about a specific **theme**, or topic.

Types of Thematic Maps

Thematic maps are useful for showing a variety of geographic information, including economic activity, natural resources, and population density. Common types of thematic maps are the point symbol map, the dot density map, and the proportional symbol map.

Before You Move On

Make Inferences Look through this textbook and identify another example of a thematic map. Why did you choose this map?

U.S. ALTERNATIVE ENERGY

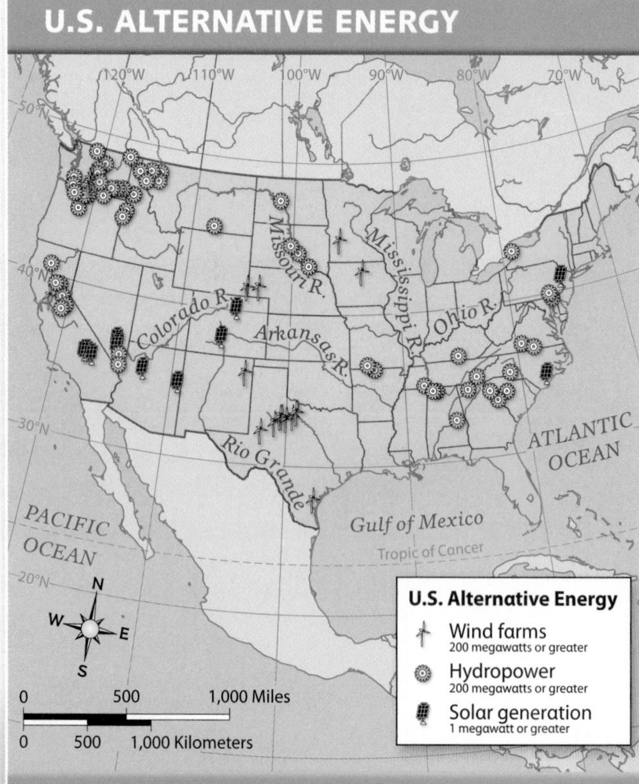

POINT SYMBOL MAP This type of map shows the location of activities at different points. For example, this map has symbols that show some sources of wind, water, and solar energy in the United States.

Critical Viewing Solar panels in the Nevada desert absorb light from the sun and turn it into energy. Why might Nevada be a good location for solar fields?

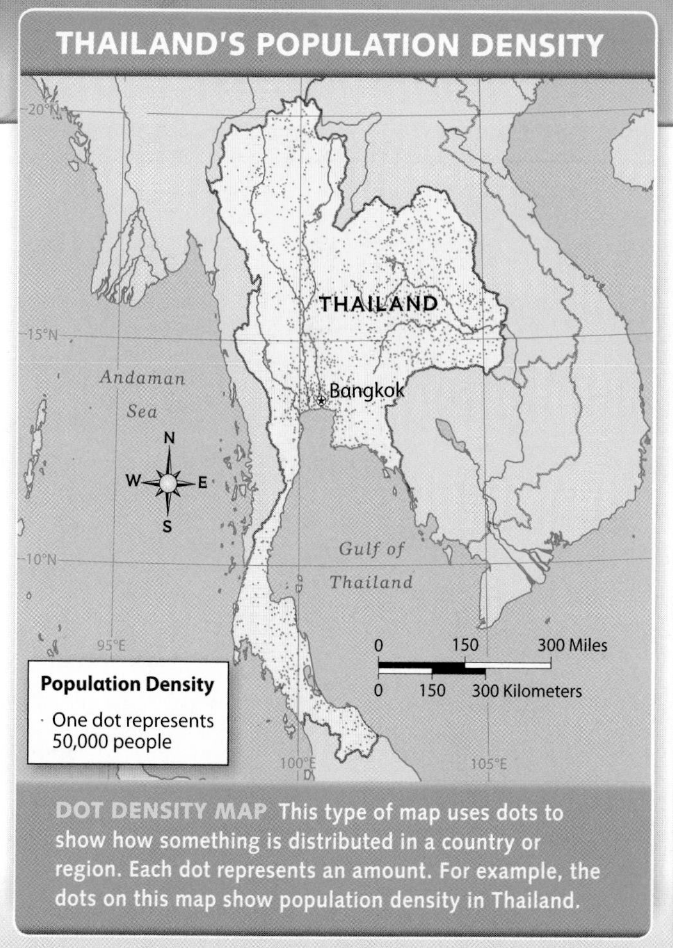

DOT DENSITY MAP This type of map uses dots to show how something is distributed in a country or region. Each dot represents an amount. For example, the dots on this map show population density in Thailand.

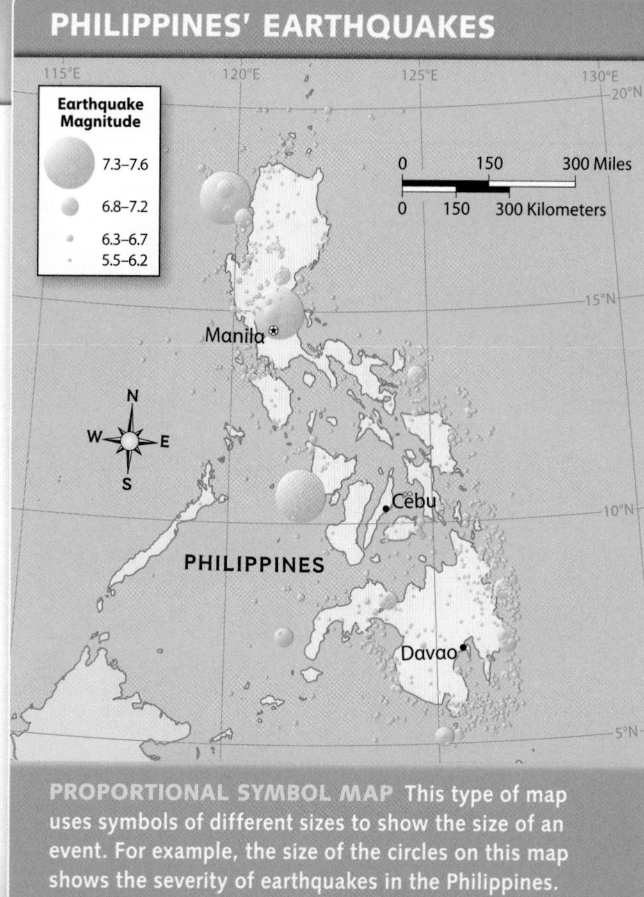

PROPORTIONAL SYMBOL MAP This type of map uses symbols of different sizes to show the size of an event. For example, the size of the circles on this map shows the severity of earthquakes in the Philippines.

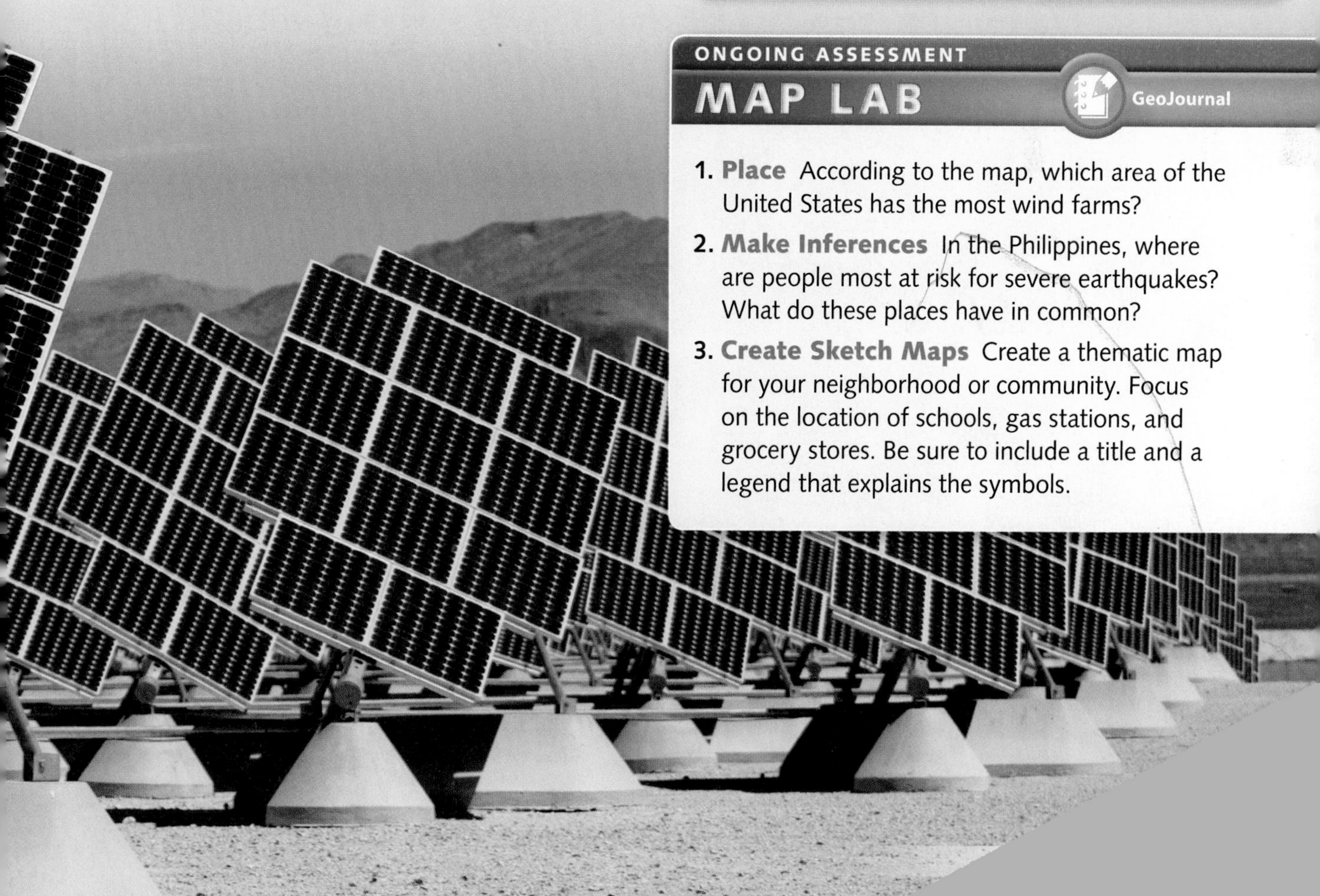

ONGOING ASSESSMENT

MAP LAB

GeoJournal

1. **Place** According to the map, which area of the United States has the most wind farms?
2. **Make Inferences** In the Philippines, where are people most at risk for severe earthquakes? What do these places have in common?
3. **Create Sketch Maps** Create a thematic map for your neighborhood or community. Focus on the location of schools, gas stations, and grocery stores. Be sure to include a title and a legend that explains the symbols.

Photo Gallery • The World At Night

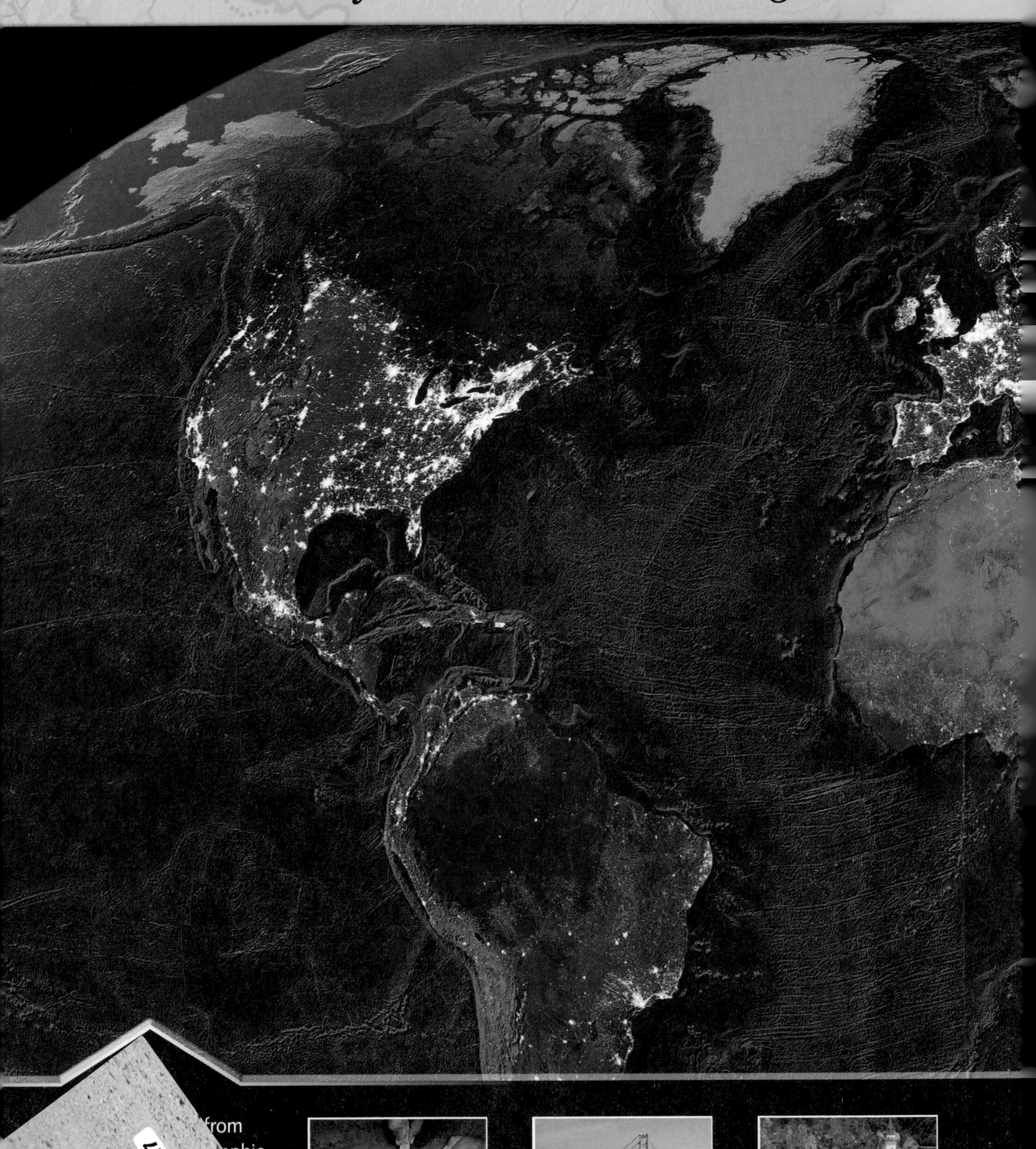

from
aphic

27
CTION 2.5

Critical Viewing This view of Earth at night was made by combining images from three satellites over a one-year period. In addition to the lighted areas, fires and natural gas burn-off are shown in orange.

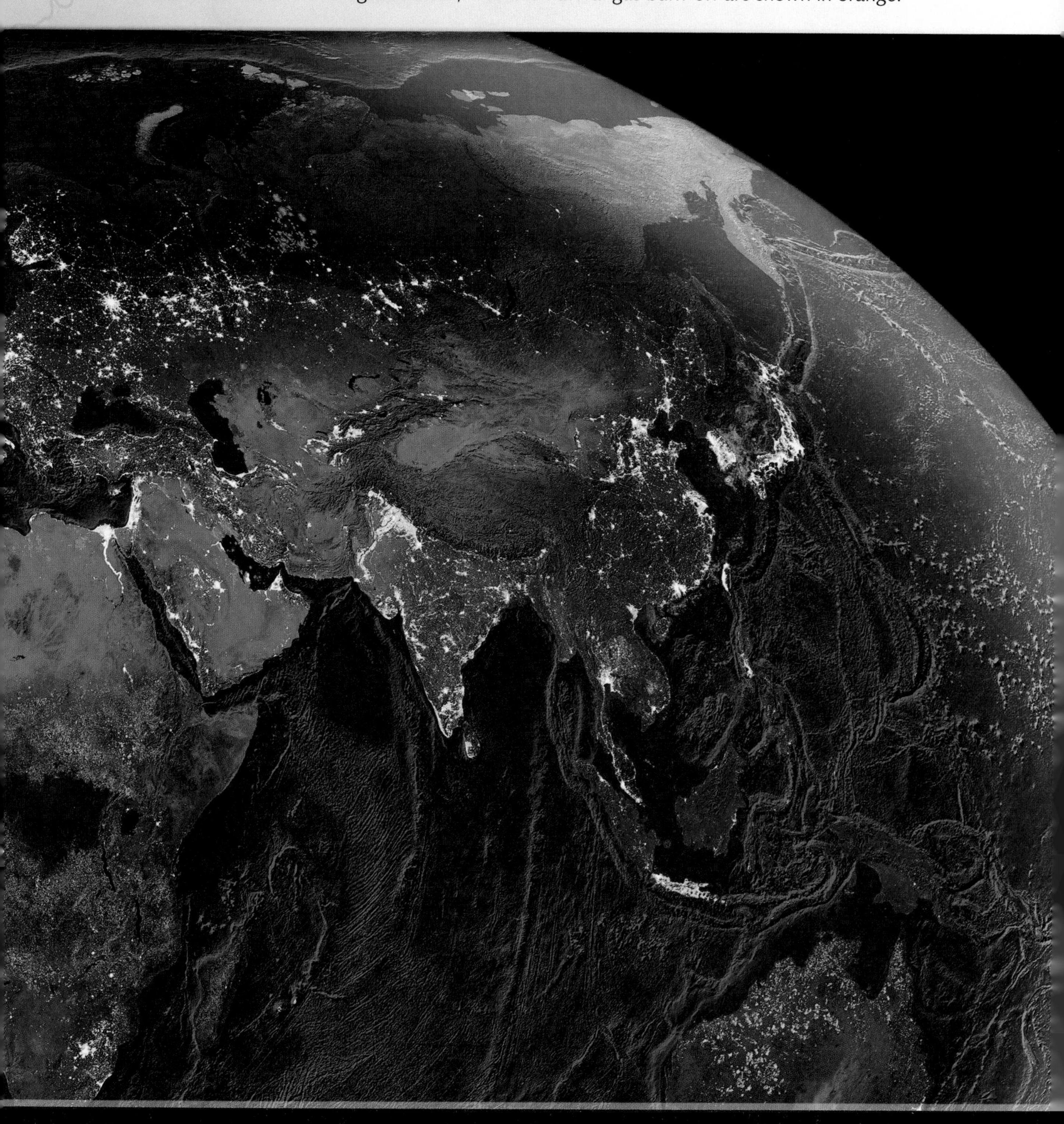

Archaeologist at work

Eastern Hemisphere, 1928

Tokyo, Japan, at night

CHAPTER 1

Review

VOCABULARY

For each pair of words, write one sentence that explains the connection between the two words.

1. absolute location; relative location

> *The absolute location of Washington, D.C., is 39° N, 77° W; its relative location is on the Potomac River.*

2. region; continent
3. latitude; longitude
4. relief; elevation
5. distort; projection

MAIN IDEAS

6. What are Geographic Information Systems? How do geographers use them? (Section 1.1)
7. How are the five themes of geography and the six essential elements similar? How are they different? (Section 1.2)
8. Why is the construction of a highway an example of human-environment interaction? (Section 1.2)
9. What are traits of a region? (Section 1.3)
10. How do latiitude and longitude help determine the absolute location of a place? (Section 2.1)
11. Is a map of the world a large-scale map or a small-scale map? Why? (Section 2.2)
12. How do ... sical maps show elevation and rel... n 2.3)

... ojections distort Earth?

... tic map would a ... how different types of ... xplain your answer.

GEOGRAPHIC THINKING

ANALYZE THE ESSENTIAL QUESTION

How do geographers think about the world?

Critical Thinking: Describe Geographic Information

15. How do geographers use spatial thinking to make sense of space on Earth's surface?
16. How would you describe New York City using the five themes of geography?
17. What is an example of a physical process, and how can it affect a region?

INTERPRET MAPS

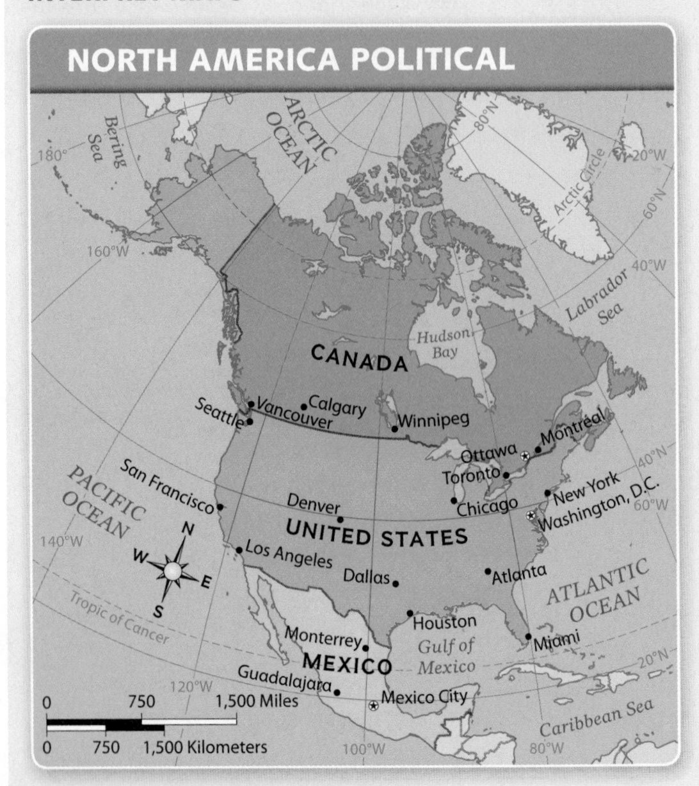

18. **Region** According to the map, what is the relative location of Canada's cities? Why do you think they are located there?
19. **Compare and Contrast** How does the relative location of Mexico's cities compare with the relative location of Canada's cities?

MAPS

ANALYZE THE ESSENTIAL QUESTION

How do people use geography?

Critical Thinking: Make Inferences

20. On a Mercator projection map, Greenland looks larger than South America. However, it is actually much smaller. How could this affect a person's understanding of land areas?

21. A geographer wants to show where economic activities are located. What kind of thematic map should the geographer use? Explain.

22. An explorer is the first to map a region that has hills, canyons, and flat areas. What type of map should the explorer create? Explain.

INTERPRET PRIMARY SOURCES

In 1953, Sir John Hunt led a group of climbers up Mount Everest, the highest mountain in the world. Read Hunt's description of climbing Everest and answer the questions.

> What makes Everest murderous is . . . its cold, its wind, and its climbing difficulties. . . . At 28,000 feet a given volume of air breathed contains only a third as much oxygen as at sea level. On the ground, even if a man were exercising violently, his lungs would need but 50 liters of air per minute. Near Everest's summit he struggles to suck in as much as 200 liters. Since he inhales his air cold and dry and exhales it warm and moist, the stress on his parched lungs and respiratory passages becomes appalling.
>
> –*National Geographic*, July 1954

23. Find Details What are the characteristics of the environment on Mount Everest?

24. Human-Environment Interaction How is this description an example of human-environment interaction?

25. ACTIVE OPTIONS

Synthesize the Essential Questions by completing the activities below.

26. Write a Travel Brochure Write a travel brochure to attract people to a place that you hope to visit. In your brochure, answer geographic questions such as the following: "Where it is located?" "In what region is it located?" "How have people made the environment more livable?" "What are the physical features of the place?" **Share your brochure with the rest of the class.**

Writing Tips
- Take notes before you begin to write.
- Write a catchy slogan that grabs the reader's attention.
- Write a list of reasons that persuades people to travel to the place.

myNGconnect.com For maps and photographs of the five themes of geography

27. Create a Digital Presentation Use presentation software to create a digital presentation on the five themes of geography. For each theme, write several bullet points. In addition, explain why each theme is useful. For each theme, select photographs from the **NG Photo Gallery** or other sources, and maps from the **Online World Atlas** that illustrate the theme. Use this format for your slides:

Region: What is a region?
- A region shares common physical and/or human characteristics.

NATIONAL GEOGRAPHIC CHAPTER 2

PHYSICAL & HUMAN GEOGRAPHY

PREVIEW THE CHAPTER

Essential Question How is the earth continually changing?

SECTION 1 • THE EARTH

KEY VOCABULARY

- solstice
- equinox
- tectonic plate
- continental drift
- plain
- plateau
- continental shelf
- butte
- erosion
- earthquake
- tsunami
- volcano
- evaporation
- condensation
- precipitation

ACADEMIC VOCABULARY

benefit, essential

TERMS & NAMES

- Ring of Fire

Essential Question What shapes the earth's varied environments?

SECTION 2 • PHYSICAL GEOGRAPHY

KEY VOCABULARY

- climate
- weather
- vegetation
- hurricane
- cyclone
- tornado
- raw material
- nonrenewable resource
- renewable resource
- habitat
- ecosystem
- marine life

ACADEMIC VOCABULARY

restore, impact

Essential Question How has geography influenced cultures around the world?

SECTION 3 • HUMAN GEOGRAPHY

KEY VOCABULARY

- culture
- civilization
- communal
- gaucho
- culture region
- kimono
- monotheistic religion
- polytheistic religion
- economy
- capital
- entrepreneurship
- free enterprise economy
- gross domestic product (GDP)
- government
- citizen
- democracy
- human rights

ACADEMIC VOCABULARY

symbol

TERMS & NAMES

- United Nations (UN)
- Universal Declaration of Human Rights

Go to myNGconnect.com for more on physical and human geography.

Scientists observe a cloud of ash erupting from Italy's Mount Etna.

1.1 Earth's Rotation and Revolution

TECHTREK
myNGconnect.com For photos and a model of the four seasons

Digital Library

Student Resources

Main Idea Earth's tilt, rotation, and revolution cause weather changes and the four seasons.

In the summer, people who live in the northern parts of Iceland, Norway, Sweden, and Finland have more than 20 hours of daylight. In the winter, these same people have more than 20 hours of darkness. These long days and nights result from Earth's tilt and revolution around the sun.

Revolution and Rotation

The solar system is formed by the sun, Earth, and seven other planets. Earth, which is the third planet from the sun, revolves around the sun at a speed of about 67,000 miles per hour. It takes one year for Earth to make one revolution.

At the same time, Earth rotates on its axis, an imaginary line that runs from the North Pole to the South Pole through the center. Each rotation takes almost one day.

Earth tilts at an angle of about 23.5°. Because of this tilt, the Northern Hemisphere receives more direct sunlight for half the year, and temperatures are warmer. During these months, the Southern Hemisphere receives less direct sunlight, and temperatures are cooler.

As Earth continues around the sun, the Northern Hemisphere faces the sun less directly and temperatures are cooler. Meanwhile the Southern Hemisphere faces the sun more directly, and temperatures are warmer. This process creates the four seasons in both hemispheres.

Summer and Winter Solstices

The exact moment at which summer and winter start is called a **solstice**. June 20 or 21 is the summer solstice in the Northern Hemisphere. It is the longest day of the year. Six months later, on December 21 or 22, the Northern Hemisphere has its winter solstice. This is the shortest day of the year.

The Southern Hemisphere is exactly the opposite. June 20 or 21 is its winter solstice and December 21 or 22 is its summer solstice.

Critical Viewing Bathers in Iceland celebrate the summer solstice with a midnight swim in a hot spring. Based on the photo, what can you tell about the summer solstice in Iceland?

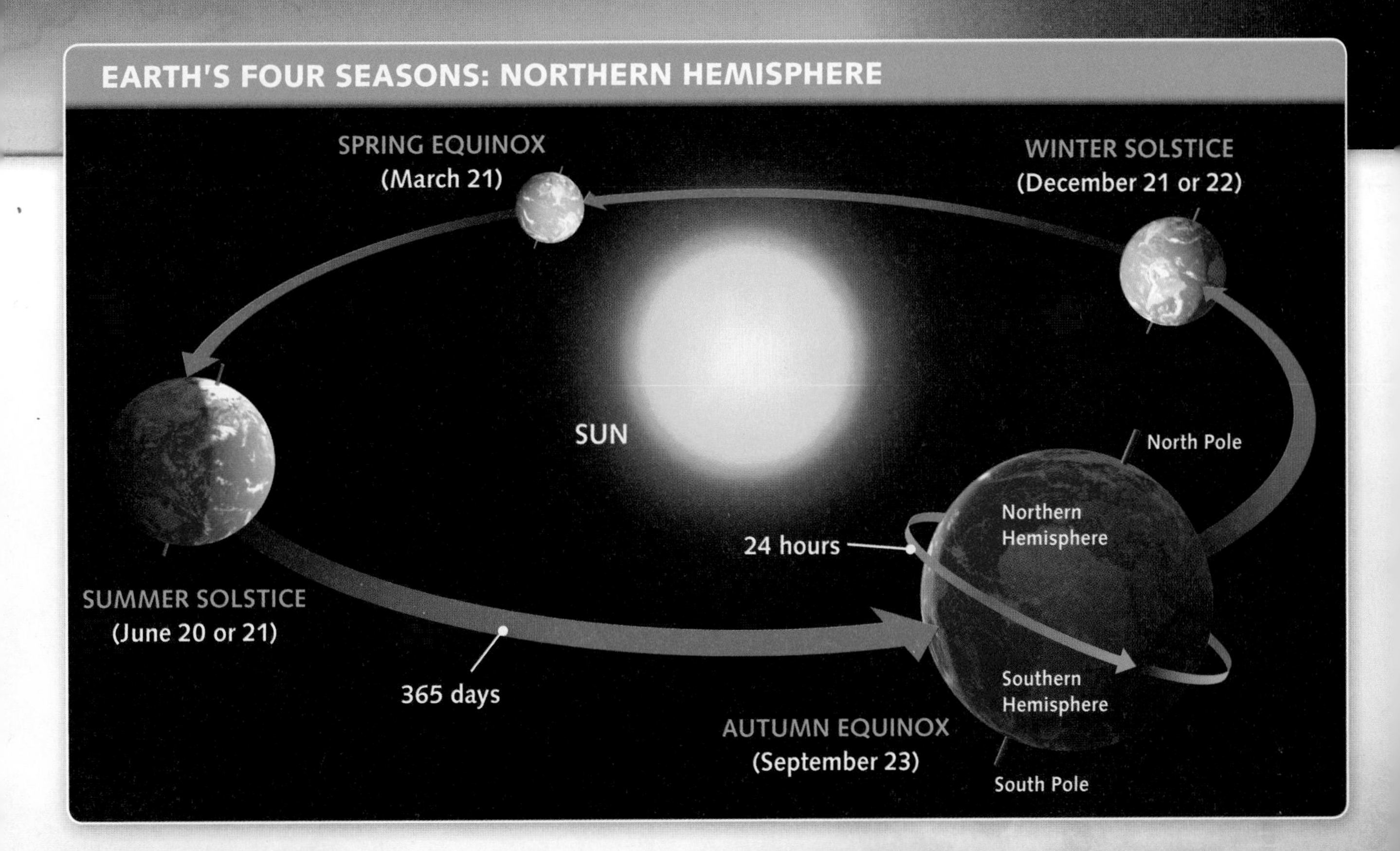

Spring and Autumn Equinoxes

The beginning of spring and autumn is called an **equinox**. Twice a year, the sun's rays hit the equator directly, and day and night are the same length. In the Northern Hemisphere, the spring equinox occurs around March 21, and the autumn equinox occurs around September 23. The Southern Hemisphere is exactly the opposite.

Before You Move On

Monitor Comprehension How do Earth's tilt, rotation, and revolution cause the seasons?

ONGOING ASSESSMENT

VIEWING LAB

GeoJournal

1. **Analyze Models** According to the model above, when do the sun's rays hit the Southern Hemisphere most directly? This is the beginning of which season?
2. **Analyze Visuals** What happens to the sun in Iceland on the day of the summer solstice? Why does this happen?
3. **Compare and Contrast** How are the spring equinox and the autumn equinox alike?
4. **Make Inferences** What happens to the length of days in the Northern Hemisphere after the spring equinox?

1.2 Earth's Complex Structure

TECHTREK
myNGconnect.com For photos and diagrams that illustrate plate tectonics

Digital Library
Student Resources

Main Idea Physical processes within Earth bring about changes on the surface.

If you could dig a tunnel to Earth's center, you would travel through several layers. Each layer would be under tremendous pressure and give off intense heat.

Earth's Layers

On your journey, you would first pass through the crust. This layer includes landmasses and the ocean floor. It is about 30 miles thick.

Next, you would come to the mantle, which consists of molten, or melted, rocks called magma. The mantle is about 1,800 miles thick and has two parts—the upper mantle and the lower mantle.

Descending even deeper, you would find yourself in the outer core, which is about 1,400 miles thick. This layer is mostly liquid, consisting of molten iron and nickel.

At the very center is the inner core. It is about 700 miles thick. It reaches a temperature of 12,000° F—hotter than the surface of the sun. The inner core is made up of iron, which remains solid because the pressure from all the layers above it is so intense.

EARTH'S STRUCTURE

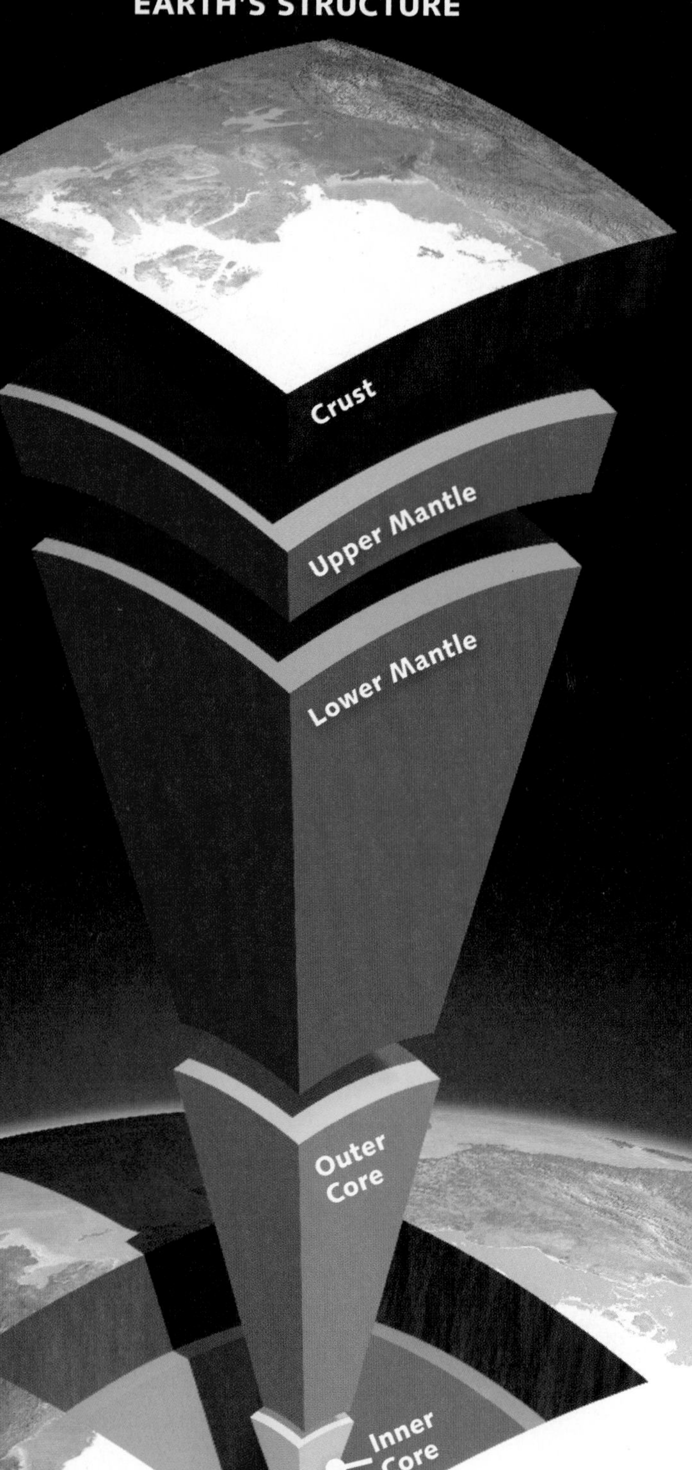

TECTONIC PLATE MOVEMENTS

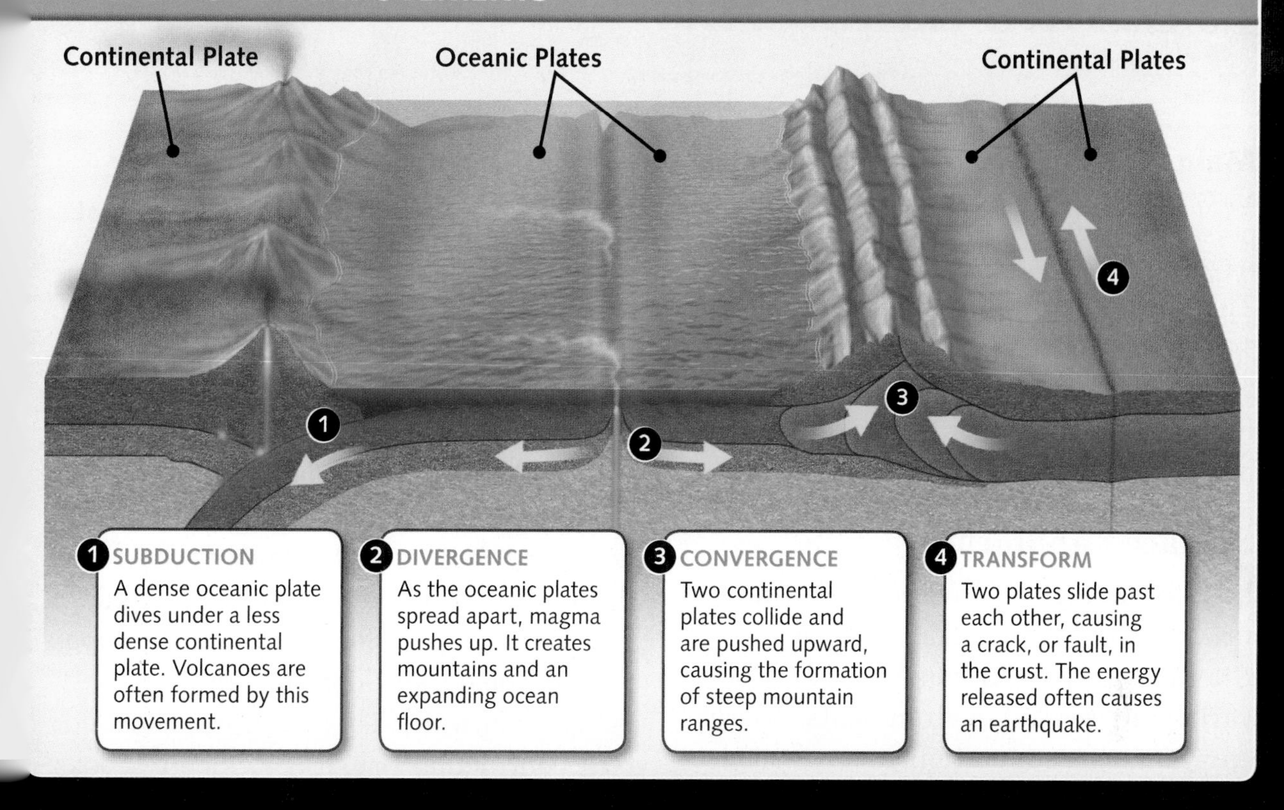

Tectonic Plates

Earth's crust is divided into sections called **tectonic plates**. The plates float on Earth's mantle. They are constantly shifting and may move up to four inches a year.

The seven continents rest on these tectonic plates. As the plates have shifted over time, the continents have moved into their current positions. This slow movement of the continents is known as **continental drift**.

Tectonic plates move in four ways, as shown in the diagram above. The enormous force of the movements and collisions creates mountains and causes earthquakes and volcanoes.

Before You Move On

Make Inferences How do the movements of tectonic plates change Earth's surface?

ONGOING ASSESSMENT

VIEWING LAB

GeoJournal

1. **Analyze Visuals** According to the diagram above, which tectonic plate movement often results in volcanoes? Which movement can cause earthquakes?
2. **Place** The Himalaya Mountains formed by convergence when the Indian plate collided with the Eurasian plate. The Indian plate is still moving almost an inch north every year. How do you predict this will affect the Himalayas?
3. **Summarize** What is the main characteristic of each layer of Earth?

1.3 Earth's Landforms

Main Idea Landforms are physical features on Earth's surface. They are continually reshaped by physical processes.

The Rocky Mountains rise more than 14,000 feet above sea level. The Grand Canyon is more than 5,000 feet deep. Both are landforms, or physical features on Earth's surface.

Surface Landforms

Landforms such as the Rocky Mountains in western North America and the Grand Canyon in Arizona provide a variety of physical environments. These environments support millions of plants and animals.

Several common landforms are found on Earth's surface. A mountain is a high, steep elevation. A hill also slopes upward but is less steep and rugged. In contrast, a **plain** is a level area. The Great Plains, for example, are flat landforms stretching from the Mississippi River to the Rocky Mountains. A **plateau** is a plain that sits high above sea level and usually has a cliff on all sides. A valley is a low-lying area that is surrounded by mountains.

Ocean Landforms

Earth's oceans also have landforms that are underwater. Mountains and valleys rise and fall along the ocean floor. Volcanoes erupt with hot magma, which hardens as it cools to form new crust.

The edge of a continent often extends out under the water. This land is called the **continental shelf**. Most of Earth's marine life lives at this level of the ocean. Beyond the continental shelf, the land develops a steep slope. Beyond the slope, before the ocean floor, the land slopes slightly upward. This landform is called the continental rise. It is formed by rocks and sediment carried by ocean currents. Together, these landforms are known as the continental margin.

Visual Vocabulary A **butte** (BYOOT) is a hill or mountain with steep sides and a flat top. These buttes in Monument Valley, Arizona, are called "the Mittens."

THE CONTINENTAL MARGIN

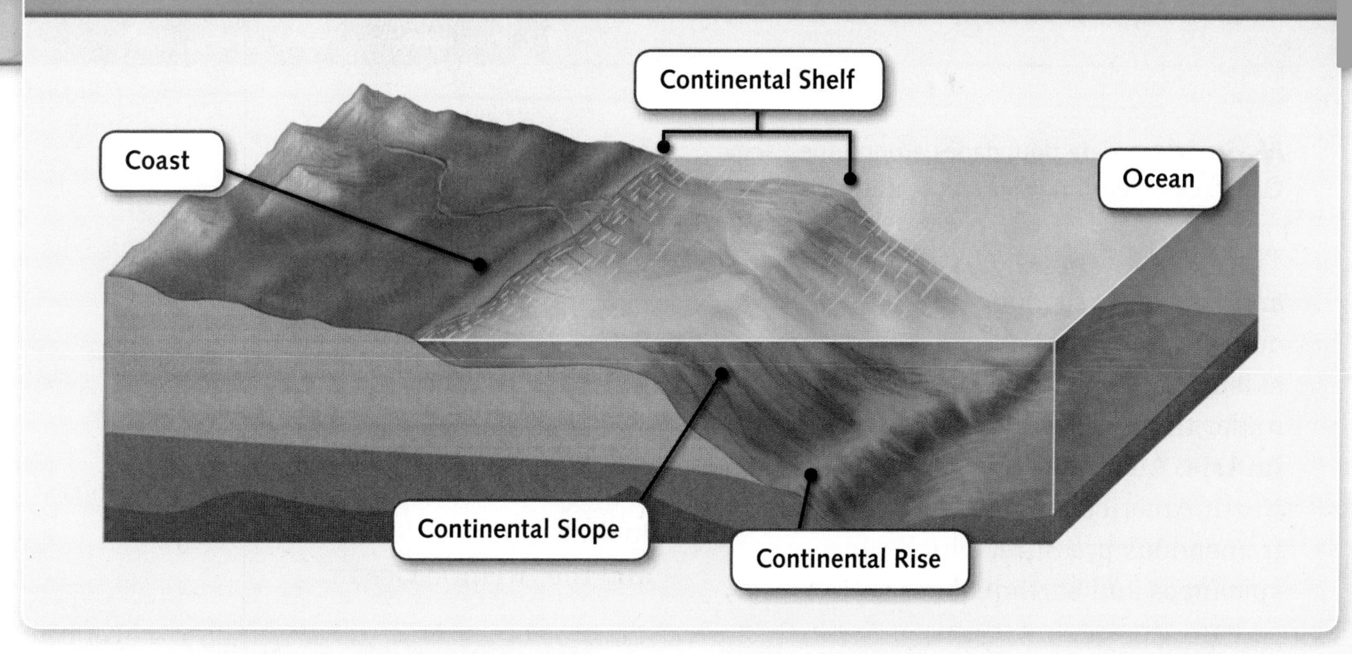

The Changing Earth

Earth is always changing, and the changes affect plant and animal life. For example, a flood can cause severe erosion, which can ruin farmers' fields. **Erosion** is the process by which rocks and soil slowly break apart and are swept away.

Erosion also results from weathering, which is when air, water, wind, or ice slowly wear away rocks and soil. The buttes in Monument Valley, Arizona, were formed in this way over a span of millions of years.

Before You Move On

Summarize How do physical processes reshape Earth's landforms?

ONGOING ASSESSMENT

SPEAKING LAB

Turn and Talk How does a landform in your community affect your daily life? Make a list of landforms you encounter every day, such as hills or plains. Then turn to a partner and talk about how one of these landforms affects your life. Develop an outline to present to the class. List your topic first. Then list your supporting details.

1.4 The Ring of Fire

Main Idea Plate boundaries around the Pacific Ocean cause earthquakes and volcanic eruptions.

The **Ring of Fire** is a circle of volcanoes and earthquakes along the rim, or outer edge, of the Pacific Ocean. It exists because a large tectonic plate under the ocean slides against plates in Asia, Australia, South America, and North America. The movements create tremendous pressure, which causes volcanoes and earthquakes.

Earthquakes

An **earthquake** is a violent shaking of Earth's crust. Many earthquakes occur along faults, which are cracks in Earth's surface. Earthquakes are common in the Ring of Fire, but they also occur in other areas on Earth. One area runs from the land around the Mediterranean Sea through East Asia. Other earthquake zones include the middle of the Arctic Ocean and the Atlantic Ocean.

THE RING OF FIRE

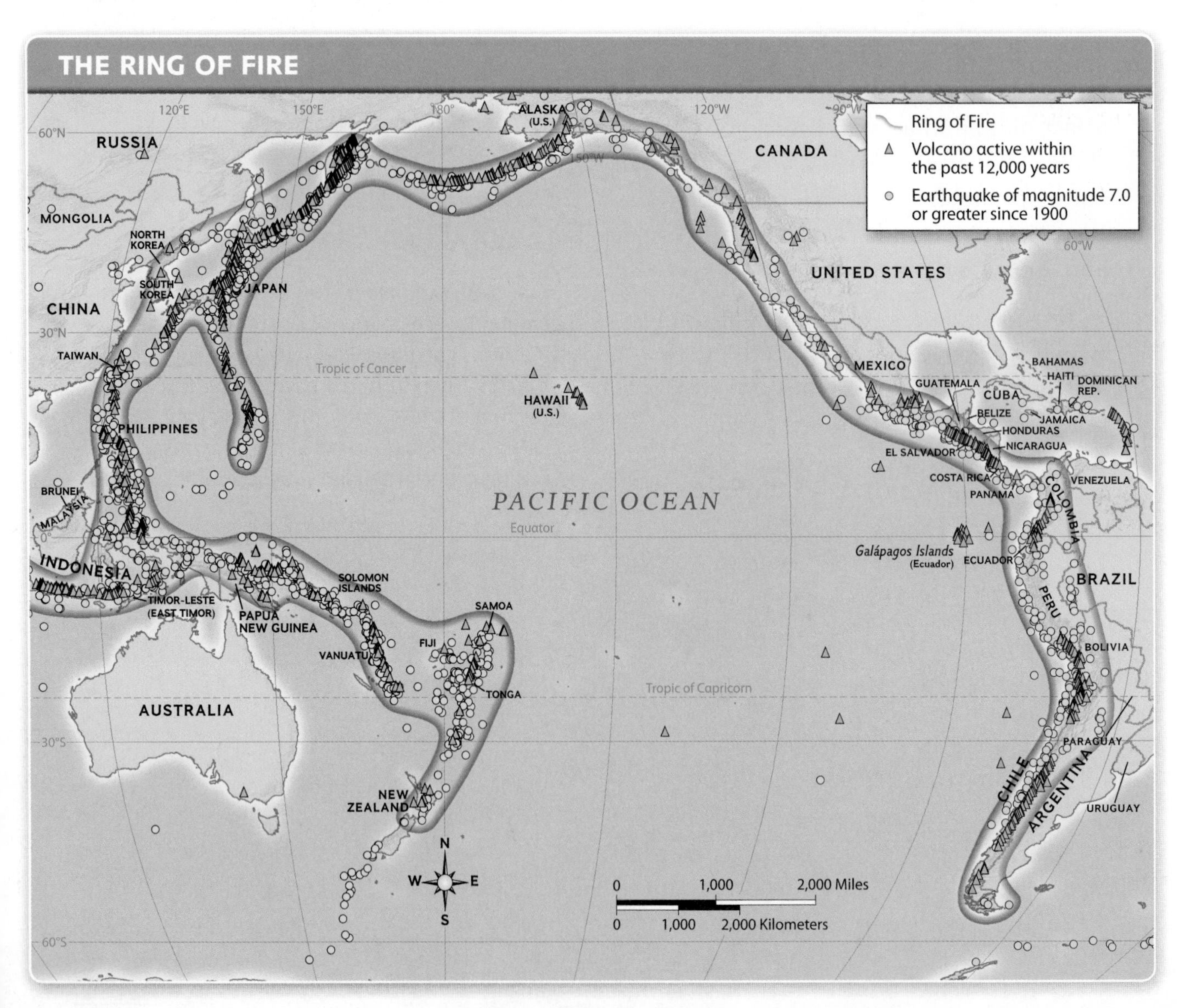

Mount St. Helens, Washington

On May 18, 1980, Mount St. Helens erupted, blasting away one side of the mountain.

Earthquakes can cause buildings, bridges, and roads to collapse. For example, in 2010, an earthquake in Haiti killed more than 200,000 people. Many people who died were trapped under buildings that collapsed.

Earthquakes beneath the ocean can cause **tsunamis**, which are large, powerful ocean waves that can cause great destruction along the coast.

Volcanoes

The Ring of Fire contains more than 75 percent of the world's volcanoes. A **volcano** is a mountain that erupts in an explosion of molten rock, gases, and ash. Lava, which is molten rock, flows down the side of the mountain.

Volcanoes can cause severe damage. In 1883, Krakatoa in Indonesia spewed ash and rock fragments over an area of 300,000 square miles. It also triggered a tsunami that killed 36,000 people. Yet volcanoes can also **benefit**, or be useful to, plant and animal life. For example, mineral-rich lava turns into fertile soil.

Scientists have learned how to predict volcanic eruptions, and engineers can design buildings that survive earthquakes. As a result, more people can live safely.

Before You Move On

Monitor Comprehension What is the Ring of Fire and why is it significant?

ONGOING ASSESSMENT

WRITING LAB

GeoJournal

1. **Make Inferences** Why do so many buildings collapse during an earthquake?
2. **Human-Environment Interaction** How do earthquakes and volcanoes affect people? How have people tried to solve these problems? Copy and complete the chart.

DISASTER	PROBLEM	SOLUTION
earthquake		
volcano		

3. **Write an Action Plan** Imagine that you live along the Ring of Fire. With a partner, write an outline of an action plan to help people survive a serious earthquake.

1.5 Waters of the Earth

TECHTREK
myNGconnect.com For photos of water and a diagram of the hydrologic cycle
Digital Library
Student Resources

Main Idea Water is essential for all forms of life on Earth.

The Mississippi River begins as a small stream in northern Minnesota. More than 2,000 miles south, it pours more than 4.7 million gallons of water per second into the Gulf of Mexico. Water flowing in rivers like the Mississippi is **essential**, or necessary, for all forms of life.

Fresh Water

The Mississippi River contains fresh water. People use fresh water to drink, cook, bathe, and irrigate crops. Early civilizations developed along rivers such as the Nile River in Egypt because of the available fresh water.

Different bodies of fresh water exist for different geographic reasons. A river is a path of water that flows from a higher elevation to a lower elevation. Streams, brooks, and creeks are like rivers, but smaller. A lake is a large body of water that is surrounded by land.

Salt Water

Salt water contains salt and other minerals. It is a major source of the world's seafood supply and a means of transportation. Oceans are large bodies of salt water. Earth's four oceans are the Atlantic, the Pacific, the Indian, and the Arctic. Continuously moving flows of water, called currents, circulate through the oceans and affect climates on land.

Seas are smaller bodies of salt water. The Red Sea, for example, lies between the Arabian Peninsula and eastern Africa.

THE HYDROLOGIC CYCLE

The hydrologic cycle is the continual movement of water from Earth's surface into the air and back again.

2 CONDENSATION
During **condensation**, cooler temperatures in the atmosphere cause the water vapor to change into droplets that form clouds.

1 EVAPORATION
During **evaporation**, the sun heats the ocean, and water vapor rises up into the atmosphere.

Before You Move On

Monitor Comprehension How is water essential for all life on Earth?

WORLD'S LONGEST RIVERS

River	Location	Length (miles)
Nile	Africa	4,241
Amazon	South America	4,000
Chang Jiang (Yangtze)	Asia	3,964
Mississippi-Missouri	North America	3,710
Yenisey-Angara	Asia	3,440

Source: *National Geographic Atlas of the World*, 8th ed.

ONGOING ASSESSMENT

DATA LAB

1. **Interpret Charts** According to the chart, which continent has two of the longest world rivers, and what are they? How do you think the two rivers have affected that continent?
2. **Interpret Models** How does the hydrologic cycle explain why rivers and lakes do not run out of water?
3. **Location** St. Louis, Missouri, is located just south of where the Missouri River flows into the Mississippi River. Why is this a good location for a major city?

2.1 Climate and Weather

Main Idea Climate and weather are different, but they both influence life on Earth.

People who live in Sacramento, California, have mild winters. When they go skiing in the nearby Sierra Nevada Mountains, they wear parkas to protect themselves from the colder temperatures. They have adapted to a different climate.

Climate Elements

Climate is the average condition of the atmosphere over a long period of time. It includes average temperature, average precipitation, and the amount of change from one season to another. For example, Fairbanks, Alaska, has a cold climate. In the winter, the temperature can reach -8°F. Yet the temperature can rise to 90°F in the summer. The city goes through changes from one season to another.

Four factors that affect a region's climate are latitude, elevation, prevailing winds, and ocean currents. Places at high latitudes, such as Fairbanks, experience more change between winter and summer. Places close to the equator have nearly the same temperature throughout the year. Places at higher elevations have generally colder temperatures than places closer to sea level.

Prevailing winds are winds coming from one direction that blow most of the time. In Florida in the summer, the prevailing winds come from the south, making a warm climate even hotter.

Ocean currents also affect climate. The Gulf Stream is a current that carries warm water from the Caribbean Sea toward Europe. Air passing over the water becomes warm and helps create a mild winter climate in England and Ireland.

Critical Viewing Skiers head to the top of Clouds Rest in Yosemite National Park, California. What does the photo suggest about the weather at this location?

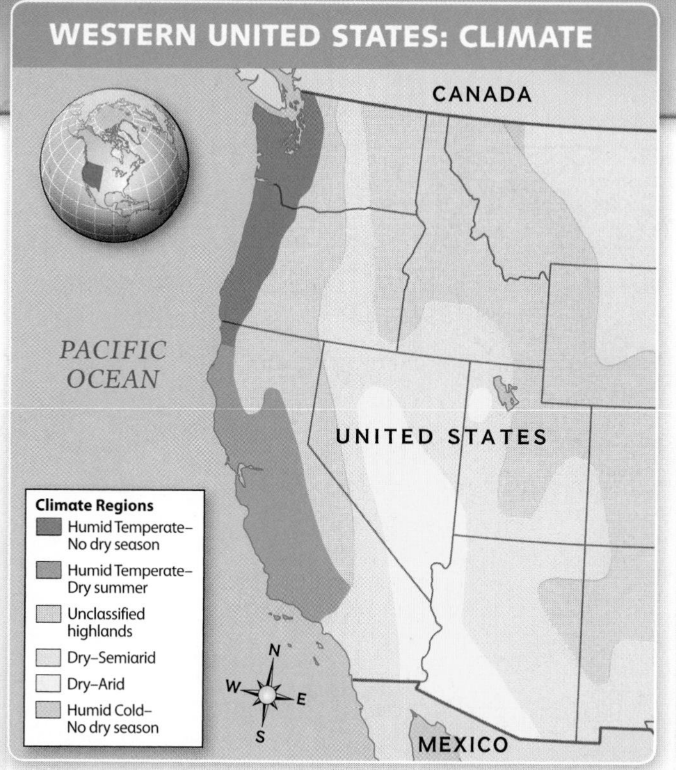

WESTERN UNITED STATES: WEATHER

Wet, stormy weather
Sunny, dry weather
Cold front
Warm front

CANADA

PACIFIC OCEAN

UNITED STATES

MEXICO

N W E S

Weather Conditions

Weather is the condition of the atmosphere at a particular time. It includes the temperature, precipitation, and humidity for a particular day or week. Humidity is the amount of water vapor in the air. If a weather forecaster says the humidity is at 95 percent, he or she means that the air is holding a large amount of water vapor.

Weather changes because of air masses. An air mass is a large area of air that has the same temperature and humidity. The boundary between two air masses is called a front. If a forecaster talks about a warm, humid front, he or she usually means that thunderstorms are headed toward the area.

Before You Move On

Monitor Comprehension What is the difference between climate and weather?

ONGOING ASSESSMENT

MAP LAB

1. **Interpret Maps** Compare the weather map and the climate map of the western United States. How do you think ocean currents and mountains affect the climate and the weather?
2. **Make Inferences** How might the climate and weather of the western United States influence everyday life there?
3. **Place** The chart shows average temperatures and rainfall for Los Angeles, California. Ask and answer questions about the data.

Los Angeles, CA	January	July	November
Temperature (°F)	57.1	69.3	61.6
Rainfall (inches)	2.98	0.03	1.13

Source: National Drought Mitigation Center

2.2 World Climate Regions

myNGconnect.com For an online map and photos of world climate regions

Maps and Graphs

Digital Library

Main Idea Geographers identify climate regions to help them understand and categorize life on Earth.

A climate region is a group of places that have similar temperatures, precipitation levels, and changes in weather. Geographers have identified 5 climate regions that are broken down into 12 subcategories. Places that are located in the same subcategory often have similar vegetation, or plant life.

Before You Move On

Make Inferences How might climate regions help geographers analyze life in a particular place?

WORLD CLIMATE REGIONS

A Dry Climates have little to no rain or snow and both hot and cold temperatures. Plant life includes shrubs and cacti.

Saguaro Cactus, Sonoran Desert, Arizona

B Humid Temperate Climates have cool winters, warm summers, and ample rainfall. Plant life includes mixed forests with evergreens and leafy trees.

Mixed Forest, Great Smoky Mountains, North Carolina

C Humid Equatorial Climates are found near the equator. They have high temperatures and rainfall all or most of the year. Plant life includes tropical plants and rain forests or grasslands with trees.

Bromeliads, Amazon rain forest, Peru

Mosses, Disko Bay, Greenland

D **Tundra** or **Ice Climates** are north of the Arctic Circle and south of the Antarctic Circle. They have long, cold winters and short summers. Plant life includes mosses or no vegetation.

Natural Park, Eastern Siberia, Russia

E **Humid Cold Climates** have cold winters, warm summers, rain, and snow. Plant life includes evergreen or deciduous (leafy) forests.

TECHTREK

Go to myNGconnect.com to explore this map with the Interactive Map Tool.

EUROPE
ASIA
AFRICA
PACIFIC OCEAN
INDIAN OCEAN
AUSTRALIA
ANTARCTICA
N
W E
S

Humid Equatorial	Humid Temperate	
No dry season	No dry season	Tundra or ice
Short dry season	Dry summer	Unclassified highlands
Long dry season	Dry winter	
Dry	**Humid Cold**	
Semiarid	Dry winter	
Arid	No dry season	

ONGOING ASSESSMENT

MAP LAB

GeoJournal

1. **Interpret Maps** What is the most common climate in northern Africa? How might this climate affect population?
2. **Compare and Contrast** How does the climate of western Europe differ from that of eastern Europe?
3. **Human-Environment Interaction** What advantages would humid temperate climates have for farming? For logging?

2.3 Extreme Weather

TECHTREK

myNGconnect.com For a map and photos of extreme weather and a diagram of tornado formation

Maps and Graphs

Digital Library

Student Resources

Main Idea Extreme weather can cause great destruction, but scientists are lessening its effect.

On August 29, 2005, Hurricane Katrina raced toward New Orleans, Louisiana. The water level in the Gulf of Mexico rose 34 feet and flooded 80 percent of the city. Thousands of people lost their homes and many businesses were destroyed.

Wild Weather

Katrina is an example of extreme weather, which is weather so powerful it can deeply affect human lives. A **hurricane** such as Katrina is a strong storm with swirling winds and heavy rainfall. Winds rotate fiercely and can reach 200 miles per hour. A hurricane is a type of **cyclone**, which is a storm with rotating winds. In the Eastern Hemisphere, a cyclone is called a typhoon.

A **tornado** is a smaller storm than a cyclone, but it has even more powerful winds that can reach 300 miles per hour. A tornado follows an unpredictable path and can rip buildings from their foundations. Tornadoes occur all over the world, but most of them form in the United States, east of the Rocky Mountains.

Other types of extreme weather are not as dangerous as cyclones or tornadoes, but they still put people at risk. A flood occurs when water covers an area of land that is usually dry. Floods often occur after a cyclone. A blizzard is a heavy snowstorm with strong winds and very cold temperatures. A drought results when the amount of rainfall drops far below the average amount. It is sometimes accompanied by a heat wave, or unusually high temperatures over a period of time.

HOW A TORNADO FORMS

Air rises from the ground into the bottom of a thunderstorm cloud.

The air begins to rotate and extends to the ground in a funnel shape.

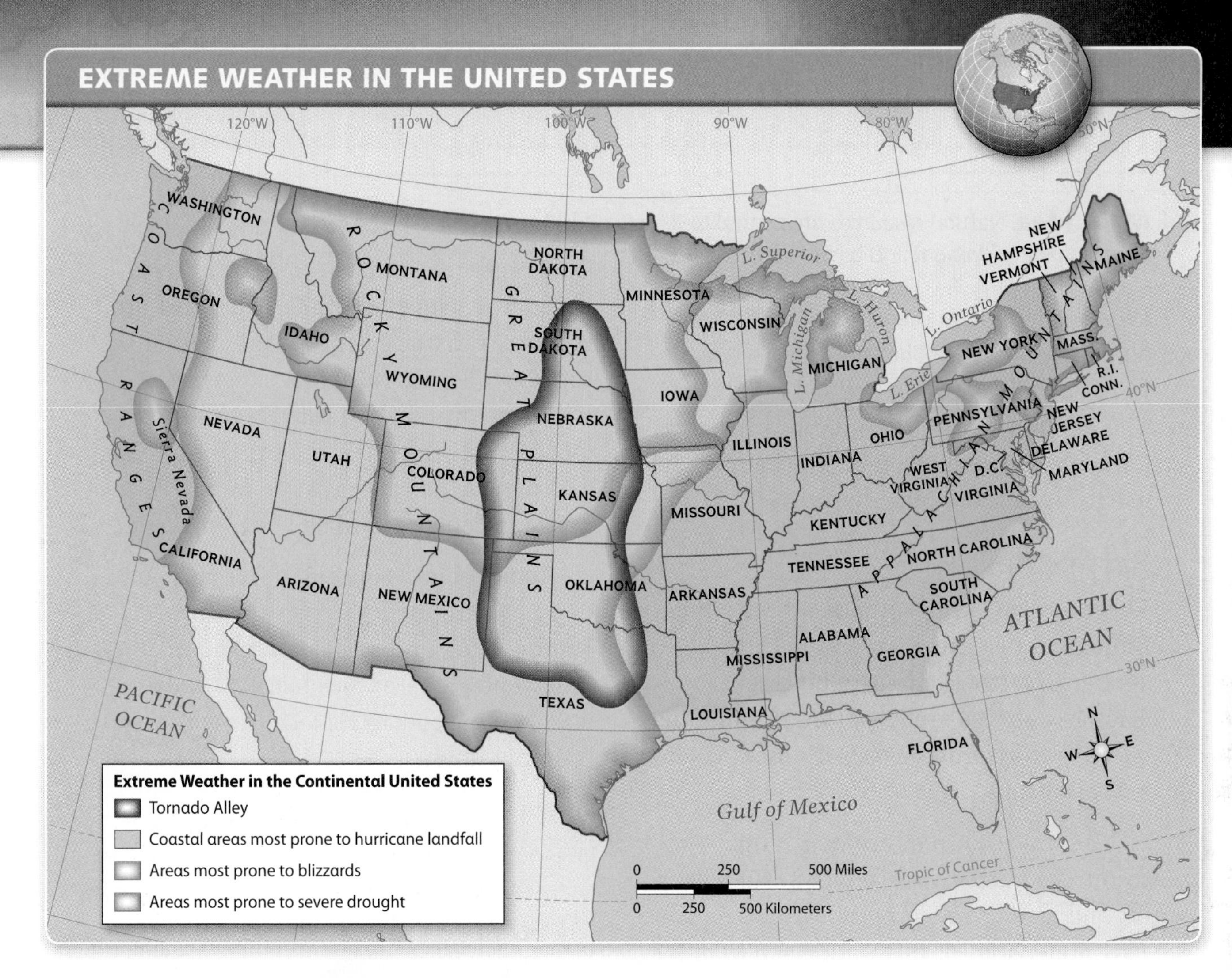

Scientific Solutions

Scientists are working to lessen the effects of extreme weather on humans. For example, they can often predict the path of a hurricane. They have also worked with engineers to design levees, or walls to hold back floodwaters. Many residents of New Orleans believe that better levees might have limited the damage caused by flooding after Hurricane Katrina.

The ability to predict tornadoes has also improved. Today, the National Weather Service uses radar and satellite imagery, as well as a network of "spotters," to track big storms. Today's instant communications technology makes it possible to broadcast warnings before a storm strikes.

Before You Move On

Summarize How are scientists helping to lessen the impact of extreme weather?

ONGOING ASSESSMENT

VIEWING LAB

GeoJournal

1. **Analyze Visuals** What happens in each stage of tornado formation?
2. **Interpret Maps** According to the map, which state is at risk for all four types of extreme weather?
3. **Draw Conclusions** Even though scientists can predict when and where tornadoes may occur, why do these storms still catch people by surprise?
4. **Human-Environment Interaction** How might a severe drought affect people who live in the Great Plains area of the United States?

2.4 Natural Resources

Main Idea Natural resources are central to economic development and basic human needs.

What materials make up a pencil? Wood comes from trees. The material that you write with is a mineral called graphite. The pencil is made from natural resources, which are materials on Earth that people use to live and to meet their needs.

Earth's Resources

There are two kinds of natural resources. Biological resources are living things, such as livestock, plants, and trees. These resources are important to humans because they provide us with food, shelter, and clothing.

Mineral resources are nonliving resources buried within Earth, such as oil and coal. Some mineral resources are **raw materials**, or materials used to make products. Iron ore, for example, is a raw material used in making steel. The steel, in turn, is used to make skyscrapers and automobiles.

Categories of Resources

Geographers classify resources in two categories. **Nonrenewable resources** are resources that are limited and cannot be replaced. For example, oil comes from wells that are drilled into Earth's crust. Once a well runs dry, the oil is gone. Coal and natural gas are other examples of nonrenewable resources.

Renewable resources never run out, or a new supply develops over time. Wind, water, and solar power are all renewable. So are trees because a new supply can grow to replace those that have been cut down.

Critical Viewing Pumpjacks pump oil at a field in California. Based on the photo, how does this action affect the land?

SELECTED NATURAL RESOURCES OF THE WORLD

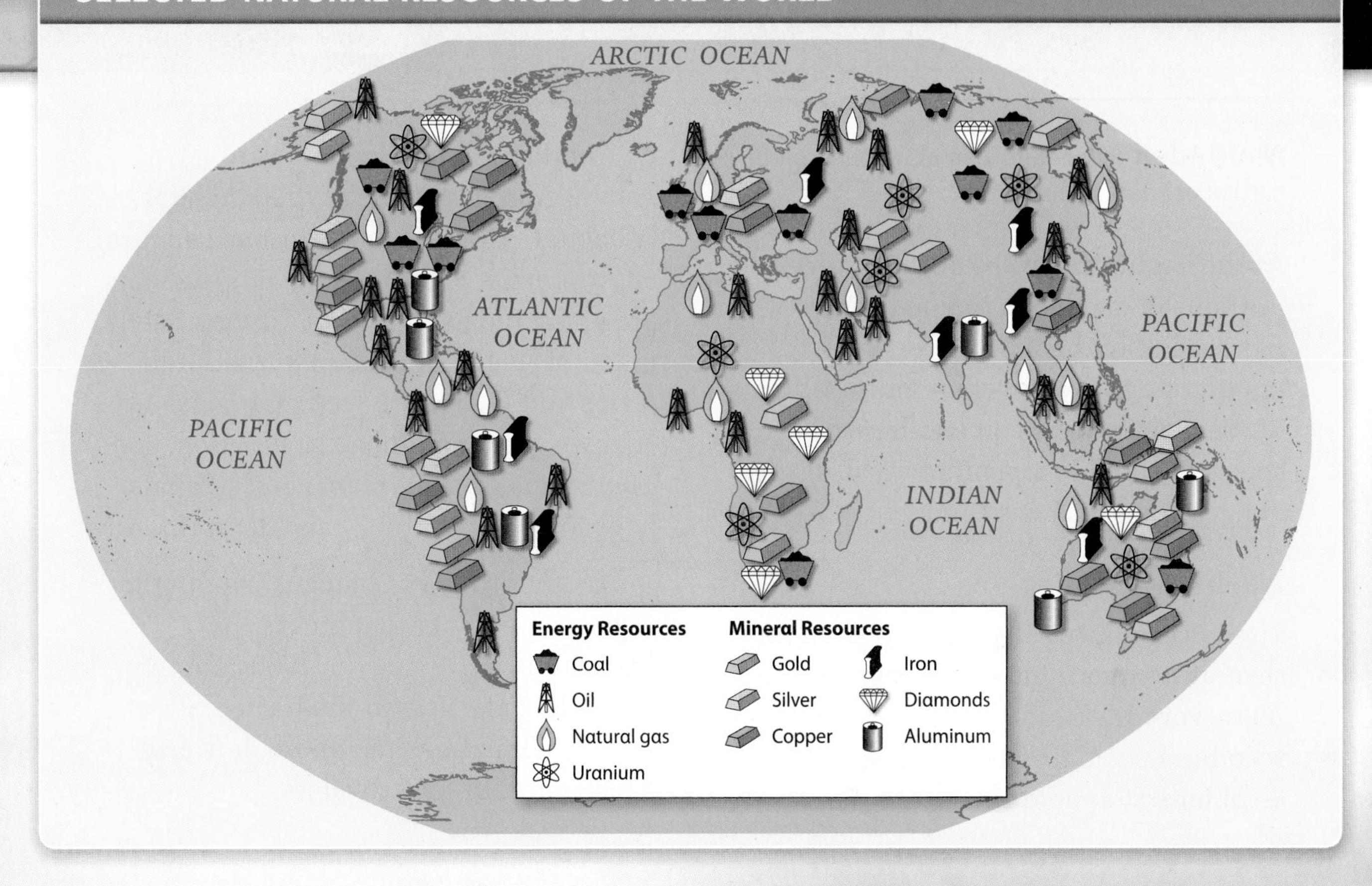

Natural resources are an important part of everyday life, yet countries with a large supply are not always wealthy. Nigeria, for example, is a major supplier of oil, but seven out of every ten Nigerians live in poverty. Japan is one of the wealthiest countries in the world—yet it must import oil from other countries.

Before You Move On

Monitor Comprehension Why are natural resources important?

ONGOING ASSESSMENT

MAP LAB

1. **Interpret Maps** Copper is in demand for electric wiring and other uses. What part of the world do you think benefits from the demand for copper? Why?
2. **Location** Where are supplies of oil found in the world? How do these supplies impact people living in these regions?
3. **Describe Geographic Information** What are examples of a biological and a mineral resource? How are these examples different?

2.5 Habitat Preservation

Main Idea Plants and animals depend on their natural habitats to survive.

At the beginning of the 20th century, millions of elephants roamed across Africa. Today the African elephant population is fewer than a half million. These elephants are an endangered species, which is a plant or an animal in danger of becoming extinct.

Natural Habitats

The African elephant is endangered for several reasons. One is the demand for their ivory tusks. Poachers, or people who hunt animals illegally, slaughtered elephants at a rapid rate in the early 1970s.

Another reason elephants are endangered is the loss of their **habitat**. A habitat is a plant or an animal's natural environment. African elephants' habitats are grasslands and forests. Unfortunately, much of this land is being turned into farms and villages to feed and house Africa's growing human population. Thousands of other plants and animals have lost their natural habitats in this way.

Another threat to habitats is pollution, or human activity that harms the environment. During the 1960s, for example, Lake Erie in the United States was a polluted habitat, and fish nearly disappeared from its waters.

Critical Viewing Elephants roam the Samburu National Reserve in Kenya. What can you tell about their natural habitat?

Habitat Loss and Restoration

The loss of habitats can destroy an entire ecosystem. An **ecosystem** is a community of plants and animals and their habitat. Earth has many different ecosystems that interact with each other. The destruction of one ecosystem affects all the others. For example, many scientists believe the destruction of rain forest habitats has led to global climate change.

People around the world have taken steps to save ecosystems and preserve natural habitats. In 1973, for example, the United States passed the Endangered Species Act, which protects the habitats of endangered species. People have also **restored**, or brought back, habitats such as forests by planting trees.

Before You Move On

Monitor Comprehension How do plants and animals lose their natural habitats?

Tiny spruce trees grow in a cut forest near Olympic National Park in the state of Washington.

ONGOING ASSESSMENT

PHOTO LAB

GeoJournal

1. **Analyze Visuals** Based on the photo of the reserve in Kenya, why might an elephant's habitat be desirable to people for farming?
2. **Describe** Based on the photo above, what steps are taken to restore forest habitats?
3. **Make Inferences** In a forest ecosystem, wolves eat deer, and deer carry the parasite that causes Lyme disease in humans. If the wolf population declines, how might it affect the number of cases of Lyme disease?
4. **Human-Environment Interaction** What is your natural habitat? With a partner, think of resources and interactions that are part of your daily life. Write a short paragraph describing your habitat.

TECHTREK
myNGconnect.com For photos of oceans and an Explorer Video Clip

Maps and Graphs

Digital Library

Exploring the World's Oceans with Sylvia Earle

Main Idea The oceans, a natural habitat for thousands of plant and animal species, face many challenges.

Life in the Ocean

As a child, Sylvia Earle loved oceans. This love continued into her adult life. In over 50 years, Explorer-in-Residence Sylvia Earle has led roughly 70 diving expeditions to explore **marine life**, or the plants and animals of the ocean. During these dives, she has seen the incredible variety of ocean life—more than 30 major divisions of animals.

However, Earle has also seen the different ways people have harmed the oceans. "Taking too much wildlife out of the sea is one way," she claims. "Putting garbage, toxic chemicals, and other wastes in is another." Earle has witnessed a huge drop in the number of fish in the ocean. She has also noted pollution's destructive **impact**, or effect, on the oceans' coral reefs. These "rain forests of the sea" house one-fourth of all marine life.

myNGconnect.com
For more on Sylvia Earle in the field today

EARTH'S "HOPE SPOTS"

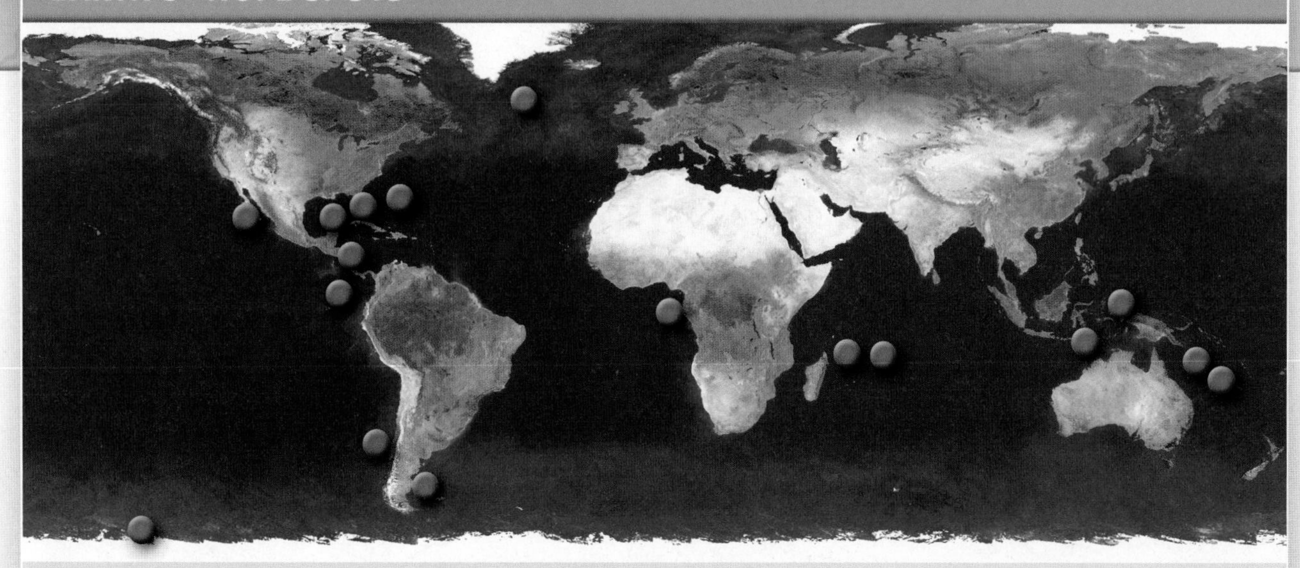

Critical Viewing This satellite map of the world shows the location of 17 "hope spots," places that are important to the overall health of Earth's oceans. What patterns, if any, do you notice about the location of these spots?

Mission Blue

Many people believe human activity has no effect on the vast oceans. They also do not understand the impact oceans have on all forms of life. "The ocean is the cornerstone of our life support system," says Earle. "Take away life in the ocean and we don't have a planet that works."

Earle is trying to educate the public. In 2009, she launched Mission Blue, a program that seeks to heal and protect Earth's oceans. One of the program's goals is to establish marine protected areas (MPAs) in endangered hot spots, or "hope spots" as she calls them. These spots are ocean habitats that can recover and grow if human impact is limited.

Voice of the Ocean

Earle's efforts to save Earth's oceans have earned her many awards and honors, including the title "Hero for the Planet." She continues to work tirelessly to protect the world's oceans.

In 2010, the Gulf of Mexico was hit with the largest oil spill in U.S. history. Earle went before Congress to testify about the impact of the spill on the Gulf's natural resources. "I really come to speak for the ocean," she began.

Before You Move On

Summarize According to Sylvia Earle, what challenges do our oceans face?

ONGOING ASSESSMENT

READING LAB

GeoJournal

1. **Monitor Comprehension** What does Sylvia Earle hope to accomplish through the Mission Blue program?
2. **Analyze Cause and Effect** How has human activity affected the world's oceans? Use a chart like the one below to list some of the causes and effects of these activities.

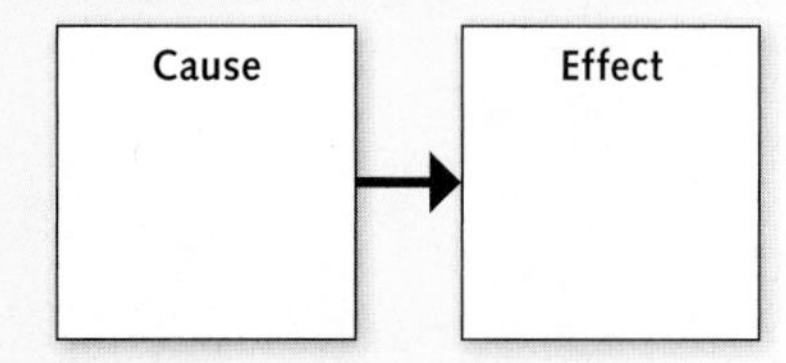

NATIONAL GEOGRAPHIC

Photo Gallery • Slip Point Tidal Pool

For more photos from the National Geographic Photo Gallery, go to the **Digital Library** at myNGconnect.com.

Grand Canyon

Kathmandu, Nepal

Japanese man with laptop

Critical Viewing A crab walks over sea urchins in a tidal pool at Slip Point in Clallam Bay, Washington. River outlets and tides create this plentiful and diverse coldwater habitat.

Maori carving

3.1 World Cultures

TECHTREK
myNGconnect.com For photos of cultures around the world
Digital Library

Main Idea The ways people speak, eat, work, play, and worship are all part of culture.

A Japanese tea ceremony is an old tradition. The host greets the guests and prepares the tea. The guests remain silent. Once the tea is poured, though, a lively conversation follows.

Expressions of Culture

The tea ceremony is an important part of Japan's culture. **Culture** is how people in a region live, behave, and think. Expressions of culture include language, religion, beliefs, and customs. Culture also includes the arts, such as music, dance, literature, theater, and film.

Culture is reflected in symbols that people recognize and respect. A **symbol** is an object that stands for something else. For example, the stars on the American flag are symbols for the 50 states.

Civilization and Culture

Culture is a main trait of civilizations. A **civilization** is a society that has a highly developed culture and technology. People in a civilization are not born knowing their culture. They learn it by watching and imitating others.

A civilization's culture affects people's lives. It guides how people meet their basic needs of food, clothing, and shelter. It also influences people's values and beliefs.

Visual Vocabulary Senegalese children gather around a **communal**, or shared, plate of food. Communal dishes are common in African cuisine.

Visual Vocabulary A **gaucho** (GOW cho), or cowboy, herds sheep in South America.

Culture Regions

Geographers study **culture regions**, areas that are unified by common cultural traits, or characteristics. For instance, some geographers group Mexico, Central America, the islands of the Caribbean, and South America in a culture region called Latin America. Many of the people in this region speak Spanish or Portuguese and practice the Roman Catholic religion. Many also share a history of Spanish colonization.

Before You Move On

Monitor Comprehension What are some expressions of culture?

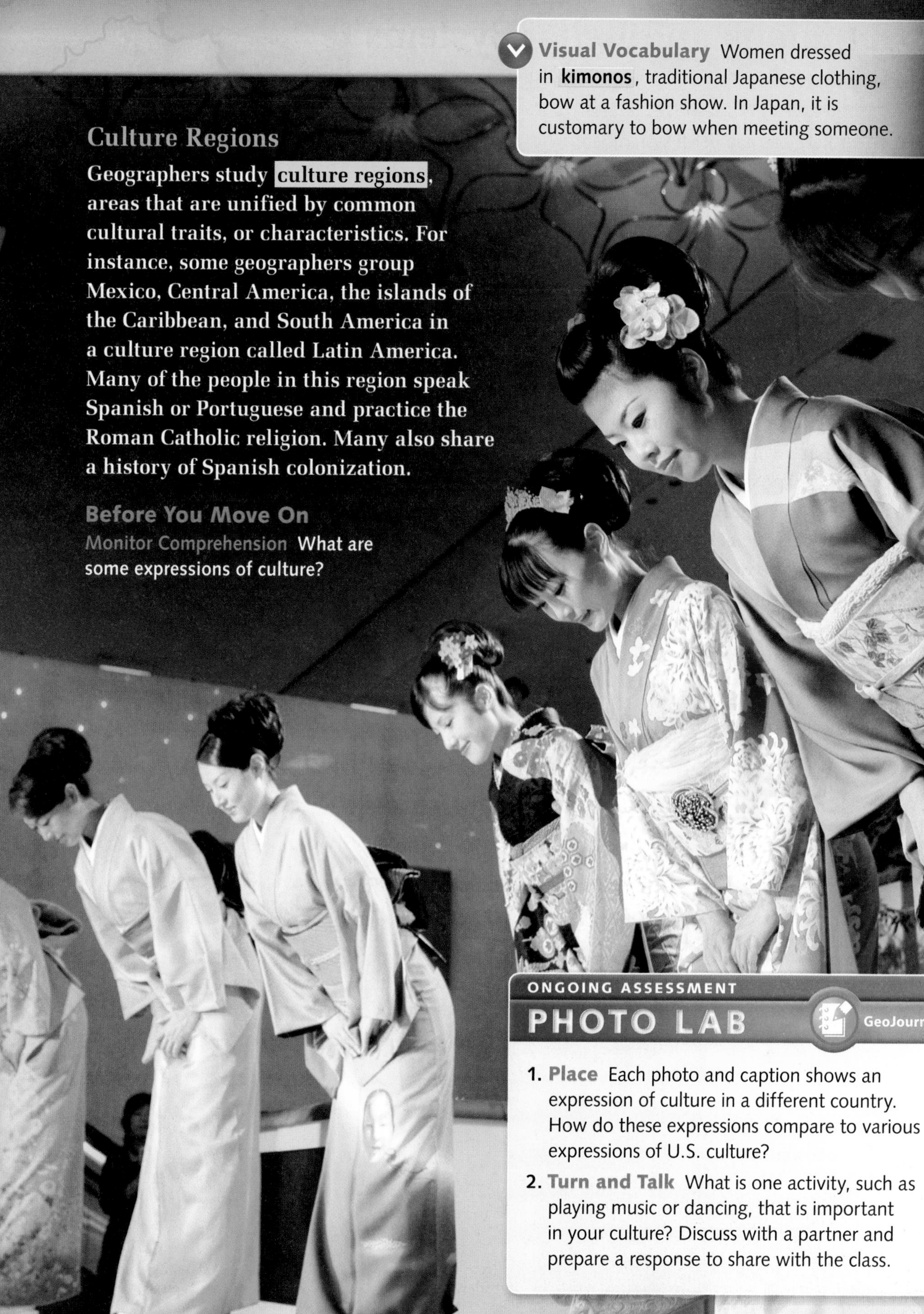

Visual Vocabulary Women dressed in **kimonos**, traditional Japanese clothing, bow at a fashion show. In Japan, it is customary to bow when meeting someone.

ONGOING ASSESSMENT

PHOTO LAB

GeoJournal

1. **Place** Each photo and caption shows an expression of culture in a different country. How do these expressions compare to various expressions of U.S. culture?
2. **Turn and Talk** What is one activity, such as playing music or dancing, that is important in your culture? Discuss with a partner and prepare a response to share with the class.

3.2 Religions and Belief Systems

TECHTREK
myNGconnect.com For an online map and photos of world religions

Maps and Graphs

Digital Library

Main Idea Religions and belief systems are important parts of cultures around the world.

A religion is a set of beliefs and practices that is often focused on one or more deities, or gods. Major world religions include Christianity, Hinduism, Islam, Buddhism, Judaism, and Sikhism.

Elements of Religion

Religion is a powerful influence that helps people answer questions such as, "What is the purpose of life?" At the center of many religions is the belief in a deity. **Monotheistic religions** are those with a belief in one deity. Christianity, Islam, and Judaism are monotheistic religions. **Polytheistic religions**, such as Hinduism, have many deities.

Every religion has a doctrine, or a set of basic beliefs. For example, Christians believe that Jesus was the Son of God.

Scriptures are sacred, or highly respected, texts that communicate the beliefs of a religion. The Bible is the sacred text for Christianity, the Koran is the sacred text for Islam, and the Torah is the sacred text for Judaism.

One of the important ways in which religion affects culture is through a code of conduct, or beliefs about right and wrong behavior. For example, the Ten Commandments are a code of conduct for Jews and Christians.

Origin and Spread of Religions

Several of the major world religions are based on the teachings of an individual. For example, Christianity grew from the teachings of Jesus Christ nearly 2,000 years ago. Other religions, such as Hinduism, grew from the beliefs of ancient peoples.

BUDDHISM

Founder Siddhartha Gautama, the Buddha
Followers 400 million
Basic Beliefs People reach enlightenment, or wisdom, by following the Eightfold Path and understanding the Four Noble Truths.

CHRISTIANITY

Founder Jesus of Nazareth
Followers 2.3 billion
Basic Beliefs There is one God, and Jesus is the only Son of God. Jesus was crucified but was resurrected. Followers reach salvation by following the teachings of Jesus.

HINDUISM

Founder Unknown
Followers 860 million
Basic Beliefs Souls continue to be reborn. The cycle of rebirth ends only when the soul achieves enlightenment, or freedom from earthly desires.

ISLAM

Founder The Prophet Muhammad
Followers 1.6 billion
Basic Beliefs There is one God. Followers must follow the Five Pillars of Islam in order to achieve salvation.

WORLD RELIGIONS

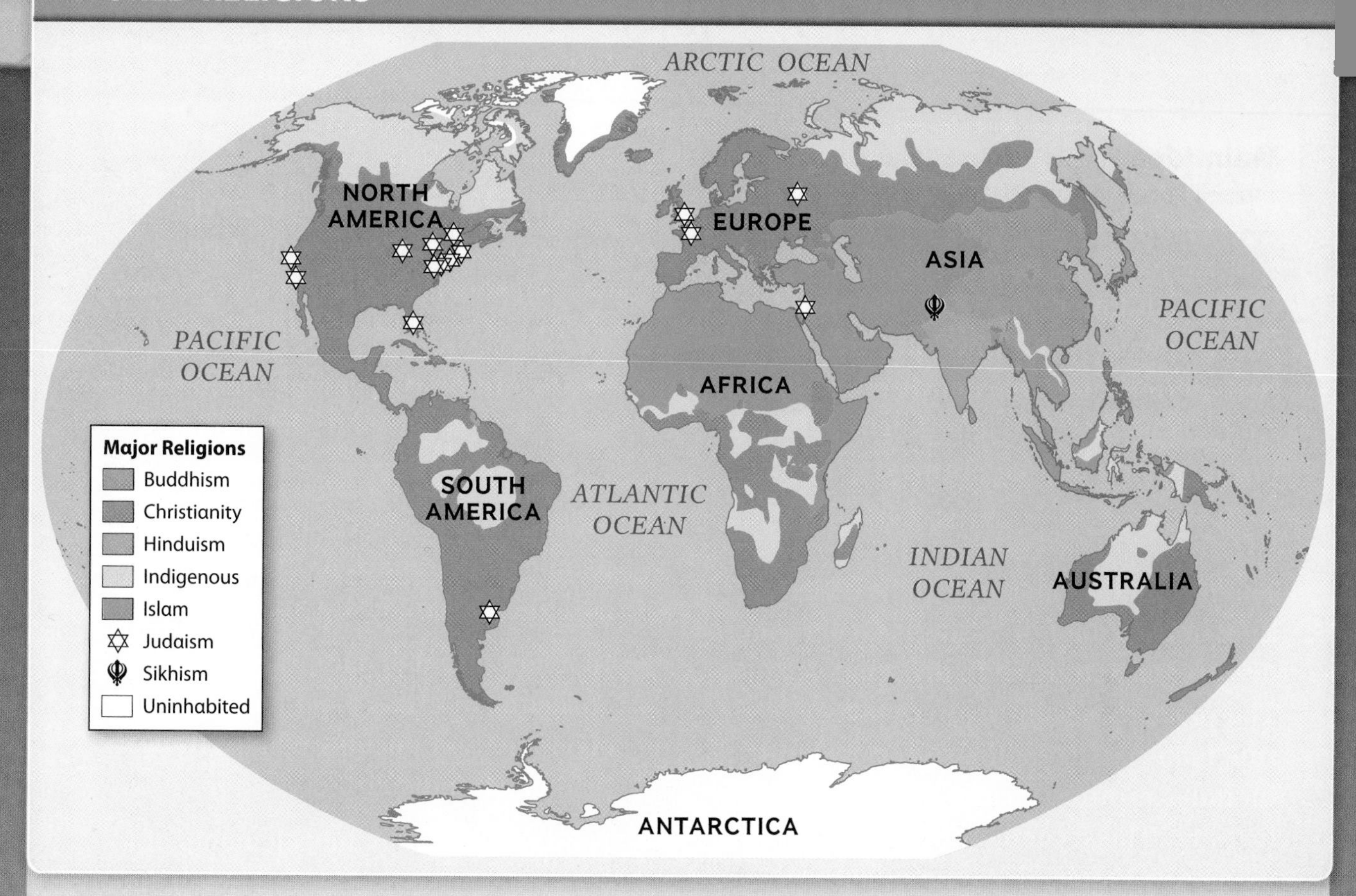

Religions have grown and spread around the world. For example, Buddhism began in India, but it spread to Japan, China, Korea, and Southeast Asia through migration and trade. Religions have also spread due to the work of missionaries, people who convert others to follow their religion.

JUDAISM

Founder Abraham
Followers 15 million

Basic Beliefs There is one God. People serve God by living according to his teachings. God handed down the Ten Commandments to guide human behavior.

SIKHISM

Founder Guru Nanak
Followers 25 million

Basic Beliefs There is one God. Souls are reborn. The goal is to achieve union with God, which a person does by acting selflessly, meditating, and helping others.

Before You Move On

Make Inferences How is religion an important part of culture?

ONGOING ASSESSMENT

READING LAB

1. **Monitor Comprehension** What is the difference between a monotheistic religion and a polytheistic religion?
2. **Compare and Contrast** How are Hinduism and Buddhism similar?
3. **Region** What major religions are found in the area around the eastern Mediterranean Sea? Why do you think they are found in this region?

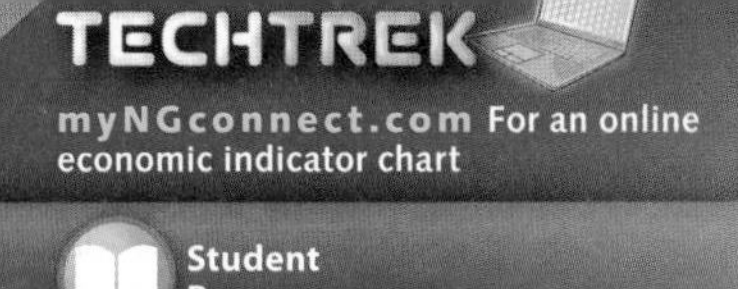

3.3 Economic Geography

Main Idea People produce, buy, and sell goods in a variety of ways.

Singapore became a trading colony of the British Empire in 1824. Today, it is an independent country, but trade is still an important part of its economy. An **economy** is a system in which people produce, sell, and buy things.

Economic Activity

The production of goods and services is known as economic activity. Geographers divide this activity into different sectors. The primary sector involves taking raw materials from the soil or water. It includes mining, farming, fishing, and forestry. The secondary sector involves using raw materials to manufacture products, such as cars. The tertiary, or third, sector includes services, such as banking and health care.

Factors of Production

Geographers study where economic activity occurs and how this activity is connected around the world. A country is more likely to have a strong economy if it has all four factors of production—land, labor, capital, and entrepreneurship. Land includes all the natural resources used to produce goods and services. Labor involves the size and education level of the workforce. **Capital** is a country's wealth and infrastructure. The fourth factor, **entrepreneurship**, involves the creativity and risk needed to develop new goods and services.

Economic Systems

Economic systems are ways in which countries organize the production of goods and services. Four main systems are found around the world:

- In a traditional economy, people trade goods and services without money.
- In a **free enterprise economy**, privately owned businesses create goods that people buy in markets. This is also called a market economy or capitalism.
- In a command economy, the government owns most parts of the economy and decides what will be produced and sold.
- A mixed economy has elements of a free enterprise and command economy.

Economic Indicators

The strength of a country's economy can be measured by several indicators, or signs. One is **gross domestic product (GDP)**. It is the total value of the goods and services that a country produces. The GDP per capita is the value of products that a country produces per person. Other indicators include income, literacy rate, and life expectancy.

Economies fall in one of two categories. Countries with high GDPs are more developed countries. Most of their economic activity is in the tertiary sector. Countries with low GDPs are less developed countries. Most of their activity is in the primary or secondary sector.

Before You Move On

Summarize What are four ways in which countries organize the production of goods and services?

ECONOMIC INDICATORS OF SELECTED COUNTRIES*

Country	Population	GDP (in U.S. dollars)	GDP Per Capita (in U.S. dollars)	Life Expectancy	Literacy Rate (percent)
Afghanistan	29.1 million	10.6 billion	366	44	28.0
Brazil	191.9 million	1.6 trillion	8,536	72	90.0
China	1.3 billion	4.5 trillion	3,422	73	93.3
Ethiopia	80.7 million	25.9 billion	321	55	35.9
Germany	82.1 million	3.7 trillion	44,525	80	99.0
Haiti	9.8 million	6.4 billion	649	61	62.1
Mexico	106.3 million	1.1 trillion	10,249	75	92.8
Singapore	4.8 million	193.3 billion	39,950	81	94.4
United States	304.3 million	14.4 trillion	47,210	78	99.0

Sources: The World Bank and the United Nations
*All figures are for 2008 with the exception of literacy rate, which is for 2007.

Critical Viewing This shipping terminal in Singapore is busy all day and all night. Based on the photo and the chart, what type of country is Singapore—a more developed or a less developed country?

ONGOING ASSESSMENT

DATA LAB

1. **Interpret Charts** Which country has the largest population? the largest GDP per capita? What can you conclude about the relationship between the two?
2. **Synthesize** What sector is probably the main source of Haiti's economic activity? Explain.
3. **Region** In the region in which you live, what is an example of an economic activity from each of the three sectors?

3.4 Political Geography

Main Idea Countries around the world have different forms of government.

How is trash collection related to the protection of free speech? Both are responsibilities of government. A **government** is an organization that keeps order, sets rules, and provides services for a society. Political geographers study boundaries between different places, where different types of government exist, and how geography affects government.

Government and Citizens

Governments govern citizens. A **citizen** is a person who lives within the territory of a government and is granted certain rights and responsibilities by that government.

Governments are either limited or unlimited. A limited government does not have complete control over its citizens. The citizens have some individual rights and responsibilities. An unlimited government has complete control over every aspect of its citizens' lives.

Critical Viewing Voters in South Africa line up for miles to vote. How important to them is the right to vote? How can you tell?

Types of Government

In the modern world, five types of government are common. The major differences among them are in the power and rights that citizens have.

- In a **democracy**, citizens elect representatives to govern them. A legislature creates laws, an executive branch carries out laws, and a judicial branch interprets laws. Citizens have many rights, such as those in the Bill of Rights of the U.S. Constitution. The United States was the first modern country to establish a representative democracy.
- In a monarchy, a king, queen, or emperor rules society. The ruler usually inherits, or is born into, the office. Citizens in an absolute monarchy, such as Saudi Arabia, have few or no rights. In a constitutional monarchy, such as the United Kingdom, the queen or king shares power with a government organized by a constitution.
- In a dictatorship, one person, the dictator, rises to power and rules society. The dictator controls all aspects of life, including education and the arts. Citizens have few or no rights. North Korea has been ruled by a dictator for more than a half century.
- In an oligarchy, a group of a few people rules society. The ruling group usually is wealthy or has military power, and citizens have few or no rights. The government of Myanmar (Burma) has been an oligarchy since 1988.

Critical Viewing Female soldiers march in a military parade in North Korea to celebrate the country's 60th anniversary. Based on the photo, what qualities are valued by North Korea's government?

- Communism is a type of command economy in which the government, controlled by the Communist Party, owns all the property. Citizens have few or no rights. Cuba has been a communist country since 1959.

Before You Move On

Summarize What are common types of government in the modern world?

ONGOING ASSESSMENT

READING LAB

GeoJournal

1. **Monitor Comprehension** Which of the five types of government are limited governments? Which are unlimited governments?
2. **Create Charts** Create a chart in which you compare the five forms of government. Use the following format:

Type of Government	Source of Power	Leader or Ruler	Citizens' Rights
democracy			

3. **Make Inferences** Look at the map of Russia in the National Geographic Atlas in the front of your textbook. How might Russia's geography affect the government's ability to rule?

3.5 Protecting Human Rights

Main Idea The United Nations adopted the Universal Declaration of Human Rights to state how all people deserve to be treated.

KEY VOCABULARY

human rights, n., political, economic, and cultural rights that all people should have

Nesse Godin was 13 years old during World War II when the Nazis occupied her town in Lithuania. Because she and her family were Jewish, they were transferred to a concentration camp. Then Nesse was sent to different labor camps, where she worked digging ditches. In January 1945, she and other prisoners were forced to march in the cold weather with little food. Many prisoners died.

The United Nations

Nesse Godin survived the events of the Holocaust, but 6 million other people did not. At the end of World War II, people vowed to never let it happen again.

In 1945, 51 countries from around the world formed the **United Nations (UN)**. The main goals of this organization were to keep peace, to develop friendly relationships among countries, and to protect people's **human rights**.

The UN established the Commission on Human Rights to decide which rights all people should have. The members of this commission came from different cultural backgrounds, but they worked together to create a "common standard of achievement for all people and all nations." On December 10, 1948, the UN General Assembly approved the commission's **Universal Declaration of Human Rights**.

The declaration has 30 articles, or sections. Article 1 states that all people should be treated with respect:

> All human beings are born free and equal in dignity and rights. They are endowed with reason and conscience and should act towards one another in a spirit of brotherhood.

Twenty-one of the articles explain political rights, such as the right to equality before the law, the right to freedom from torture, and the right to take part in government. Six of the articles address people's economic and cultural rights, such as the right to work, the right to education, and the right to participate in the cultural life of the community.

The Impact of Human Rights

The declaration has had an impact, or effect, on people and governments. For example, during the 1960s–1980s, countries around the world pressured the Republic of South Africa to grant human rights to its non-white population. Many countries refused to trade with South Africa, and the country was barred from participating in the Olympic Games from 1964–1990.

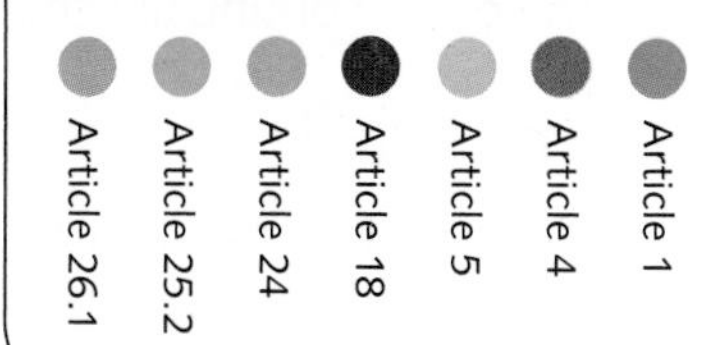

Motherhood and childhood are entitled to special care and assistance.

No one shall be held in slavery or servitude ...

All human beings are born free ...

Everyone has the right to freedom of thought, conscience, and religion ...

No one shall be subjected to torture or to cruel, inhuman, or degrading treatment or punishment.

Everyone has the right to education.

Everyone has the right to rest and leisure ...

In 1994, South Africa finally gave in to the pressure and held elections in which all people could vote. The people elected Nelson Mandela, a leader of the African population, as president. This action showed the power of human rights.

Before You Move On

Monitor Comprehension According to the Universal Declaration of Human Rights, how do all people deserve to be treated?

ONGOING ASSESSMENT

WRITING LAB

1. **Form and Support Opinions** Choose one of the rights from the Universal Declaration of Human Rights that are listed above. In a paragraph, explain what this right means and why it is important.
2. **Write Reports** Select a news story in the newspaper or on television. In a short report, explain how the story shows the importance of protecting human rights.

CHAPTER 2

Review

VOCABULARY

For each pair of vocabulary words, write one sentence that explains the connection between the two words.

1. solstice; equinox

Summer begins at the solstice while spring begins at the equinox.

2. plateau; continental shelf
3. renewable resource; nonrenewable resource
4. habitat; restore
5. free enterprise economy; democracy

MAIN IDEAS

6. How does Earth's tilt affect the seasons? (Section 1.1)
7. What layers would you pass through on a journey to Earth's center? (Section 1.2)
8. How does erosion change Earth's surface? (Section 1.3)
9. How have scientists tried to lessen the impact of earthquakes and volcanoes? (Section 1.4)
10. In what ways is climate different from weather? (Section 2.1)
11. How is a humid equatorial climate different from a humid temperate climate? (Section 2.2)
12. What is an example of a renewable resource and a nonrenewable resource? (Section 2.4)
13. What are endangered species, and what factors threaten them? (Section 2.5)
14. What are four ways in which culture affects people's lives? (Section 3.1)
15. How is a free enterprise economy different from a command economy? (Section 3.3)
16. What are the common types of government in the modern world? (Section 3.4)

THE EARTH

ANALYZE THE ESSENTIAL QUESTION

How is the earth continually changing?

Critical Thinking: Make Inferences

17. What impact do the changing seasons have on how farmers grow food?
18. Why did early civilizations develop along rivers?
19. How does the hydrologic cycle return water to Earth?

PHYSICAL GEOGRAPHY

ANALYZE THE ESSENTIAL QUESTION

What shapes the earth's varied environments?

Critical Thinking: Draw Conclusions

20. How would plants in humid cold climates and humid equatorial climates differ?
21. What factors might prevent a country rich in natural resources from using them effectively?

INTERPRET CHARTS

U.S. AND WORLD ENDANGERED SPECIES

Group	United States	Other Countries	Total Number
Mammals	71 species	255 species	326 species
Reptiles	13 species	66 species	79 species
Fish	74 species	11 species	85 species
Birds	76 species	184 species	260 species

Source: U.S. Fish and Wildlife Service

22. **Analyze Data** What percentage of endangered animals in the United States are mammals? What percentage in other countries are mammals?
23. **Human-Environment Interaction** What steps might wildlife conservationists take to reduce the number of endangered species?

HUMAN GEOGRAPHY

ANALYZE THE ESSENTIAL QUESTION

How has geography influenced cultures around the world?

Critical Thinking: **Find Main Ideas and Details**

24. What are some characteristics of American culture?
25. In what ways do different religions spread?
26. What is an example of each sector of economic activity?
27. What are two examples of different ways in which governments govern?

INTERPRET MAPS

CANADA'S BIOLOGICAL RESOURCES

28. **Place** Are Canada's resources mostly in the north or the south? Why do you think this is the case?

ACTIVE OPTIONS

Synthesize the Essential Questions by completing the activities below.

29. **Write a Public Service Announcement (PSA)** Write a PSA about what to do during an extreme weather emergency in your community. Focus on one type of weather event that is common in your area. Explain the supplies that a household should have. Provide an escape route or hiding place. Also, explain how people should behave during the emergency. **Share your PSA with the class.**

Writing Tips

- Do research to find the advice of weather experts.
- Take notes and organize your information to make it clear to readers.
- Use a tone that is calm and shows that you are informed.

Go to **Student Resources** for Guided Writing support.

TECHTREK myNGconnect.com For photographs of elements of culture

30. **Create a Visual Overview** Choose a part of culture that interests you, such as music, dance, language, or sports. Use photos from the **Digital Library** or from other online or print sources to create a visual introduction to that aspect of culture around the world. Show similarities and differences among countries around the world. Copy and complete the following organizer to organize your ideas:

Element of Culture: ____________

Examples: ____________

Similarities: ____________

Differences: ____________

Climate

TECHTREK

myNGconnect.com For climographs and research links about U.S. climates

 Maps and Graphs

 Connect to NG

In this unit, you learned that climate is the average condition of the atmosphere in a region over a long period of time. Climate is determined by the latitude of a location, as well as by ocean currents, air currents, and elevation.

Two important factors of climate are temperature and precipitation. These factors determine the length and timing of the growing season and influence the types of economic activities in a region. For instance, a country with year-round warm temperatures and precipitation probably has a better agricultural industry than a country with low temperatures and a short rainy season.

Compare

- Brazil
- Russia
- United States

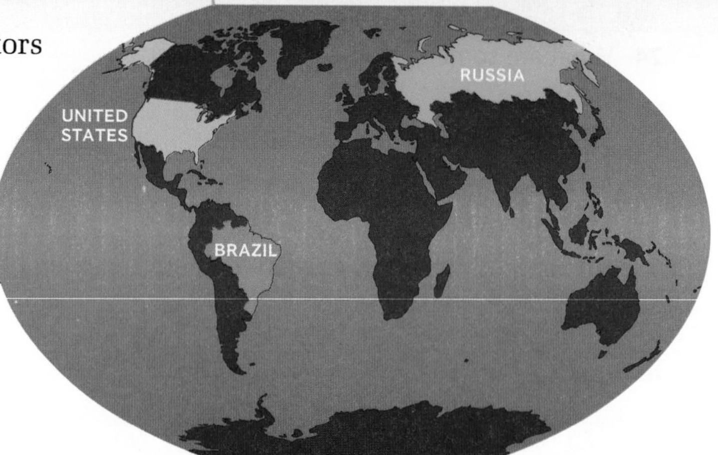

CLIMATE MAPS AND CLIMOGRAPHS

A climate map provides an overview of the climate in a region. However, sometimes more specific information about a city or country is needed. Geographers use a special tool called a **climograph** (climate + graph) to graphically show a range of temperature and precipitation in a place over a period of time. A climograph includes a bar graph that shows the amount of precipitation for a location. Average monthly temperature is shown by a line connecting 12 points, one for each month of the year.

Climographs are useful in comparing the climates of two different locations. They help geographers and others better understand the effects of climate on human activities in those locations.

INTERPRET CLIMOGRAPHS

Look at the climographs for the cities of Belem, Brazil, and Omsk, Russia, on the opposite page. The months of the year are listed across the bottom of each climograph. A scale on the left vertical axis measures precipitation, or rainfall, in inches. A scale on the right vertical axis measures temperature in degrees Fahrenheit (°F).

The bars on the climographs show the rainfall for Belem, Brazil, and Omsk, Russia. The lines connecting the dots show the range of temperatures for each location.

Study the data in the climographs to analyze the climate in Belem and Omsk. Then answer the questions at right.

AVERAGE MONTHLY TEMPERATURE AND PRECIPITATION

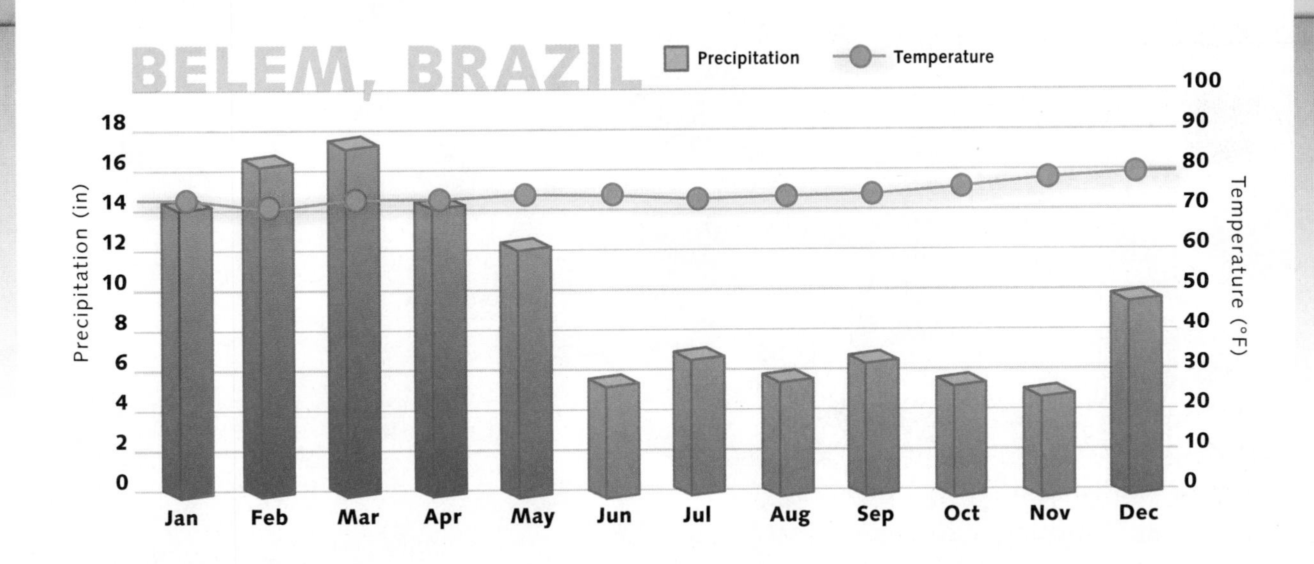

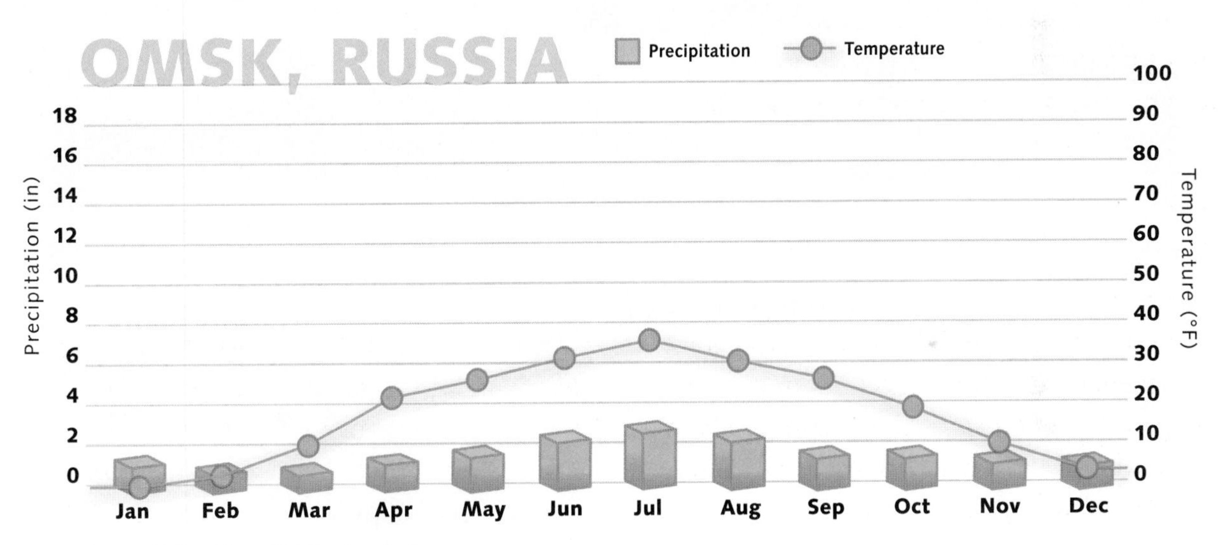

Source: National Drought Mitigation Center, University of Nebraska–Lincoln

ONGOING ASSESSMENT

RESEARCH LAB

GeoJournal

1. **Interpret Graphs** How would you describe the range of temperatures in Belem? What is the average temperature in Omsk in January? In July?
2. **Make Inferences** Most crops need water and warm temperatures in order to grow. Based on this information, what can you tell about the growing season in each city?

Research and Create a Climograph Choose a city in the United States and do research to find the average temperature and rainfall for each month of the year. Record the data in a chart. Use the data to create a climograph for that city. Then, with a partner, write three questions to help someone analyze and compare data on your two climographs.

Active Options

TECHTREK
myNGconnect.com
For writing templates

Student Resources

Magazine Maker

ACTIVITY 1

Goal: Extend your understanding of the environment.

Compose a Top Ten List

For over 40 years, countries around the world have celebrated Earth Day. The purpose of the day is to appreciate our planet and to raise awareness of the need to protect the environment. With your classmates, create a Top Ten list of actions students in your school can plan to do to show appreciation for Earth on the next Earth Day. See if you can get permission to post copies of your list around the school to share with other students.

A reforestation event in the Philippines

ACTIVITY 2

Goal: Learn about culture through religious architecture.

Create a Catalog of Religious Architecture

Some of the world's greatest architecture has been inspired by religious beliefs. Use the **Magazine Maker** to create a visual catalog of examples of architecture representing several religions around the world. Include important information in your catalog, such as the name of the architect, the name of the structure, and the year the structure was created.

ACTIVITY 3

Goal: Research the history of human migration.

Write a Migration History

Explorer Spencer Wells is researching the history of human migration. Interview a friend to find out more about his or her migration history. Include these questions:

- When did your ancestors arrive in the United States?
- Why did they come?
- What did they do when they arrived?

Then write an account of your friend's migration history.

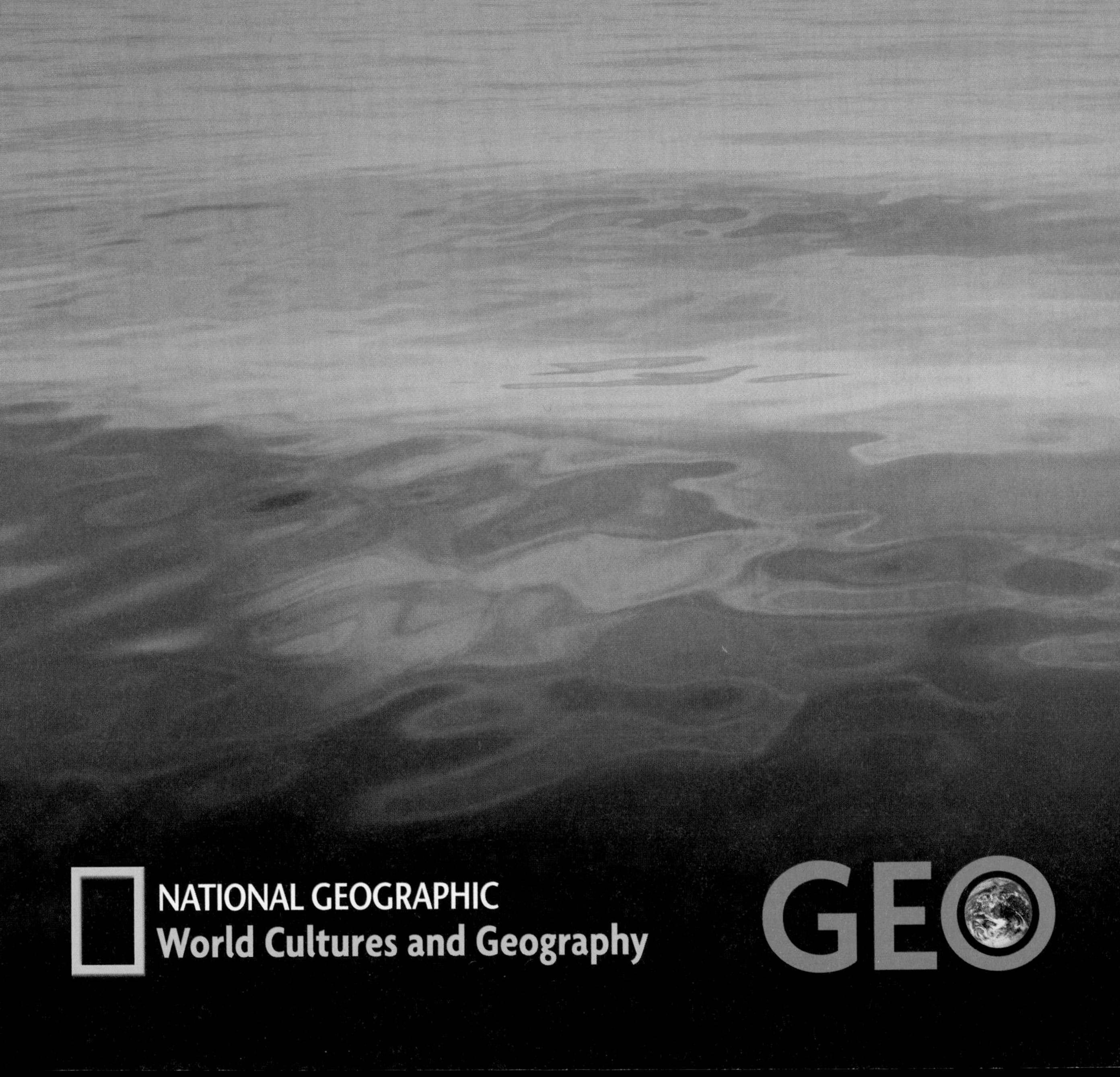
NATIONAL GEOGRAPHIC
World Cultures and Geography
GEO

EXPLORE NORTH AMERICA WITH NATIONAL GEOGRAPHIC

MEET THE EXPLORER

NATIONAL GEOGRAPHIC

By analyzing ancient DNA, Emerging Explorer Beth Shapiro traces the decline of bison in North America. She has proved the decline was not caused by hunters but by an ice age 20,000 years ago.

INVESTIGATE GEOGRAPHY

Niagara Falls attracts millions of visitors every year. The waters also generate plenty of low-cost electricity. For nearly 50 years, the United States and Canada have shared Niagara's power production, along with a strong desire to preserve its unique setting.

STEP INTO HISTORY

The U.S. Capitol in Washington, D.C., houses the legislative branch of government in two chambers of Congress. It symbolizes a federal republic, a federation of states with elected representatives. Mexico has also practiced this form of government since 1824.

CONNECT WITH THE CULTURE

Tourists peer over the western rim of the Grand Canyon, in northwest Arizona. The Skywalk, completed in 2007, is owned and operated by the Hualapai tribe.

NORTH AMERICA GEOGRAPHY & HISTORY

PREVIEW THE CHAPTER

Essential Question What are the significant physical features of North America?

SECTION 1 • GEOGRAPHY

KEY VOCABULARY

- contiguous
- temperate
- glacier
- export
- drought
- commercial agriculture
- cordillera
- rain shadow effect
- dam
- peninsula
- subsistence farming
- aquifer
- cenote
- sustainable

ACADEMIC VOCABULARY

modify

TERMS & NAMES

- Great Plains
- Great Lakes
- Grand Canyon
- Mexican Plateau

Essential Question How did the United States and Canada develop as nations?

SECTION 2 • U.S. & CANADIAN HISTORY

KEY VOCABULARY

- colonize
- plantation
- fortify
- missionary
- tax
- revolution
- constitution
- amendment
- pioneer
- industrialization
- transcontinental
- abolition
- secede
- civil war
- alliance
- neutrality
- dictator
- terrorism

ACADEMIC VOCABULARY

adapt, protest

TERMS & NAMES

- Declaration of Independence
- Bill of Rights
- Louisiana Purchase
- Manifest Destiny
- Trail of Tears
- Emancipation Proclamation
- Gettysburg Address
- Reconstruction
- Pearl Harbor
- Holocaust
- Cold War

Essential Question How have various cultures influenced Mexico's history?

SECTION 3 • HISTORY OF MEXICO

KEY VOCABULARY

- civilization
- hieroglyphics
- empire
- tribute
- conquistador
- epidemic
- republic
- exile
- annexation
- reform
- land reform
- tyranny

ACADEMIC VOCABULARY

oppose

TERMS & NAMES

- Olmec
- Maya
- Aztec
- Hernán Cortés
- Montezuma
- Miguel Hidalgo
- Santa Anna
- Alamo
- Mexican Cession
- Gadsden Purchase

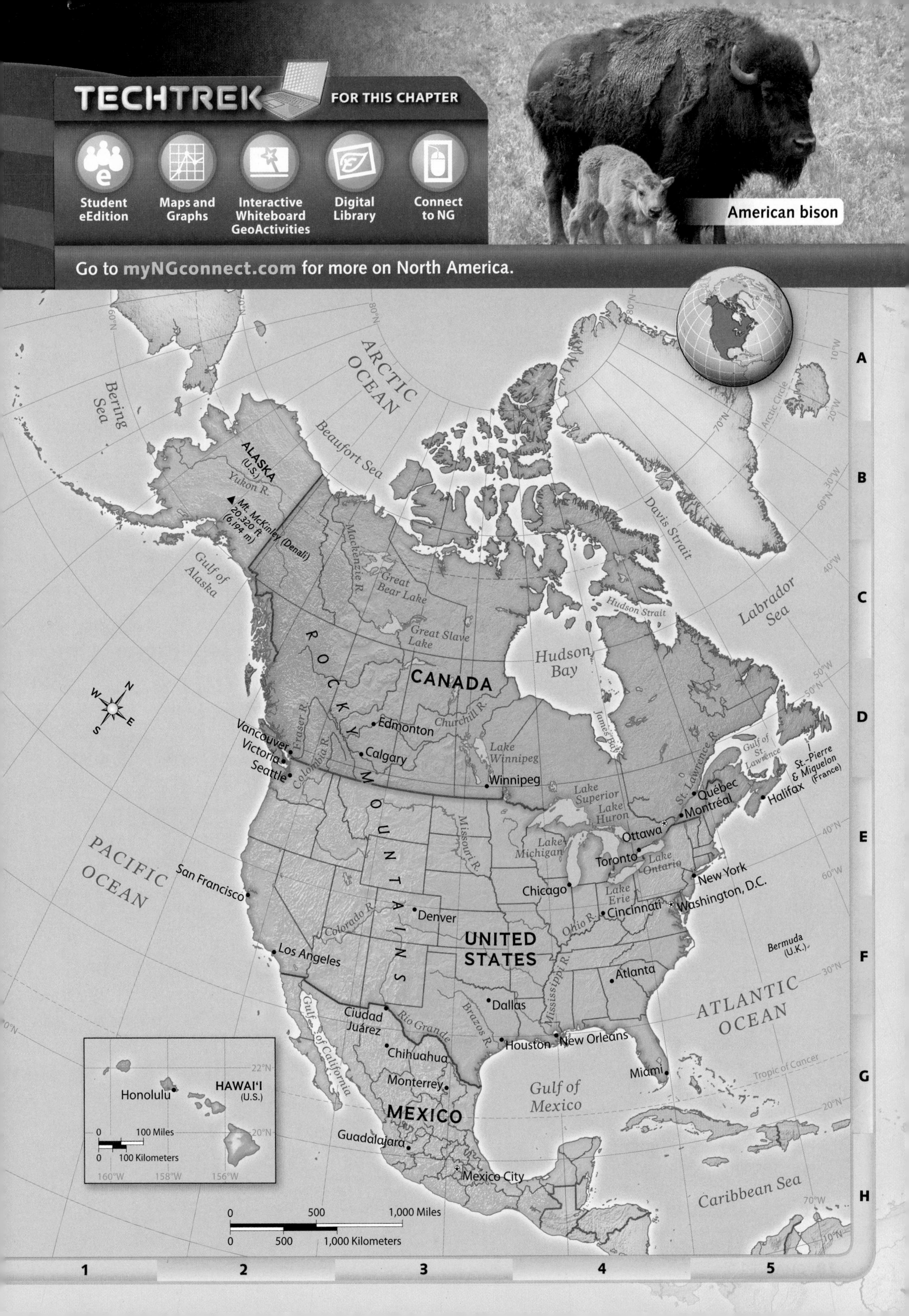
TECHTREK FOR THIS CHAPTER
Student eEdition
Maps and Graphs
Interactive Whiteboard GeoActivities
Digital Library
Connect to NG
American bison
Go to myNGconnect.com for more on North America.
ARCTIC OCEAN
Bering Sea
Beaufort Sea
ALASKA (U.S.)
Yukon R.
Mt. McKinley (Denali) 20,320 ft (6,194 m)
Gulf of Alaska
Mackenzie R.
Great Bear Lake
Great Slave Lake
Davis Strait
Labrador Sea
Hudson Strait
Hudson Bay
ROCKY MOUNTAINS
CANADA
Churchill R.
Edmonton
Calgary
Vancouver
Victoria
Seattle
Fraser R.
Columbia R.
Lake Winnipeg
Winnipeg
James Bay
St. Lawrence R.
Gulf of St. Lawrence
St.-Pierre & Miquelon (France)
Québec
Montréal
Halifax
Lake Superior
Lake Huron
Lake Michigan
Ottawa
Toronto
Lake Ontario
New York
Missouri R.
PACIFIC OCEAN
San Francisco
Chicago
Lake Erie
Washington, D.C.
Cincinnati
Ohio R.
Denver
Colorado R.
UNITED STATES
Los Angeles
Bermuda (U.K.)
Atlanta
Mississippi R.
Dallas
Brazos R.
ATLANTIC OCEAN
Ciudad Juárez
Rio Grande
Gulf of California
Chihuahua
Houston
New Orleans
Miami
Monterrey
Gulf of Mexico
Tropic of Cancer
MEXICO
Guadalajara
Mexico City
Caribbean Sea
HAWAI'I (U.S.)
Honolulu
100 Miles
100 Kilometers
0
500
1,000 Miles
1,000 Kilometers
Arctic Circle
A
B
C
D
E
F
G
H
1
2
3
4
5

1.1 Physical Geography

TECHTREK
myNGconnect.com For online maps of North America and Visual Vocabulary

 Maps and Graphs

 Digital Library

NORTH AMERICA PHYSICAL

Main Idea North America has a wide variety of landforms, bodies of water, and climates.

North America stretches from the cold arctic of northern Canada to the warm tropics of Mexico. At the center lie the 48 **contiguous** United States, which means they are all connected in one block. The state of Alaska is in the northwest of the continent. The islands of the state of Hawaii are in the Pacific Ocean.

Highlands, Plains, and Plateaus

Land elevation in North America generally rises from east to west, though the east has some highlands, or areas of hills and mountains. East of the Rocky Mountains lie the **Great Plains**. Plains are flat areas of land, which make up most of the center of North America. Plains also appear near coasts. Plateaus—flat lands of high elevation—are located between mountains in the western United States and central Mexico.

Rivers and Lakes

Major cities developed along the region's numerous rivers, such as Cincinnati on the Ohio River, New Orleans on the Mississippi River, and Juárez (HWA rez) on the Rio Grande. The St. Lawrence River provides a waterway from the Atlantic Ocean to the **Great Lakes**. Combined, these five lakes form the largest body of fresh water in the world. Four of the lakes provide a physical boundary between the United States and Canada, just as the Rio Grande is a natural border between the United States and Mexico.

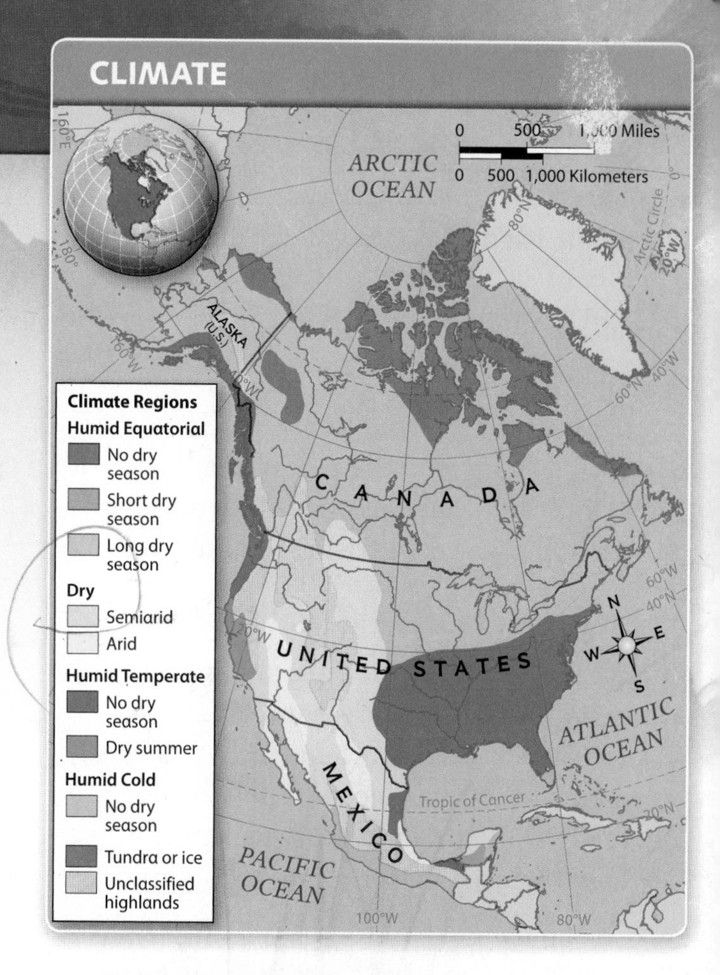

A Variety of Climates

This region includes **temperate**, or mild, climates as well as extremes of cold and heat. Parts of the far north are covered by **glaciers**, or large masses of ice and packed snow. Much of northern Mexico is warm and dry, and Mexico's tropical south has warm climates and rainfall all year.

Before You Move On

Monitor Comprehension Describe the region's major landforms, bodies of water, and climates.

ONGOING ASSESSMENT

MAP LAB

GeoJournal

1. **Interpret Maps** According to the physical map, how might goods be moved from Lake Michigan to the Atlantic Ocean?
2. **Region** According to the climate map, what types of climate occur in both Mexico and Alaska?

1.2 The Great Plains

TECHTREK
myNGconnect.com For a map and photos of the Great Plains in North America

Maps and Graphs

Digital Library

Main Idea The Great Plains of the United States and Canada are a rich agricultural region with valuable energy resources.

The Great Plains region runs through the center of the continent. The crops grown there feed the population of North America with enough left to **export**, or send to other countries for aid or profit.

Farming on the Great Plains

The Great Plains are well suited to agriculture for two reasons. First, the soil is rich with nutrients, so it produces bountiful crops. Second, the climate on the Great Plains is temperate and the area usually has a plentiful amount of rain.

Some years, rainfall is below normal for a long period of time, causing **drought** that can kill crops. In the 1930s, for example, the region experienced a drought that lasted several years. Plowing of native prairie grasses that helped hold the soil contributed to soil erosion, and the Plains became known as the "Dust Bowl." The persistent winds in the area stirred up great clouds of dry soil, or dust. Today, these winds can be a source of power.

Much of the original prairie land has been replaced by fields of wheat, corn, and other grains. These crops are grown on huge farms where planting and harvesting are done by machine. Such large, highly productive farms are typical of **commercial agriculture**, or the business of producing crops to sell.

Rivers such as the Missouri and the Mississippi transport goods from the Great Plains to lowland areas. Grain from the Canadian plains is moved by rail to the Atlantic Ocean and by ship on the Great Lakes and St. Lawrence River.

Energy Resources

The Great Plains in the United States and Canada are home to major deposits of oil and natural gas. The main oil fields in the United States are found in the southern part of the Great Plains, from Kansas to Texas. Texas also has many offshore oil fields in the Gulf of Mexico.

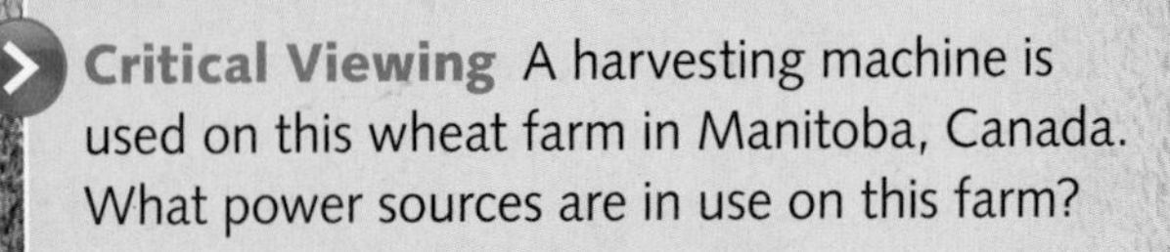
Critical Viewing A harvesting machine is used on this wheat farm in Manitoba, Canada. What power sources are in use on this farm?

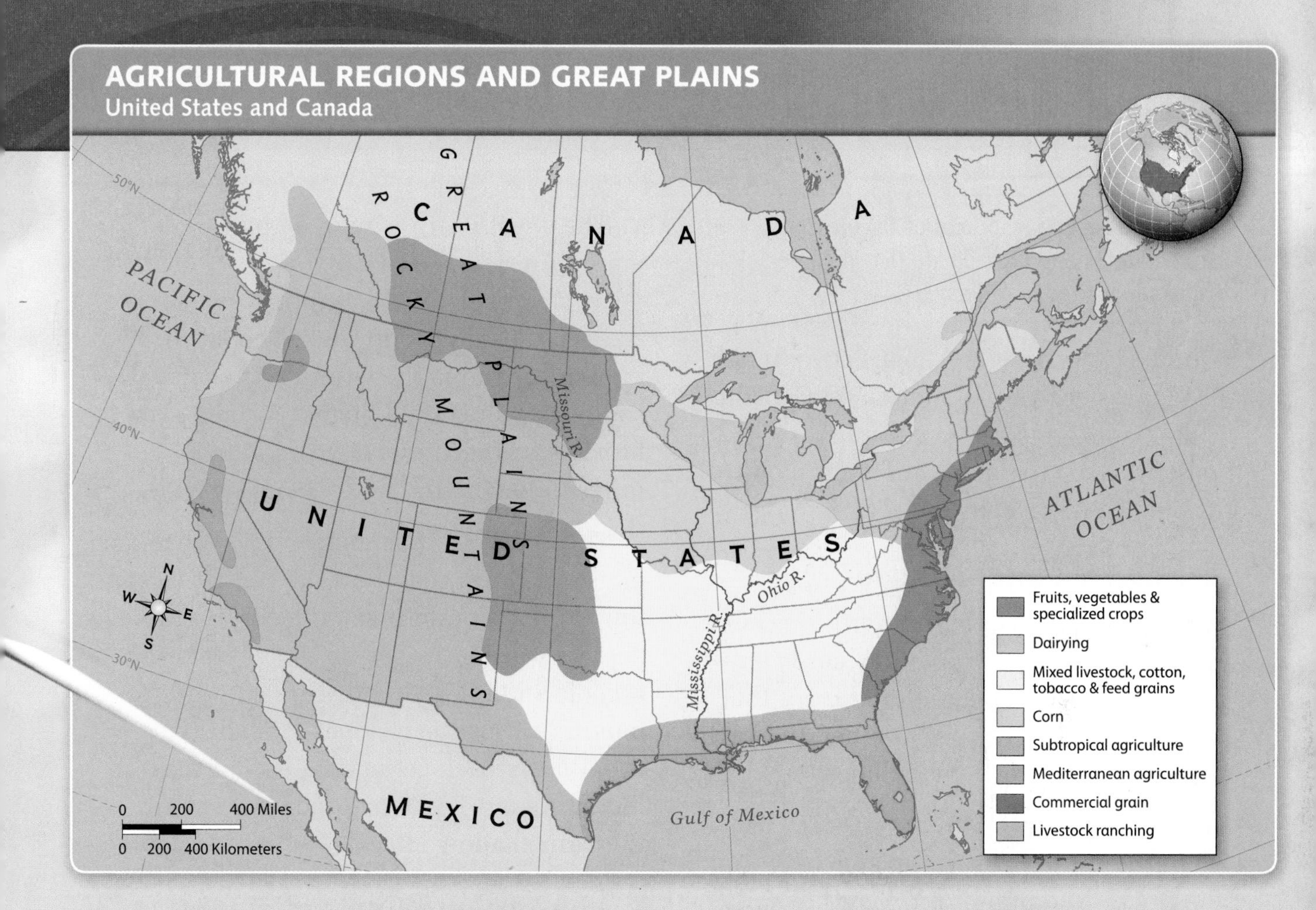

Commercial drilling in the Gulf of Mexico is risky. Oil deposits lie beneath layers of salt that can shift and cause an underwater earthquake. In 2010, human error caused an explosion of a large deepwater drilling structure, resulting in a major oil spill. The spill harmed wildlife and threatened the economy of the region.

Energy resources from the Great Plains are important to the United States, where more energy is used than is produced. High winds in the plains may be utilized as an alternative energy source.

Before You Move On

Summarize What makes the Great Plains an important area for agriculture and energy?

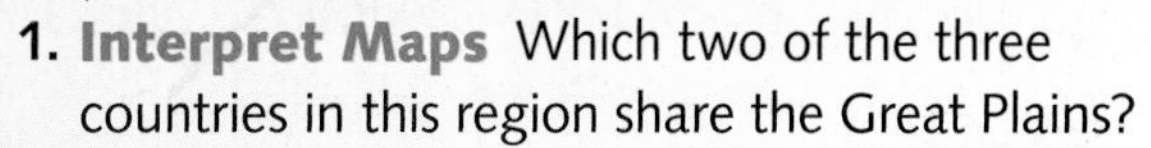

ONGOING ASSESSMENT

MAP LAB

1. **Interpret Maps** Which two of the three countries in this region share the Great Plains?
2. **Make Inferences** Look at the climate map in Section 1.1 and the map above. What reason can you suggest for the differences in climate across the Great Plains?
3. **Movement** What methods might be used for transporting goods out of the Great Plains? Use evidence from the map to support your answer.

1.3 Western Mountains and Deserts

TECHTREK

myNGconnect.com For a map and photos of the western United States and Canada

Maps and Graphs

Digital Library

Main Idea The resources of the western United States and Canada are rich in some areas and limited in others.

Mountains and high plateaus cover much of the western United States and Canada. These landforms create natural barriers to western coastal regions and offer plentiful resources in some areas.

Landforms and Climate

The main landform in the western part of the region is the cordillera. A **cordillera** is a system of several mountain ranges that often run parallel to one another. In North America, the cordillera includes the Rocky Mountains and the Sierra Nevadas.

In the United States, the area between the Rocky Mountains and Sierra Nevada Mountains is the Great Basin. A basin is a depression in the surface of the land. The Great Basin is a desert, a dry, often sandy area with little rainfall or plant life.

The Great Basin is marked by smaller mountain ranges and canyons, which are deep, steep-sided valleys formed when rivers cut through soft rock. The best known canyon is the **Grand Canyon** in the southwestern United States. Formed over hundreds of millions of years, it is 277 miles in length and up to 18 miles wide.

The Great Basin is mostly dry. Warm, moist air flows east from the Pacific Ocean toward the mountains of the cordillera. As this warm air rises up the mountains, it cools and releases moisture on the mountains' western slopes. The air that eventually reaches the land east of the mountains is dry. This process is called the **rain shadow effect**.

Resources and Conservation

Varying climates contribute to a varied supply of resources. The Great Basin and the mountain ranges that surround it contain important mineral deposits.

Critical Viewing A group of horses follows hikers in the Great Basin. What landforms can you identify in the photo?

Areas of southwestern Canada hold reserves of natural gas. This area's heavy rainfall and many lakes allow for the use of water power to provide electricity.

Water for human use is in short supply in the southwestern United States. The population there has grown rapidly in recent years, and the demand for water has increased in this already dry area. A **dam**, which is a barrier that controls the flow of water, can help solve the problem of water shortage. However, dams can also cause problems such as excess soil erosion. The Hoover Dam on the Colorado River makes use of water power to supply electricity, irrigation, and drinking water to parts of Arizona, Nevada, and southern California.

Before You Move On

Make Inferences What does the shortage of certain resources mean for the people who live in the West?

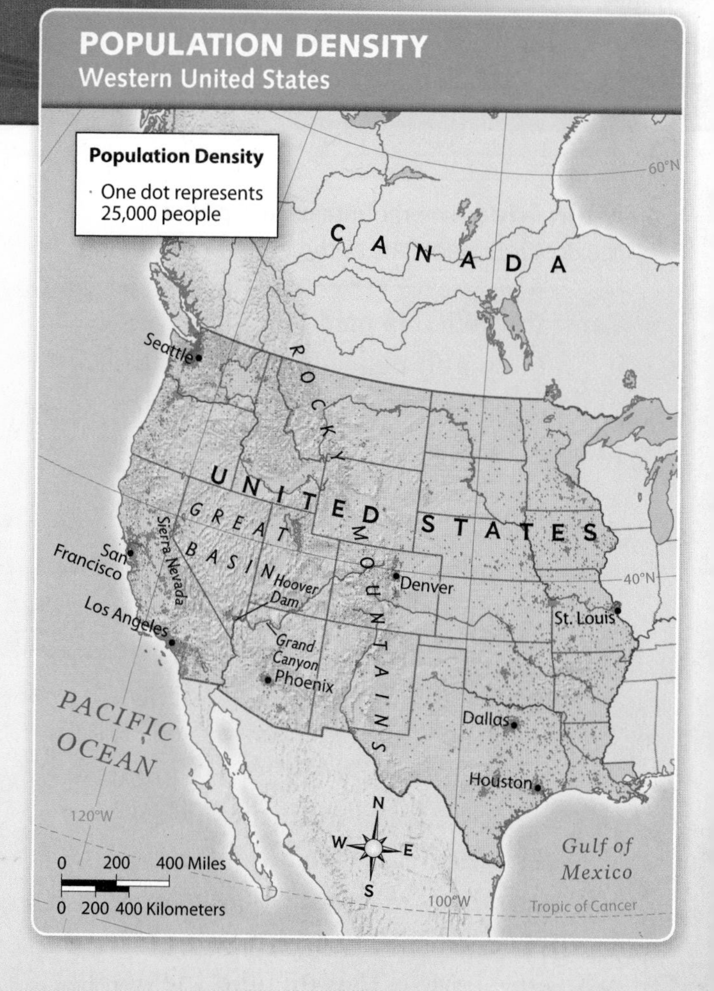

ONGOING ASSESSMENT

PHOTO LAB

1. **Describe Geographic Information** Turn to a classmate and brainstorm words to describe the photo. What information in the lesson supports your description?
2. **Interpret Maps** Where is population density highest and lowest in this part of the region?
3. **Analyze Visuals** What physical features shown in the photo might explain the population distribution in this region?

1.4 Mexico's Mountains and Plateaus

myNGconnect.com For a map and photos of Mexico's resources

Maps and Graphs

Digital Library

Main Idea The mountains and plateaus of Mexico are important to the country's economy.

Central Mexico is made up mainly of mountains and plateaus. These landforms yield rich resources that contribute to Mexico's economy.

Landforms and Climate

Mexico includes two **peninsulas**, narrow strips of land that extend out into a body of water: the Baja (BAH hah) in western Mexico and the Yucatán (yoo kah TAHN) in the southeast. Mexico is shaped like an upside-down triangle with the mountains of the Sierra Madre running along each side. The **Mexican Plateau** lies between the two ranges of the Sierra Madres. Mexicans call the southern area of this plateau the Mesa Central, and the northern area the Mesa del Norte. A mesa—the Spanish word for "table"—is a high, flat plateau. Mexico's highest mountain, volcano Pico de Orizaba, rises at the southern edge of the plateau.

Mexico City, the capital, is on the Mesa Central. The city is home to more than 20 million people, almost 20 percent of the population of Mexico. Some areas of the Mesa Central have been subject to volcanic activity, which results in rich, volcanic soil. This soil helps to produce crops that are important to Mexico's economy, such as sugarcane, corn, and wheat.

Northern Mexico sits in the temperate zone, and the southern half lies in the tropics. On the high Mesa Central, climate is **modified**, or made less extreme, by a higher elevation. Temperatures there are cooler than along lower coastal areas.

Resources and Agriculture

Mexico's mountains hold resources such as copper, silver, and zinc. However, the richest resource is the oil found in and around the Gulf of Mexico. Over three million barrels are produced each day.

Agriculture is also important to the country. Farmers in the north grow cotton, wheat, and fruit, and also raise cattle. To the south, farms produce sugarcane, coffee, and tropical fruits. Many rural Mexicans live by **subsistence farming**, growing just enough food to feed their families. This type of farming occurs mostly in the southern highlands.

Before You Move On

Monitor Comprehension Why is the Mexican Plateau important to the country's economy?

Critical Viewing At a height of more than 18,000 feet, Pico de Orizaba is the highest mountain in Mexico. What words would you use to describe this mountain?

Volcano Pico de Orizaba, Mexico

ONGOING ASSESSMENT

MAP LAB

GeoJournal

1. **Location** Based on the map, where are most of Mexico's oil reserves?
2. **Interpret Maps** Using directional words, describe where most cattle is raised in Mexico.
3. **Make Inferences** On the map, move your finger along the areas with sugarcane. What type of land is best for growing sugarcane?

TECHTREK
myNGconnect.com For photos of the Yucatán and an Explorer Video Clip
Digital Library

Exploring the Yucatán with Sam Meacham

myNGconnect.com
For more on Sam Meacham in the field today

Main Idea Exploring untapped resources can help solve the problem of scarce water in dry areas of Mexico.

Sacred Wells

Mexico's Yucatán Peninsula has both wet and dry seasons. Because of the peninsula's physical geography, very little surface water is available. National Geographic Grantee Sam Meacham is helping to find ways to gain access to this valuable resource.

The peninsula is composed mostly of limestone. Water from rainfall seeps through pores, or openings, in the limestone and flows underground. The weight of this groundwater creates cracks underground. Over time, the cracks expand into caves. These underwater caves function as **aquifers**, or layers of rock beneath the earth that contain water. The Yucatán Peninsula has a dry surface, so the water contained in these aquifers is critical to the water supply.

Sometimes the surface rock collapses to reveal the underground pools. These exposed pools are called **cenotes** (se NO tayz)—Mayan for "sacred wells." Meacham's team explores the cenotes, which provide entry to underground rivers and underwater caves that contain substantial water resources. One cave that Sam's team explored has 112 miles of underwater passages.

Saving an Important Resource

Meacham's team is helping to reveal not only the importance of this groundwater, but also a potential threat to its safety. The Yucatán Peninsula is a popular tourist destination.

Visitors come to enjoy the white sand beaches of the Caribbean coastline on the eastern side of the peninsula. While income from tourism is needed to build Mexico's economy, Meacham fears that high levels of tourism can be a threat to the aquifer. Tourist activities produce garbage and sewage. If this material is not disposed of properly, the water that collects in the cenotes can become polluted, and therefore unusable.

Meacham leads a nonprofit organization devoted to protecting the Yucatán aquifer. His work exploring and mapping underwater caves will provide access to scientists who want to study these reserves of water. Access to the water from this underground reserve can help Mexico's economy grow in a way that is **sustainable**, or based on preserving resources rather than using them up.

Before You Move On

Summarize How does Sam Meacham's work help preserve the water supply?

Critical Viewing This cenote is in the Yucatán. Would you find exploring cenotes appealing? Why or why not?

ONGOING ASSESSMENT

VIEWING LAB

GeoJournal

1. **Analyze Visuals** Go to the **Digital Library** and watch the Explorer Video Clip. What images from the video show the value of Meacham's work?
2. **Make Inferences** Why is Meacham's work important for the Yucatán Peninsula?
3. **Human-Environment Interaction** What impact does human behavior have on the environment in the Yucatán?

2.1 Exploration and Colonization

myNGconnect.com For a map and images of European colonies

Main Idea European colonization of North America brought settlers from several countries and permanently changed life on the continent.

In the late 1400s, many Native American groups lived on the continent. Each group **adapted**, or adjusted, to the environment in which it lived. For example, Native Americans in the eastern woodlands hunted deer and farmed the land. In the West, the Lakota hunted the large herds of American bison, a type of buffalo that grazed on the Great Plains.

Europe Meets Native America

In 1492, Christopher Columbus sailed west from Spain and reached islands in the Caribbean Sea. This voyage was part of a period of European exploration of the Americas. Explorers wanted to find riches and claim new lands for their rulers.

The lives of Native Americans were permanently changed by the arrival of Europeans. Settlers carried diseases such as smallpox, which the native people could not fight. Disease killed a large percentage of certain Native American populations. The spread of settlements also pushed many Native American groups off their traditional lands.

By the late 1500s, European countries had begun to **colonize** the area, or build settlements and develop trade in lands that they controlled. Spain formed the colony of St. Augustine in present-day Florida. In the 1600s, the British settled Jamestown in Virginia and Plymouth in Massachusetts. New Sweden was established in what is now Delaware, and New France arose along the St. Lawrence River in what is now Quebec, Canada. These colonies were mostly small settlements. Most people farmed, traded furs, or did craft work.

The European powers began to compete for land. The British and French fought more than once over their colonies. In 1763, the British would eventually gain control of New France and become the major colonial power north of Mexico.

Slavery in the Colonies

Over time, the British colonies thrived. In the south, huge **plantations**, or large farms that grow crops for profit, needed more labor than farmers' families could provide. Thousands of Africans were forced into slavery and brought across the Atlantic to the colonies.

Cathedral Basilica of St. Augustine, Florida

1550

1565 Spain forms colony in St. Augustine, Florida.

1600

1607 English establish colony in Jamestown, Virginia.

1608 French establish colony in Quebec, Canada.

EUROPEAN COLONIES AND NATIVE AMERICAN GROUPS, c. 1650

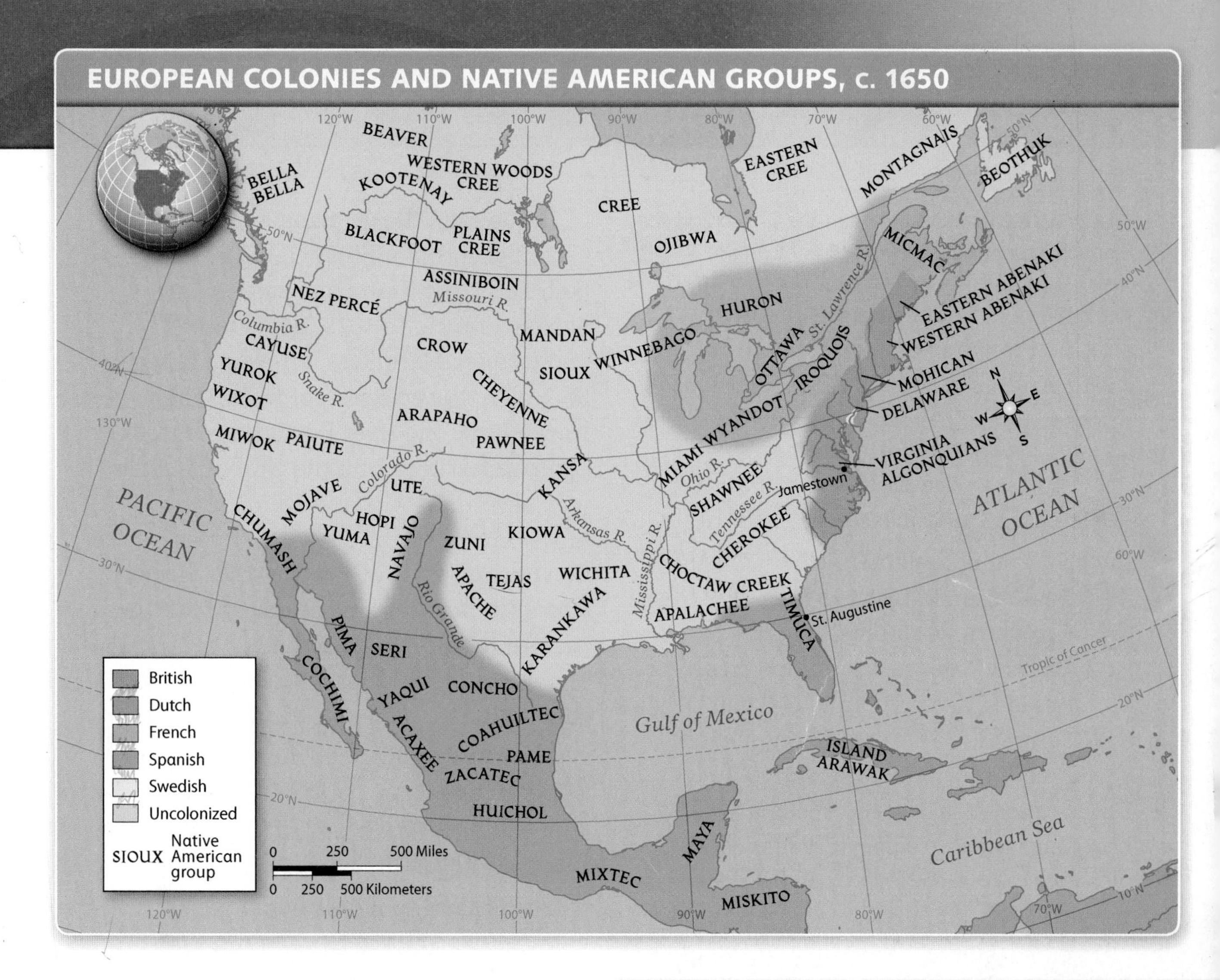

As production of cotton increased on plantations in the South, the population of slaves in these states grew rapidly. The economy of the Southern colonies became more and more dependent on slave labor.

Before You Move On

Summarize How did European colonization change life in North America?

ONGOING ASSESSMENT

MAP LAB

1. **Identify** According to the map, where were most of the European colonies located by 1650?
2. **Interpret Maps** In 1650, which Native American groups might have been most affected by Europeans? Why?
3. **Location** Why did British colonies in the South have the greatest number of slaves?

1620
English settlers set up colony in Plymouth, Massachusetts.

1650

1664
Dutch colony Beverwyck is renamed Albany, New York.

British settlers arrive in Plymouth, Massachusetts.

1700

1730s
Swedish colony, present-day Wilmington, Delaware, becomes successful market town.

2.2 Settling Quebec

TECHTREK
myNGconnect.com For a map and images of New France

Maps and Graphs

Digital Library

Main Idea Conflict between the British and the French shaped the development of Canada.

As European colonies grew, the British and the French competed for land in North America. Their conflict had a major influence on the development of Canada as a nation.

Founding New France

In 1534, French explorer Jacques Cartier (kar tee AY) sailed into the Gulf of St. Lawrence and claimed the area as New France. Later he sailed down the St. Lawrence River as far as present-day Montreal. This exploration opened a profitable fur trade with Native Americans. French traders exchanged European goods for beaver pelts, which were used to make hats that were popular in Europe.

In the early 1600s, Samuel de Champlain (sham PLAYN) built Quebec (keh BEK), the first major settlement of New France. In 1672, Louis Jolliet and Jacques Marquette left New France to explore the Mississippi River. They learned that the river flowed through Spanish territory into the Gulf of Mexico.

Life in New France

Most people in New France were farmers. However, farming was more difficult there than in the British colonies farther south. The soil was not as productive, and the colder climate made the growing season shorter. As a result, New France never had a large population. By the early 1700s, New France had far fewer people than the British colonies in North America.

The French generally got along better with Native Americans than did the British. Certain French groups, such as *voyageurs* and missionaries, had positive contact with some Native Americans in New France.

Critical Viewing Walls were built around Quebec City to **fortify**, or strengthen, the city. Why do you think these were needed?

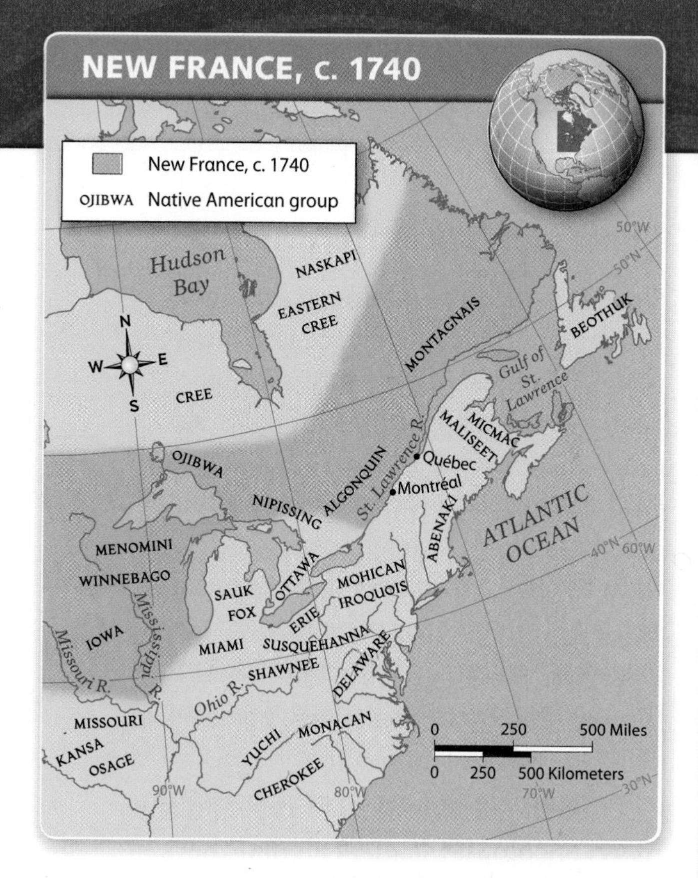

French explorer Samuel de Champlain

Voyageurs were adventurous men who traveled the area to trade with Native Americans for furs. **Missionaries** were sent by the Roman Catholic Church to convince Native Americans to accept Christianity. This contact was often cooperative and usually peaceful.

The British Take Control

The British and French were rivals for power in the region. War erupted in 1754 over whether a specific area of the upper Ohio River belonged to Britain or France. A bigger issue at the time, though, was which culture—British or French—would gain a stronger hold in North America. With greater military and financial resources, the British conquered Quebec and in 1760 gained control of the rest of New France.

With tension growing in the American colonies, Britain needed loyalty from its French subjects in Quebec. In 1774, the British passed the Quebec Act, which strengthened elements of French culture.

The Quebec Act established a French system of law and allowed the mostly Catholic population freedom to practice their religion. After the American Revolution, some people loyal to Britain migrated to Quebec from the United States. As a result, Quebec has both French and English influences.

Before You Move On

Make Inferences Why were there elements of French culture in Quebec after British conquest?

ONGOING ASSESSMENT

MAP LAB

1. **Interpret Maps** Look at the map in Section 2.1. Where was New France in relation to the British colonies?
2. **Compare** Based on maps in 2.1 and 2.2, how did the two countries' colonies compare in size?
3. **Draw Conclusions** Why were the British able to defeat the French?

2.3 Revolution and Independence

myNGconnect.com For a map and images of the events of the Revolution

Maps and Graphs

Digital Library

Main Idea The American colonists fought for and won independence from Britain.

As the British colonies in America grew, conflicts arose between the colonists and the government of Britain about how to govern the new land.

Trouble in the Colonies

In the 1760s, the British government passed several laws taxing the colonists. A **tax** is a fee demanded by the government to pay for public services. These taxes angered the colonists, who had no representation in British government. The colonists **protested**, or objected to, British control. In 1773, in an event that came to be known as the Boston Tea Party, angry colonists dumped a shipment of tea from Britain into Boston Harbor because they felt the tea was unfairly taxed.

In response, Britain enacted measures meant to punish the colonies. These actions led some colonists to become more determined to govern themselves. Many were ready to overthrow the government and replace it with one of their own making. In other words, some colonists were ready for a **revolution**.

The American Revolution

On April 19, 1775, violence erupted in Lexington and Concord, Massachusetts, when British soldiers were sent to destroy military supplies of colonial rebel groups. One of these rebels was Paul Revere. He alerted the Minutemen, colonists who were standing ready to fight at only a minute's warning, that the British were coming. The trouble between Patriots—colonists who wanted American independence—and Britain was now an armed conflict.

Over the next year, fighting continued. In 1776, colonial leaders signed the **Declaration of Independence**, written by Thomas Jefferson and claiming rights for America as an independent country.

Benjamin Franklin, a colonial political leader, traveled to France to help convince the French government to support the Patriot cause against the British. France decided to loan money to the Patriots and to send troops and ships. Marquis de Lafayette, a French nobleman, became an officer in the Patriot army and helped General George Washington.

Tax stamp issued by the British government for use in the colonies

1760

1765 British Parliament passes Stamp Act, creating a new tax; colonists protest.

1767 Parliament enacts new taxes, resulting in more protests.

1770

1773 Colonists protest tea tax in the "Boston Tea Party."

Liberty Bell rung for the first time on July 8, 1776

General Washington led the Patriots to final victory at Yorktown, Virginia, where British troops surrendered in 1781. The two sides signed the Treaty of Paris in 1783, which was a formal agreement that resolved the issues between Britain and the United States. The Treaty of Paris recognized American independence. It also set the borders of the United States from the Atlantic Ocean to the Mississippi River and from Canada to the northern border of Florida. The American Revolution had created the United States of America.

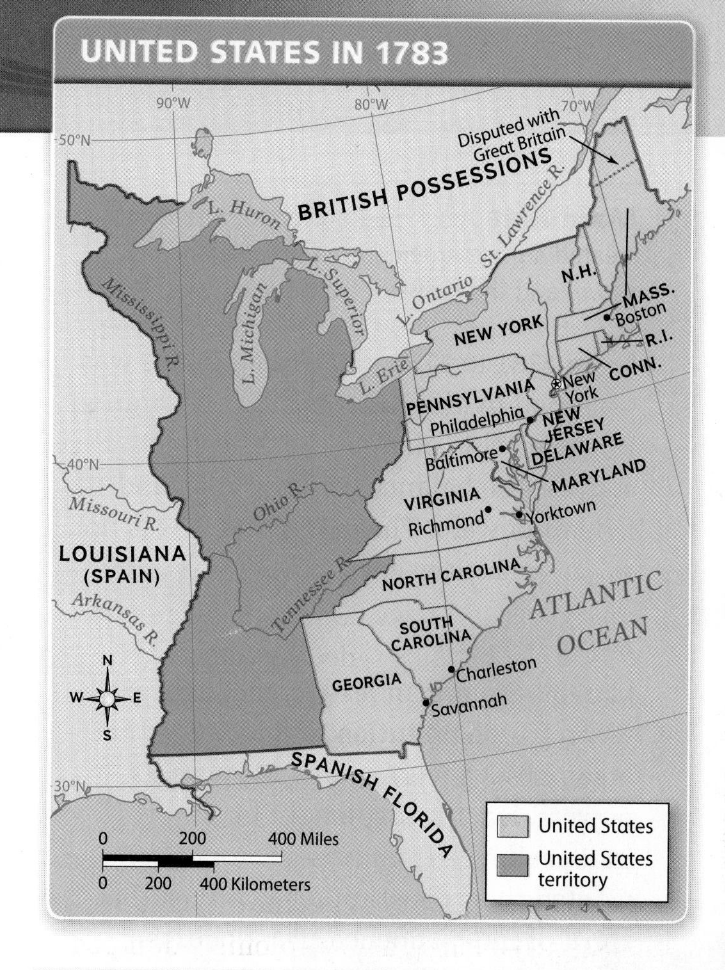

Lasting Effects

The Revolution had worldwide impact. Assistance from the French had helped Americans win the war. However, France suffered economic trouble because of the aid it had provided, and this contributed to its own revolution. The American example also helped inspire revolutions in Haiti and Central America. Finally, the Revolution also affected Canada. Many who were still loyal to Britain moved to Canada, increasing British presence there.

Before You Move On

Summarize What factors contributed to the Americans' victory over the British?

ONGOING ASSESSMENT

READING LAB

1. **Make Inferences** Why would an action such as the Boston Tea Party be an effective way to show disagreement?
2. **Summarize** In what ways were colonists ready for a revolution?
3. **Analyze Cause and Effect** In what ways did the Revolution influence other nations?

Replica of General Washington's Patriot Army uniform

2.4 U.S. Constitution

myNGconnect.com For images and a diagram of the U.S. government

Main Idea After the Revolution, Americans created a government based on a balance of power and the rights of the people.

From 1781 to 1789, the United States was governed by the Articles of Confederation. However, this plan did not effectively address the balance between state and federal powers. The new country was not as united as it needed to be.

A New Plan

In 1789, American leaders met to begin writing a **constitution**, a document that organizes a government and states its powers. Having just fought for and won independence from Britain, Americans wanted their constitution to reflect that spirit of independence. The document needed to grant certain freedoms to individuals and limit government control. Differing concerns of northern and southern states also had to be addressed.

Officially approved in 1789, the Constitution of the United States has become a model for other countries. It establishes a government based on five principles, shown in the graphic (right). The federal government consists of three branches, shown in the diagram below.

Bill of Rights

One reason the U.S. Constitution has lasted for so many years is that it can be changed, which makes it a living document. It allows for **amendments**, which are formal changes to a law.

The first ten amendments to the U.S. Constitution make up the **Bill of Rights**. The rights and freedoms it guarantees are summarized below.

1. **Freedom of religion, speech, press, assembly, and petition**
2. **Right to bear arms**
3. **Freedom from quartering of troops**
4. **Freedom against unreasonable search and seizure**
5. **Freedom from self-incrimination (testifying against yourself) and right to due process**
6. **Right to a speedy trial and to confront accusers**
7. **Right to a trial by jury**
8. **Freedom from cruel and unusual punishment**
9. **Protection of rights not named in the Constitution**
10. **Powers not given to the federal government are given to states and people**

Before You Move On

Make Inferences Why did Americans want to ensure that the U.S. Constitution limited the powers held by their own government?

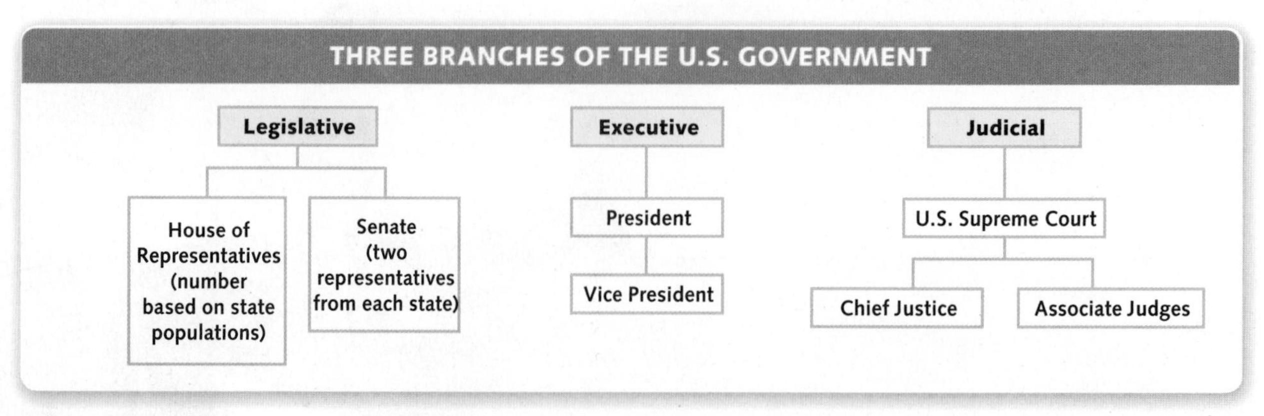

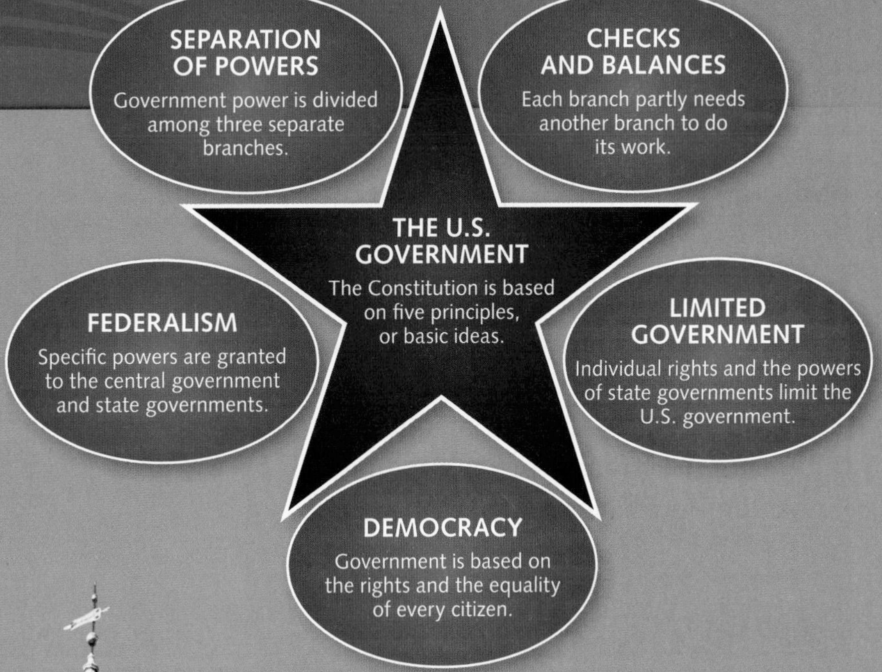

SEPARATION OF POWERS
Government power is divided among three separate branches.

CHECKS AND BALANCES
Each branch partly needs another branch to do its work.

THE U.S. GOVERNMENT
The Constitution is based on five principles, or basic ideas.

FEDERALISM
Specific powers are granted to the central government and state governments.

LIMITED GOVERNMENT
Individual rights and the powers of state governments limit the U.S. government.

DEMOCRACY
Government is based on the rights and the equality of every citizen.

The U.S. Constitution was written in Independence Hall in Pennsylvania.

ONGOING ASSESSMENT

SPEAKING LAB

GeoJournal

Pose and Answer Questions With a partner, state questions to ask the Constitution writers. Use information from the lesson for answers. If you cannot answer a question, check other sources.

2.5 Expansion and Industrialization

TECHTREK
myNGconnect.com For a map and images of expansion and industrialization

Maps and Graphs
Digital Library

Main Idea During the 1800s, the United States expanded its territory and industries.

The treaty that settled the Revolution expanded U.S. territory to include land east of the Mississippi River. In 1803, President Thomas Jefferson doubled the size of the country by making the **Louisiana Purchase.**

Settling the West

Some Americans expected that the United States would expand its territory all the way to the Pacific Ocean. Many believed it had the right to do so. This idea came to be known as **Manifest Destiny**.

In 1804, Army officers Meriwether Lewis and William Clark were hired to lead an exploration of the newly purchased land. The expedition set out from St. Louis and traveled to the Pacific Ocean and back. Its arrival at the Columbia River in Oregon ▶ was important because it helped establish a U.S. claim as far west as the Pacific Ocean.

Starting in the 1840s, thousands of Americans became **pioneers**, or settlers of new land. Their westward journey was marked by rough lands, deep rivers, the threat of disease, and even possible attack by Native Americans. The trails they used, such as the Sante Fe and Oregon Trails, still show deep ruts from wagon wheels.

Settlers' thirst for land also led to the forced removal of Native Americans. The Indian Removal Act of 1830 relocated tribes to the West. Some tribes in the southeast had highly developed farming and governments, and no interest in land to the West. If tribes refused to leave, U.S. troops forced them. The Cherokee, who tried to negotiate to keep their land, were eventually forced into a strenuous 116-day journey to Oklahoma. This route became known as the **Trail of Tears**.

Critical Viewing Camps like this one were set up for workers along the construction of the transcontinental railroad. What would be the advantages and disadvantages of being a worker on this project?

Industrialization

After independence, industry grew quickly in the east, especially textiles. By 1813, for example, the production of cloth was completely mechanized, or made by machines rather than by hand. The town of Lowell, Massachusetts, with its early textile mills, was the first U.S. town to be planned around an industry.

Industrialization—the shift to large-scale production—expanded throughout the 19th century. The construction of the Erie Canal in 1825 provided a water route from the Atlantic to the Great Lakes. As a result, New York City became a major American port. Railroads allowed goods and people to travel even greater distances, especially the **transcontinental** railroad. Completed in 1869, this railroad crossed the entire continent. Movement to the West was faster and easier for Americans than it had ever been before.

Before You Move On

Monitor Comprehension In what ways did the United States expand its territory and industries during the 1800s?

ONGOING ASSESSMENT

READING LAB

1. **Summarize** What is Manifest Destiny? What effect did it have on the United States?
2. **Location** How did the Indian Removal Act of 1830 affect tribes in the southeastern states?
3. **Make Inferences** How did the construction of canals and railroads benefit manufacturers?

2.6 Civil War and Reconstruction

myNGconnect.com For a map and images of the Civil War

Maps and Graphs

Digital Library

Main Idea Differences between Northern and Southern states led to the Civil War.

As the United States grew in the early 1800s, a powerful division arose within the country. Southerners wanted slavery to be allowed in new territories in the West. However, **abolition**, or the movement for ending slavery, was growing in the North.

Causes of the Civil War

Abraham Lincoln was elected president in 1860. Southerners, who depended on slave labor to run their plantations, feared the new president would try to end slavery in the South. As a result, 11 Southern states **seceded**, or formally withdrew, from the Union in 1860 and 1861. These states formed the Confederate States of America. Richmond, Virginia, became the capital and Jefferson Davis was president.

Lincoln declared the states in rebellion and vowed to reunite the Union. Soon after, Confederate troops fired upon Fort Sumter, a federal fort located in South Carolina. This event was the beginning of the Civil War in the United States. A **civil war** is war between opposing groups of citizens in the same country.

Conduct of the War

The war lasted four years, from 1861 to 1865. The Confederates, led by General Robert E. Lee, won many early battles. However, the Union had a larger military, a stronger economy, and more resources.

In the midst of the war, on January 1, 1863, the **Emancipation Proclamation**, which freed all slaves in Confederate territory, became effective. The same year, Lincoln delivered the **Gettysburg Address** to honor those who died there in a key battle. These two events helped to highlight a moral purpose to the war—freedom and equality. Eventually, the resources and economy of the North proved too strong, and the Confederacy surrendered in 1865.

Reconstruction After the War

Racial tensions, desperate poverty, and hostility toward the U.S. government lingered in the South after the war. These problems presented challenges to **Reconstruction**, the effort to rebuild and reunite the states as one nation.

Although former slaves had been granted new rights, many white Southerners did not allow them new freedoms.

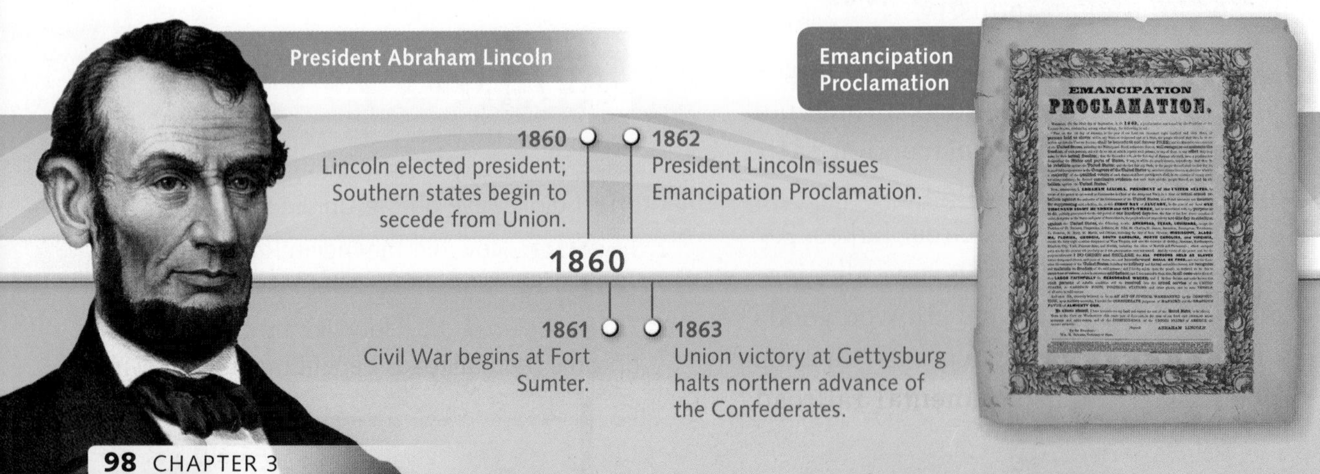

CIVIL WAR IN THE UNITED STATES

The government had to send federal troops to protect former slaves. By 1877, Reconstruction had abruptly ended. Almost a century passed before the federal government would again protect the civil rights of African Americans.

Before You Move On

Summarize What differences between Northern and Southern states led to the Civil War?

ONGOING ASSESSMENT

MAP LAB

GeoJournal

1. **Interpret Maps** How did the number of Union and Confederate states compare?
2. **Location** Based on the map, where were most battles fought?
3. **Make Inferences** Which side was most likely to suffer damage from the war? Why?

1864
Union captures Atlanta; Lincoln is re-elected president.

1865

April 11, 1865
Confederacy surrenders.
April 14, 1865
President Lincoln is shot.

A schoolroom where freed black slaves learned to read

1870

1877
Reconstruction ends.

2.7 World Conflict

TECHTREK
myNGconnect.com For photos and Guided Writing support

Digital Library

Student Resources

Main Idea In the 20th and 21st centuries, the United States became increasingly involved in world affairs and conflicts.

As the United States grew in population and prosperity, so did its economic power. This new strength made it difficult for the country to remain uninvolved in world issues and conflicts.

World War I

Due to growing conflicts between major powers in Europe, war broke out there in 1914. Russia, France, and Britain—the Allies—fought the Central Powers, which were led by Germany. The United States had economic **alliances**, or partnerships, with France and Britain. Nevertheless, the country stood firm in its **neutrality**, or refusal to take sides, and worked to bring about peace in Europe. When German submarines sank the Lusitania, a ship carrying U.S. passengers, the United States could no longer remain neutral. In 1917, America joined the Allies and helped them win the war. The terms of the peace treaty signed after the war were thought by many Germans to be unfair.

World War II and the Cold War

After a period of prosperity following World War I, the U.S. stock market crashed in 1929. The crash set in motion a series of events that led to a worldwide economic downturn called the Great Depression. In 1933, Adolf Hitler rose to power with promises to restore Germany's political and economic strength. He became a **dictator**, a ruler with complete control, and he planned to conquer all of Europe.

Critical Viewing A woman works in an engine shop in 1942. Why would women be doing this kind of work at this time?

Hitler's invasion of Poland in 1939 started World War II. The United States again stayed neutral, until the Japanese bombed **Pearl Harbor**, a U.S. naval base in Hawaii, on December 7, 1941. The next day, the United States joined Britain, France, and the Soviet Union against the Axis Powers, led by Germany, Italy, and Japan.

After years of fighting, Germany surrendered in May, 1945, but Japan fought on. In August that year, U.S. planes dropped atomic bombs on Japan. Bombs hit Hiroshima on August 6, and Nagasaki on August 9. Both cities were completely destroyed, and as many as 140,000 civilians were killed. Japan surrendered on September 2, 1945.

ONGOING ASSESSMENT

WRITING LAB

1. **Find Main Idea and Details** What role did the United States play in World Wars I and II?
2. **Summarize** What ideas did the United States promote during the Cold War?
3. **Write a Paragraph** Write a short paragraph that describes U.S. involvement in world affairs in the 1900s and beyond. Go to **Student Resources** for Guided Writing support.

The war had a great human cost. The **Holocaust**—Hitler's organized murder of Jews and other groups—killed an estimated six million people. Overall, an estimated 50 million people were killed, primarily in Europe and the Soviet Union.

Even though the United States and the Soviet Union had been allies, the two countries had political differences. They engaged in a **Cold War**, a long period of political tension without fighting. The United States promoted global democracy and freedom, while the Soviet Union promoted communism, in which the state controls all parts of the economy. The Cold War ended in 1991 when the government of the Soviet Union broke apart.

Terrorism and Modern Conflict

As the 21st century approached, a new type of warfare was becoming more common: **terrorism**. Terrorists use violence to achieve political results.

On September 11, 2001, terrorists attacked the United States. They hijacked four passenger planes and crashed them into civilian and military targets, killing over 3,000 people. The United States invaded Afghanistan where the terrorist group originated. The campaign to defeat terrorism led to a U.S. operation in 2011 that resulted in the death of Osama bin Laden, believed to have planned the September 11 attack.

Before You Move On

Summarize In what ways did the United States become increasingly involved in world conflict in the 20th and 21st centuries?

Rescue workers provide medical aid in New York City on September 11, 2001.

3.1 The Maya and the Aztecs

myNGconnect.com For maps of the Maya and Aztecs and photos of artifacts

Main Idea The Mayan and Aztec civilizations made important cultural contributions to Mexico.

Present-day Mexico was settled several thousand years earlier than the United States. About 11,000 years ago, Native American groups lived in the Valley of Mexico, the area around modern Mexico City. They survived by hunting and gathering plants to eat. Around 7,000 years ago, the settlers began growing maize, or corn, which was a native grass. High yields from these crops allowed the population to grow.

The Maya

By about 1000 B.C., an organized society called the **Olmec** lived along the southern coast of the Gulf of Mexico. Their culture had a strong influence on later cultures in Mexico, such as the Maya. The **Maya** lived on the present-day Yucatán Peninsula in Mexico and in northern Central America.

Around 100 B.C., the Maya began to develop into a **civilization**, a society with highly developed culture, politics, and technology. Evidence the Maya left behind provides information about their culture, such as their system of writing.

The Maya used **hieroglyphics**, a system of writing that consisted mostly of pictures and symbols—or hieroglyphs—as characters. The recorded history they left behind in hieroglyphic paintings reveals a highly developed written language. The Maya also studied the sun, moon, stars, and planets, which allowed them to develop an accurate calendar. They used the calendar to mark dates that were important in their religion.

After about A.D. 900, Mayan civilization apparently declined. Historians do not fully understand why this happened. Possible theories include violent conflict between cities, overpopulation, or the overuse of land available for farming.

Critical Viewing Tenochtitlán was built on islands within a lake. What might be an advantage of building the Aztec capital city on islands rather than on mainland?

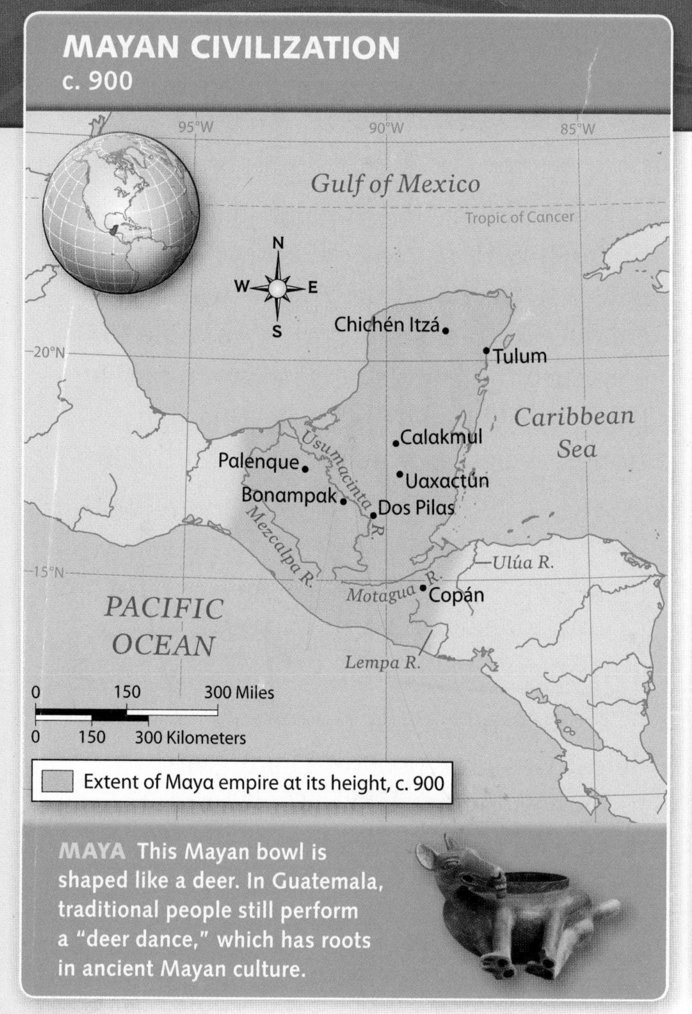

MAYA This Mayan bowl is shaped like a deer. In Guatemala, traditional people still perform a "deer dance," which has roots in ancient Mayan culture.

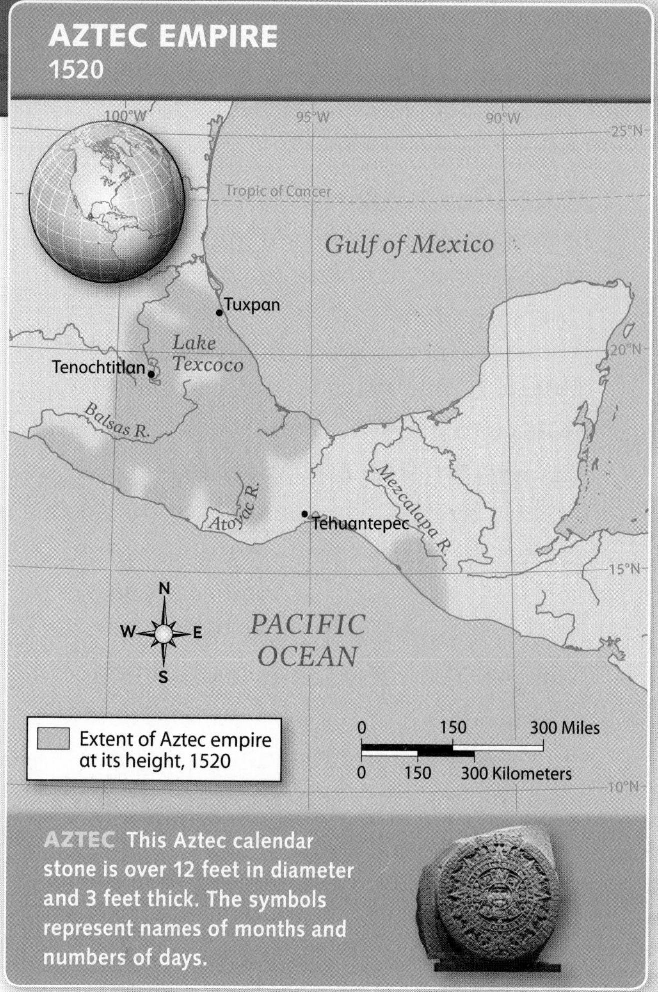

AZTEC This Aztec calendar stone is over 12 feet in diameter and 3 feet thick. The symbols represent names of months and numbers of days.

The Aztecs

After the Mayan decline, other native groups gained power in Mexico. The dominant group was the **Aztecs**, who settled in the area of modern Mexico City around A.D. 1325. The Aztec people built a city called Tenochtitlán (teh nohch teet LAHN) on islands in Lake Texcoco. They constructed human-made islands called *chinampas*, or "floating gardens," on the lake to give them more area to grow food.

The Aztecs built a broad **empire**—an extensive group of peoples governed by one ruler—through military conquest of neighboring lands. They gained further power by enslaving conquered people and using their labor to build more cities. Rulers also collected **tribute**, or fees, in the form of money, crops, or other goods.

The Aztec Empire continued to thrive until Spanish explorers arrived in the 1500s, which changed life in the region.

Before You Move On

Monitor Comprehension What cultural contributions did the Maya and the Aztecs make?

ONGOING ASSESSMENT

MAP LAB

1. **Interpret Maps** Use directional words to describe the location of the Maya and Aztecs.
2. **Make Inferences** What might account for so little overlap in the land areas of each group?
3. **Analyze Visuals** What does the circular shape of the Aztec calendar suggest about the Aztec idea of a year?
4. **Summarize** How was the Aztec Empire able to gain so much power?

3.2 The Conquistadors

myNGconnect.com For images of warriors and conquistadors

Digital Library

Main Idea The Spanish conquest of the Aztec Empire had a lasting effect on the native populations of Mexico.

After Columbus's voyages, Spain seized control of many islands in the Caribbean in the early 1500s. From there, Spanish explorers continued to look for new lands. In their travels, they heard tales of wealth among the Aztecs and set out to obtain it.

The Conquest of Mexico

Late in 1518, **Hernán Cortés** landed in Mexico with a force of about 500 soldiers. Cortés was a **conquistador**, a Spanish soldier-explorer. He was determined to conquer more of the Americas. Cortés quickly realized that the Aztecs were widely hated by other native groups—and that the rumors of Aztec wealth were true.

The Aztec ruler, **Montezuma**, welcomed Cortés and his men to Mexico's capital, Tenochtitlán. Within three years, however, Cortés had killed Montezuma and captured the city. He ordered the city burned and built a new capital—Mexico City—on its ashes. By 1525, Spanish control reached as far south as Central America.

Spanish Advantage

Cortés was a bold, skillful leader. The Spanish had guns, swords, and armor, all superior to the Aztecs' weapons. The Spanish also used horses and dogs in battle. The Aztecs were slow to act at first because they believed Cortés was a god. When their leader, Montezuma, was killed by Cortès in 1520, the Aztec armies fell apart in confusion.

Native Mexican people provided the Spanish with further advantage. Many native tribes hated the Aztecs who had conquered them, and Cortés was able to convince thousands of warriors in native tribes to fight along with the Spanish troops. A native woman called Malinche served as an interpreter and guide for the Spanish. Her knowledge of native languages was important for successful interactions between the Spanish and other native tribes.

One of the biggest reasons for the Aztec defeat was unexpected. The Spanish unknowingly brought over diseases that the natives had never encountered and could not resist.

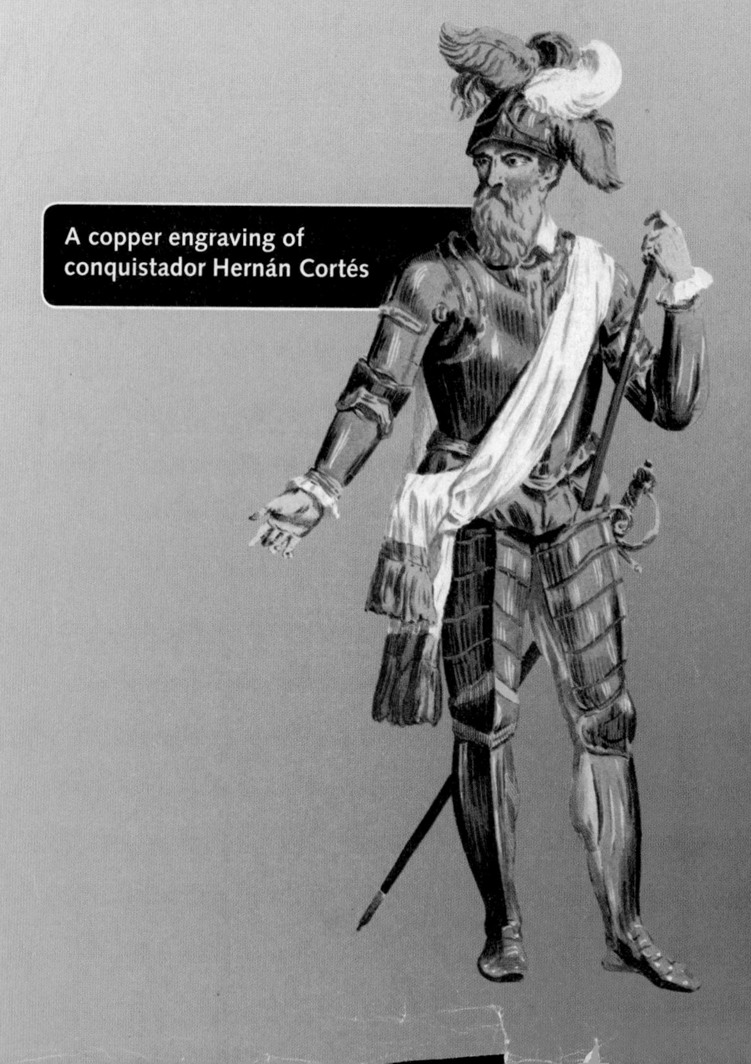

A copper engraving of conquistador Hernán Cortés

Illnesses like measles and smallpox became epidemics. An **epidemic** is the outbreak of a disease that spreads widely. These diseases infected and killed huge numbers of the native population. The deaths weakened the Aztecs' ability to resist the Spanish conquest.

Results of the Conquest

From their base in Mexico, the Spanish armies conquered all of Central America. They went on to conquer most of South America, building their own vast empire. European diseases continued to wipe out native populations for many years.

Natives who survived were enslaved by the Spanish and forced to work on farms and in silver mines. Using slave labor, the Spanish mined vast wealth from the large amounts of silver in the mountains of Northern Mexico.

Spain's success and sudden wealth had another effect. It drove Spain's rivals in Europe to send explorers to the Americas in search of wealth. These European powers also created new empires there.

Before You Move On

Monitor Comprehension What did the Spanish conquest mean for the Aztecs and other native populations of Mexico?

An engraving of Aztec ruler Montezuma

ONGOING ASSESSMENT

VIEWING LAB

GeoJournal

1. **Compare and Contrast** Look at the engravings of Hernán Cortés and Montezuma. What similarities and differences do you notice? Copy the Venn Diagram and complete it with details from the images to compare and contrast the two leaders.

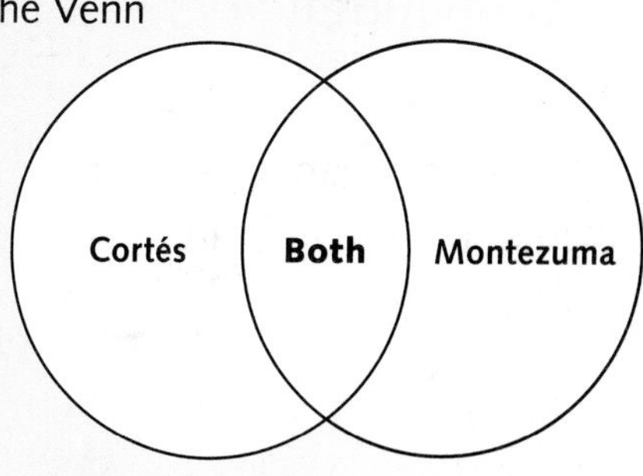

2. **Analyze Visuals** Based on what you see in the engravings, what features of Cortés and Montezuma suggest their position as leaders?
3. **Summarize** What factors contributed to the Spanish army's ability to conquer the Aztecs?
4. **Analyze Visuals** Based on what you see in the illustrations, what would you expect to have happened when these two fighting men met in battle? Why?

3.3 Mexican Independence

Main Idea Conflict between social classes in colonial Mexico led to a movement for Mexican independence from Spain.

Because of the conquistadors, Mexico became a major part of Spain's colonial empire. The empire included Central America and Spanish-held areas in the Caribbean and present-day United States.

Colonial Mexico

In the early 1800s, the society of colonial Mexico was divided into social classes. At the top were people born in Spain, called *peninsulares* (pen in soo LAH rayz). They held the highest offices in the government and the church. Next were *criollos* (kree OH yohz), people of Spanish background born in Mexico. Next came mestizos (mes TEET sohz), people of mixed Spanish and native heritage. Native Americans and slaves were the lowest class of society.

Independence

In 1810, **Miguel Hidalgo**, a Catholic priest, led a revolt against the *peninsulares.* He led an army of almost 100,000 men, including many mestizos. His army killed many *peninsulares* and *criollos* and captured several of Mexico's cities.

Government troops resisted Father Hidalgo's army, and in 1811, the troops defeated the priest's army and seized him. On July 30, 1811, they executed him, but his rebellion inspired others in Mexico to continue the fight.

José Morelos, a priest and revolutionary, took over leadership of the Mexican independence movement. In November 1813, he and his followers declared independence from Spain. They drafted a constitution that made Mexico a **republic**, a government of elected officials. The constitution also guaranteed freedom, equality, and security to Mexico's citizens. However, Spain refused to accept this constitution, and Spanish troops refused to surrender. For nearly two years they pursued Father Morales's army throughout southern Mexico. He was captured by Spanish troops and executed in 1815.

The Struggle Continues

Small groups of rebels continued to fight. In 1820, one leader united them—Colonel Augustín de Iturbide (ee toor BEE day). Early in 1821, he issued a plan that included Three Guarantees. First, Mexico would be independent of Spanish control.

Father Miguel Hidalgo

1800s
Colonial Mexico class system solidified

1800

1810
Miguel Hidalgo leads a revolt against the *peninsulares.*

1810

1811
Hidalgo is executed.

Second, *peninsulares* and *criollos* would be equal to each other under the law. Third, Roman Catholicism would be the only religion in the country.

Iturbide's army rapidly defeated Spanish troops in Mexico. In 1821, Spain signed a treaty granting independence to Mexico. In 1822, with a tremendous thirst for power, Iturbide crowned himself the emperor of Mexico. His reign was a disaster. Even though he had great success leading the rebels to defeat Spain, as emperor he could not unify the country. One of his generals, Antonio López de **Santa Anna**, rebelled against him. In 1823, Iturbide gave up his crown and was **exiled**, or forced to leave the country.

Santa Anna's new government wrote the Constitution of 1824. That year, the United States, Britain, and other countries recognized Mexico as an independent country. However, Spain continued to fight for control of Mexico. Santa Anna's troops finally defeated Spanish forces in 1830. That same year, Mexico's people elected Santa Anna as their president.

Before You Move On

Summarize How did the colonial structure in Mexico contribute to demands for independence?

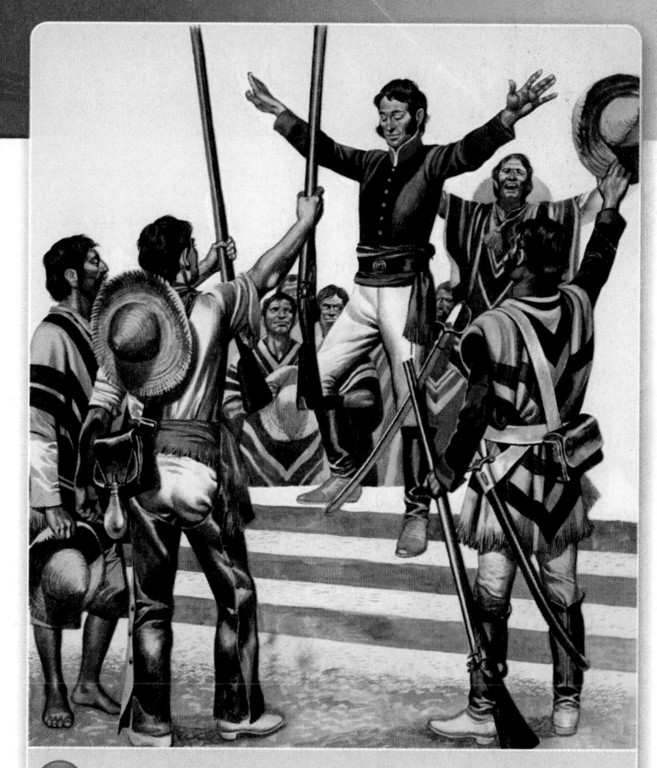

Critical Viewing A crowd surrounds Colonel Iturbide before he crowns himself emperor. What details in the illustration show that the Mexican people were in support of Iturbide?

ONGOING ASSESSMENT

READING LAB

GeoJournal

1. **Make Inferences** Do you think the Catholic Church was on the side of Spain or the native people? Explain your answer.
2. **Categorize** Identify the classes of society in Mexico from top to bottom.
3. **Draw Conclusions** How was General Iturbide able to both unite and divide the country?

1820

1821 Spain grants Mexico's independence.

1822 Iturbide proclaims himself emperor of Mexico.

1823 Santa Anna and others force Iturbide out of power.

1824 Mexico adopts a new constitution as a republic.

1830

Mexican general and political leader Antonio López de Santa Anna

3.4 The U.S.-Mexican War

TECHTREK
myNGconnect.com For maps and images from the U.S.-Mexican War

Maps and Graphs

Digital Library

Main Idea As a result of the U.S.-Mexican War, Mexico lost some territory to the United States.

When Santa Anna became president of Mexico, the country included land from present-day Texas to California and north to Utah. This territory soon faced a crisis.

The Texas Revolution

To encourage new settlers in the area of Texas, Mexico offered land grants to Americans to set up colonies there, and they had begun arriving in 1821. The American population grew so quickly that in 1830 Mexico passed a law blocking further settlement. This move started a growing hostility between the Mexican government and the settlers in Texas.

In 1835, Texans began to rebel against Mexican forces. In response, President Santa Anna led a large army to Texas to end the rebellion. In 1836, in a battle that lasted 13 days, Mexican forces killed almost all of the approximately 200 Texan rebels defending a fort called the **Alamo**. Despite this defeat, Texans continued their fight for independence. The next month, under the leadership of General Sam Houston, the Texan army defeated Santa Anna's troops at the Battle of San Jacinto.

War with the United States

After winning independence from Mexico, Texans created the Republic of Texas. For nine years, the country struggled with debt, disputes with Mexico, and violence between settlers and Native Americans. In 1845, Texas joined the United States through **annexation**, or adding territory.

Texas and the United States claimed that the southern border was the Rio Grande, but Mexico said the border was the Nueces River, farther north. In early 1846, American troops were sent to occupy the disputed area between the two rivers. When Mexican and U.S. troops fought near the Rio Grande in May 1846, the United States declared war on Mexico.

Santa Anna's troops battled the U.S. army but were defeated after two years. As a result, in 1848 Mexico gave up the area from Texas to California. This land became known as the **Mexican Cession**. In 1853, Santa Anna sold further Mexican land to the United States in what came to be known as the **Gadsden Purchase.** In Mexico, many people **opposed**, or objected to, this decision.

General Sam Houston

The Alamo

1820

1830 Mexico blocks further American settlement of Texas, limits rights of Texans.

1836 Texans revolt and win independence, but Mexico refuses to recognize it.

1840

1845 United States annexes Texas.

1846 War between Mexico and United States begins.

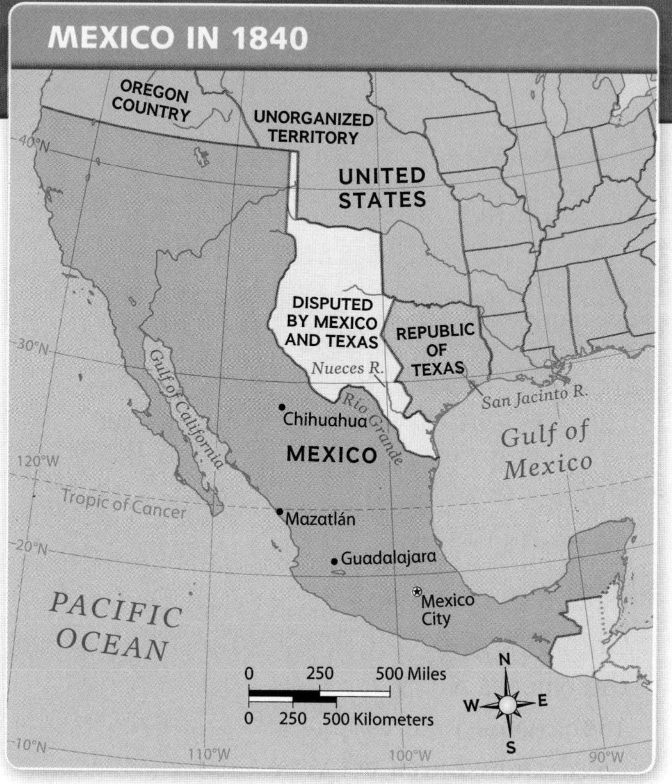

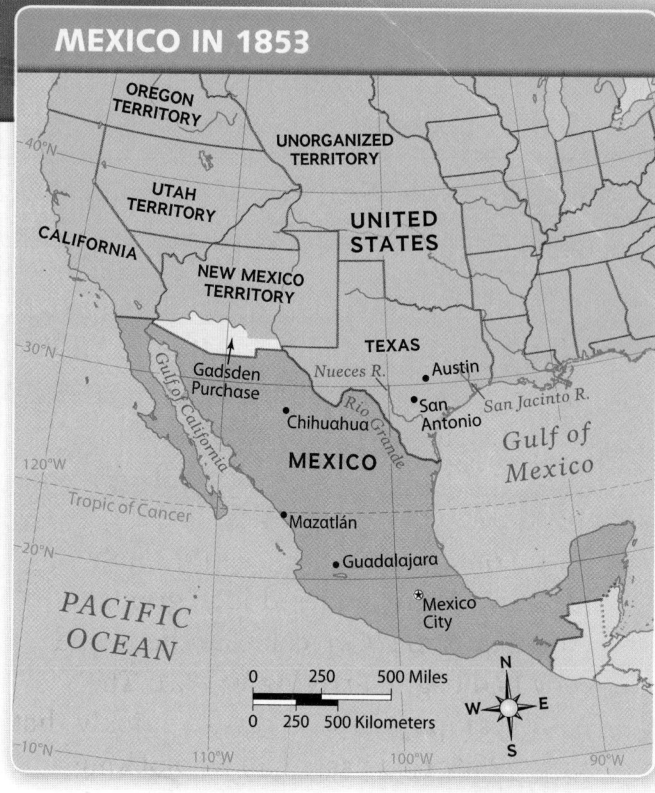

La Reforma—The Reform

Due to growing opposition to Santa Anna and his policies, he was forced out of leadership in 1854. New leaders wanted to change Mexican society. In 1857, President Benito Juárez proposed major **reforms**, or changes, to promote social equality. Religious and military leaders refused to accept these reforms, and the country entered a long period of civil unrest and violence. Conflicts between those who wanted reform in Mexico and those who did not continued throughout the 1860s.

Before You Move On

Make Inferences Why would the Republic of Texas have agreed to become part of the United States after winning independence?

ONGOING ASSESSMENT

MAP LAB

1. **Location** Find the Nueces River and the Rio Grande on the map. Describe the distance between the two rivers.
2. **Compare and Contrast** Based on the two maps, what territories did the United States gain after war with Mexico?
3. **Analyze Cause and Effect** How did the dispute over the border between Texas and Mexico lead to war?

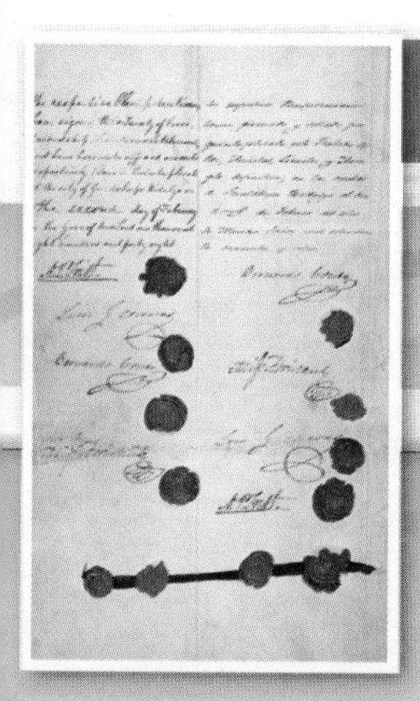

Treaty of Guadalupe Hidalgo, with seals pressed in red wax

1848 War between Mexico and United States ends; United States gains Mexican Cession.

1853 In Gadsden Purchase, Mexico sells parts of modern Arizona and New Mexico to United States.

1858 Reformer Benito Juárez becomes president of Mexico.

1860

DE MEXICO S.A. PAGARA TA PESOS VISTA AL PORTADOR SERIE LT

President Benito Juarez image on Mexican banknote

3.5 The Mexican Revolution

TECHTREK

myNGconnect.com For writing templates

Student Resources

The U.S.-Mexican War and the resulting civil conflict had weakened Mexico's economy and government. After Juárez's death, General Porfirio Díaz became dictator. Díaz's policies mainly benefited the wealthy.

The Mexican Revolution was partly a fight for **land reform**, in which large estates would be broken up and the land would be given to the poor. Emiliano Zapata and Pancho Villa led the struggle, which lasted from 1910 to 1920 and killed more than one million Mexicans.

General Porfirio Diaz

DOCUMENT 1

Plan of Ayala

Zapata issued the Plan of Ayala in 1911 to state goals for the revolution.

[W]e who undersign . . . declare solemnly . . . . . to end the tyranny [harsh government] which oppresses us. . .

[T]he pueblos [villages] or citizens who have the titles [to property being held by the government] . . . will immediately enter into possession of that real estate [property] . . . .

[T]he immense majority of Mexican [settlements] and citizens are owners of no more than the land they walk on [and] suffer . . . the horrors of poverty without being able to improve their social condition in any way.

CONSTRUCTED RESPONSE

1. What does the plan say about conditions in Mexico?

DOCUMENT 2

Constitution of 1917

One important goal of the constitution was to bring about changes in land ownership.

Article 27. Ownership of the lands and waters within the boundaries of the national territory is vested [included] originally in the Nation, which has . . . the right to transmit [give] title thereof to private persons . . . .

[N]ecessary measures shall be taken to divide up large landed estates; to develop small landed holdings [smaller farms] in operation; . . .

to encourage agriculture . . . and to prevent the destruction of natural resources. . . The rights of small landed holdings in operation [will be] respected at all times.

CONSTRUCTED RESPONSE

2. How might the division of large estates into small holdings change Mexico?

[T]HE IMMENSE MAJORITY OF MEXICAN . . . CITIZENS ARE OWNERS OF NO MORE THAN THE LAND THEY WALK ON.

— PLAN OF AYALA, 1911

Pancho Villa

DOCUMENT 3 

Photo of Pancho Villa

Pancho Villa (VEE yuh), the son of a poor farm laborer, became a folk hero in Mexico. He gained popularity by stealing from the rich in order to give to the poor. Villa's knowledge of the physical geography of northern Mexico helped him avoid being captured.

CONSTRUCTED RESPONSE

3. What does this photo of Villa suggest about him as a leader?

ONGOING ASSESSMENT

WRITING LAB  GeoJournal

DBQ Practice Think about the Plan of Ayala, the Constitution of 1917, and the photograph of Pancho Villa. What do these show about the conditions that led to the Mexican Revolution?

Step 1. Review the description of the revolution and the statements from the Plan of Ayala and the Constitution. Reread the biographical information about Villa and look at the photograph.

Step 2. On your own paper, jot down notes about the main ideas expressed in each document.

Document 1: Plan of Ayala

Main Idea(s) ______________________

Document 2: Constitution of 1917

Main Idea(s) ______________________

Document 3: Photo of Pancho Villa

Main Idea(s) ______________________

Step 3. Construct a topic sentence that answers this question: What do the statements in the Plan of Ayala and the Mexican Constitution and the photograph of Pancho Villa show about the conditions that led to the Mexican Revolution?

Step 4. Write a detailed paragraph explaining what the documents tell you about the conditions that led to the Mexican Revolution.

CHAPTER 3

Review

VOCABULARY

For each pair of vocabulary words, write one sentence that explains the connection between the two words.

1. temperate; glacier

> Many areas of North America have a temperate climate, but others are so cold they are covered by glaciers.

2. export; commercial agriculture
3. colonize; plantation
4. abolition; secede
5. civilization; hieroglyphics
6. empire; tribute

MAIN IDEAS

7. How does the geography of eastern and western North America differ? (Section 1.1)
8. What are important resources of the Great Plains? (Section 1.2)
9. What resources are available from Mexico's mountains and plateaus? (Section 1.4)
10. Why were European countries interested in settling North America? (Section 2.1)
11. Why has the U.S. Constitution been able to last so long? (Section 2.4)
12. What were the causes and results of the Civil War? (Section 2.5)
13. How did the United States change in the late 1800s? (Section 2.6)
14. What events led to U.S. involvement in three world conflicts? (Section 2.7)
15. What were the cultural achievements of the Maya? (Section 3.1)
16. Why were the Spanish able to conquer the Aztec Empire? (Section 3.2)

GEOGRAPHY

ANALYZE THE ESSENTIAL QUESTION

What are the significant physical features of North America?

Critical Thinking: Compare and Contrast

17. In what ways is the physical geography of Mexico different from that of the United States and Canada? In what ways is it similar?
18. What energy resources are found in all three countries of North America?
19. How are the western United States and the Yucatán similar in terms of resources?

U.S. & CANADIAN HISTORY

ANALYZE THE ESSENTIAL QUESTION

How did the United States and Canada develop as nations?

Critical Thinking: Make Inferences

20. How did Reconstruction attempt to address issues that resulted from the Civil War?
21. Which groups suffered as a result of U.S. expansion west? Which groups gained?
22. Why did the population of the United States grow faster than that of Canada?

INTERPRET CHARTS

SETTLING QUEBEC	
1534	Jacques Cartier claims St. Lawrence River Valley for France.
1608	Samuel de Champlain founds colony of Quebec for France.
1671	France claims the Great Lakes and the Mississippi River.
1759	British capture Quebec during Seven Years' War.

23. **Interpret Charts** What land and water claims did France make in the 1500s and 1600s?
24. **Make Inferences** What can you infer about the French explorations in the Americas?

HISTORY OF MEXICO

ANALYZE THE ESSENTIAL QUESTION

How have various cultures influenced Mexico's history?

Critical Thinking: Summarize

25. Why did the Spanish place the capital of colonial Mexico where they did?
26. What are the social divisions that led to the Mexican Revolution?
27. How did this social class structure lead to Mexican independence from Spain?

INTERPRET MAPS

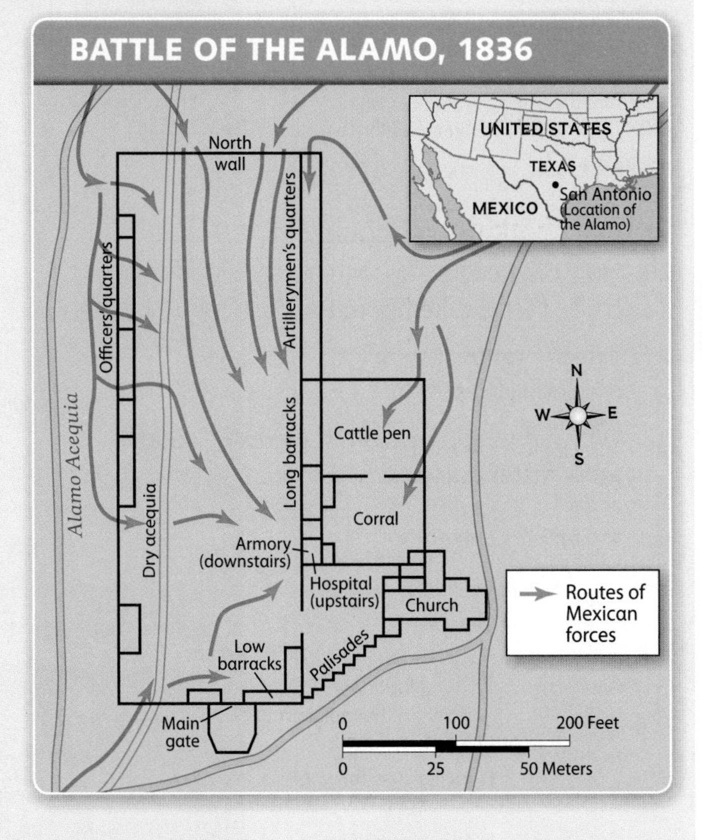

28. **Interpret Maps** Which areas of the Alamo appear to have been avoided by Mexican forces?
29. **Make Inferences** Which part of the Alamo did Mexican forces not use as entry? Why might this be?

ACTIVE OPTIONS

Synthesize the Essential Questions by completing the activities below.

30. **Write a Comparison Speech** Write a speech that compares the groups that were part of settling the United States, Canada, and Mexico. Your speech should point out similarities and differences based on the groups that settled those three countries. Use the tips below to help you prepare the speech. **Offer to deliver your speech to other classes who are studying North America.**

Writing Tips
- Identify the groups that settled each country and think about their similarities and differences.
- Decide whether you think the groups are more similar than they are different.
- Write that conclusion as your main idea.
- Support your main idea with details.

31. **Conduct Internet Research** Choose a city in one of the three countries. Use research links at **Connect to NG** to gather information about the history of that city.

    Find out when and how the city was settled, how it has grown over time, and what groups contributed to its history. Then choose one of the following options to show what you found:

    - Construct a time line that shows the sequence of events that make up the city's history. Illustrate your time line.
    - Draw a diagram of the city, including its historical landmarks, such as statues or other historical markers. Also label the locations of historical events. The labels should explain why that landmark or location is important. Add your time line or diagram to a class exhibit on North America.

CHAPTER 4

NORTH AMERICA TODAY

PREVIEW THE CHAPTER

Essential Question What issues do the United States and Canada face today?

SECTION 1 • FOCUS ON THE UNITED STATES & CANADA

KEY VOCABULARY

- diversity
- indigenous
- immigrant
- tolerance
- mass media
- smartphone
- global
- mobile
- manufacturing
- fiber optic
- recession
- petroleum
- ethanol
- hybrid
- wind turbine
- naturalization
- due process

ACADEMIC VOCABULARY

dynamic

TERMS & NAMES

- Internet

Essential Question How has globalization affected Mexico?

SECTION 2 • FOCUS ON MEXICO

KEY VOCABULARY

- descendant
- ancestry
- mural
- artifact
- globalization
- nationalize
- economic sector
- multi-party democracy

ACADEMIC VOCABULARY

reform

TERMS & NAMES

- North American Free Trade Agreement (NAFTA)
- Institutional Revolutionary Party (PRI)

New York City's Empire State Building stands out against the Hudson River in the background. Manhattan, a densely populated area of the city, is a major economic center in the global economy.

1.1 North America's Cultural Diversity

TECHTREK
myNGconnect.com For a chart of languages in North America
Student Resources

Main Idea The opportunities and freedoms offered in the United States and Canada have attracted a diverse population.

Diversity, or variety, is a major feature of U.S. and Canadian culture. Both countries have a diverse mix of races, languages, religions, and nationalities.

Diversity in the United States

Early cultures included **indigenous**, or native, tribes, European settlers, and Africans who had been forced onto the continent as slaves. After independence, immigrants came from many other parts of Europe seeking freedom and opportunity. An **immigrant** is a person who takes up permanent residence in another country.

Many historians describe immigration as a series of waves, starting in the 1600s with European colonists. From 1820 to 1870, a second wave brought more than seven million people, many from Ireland and Germany, fleeing poor conditions. Immigrants also came from China, seeking opportunity in the American West. Starting in 1880, a third wave, mostly from parts of southern and eastern Europe, brought 23.5 million more. Changes to U.S. immigration laws in the 1960s brought a fourth wave, many from Asia and the Caribbean. Today, most immigrants come from Mexico, China, the Philippines, and India.

Immigrants bring not only their culture, but also their skills and ambitions, and can play a major role in economic growth. In fact, many immigrants come here for that reason. In the United States they have the freedom to start a business.

Critical Viewing Street signs in Chicago's Chinatown are posted in two languages. In what way would signs like this be helpful to immigrants?

American culture has come to be defined by its diversity. **Tolerance**—acceptance of different views—is an important American value.

Diversity in Canada

Native groups such as the Inuit have been in Canada for thousands of years and are still there today. Historically, Canada has encouraged immigration in order to attract workers to help build its economy.

In the 1600s and 1700s, British and French colonies in Canada had attracted large numbers of immigrants from France, Ireland, and Britain. After World War II, many Europeans who had lost their homes moved to Canada. Immigrants from Asia and Latin America began to arrive in the late 1900s. However, most people in Canada today still have European roots.

Before You Move On

Summarize Why are the cultures of the United States and Canada so diverse?

SPEAKERS OF MAJOR LANGUAGES, UNITED STATES AND CANADA

Language	United States	Canada
Arabic	845,396	261,640
Chinese	2,600,150	1,102,065
English	228,699,523	17,882,775
French	1,305,503	6,817,655
German	1,109,216	450,570
Italian	753,992	455,040
Polish	593,598	211,175
Portuguese	731,282	219,275
Spanish	35,468,501	345,345
Tagalog (Filipino)	1,513,734	235,615

Sources: 2009 American Community Survey; Statistics Canada, Census of the Population, 2006

People gather in Times Square in New York for a New Year's celebration.

ONGOING ASSESSMENT

LANGUAGE LAB

GeoJournal

1. **Compare and Contrast** After English, what is the next major language spoken in the United States and in Canada?
2. **Place** What do the top two languages spoken in Canada suggest about its history?
3. **Make Predictions** Which languages do you predict will have a higher number of speakers in the United States in future years? Why?

1.2 The Media Culture

myNGconnect.com For photos and a map on media culture

Main Idea The movement of ideas through various media has shaped communication and cultures in the United States and Canada.

Mass media refers to communication that reaches large audiences. Traditionally this meant newspapers, radio, and television. Today, mass media also includes the rapid and nearly continuous flow of information through personal electronic devices.

Print Media to Electronic Media

During the 1800s, newspapers provided information that often influenced public opinion. By the 1920s, radio provided news and entertainment. It was the most important source of news during World War II. President Franklin Roosevelt used radio for his "fireside chats," in which he shared information about the war with the American public. By this time it was clear that mass media was a powerful tool.

In the 1950s, many U.S. households acquired their first television sets, and this medium surpassed radio in its popularity as a source for news and entertainment. In 1963 the entire nation viewed film footage of the assassination of President Kennedy. In the same decade television became an important medium for witnessing the Vietnam War and the protests against it.

In the 1990s, new media technologies exploded in popularity. Cell phones connected people around the world, even if they were on the move. The **Internet**, a communications network, gave people the power to access huge amounts of information from a variety of sources. It also allowed individuals to distribute their own media, such as blogs and videos, to a mass audience. With so much information available electronically, people rely less on print publications.

Critical Viewing Mobile media today combines communication and visual technology. Which details in the photo best illustrate American culture?

American Gothic
Grant Wood
Art Institute of Chicago

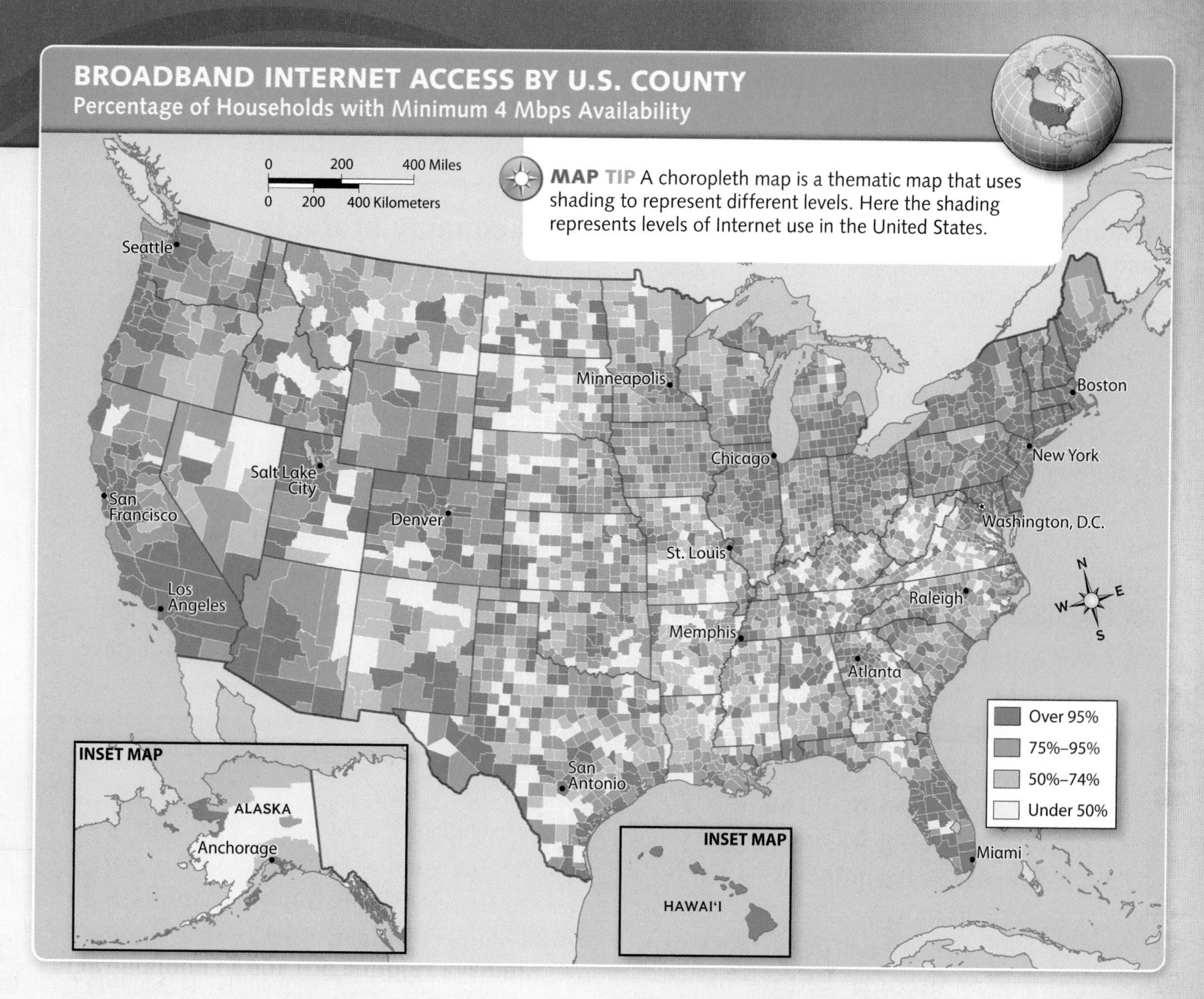

Global and Mobile Media Today

Today, texting, instant messaging, and social networking allow fast and easy communication between people around the world. Much of this is done with **smartphones**, which combine communication and software applications into one handheld device. Media technology encourages a culture that is **global**, or worldwide. Access to information is available from anywhere in the world. Media culture forms around shared ideas or interests, rather than physical location. Most technology is now **mobile**, which means it moves along with the people who use it. For many people, new media has entirely replaced traditional methods of communication.

Before You Move On

Make Inferences Why do you think mass media has such a strong influence on culture in the United States and Canada?

ONGOING ASSESSMENT

MAP LAB

1. **Region** Based on what you know about population density in the United States, how would you describe the pattern of Internet access across the country?
2. **Compare and Contrast** Find where you live on the map. How does the level of high speed Internet access in your county compare to the rest of the country? Why might this be?
3. **Draw Conclusions** What effect do you think widespread availability of the Internet has had on print publications? Why?

1.3 The Changing Economy

TECHTREK
myNGconnect.com For photos of manufacturing and current events

Connect to NG

Digital Library

Main Idea The economies of the United States and Canada are changing due to new technology and worldwide trends.

The Canadian and U.S. economies differ in some ways. For example, businesses are less regulated in the United States. Canada's government plays a larger role in its economy by owning some businesses and providing health care for all citizens. The economies of both countries have changed dramatically over the past 150 years.

The Economy of the Past

In the early 1800s, most Americans and Canadians worked in industries in which people make a living directly from the land. Most people farmed, worked in mines, fished, or logged timber.

During the mid-1800s, the Industrial Revolution, which began in Europe, swept the United States and Canada. **Manufacturing**, or using machines to make raw materials into a usable product, became an important part

Critical Viewing Automated robot arms weld trucks in a Michigan assembly plant. What details in the photo illustrate this automation?

of the economy. In vast factories, workers manufactured products such as iron and steel. The manufacturing industry led the economy from the 1870s until the 1950s.

During the 1950s, the service industry expanded. Workers in this industry provide a service rather than make a product. Jobs in areas such as health care, education, entertainment, banking, and retail are all part of the service industry.

The Information Economy

One important similarity between the U.S. and Canadian economies is that they are **dynamic**; that is, they change quickly. Part of the reason for rapid change is the availability of new computer and information technologies.

In 1974, a tiny company called MITS built the first personal computer, the Altair. It had limited memory but worked well. The only problem was that you had to build it yourself from a kit! Today, more than 8 out of every 10 U.S. and Canadian households have computers, and **fiber optics** uses digital technology to transmit voice, text, and visual messages. These types of advancements have shifted the economy in a new direction.

Today's technologies allow people to store, organize, and retrieve enormous amounts of financial information with a few mouse clicks. New technologies also can bring about more jobs in the communication and information industry. Faster communication across greater distances helps support a global economy, unlimited by physical distance.

Visual Vocabulary **Fiber optics** is a method of sending light through glass fibers. Fiber optic cables can transmit digital code quickly across great distances.

Global Recession

In 2007, the economy faced a **recession**, or a slowdown in economic growth. Lending practices from earlier in the decade led banks to grant large loans to consumers and businesses—loans they could not pay back. These unpaid debts led to a global financial crisis, including the collapse of some major worldwide banks. A recession is officially over after six consecutive months of economic growth. The U.S. economy reached this mark in early 2009.

Before You Move On

Make Inferences In what ways might new technologies bring about more jobs?

ONGOING ASSESSMENT

PHOTO LAB

1. **Analyze Visuals** What do you expect to see on an assembly line that is missing from the photo? How might this technology affect jobs?
2. **Synthesize** What similarities and differences do you notice in the photos?
3. **Make Inferences** In what ways can new technologies affect a country's economy?

1.4 Finding New Energy Sources

Main Idea The growing demand for nonrenewable fossil fuels for energy has led to exploration into renewable, alternative energy sources.

KEY VOCABULARY

petroleum, n., the raw material used to create oil products

wind turbine, n., an engine powered by wind to generate electricity

ethanol, n., a fuel obtained from plants that can be used alone or blended with gasoline

hybrid, n., a vehicle that can run using either an electric motor or a gas-powered engine

In April 2010, the *Deepwater Horizon* oil drilling rig in the Gulf of Mexico exploded. The explosion opened a hole in a well drilled in the seafloor. Oil gushed from that hole every day for nearly three months, leaking nearly 5 million barrels of oil. Before this accident, the worst U.S. oil disaster at sea was the grounding of the *Exxon Valdez* tanker in 1989. That disaster lost around 260,000 barrels of oil—a large amount, but far smaller than the amount of oil leaked after the explosion of the *Deepwater Horizon.*

Oil Supply and Demand

The United States runs on energy, and oil provides about a third of that energy. **Petroleum** is a nonrenewable resource, meaning that it will eventually run out. However, no one knows exactly how much petroleum might be left in reserves that lie deep beneath the earth's surface.

Experts try to predict when the United States will use up all of its own oil supply, and they don't all agree. However, they do agree that the demand for oil is growing, rapidly shrinking the supply of this nonrenewable resource. Some experts say that as early as 2020, there won't be enough oil in the United States for everyone who wants it.

Today, the United States depends on other countries for more than two-thirds of its oil. As demand rises and supply falls, the price of oil will rise. Rising oil prices can slow economic growth and drive countries into competition for oil.

Before You Move On

Summarize What are some of the effects of an increasing demand for oil in the United States?

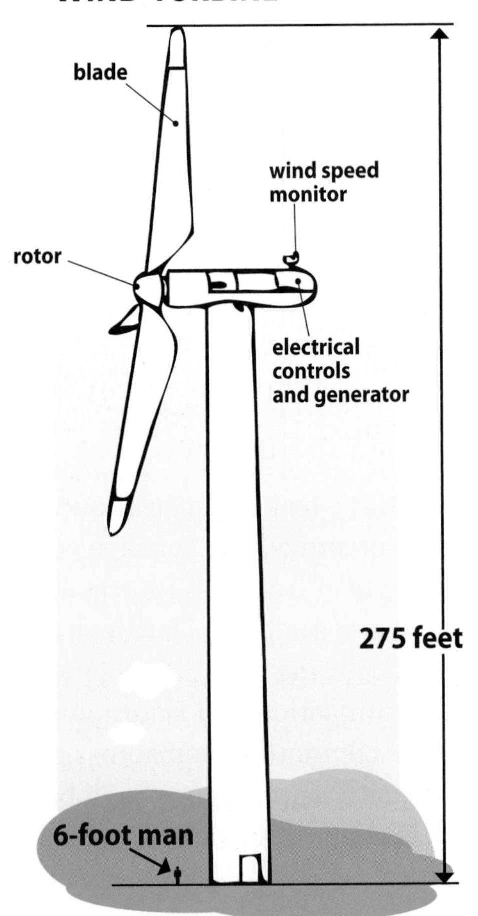

Alternative Solutions

Energy sources other than oil include nuclear power, **ethanol**, solar power, and wind power. However, each has issues that make it less than ideal. Nuclear power plants take a long time to build and are very expensive. They also create radioactive waste. Making ethanol from grains means some agricultural production is used for fuel instead of food. Solar panels and wind turbines cannot always produce electricity on demand.

Auto manufacturers are now making more electric and **hybrid** vehicles. Hybrids use fuel cells to make energy. Scientists hope to use common gases like hydrogen and oxygen in these cells. If so, the supply of energy will be cheap and almost unlimited.

COMPARE ACROSS REGIONS

Innovation in Other Countries

Countries around the world are using a variety of energy innovations. For example, Brazil has been producing ethanol since the 1970s and is a leading producer of this fuel. Britain has banned conventional light bulbs in favor of bulbs with lower wattage, which use less energy. Nuclear power supplies more than 76 percent of France's electricity. Denmark's coal plants produce both electricity and hot water, making them more efficient. Some nations, such as Iceland and New Zealand, use geothermal energy, heat produced within the earth. These alternatives are some energy-efficient solutions and illustrate new ways of looking at energy production.

Visual Vocabulary A **wind turbine** is an engine that is powered by wind in order to generate electricity. A group of wind turbines placed together is a wind farm.

Before You Move On

Make Inferences In what ways might the use of alternative energies in other countries affect energy use in the United States?

ONGOING ASSESSMENT

READING LAB

1. **Make Inferences** Which alternative energy sources depend on physical geography, and in what ways?
2. **Summarize** What issues prevent certain alternative energy sources from being used?
3. **Form and Support Opinions** Which alternative energy source do you think is most promising? Write a paragraph stating your choice. Support your opinion with evidence from the lesson.

1.5 Citizens' Rights and Responsibilities

TECHTREK
myNGconnect.com For photos and a chart on citizens' rights
Digital Library
Student Resources

Main Idea Citizens in the United States and Canada enjoy many rights and also have certain responsibilities.

The governments of the United States and Canada share many similarities. Leaders are elected by voters. Government power is shared by the national government and smaller units—states in the United States and provinces in Canada. Laws are made by a legislative body. Perhaps most importantly, all citizens in these countries have rights and responsibilities.

Government Structures

Both governments consist of three branches—the executive, the legislative, and the judicial—each with specific powers. The legislative, or law-making, branches are the U.S. Congress and the Canadian Parliament. In the United States, the head of the executive branch is the president, elected by the people. In Canada, by contrast, the chief executive is the prime minister, who is chosen by the party with the most seats in Parliament.

Critical Viewing A young man registers to vote at the University of Texas. What details in the photo convey the importance of voting?

Rights and Responsibilities

In both countries, all citizens enjoy basic rights, such as the right to citizenship, equality, and fair treatment. People become citizens by being born in the country or by **naturalization**, the process that someone born in another country follows to become a citizen.

The Bill of Rights, which is the first ten amendments to the U.S. Constitution, and Canada's Charter of Rights guarantee basic freedoms. Among these are freedom of speech and religion, and certain legal rights. For example, citizens have the right to **due process**, or specific rules that authorities must follow. Canada's charter also identifies two official languages, English and French. Citizens in Canada have the right to be educated in either one.

The citizens of the United States and Canada enjoy a great deal of freedom. They can live where they wish and speak their own languages. They can express their opinions freely and practice any religion they choose. They can gather together and speak out against their government. Many people who live in these two countries immigrated from countries that did not allow these basic freedoms.

In both countries, citizens with rights also have responsibilities. The right to vote brings with it the responsibility to cast a vote fairly, through formal elections. Legal rights bring about the responsibility to obey laws and to serve on a jury, which is a group of fellow citizens who listen to both sides of a legal matter and decide whether laws were broken.

CITIZENS' RIGHTS

UNITED STATES

Protection of rights not guaranteed by the Constitution

Right to bear arms

UNITED STATES & CANADA

Right to a speedy trial

Right to trial by jury

Right to vote

Freedom of speech, religion, press, assembly

Protection against unreasonable search

Right to have a lawyer

CANADA

Right to be educated in a minority language (French)

Right to be protected against discrimination on basis of race or gender

Some responsibilities are not identified in any document, but are just as important as those required by law. For example, citizens are expected to treat each other fairly. The United States and Canada are home to a wide variety of races, religions, and customs. Citizens in diverse societies like these have the responsibility to accept one another's differences.

Before You Move On

Make Inferences Why is responsibility an important part of having rights and freedoms?

ONGOING ASSESSMENT

SPEAKING LAB

GeoJournal

1. **Make Generalizations** Based on the graphic above, which document, the Bill of Rights or Canada's Charter of Rights, provides stronger protection for rights? How?
2. **Express Ideas Through Speech** Write and deliver a speech that explains your ideas about which citizens' right and responsibility you think is most important in a society, and why.

2.1 Daily Life in Mexico

myNGconnect.com For photos of Mexico's daily life

Digital Library

Main Idea Life in Mexico today reflects a blend of traditional and modern elements and Native American and Spanish cultures.

Much of Mexican life is modern, but many Mexican people hold a deep respect for the past. Mexican culture blends contributions from the Native American and Spanish cultures that shaped its history.

A Blending of Cultures

Native people of Mexico had created successful empires before the Spanish conquest in 1521. **Descendants**, or future generations, of these indigenous people still make up a large part of Mexico's population. The largest population group is the mestizos, who are Mexicans of mixed **ancestry**, or heritage. This ancestry can include native, European (particularly Spanish), and African family roots. Mixed ancestry is a source of pride among mestizos in Mexico.

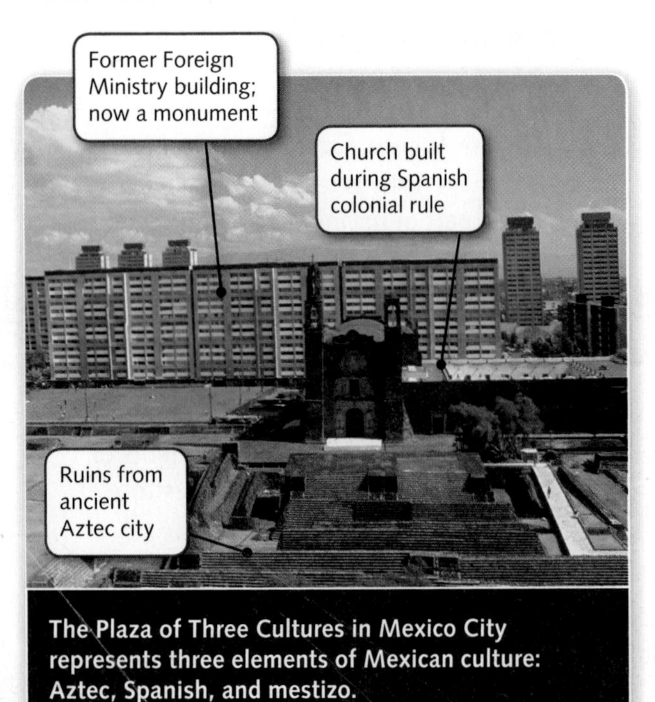

The Plaza of Three Cultures in Mexico City represents three elements of Mexican culture: Aztec, Spanish, and mestizo.

The blend of cultures is seen in the many languages and religions of Mexico. While nine out of ten Mexicans speak only Spanish, the country's official language, the rest speak one of more than 50 Native American languages. Similarly, more than 90 percent of Mexico's people follow the traditional Roman Catholic religion. However, many Native Americans practice traditional religions, and many Mexican customs show the blending of cultures. For example, a celebration called the Day of the Dead, which honors one's ancestors, mixes native and Catholic traditions.

Many Mexican artists have blended native and European culture. Diego Rivera and David Alfaro Siqueiros (sih KAY rohs) used European style to paint **murals**, or large paintings on walls, that highlight Native American culture in Mexico.

Modern and Traditional Life

Today, life in Mexico is a balance of the traditional and modern. Once largely rural, Mexico is now an urban nation. More than three-quarters of its people live in cities. The capital, Mexico City, is one of the world's largest cities, and is home to one-fifth of Mexico's population.

Mexico embraces past and present. The country has preserved sites that include Aztec and Mayan art and **artifacts**, or tools and other ornaments that show something about how a culture lived. The National Palace in Mexico City, built to house colonial viceroys, is still used as a government building. Mexico also has bold, modern architecture, such as the Lighthouse of Commerce in Monterrey.

Mexican cooking continues to include staple foods of its indigenous people: corn, beans, and squash. Corn is ground to make *tortillas*, a flat bread served with many meals. Traditional foods now must fit into a more modern lifestyle. Families used to share a large midday meal, but shorter lunch hours and long commutes for modern workers make it difficult to keep up this tradition.

Some Mexican people follow traditional practices as a way of life. This is especially true among rural people. For example, *vaqueros*, or cowboys, still work on some ranches in northern Mexico, using their traditional skills.

Before You Move On

Summarize In what ways is Mexican culture a blend of influences?

ONGOING ASSESSMENT

PHOTO LAB

1. **Synthesize** What details in the photos reveal the importance of tradition in Mexico?
2. **Place** In what ways does the Plaza of Three Cultures represent Mexican culture?
3. **Make Inferences** Why might the mixed cultural heritage of the *mestizos* be a source of pride in Mexico?

Critical Viewing A quinceañera (keen sin NYAR ah) is a traditional celebration of a girl's 15th birthday. It is both a religious and a social event. What details in the photo remind you of other ceremonies?

2.2 The Impact of Globalization

myNGconnect.com For photos and a graph of Mexico's Gross Domestic Product

Maps and Graphs

Digital Library

Main Idea Mexico has a developing economy that faces challenges as a result of global trends.

As parts of Mexico are modernizing, the country's economy is growing. However, **globalization**, the development of a world economy based on free trade and the use of labor from other countries, has presented economic challenges.

A Drive for Growth

For much of the 20th century, some major industries in Mexico were **nationalized**, or placed under government control. All income from a nationalized industry goes to the government. The profitable oil industry, for example, was nationalized in 1932. For periods of time during the 1900s, the government also controlled Mexico's banking, transportation, and telecommunications systems.

In the 1980s, in an effort to improve the country's economy, Mexico released some control of these industries to private and foreign investors. Another factor of Mexico's economic growth was the adoption of the **North American Free Trade Agreement (NAFTA)** in 1994, which removed many barriers to trade. Since then, Mexican trade with the rest of North America—mostly with the United States—has increased nearly 300 percent. However, NAFTA's critics say the agreement unfairly favors commercial agriculture over small-scale farmers.

Three Economic Sectors

Economic activity in Mexico takes place in three **economic sectors**, or subdivisions of the economy: agriculture, manufacturing, and service. The agricultural sector includes large-scale production for export.

Critical Viewing This worker cleans molds that are used to make shoe soles in a factory in Leon, Mexico. What skills would be required of factory workers?

Major exports include tropical fruits, coffee, sugarcane, and cotton. Silver is also an important mineral resource for Mexico's export income in this sector. Export income also comes from oil sold on world markets.

Manufacturing in Mexico includes automobiles, food processing, and metal products. Heavy foreign investment followed the adoption of NAFTA, which increased manufacturing activity. This investment led to more *maquiladoras*, foreign-owned factories where parts made elsewhere are put together into finished goods and then exported worldwide.

The dollar amount provided to the GDP (Gross Domestic Product) by service industries increased by a factor of more than 25 between 1970 and 2009. This sector includes services such as banking and transportation. Mexico's warm climate and cultural treasures attract tourists from around the globe, making tourism another big part of the service economy.

MEXICO'S GDP
(IN US$ BILLIONS)

Services
Manufacturing
Agriculture

1 □ = $5 billion

Source: World Bank

1970
1990
2009

Economic Challenges

Mexico still faces challenges. Economic growth hasn't happened for most poor people. Many Mexicans have difficulty finding work and migrate to the United States. Some Mexican-owned businesses have transferred jobs to countries where workers are paid less.

Before You Move On

Make Inferences In what ways has globalization affected Mexico's economy?

ONGOING ASSESSMENT

DATA LAB

1. **Analyze Data** Based on the graph, which economic sector has grown the most?
2. **Synthesize** How might NAFTA account for the growth in agriculture between 1990 and 2009?
3. **Region** Why is tourism important to Mexico?

2.3 Reaching Toward Democracy

TECHTREK
myNGconnect.com For a graph and photos on democracy in Mexico

Maps and Graphs

Digital Library

Main Idea Mexico has made progress toward democracy but still faces serious obstacles.

Mexico faces not only economic challenges but also political ones. The country's political history is a factor in today's struggles toward democracy.

Stability Without Democracy

The biggest limit on democracy came from the dominance of the **Institutional Revolutionary Party (PRI)**, which controlled Mexico's government from 1929 to 2000. The PRI formed as a result of major political instability that included the assassination of Mexico's newly elected president in 1928. The party's goal was to enact the reforms that were fought for during the Mexican Revolution.

Multi-Party Democracy

Economic and political problems during the 1990s weakened the PRI's hold on power in the Mexican states. In 1994, the Zapatista Liberation Army staged a violent rebellion in response to new economic policies that were harmful to poor, indigenous people. This rebel group seized control of several Mexican states.

In 1996, the government enacted election **reforms**, or changes aimed at correcting problems. These reforms made voting more fair and made it easier for parties other than the PRI to run candidates. Elections between candidates from the PRI and other parties were closer each year. In 2000, the election of Vicente Fox of the National Action Party (PAN) broke 71 years of PRI rule.

Fox was appealing to many Mexican people. He promised to improve the economy, clean up government corruption, and resolve disputes with rebellious political groups, particularly the Zapatistas. The results of Fox's presidency were complex. He had raised the expectations of many Mexican people who were suffering economically. However, the PRI still controlled Congress. Therefore, Fox's reform efforts moved slowly and many Mexicans who originally supported Fox were disappointed.

Critical Viewing Felipe Calderón greets supporters after being declared winner of the 2006 presidential election. How would you describe the emotions of the crowd?

Perhaps the most important result of Fox's presidency was that it brought about **multi-party democracy**, a political system in which elections include candidates from more than one party. Presidents in Mexico serve only one six-year term, so the 2006 election included two new candidates: Felipe Calderón of the PAN and López Obrador from the leftist party of the Democratic Revolution, which had become another important political voice in Mexican politics. Calderón was declared winner, but with less than a one percent lead over Obrador. This election further proved that multi-party democracy was in place in Mexico.

Challenges Ahead

One of the obstacles to Mexico's democracy comes from the illegal drug trade, which is controlled by several drug cartels. The activities of the cartels brought about heavy violence in the early 2000s, especially along the U.S.-Mexico border. In 2009, President Calderón sent troops to the area to control the violence, but the cartels still challenge the authority of Mexico's government. Democracy is also challenged by the Zapatistas' ongoing control in several Mexican states.

A positive trend in Mexico is its increasing literacy rate. By the start of the 21st century, nearly 90 percent of Mexico's population was literate, or able to read and write. A high literacy rate is one factor that helps a society be more productive and contribute to the country's economy. It is also an important element of a democracy, because people need to be informed to take part in decision-making.

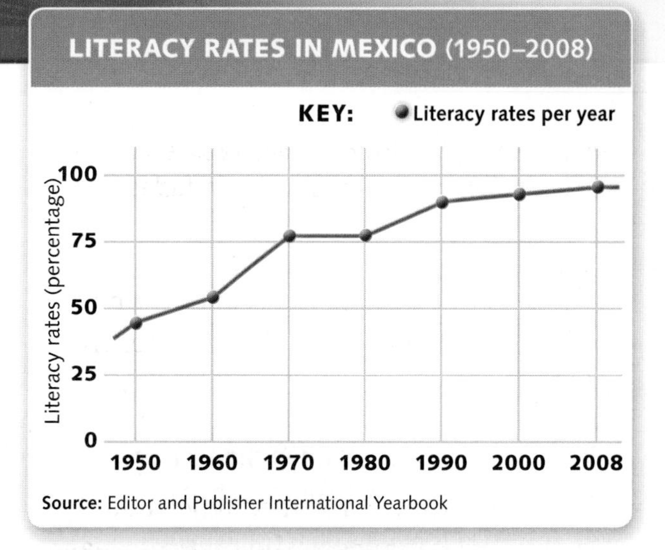

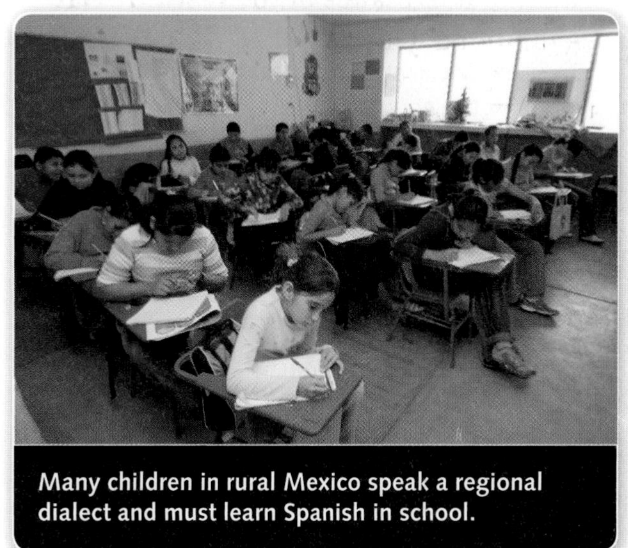
Many children in rural Mexico speak a regional dialect and must learn Spanish in school.

Before You Move On

Monitor Comprehension How has the Mexican government become more democratic?

ONGOING ASSESSMENT

READING LAB

1. **Identify Problems and Solutions** What problems does Mexico's government now face? How might it solve them? Explain your answer.
2. **Interpret Graphs** During which 10-year period did Mexico's literacy rate increase the most? When did it increase the least?
3. **Draw Conclusions** How might Mexico's growing literacy rate help achieve democracy?

CHAPTER 4

Review

VOCABULARY

On your paper, write the vocabulary word that completes the following sentences.

1. North America's ________ shows in its mix of races, languages, and religions.
2. With new media able to reach people all over the world, culture has become ________.
3. A ________ occurs when the economy shows a sharp decline in business activity.
4. A vehicle that runs on a combination of electric and gas power is called a ________.
5. Some Mexican artists create ________ that show the country's native cultures.
6. When an industry is ________, it is placed under government control.

MAIN IDEAS

7. How did the United States and Canada become so diverse? (Section 1.1)
8. How has mobile technology affected cultures in the United States and Canada? (Section 1.2)
9. How did the information age transform the U.S. and Canadian economies? (Section 1.3)
10. Why are researchers looking at alternative sources of energy? (Section 1.4)
11. How do immigrants from other lands gain the rights and responsibilities of American and Canadian citizenship? (Section 1.5)
12. What are two examples of the continuing influence of Spanish culture on Mexico? (Section 2.1)
13. What economic success has Mexico seen in recent decades? (Section 2.2)
14. Why is it important for a country struggling toward democracy to have more than one political party? (Section 2.3)

FOCUS ON UNITED STATES & CANADA

ANALYZE THE ESSENTIAL QUESTION

What issues do the United States and Canada face today?

Focus Skill: Identify Problems and Solutions

15. What do you think is the single biggest challenge facing the United States? Why do you think so?
16. What do you think is the single biggest challenge facing Canada? Why do you think so?
17. Choose one of the challenges you identified in the previous two questions. What do you think the country should do about that challenge? Why would those steps help?

INTERPRET TABLES

UNITED STATES TRADE BALANCE ($ BILLIONS)

	2006	2007	2008
Total	-760.4	-701.4	-696.0
With main trading partners			
Canada	–71.8	–68.2	–78.3
China	–234.1	–258.5	–268.0
Germany	–89.1	–94.2	–43.0
Japan	–89.7	–84.3	–74.1
Mexico	–64.5	–74.8	–64.7

Source: 2010 Statistical Abstract of the United States, Table 1264 and Table 1271

18. **Analyze Data** A negative trade balance means that a country imports more than it exports. What trend do you see in the overall trade balance from 2006 to 2008?
19. **Synthesize** If the United States eliminated its trade deficit with Canada and Mexico, would that have much effect on the overall trade balance? Why or why not?

FOCUS ON MEXICO

ANALYZE THE ESSENTIAL QUESTION

How has globalization affected Mexico?

Focus Skill: Summarize

20. How has the balance of rural and urban populations shifted in Mexico?
21. What factors have contributed to economic growth in Mexico?
22. What challenges come along with economic growth in Mexico?
23. What qualities will help Mexico become a successful democracy and take part in the global economy?

INTERPRET MAPS

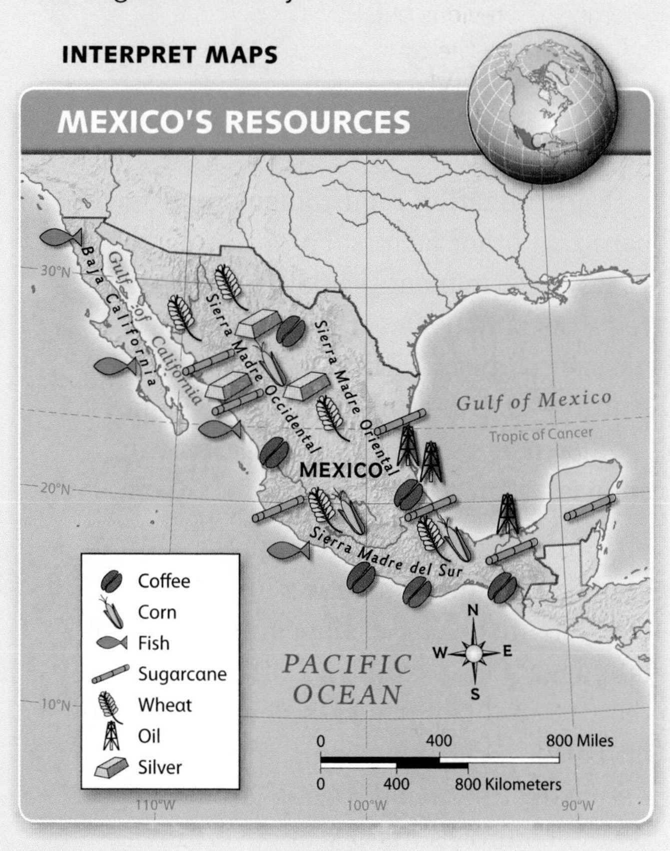

24. **Interpret Maps** Which natural resource is found only in the mountains? Which resource besides fish and oil is found on coastal plains?
25. **Synthesize** Which resources on the map are important to Mexico's success in the global economy? Explain why.

ACTIVE OPTIONS

Synthesize the Essential Questions by completing the activities below.

26. **Write a Speech** Imagine that you are president of the United States preparing to visit either Mexico or Canada. In a speech to that country's people, you want to emphasize the cooperation and good relations between your two countries. Choose one of those countries and then write and deliver a two-minute speech that a president might deliver. Use the following tips to help you write. **After all the speeches are delivered, discuss the ideas that were presented and evaluate which are best.**

Writing Tips

- Identify a clear theme.
- Use details in history and the current situation to support the idea that the two countries cooperate well.
- State your goals for relations in the future.

TECHTREK **myNGconnect.com For research links on North America today**

27. **Create a Multimedia Presentation** Make a chart like the one below. Use it to plan a multimedia presentation on one of the three countries in this chapter. Go to **Connect to NG** and other online sources to gather facts and photos.

Country	
Population	
Culture	
Economy	
Government	
Challenges	

Becoming a Citizen

myNGconnect.com For graphs of U.S. naturalizations

Naturalization is the process by which someone who was not born in a country can become a citizen of that country. Every country spells out in detail who can or cannot become a citizen and what steps must be taken as part of the naturalization process.

Requirements for naturalization in most countries include specific guidelines in several areas: residency (the amount of time spent living in the country); language proficiency (the ability to speak the country's official language); cultural and historical knowledge of the country (usually a written test); and character, or the kind of person you are (usually determined by a background check).

Compare

- China
- Cuba
- Ethiopia
- India
- Iran
- Mexico
- Ukraine
- United States

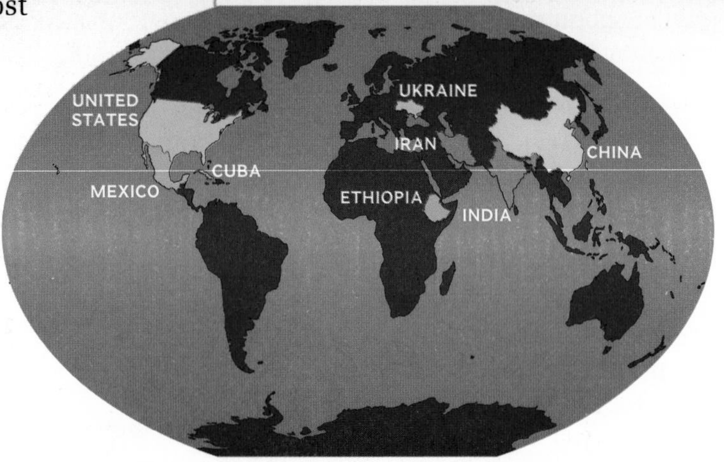

NATURALIZATION AND CITIZENSHIP

Most countries require a person who wishes to become a citizen to first gain permanent residency status before even applying for naturalization. A permanent resident is allowed to live and work in a country, but legally is still a citizen of the home country.

Permission for a non-citizen to immigrate and live as a permanent resident in a country can be very difficult to obtain. Eligibility may depend on the need a country has to fill certain categories of jobs, or the country's political situation. Permission for permanent residency also may be limited by a variety of requirements. Background checks for any illegal activity or criminal conviction are typical and can disqualify a candidate for both the right to immigrate and the right to apply for naturalization.

Naturalization requirements may vary based on other factors. For example, the years of permanent residency a country requires may be fewer for a person who has served or is willing to serve in that country's military. Permanent residency requirements are usually lower for the spouse, or married partner, of a citizen.

Most countries require a person applying for naturalization to be a competent speaker of the country's official language. Many countries offer and even require applicants to complete a series of citizenship classes. People applying for naturalization usually have to pass a test on the country's history or government, in addition to proving their language proficiency. The final step in obtaining citizenship is often a naturalization ceremony, in which new citizens take an oath, or make a formal promise to live by their new country's laws and customs.

U.S. NATURALIZATION*

BY THE NUMBERS

570,442
Number of applications filed for U.S. naturalization, 2009

109,813
Number of denied applications for U.S. naturalization, 2009

$680
U.S. naturalization application filing fee

39
Median age at naturalization, 2000–2008

8
Median years as permanent resident, 2000–2008

20%
Percentage of naturalized citizens with a Bachelor's degree, 2000–2008

$57,030
Median household income of naturalized citizens, 2000–2008

U.S. MILITARY NATURALIZATIONS

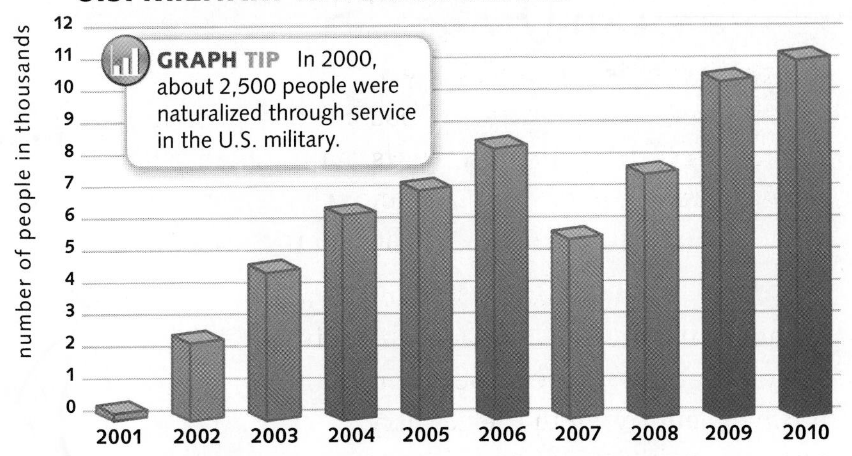

U.S. NATURALIZATIONS

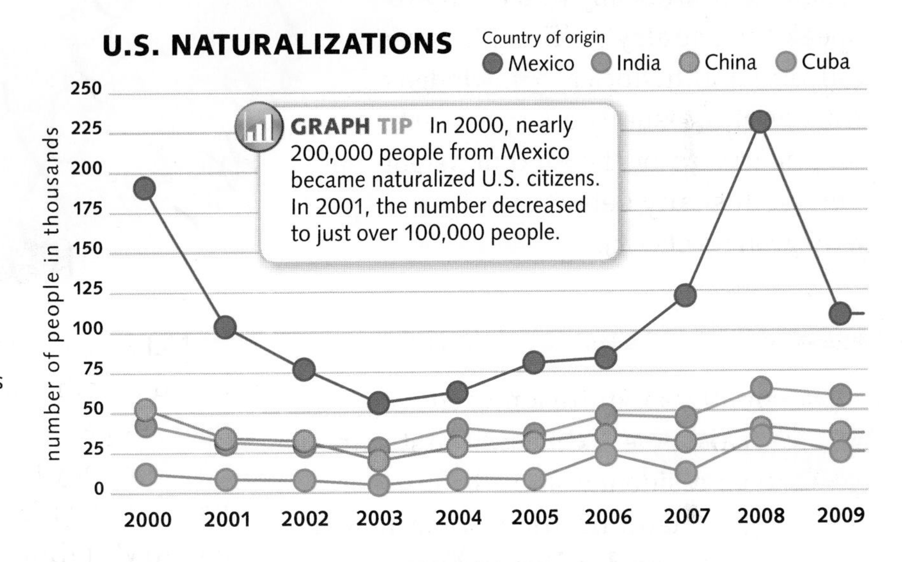

***Sources:** U.S. Department of Homeland Security; Congressional Research Service

ONGOING ASSESSMENT

RESEARCH LAB

GeoJournal

1. **Draw Conclusions** Why would a country require knowledge about its history, government, and culture to gain citizenship?
2. **Make Predictions** What would you predict about the number of U.S. military naturalizations in 2011? Use evidence from the U.S. Military Naturalizations graph to explain your answer.
3. **Interpret Graphs** What is similar about the number of U.S. naturalizations from all four countries from 2000 to 2001? What about from 2007 to 2008?

Research and Create Graphs Do research to find out the number of people from Ethiopia, Iran, and Ukraine who became U.S. naturalized citizens each year from 2000 to 2009. Create a line graph like the one above to show your data. Then write a short paragraph explaining the changes in the numbers throughout the time period. For example, in which year(s) did the number go down for all three countries? In which year(s) did it increase? What inferences can you draw from the data?

Active Options

TECHTREK

myNGconnect.com For research links on national parks

Connect to NG

Magazine Maker

ACTIVITY 1

Goal: Evaluate important physical structures.

Write Seven Wonders List

Since ancient times, people have made "Seven Wonders" lists of magnificent structures around the world, such as the Great Pyramids in Egypt or the Statue of Zeus in Greece.

Create such a list for the continent of North America. Review the photos in the **Digital Library** and use the research links to help you decide which structures belong on the list. Your list can include both natural structures and structures built by humans. For each item on your list, write the reasons why the structure deserves to be named as one of the Seven Wonders of North America.

Mount Rushmore in South Dakota

ACTIVITY 2

Goal: Research a national park.

Conduct a Walk in the Park

The U.S. National Park Service protects about 400 parks, and Parks Canada protects over 40 national parks. Choose a national park in North America from the list below or from another source. Do research and use the **Magazine Maker** to create a "show and tell" route through the park to show what makes it so important and worth preserving.

National Parks

- Denali, AK
- Everglades, FL
- Glacier, MT
- Yosemite, CA
- Zion, UT
- Yellowstone, WY, MN, ID
- Glacier, British Columbia, Canada
- Prince Edward Island, Canada

ACTIVITY 3

Goal: Learn about National Geographic Explorers.

Explore the Explorers

With a group, find out more about National Geographic Explorers Beth Shapiro or Sam Meacham. Research the explorer's background and how he or she became interested in the work he or she does today. Find out more details about the explorer's current project and possible future plans. Then create a poster that highlights the explorer and his or her work.

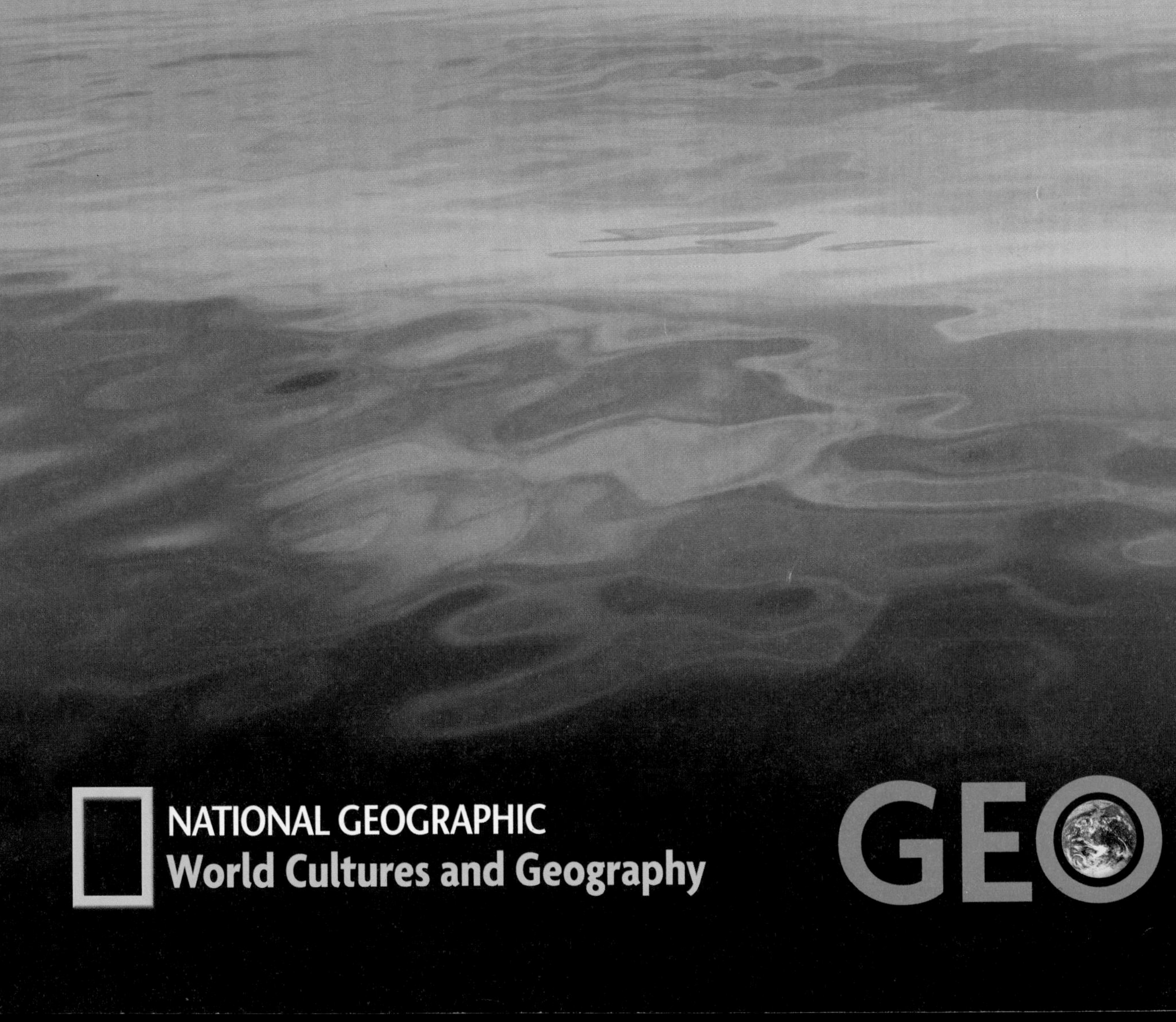
NATIONAL GEOGRAPHIC
World Cultures and Geography
GEO

Explore Central America & the Caribbean with NATIONAL GEOGRAPHIC

MEET THE EXPLORER

NATIONAL GEOGRAPHIC

Emerging Explorer Ken Banks created a communication tool for farmers in El Salvador. They can discuss crop pricing and remain competitive using an inexpensive cell phone.

INVESTIGATE GEOGRAPHY

Central America was once entirely covered by rain forests. Farmers have cleared many of these hot, tropical areas for cattle ranching and sugar plantations. The remaining rain forests still contain many plants and tropical birds, including this toucan.

STEP INTO HISTORY

Toussaint L'Ouverture led a slave revolt in Santo Domingo, later called Haiti, and laid the groundwork for its independence in 1804. The former slave was a brilliant general, defeating the powerful French army.

Go to myNGconnect.com for maps of Central America and the Caribbean.

CONNECT WITH THE CULTURE

Brightly colored, handwoven Guatemalan textiles, like the young girl wears in this photo, reflect the country's traditional Mayan roots.

CHAPTER 5

Central America & the Caribbean GEOGRAPHY & HISTORY

PREVIEW THE CHAPTER

Essential Question How has physical geography been a positive or negative influence on the economy of the region?

SECTION 1 • GEOGRAPHY

KEY VOCABULARY

- isthmus
- coastal plain
- rain forest
- archipelago
- tectonic plate
- seismic
- ecosystem
- deforestation
- fertile
- tourism
- canopy
- extinction
- poacher

ACADEMIC VOCABULARY

critical

TERMS & NAMES

- Caribbean Sea

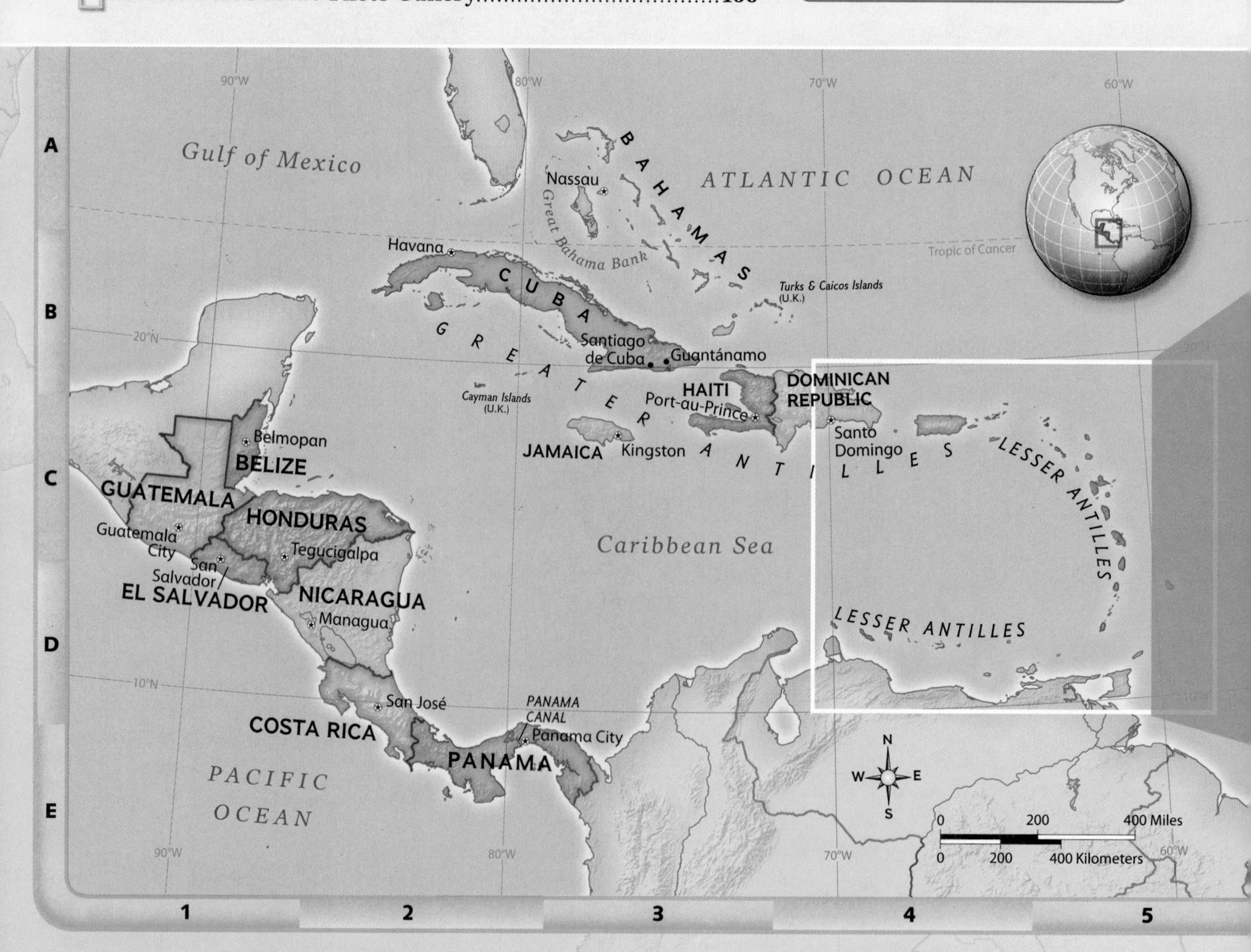

TECHTREK FOR THIS CHAPTER

Student eEdition | Maps and Graphs | Interactive Whiteboard GeoActivities | Digital Library | Connect to NG

Go to myNGconnect.com for more on Central America and the Caribbean.

lemur frog, Costa Rica

Essential Question How have economic resources influenced the history of the region?

SECTION 2 • HISTORY

KEY VOCABULARY

- cash crop
- scarcity
- triangular trade
- multitude
- staple
- viceroy
- province
- harbor
- dictator
- commonwealth

ACADEMIC VOCABULARY

exploit

TERMS & NAMES

- Columbian Exchange
- Toussaint L'Ouverture

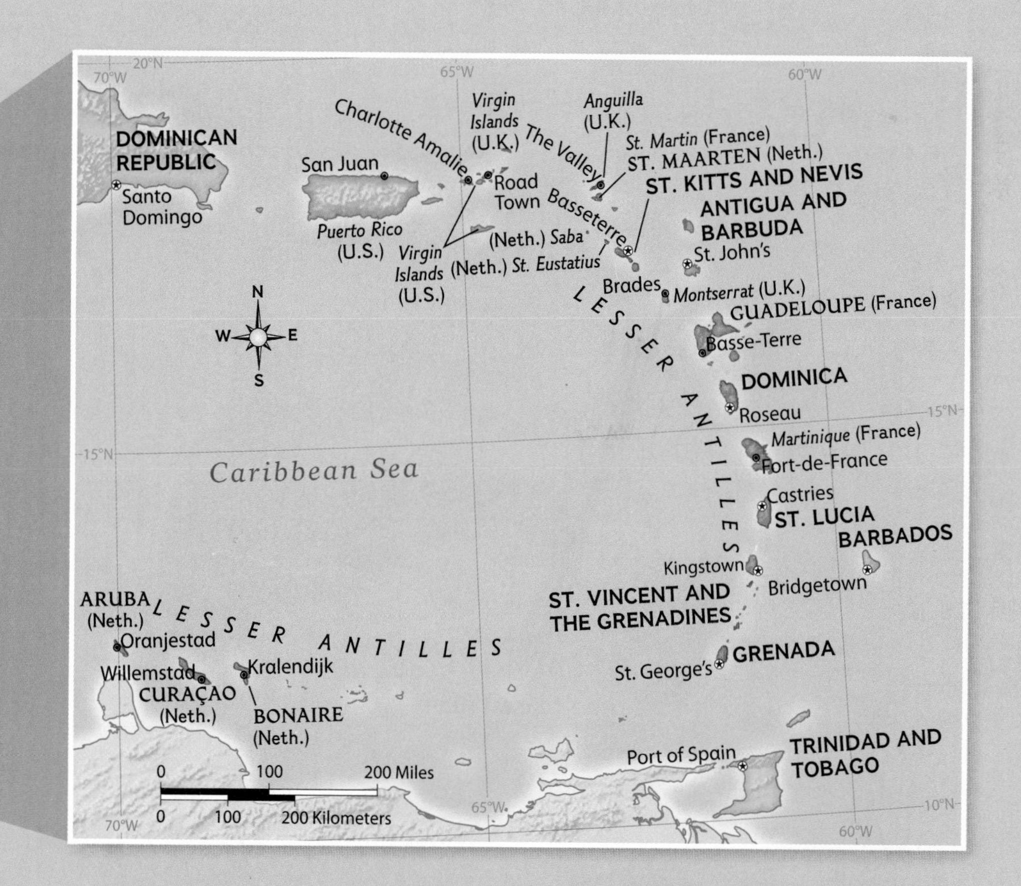

A B C D E
6 7 8 9

1.1 Physical Geography

TECHTREK
myNGconnect.com For online maps of the region and Visual Vocabulary
Maps and Graphs
Digital Library

CENTRAL AMERICA AND THE CARIBBEAN PHYSICAL

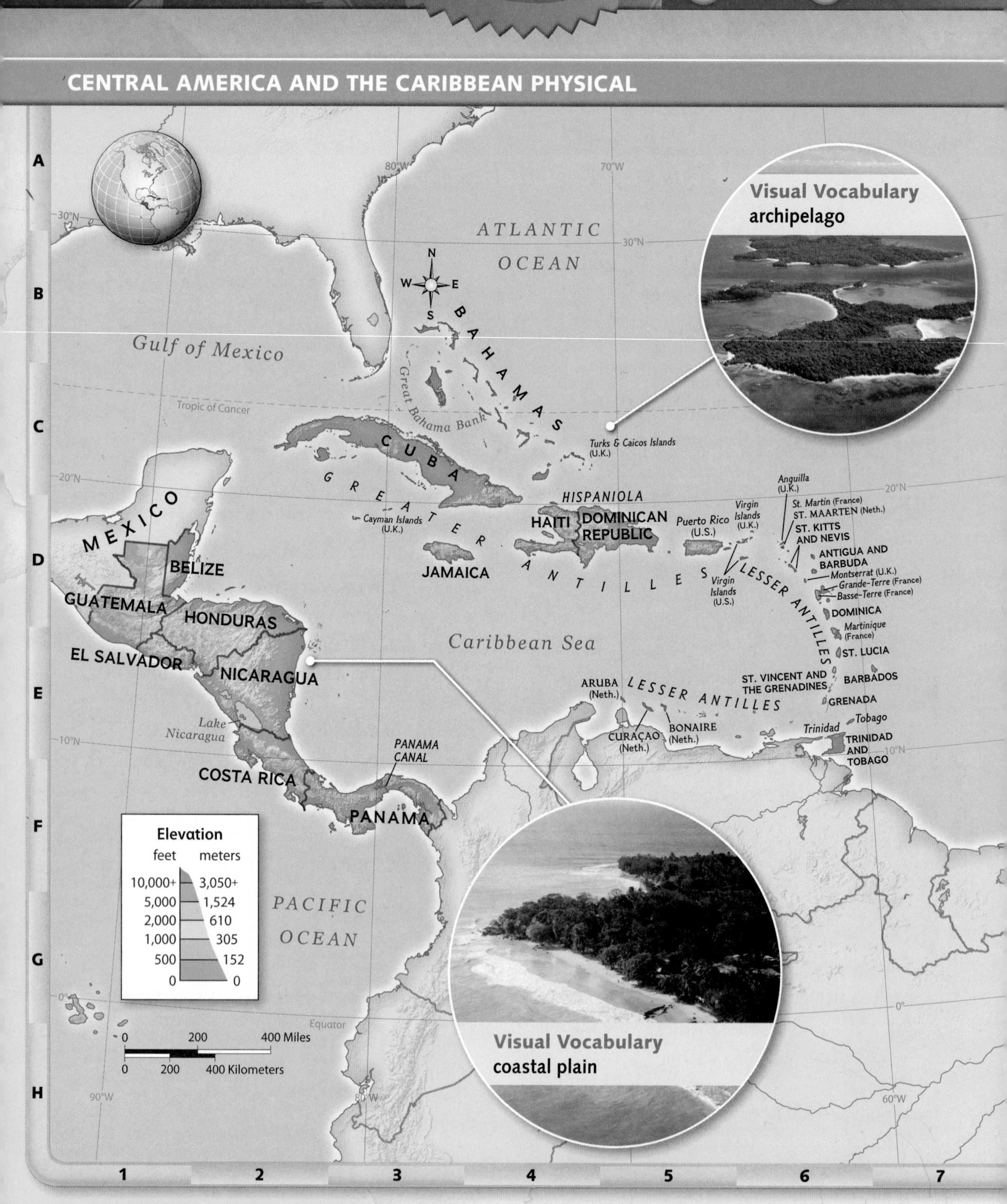

CLIMATE

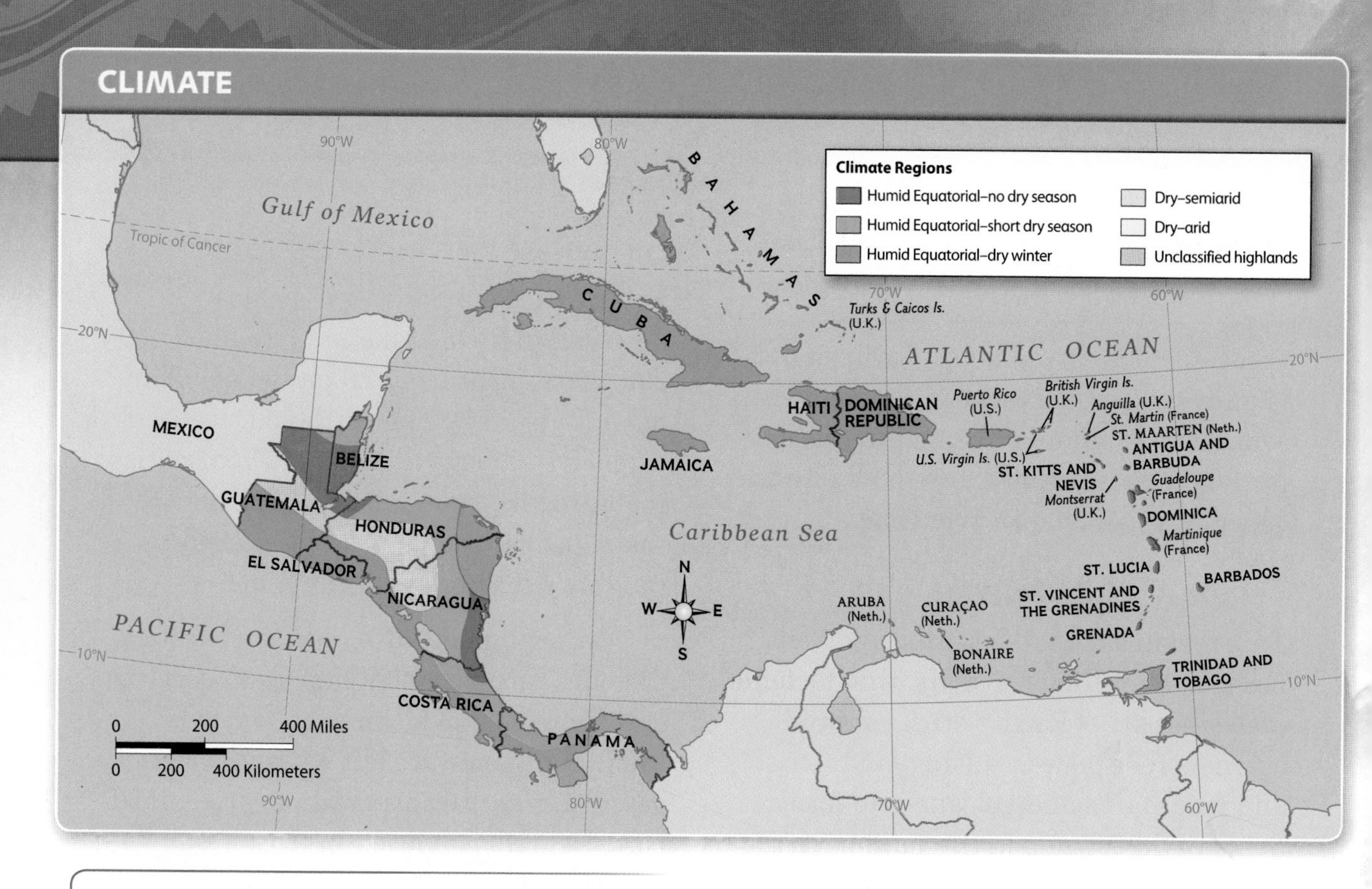

Main Idea Mountains, coastal plains, and rain forests are the main landforms that support the region's economy.

Central America and the Caribbean islands are located between the continents of North America and South America. The physical geography of the region supports the agriculture and tourism that are so valuable to the region's economy.

Central America

Central America is an **isthmus**, a narrow strip connecting two large land areas. A range of volcanic mountains spreads across the seven countries of the region. The climate in the mountains is cool and the rich volcanic soil is ideal for growing coffee beans, an important export. The **coastal plains**, which are the lowlands next to the seacoast, and the tropical **rain forests**—heavily wooded forests that may receive more than 100 inches of rain per year—provide resources that boost the economy.

The Caribbean Islands

The Caribbean islands curve in an **archipelago**, or chain of islands, between the Atlantic Ocean and the **Caribbean Sea**. Sugarcane, grown on coastal plains, is the leading crop. The climate is mild in winter and hot in summer, and attracts visitors year-round.

Before You Move On

Summarize In what ways do the main landforms of the region support the economy?

ONGOING ASSESSMENT

MAP LAB

1. **Location** Based on the physical map, where are the mountainous regions of Central America located?
2. **Interpret Maps** What is the climate like on the Central American and Caribbean coasts?
3. **Turn and Talk** Discuss with a classmate the differences in climate in the mountains and on the coastal plains. Where would you like to live?

SECTION 1 GEOGRAPHY

1.2 Earthquakes and Volcanoes

Main Idea Earthquakes and volcanoes affect daily life and economic activities in the region.

Central America and many Caribbean islands were underwater millions of years ago. The formation of the mountains and islands in this region shows how physical geography can change over time.

Tectonic Plate Movements

This region sits on the Caribbean Plate. For tens of millions of years, this **tectonic plate**—a part of Earth's crust—moved against other plates, causing land to rise and mountains and volcanoes to form. Some of the islands in the Caribbean are really just the tips of ancient volcanoes created by tectonic plate movement.

The location of the islands on these plates is significant. Today, plates slowly continue to move, crashing into or sliding over or under each other. This movement causes earthquake, or **seismic**, activity and volcanic eruptions. An eruption in Montserrat in 1996 caused two-thirds of the population to flee their homes.

Impact of Earthquakes

In 2010, Haiti experienced its most destructive earthquake in more than 200 years, measuring 7.0 on the moment magnitude scale. The earthquake caused extensive damage to many of Haiti's cities, including its capital, Port-au-Prince. More than 200,000 people were killed and roughly a million were left homeless.

The earthquake also destroyed **critical** systems, the extremely important services and supplies that a community needs. Many Haitians were left without water, gas, electricity, transportation, or medical care. Donations poured in from around the world in an international effort to help Haiti recover and rebuild.

Before You Move On

Summarize What causes earthquakes and volcanoes in this region, and what are some of the effects?

Critical Viewing This courthouse was buried in a landslide caused by a volcano on Montserrat. What does this photo suggest about volcanoes?

PLATES AND VOLCANOES

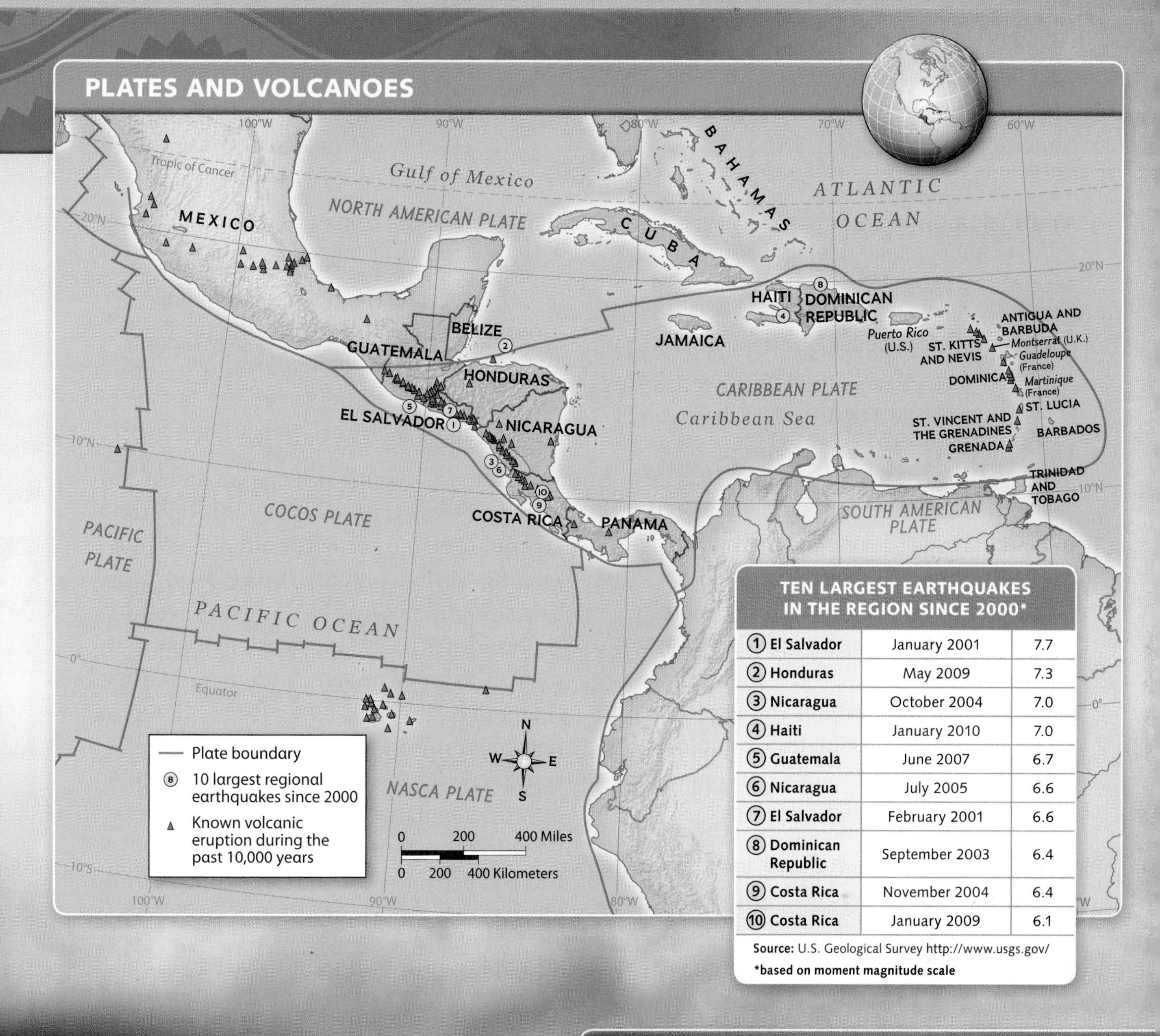

TEN LARGEST EARTHQUAKES IN THE REGION SINCE 2000*

① El Salvador	January 2001	7.7
② Honduras	May 2009	7.3
③ Nicaragua	October 2004	7.0
④ Haiti	January 2010	7.0
⑤ Guatemala	June 2007	6.7
⑥ Nicaragua	July 2005	6.6
⑦ El Salvador	February 2001	6.6
⑧ Dominican Republic	September 2003	6.4
⑨ Costa Rica	November 2004	6.4
⑩ Costa Rica	January 2009	6.1

Source: U.S. Geological Survey http://www.usgs.gov/

***based on moment magnitude scale**

ONGOING ASSESSMENT

MAP LAB

1. **Interpret Maps** What is significant about Central America's location near where the Caribbean Plate and the Cocos Plate meet?
2. **Location** According to the map, where do the most and the fewest volcanic eruptions happen in Central America? Explain.
3. **Make Inferences** Trace the boundary of the Caribbean Plate. Why are the land masses clustered around the edges of the plate?

1.3 Rain Forests of Central America

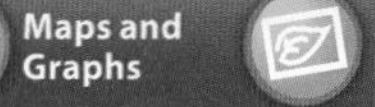

TECHTREK
myNGconnect.com For an online map and photos of the rain forest
Maps and Graphs
Digital Library

Main Idea Central American rain forests are an important economic resource for the region.

Central America's rain forests cover a large portion of the region. Countries in the region are working to save rain forests and grow economically at the same time.

The Importance of Rain Forests

Central American rain forests have tall trees with broad leaves, and grow in tropical areas with heavy rainfall. A rain forest is an individual **ecosystem**, a place where plants and animals rely on the environment to survive. Many rare species make their homes in rain forests. The quetzal (ket SAHL), for example, is the national symbol of Guatemala. It makes a nest in rain forest trees.

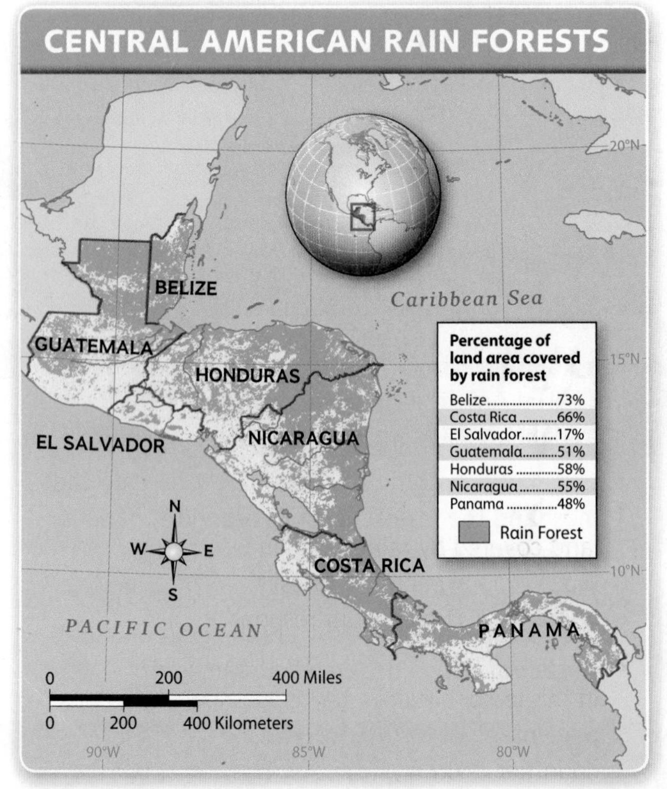

Rain Forest Destruction

In the 1500s, rain forests covered most of Central America. They thrived for hundreds of years until the 1900s. From 1990 to 2005, Nicaragua, Honduras, Guatemala, and El Salvador lost 14 to 30 percent of their rain forest land.

There are several reasons for forest loss, or **deforestation**. Many forests are cleared for commercial farmland and for grazing cattle to supply the beef industry. Trees provide valuable timber, or wood for building. Rain forests are usually located in rural areas, where people use the land to survive. Trees are cut down for firewood and for small farms. However, rain forest soil is not very **fertile**, or able to produce plentiful crops. Rural farmers need to clear more land after nutrients in the soil are used up. The forests are being cleared faster than they can regrow.

The Future

Many countries encourage rain forest tourism. **Tourism**—the travel industry—brings income to the region and provides jobs so rural people do not need to depend on rain forest land for farming. Certain farming methods can also prevent further damage to rain forests. Shade coffee, for example, grows under the protection of trees. This crop can grow on the same soil as other crops, such as beans. This practice preserves soil quality.

Before You Move On

Make Inferences How do Central America's rain forests contribute to the economy?

Visual Vocabulary **Canopy** refers to the tops of trees that create a roof over the rain forest. Many visitors take rain forest canopy tours.

Critical Viewing A tourist ziplines through the rain forest in Honduras. A cable is strung at an angle between two points, and the rider uses a pulley to "zip" from the top of the line to the bottom. What would you see by exploring the canopy of the rain forest that would be different from exploring at ground level?

ONGOING ASSESSMENT

DATA LAB

GeoJournal

1. **Analyze Data** Based on the percentages of land covered by rain forest, in which country would you expect rain forest tourism to be most important? Explain your response.
2. **Create Graphs** Using data shown on the map in this lesson, make a bar graph showing the percentage of forested land in Central American countries. Which method of showing the data do you find the most clear, and why?

Saving Sea Turtles with José Urteaga

Main Idea People can save their region's valuable resources by revisiting traditional ideas and finding new economic solutions.

Identifying the Problem

myNGconnect.com
For more on José Urteaga in the field today

In the early 1980s, sea turtles in Nicaragua began to disappear. Their decline in population was unusual, considering that they had outlived dinosaurs. For more than 100 million years, sea turtles migrated to Nicaragua's beaches to lay eggs, sometimes hatching 600,000 babies at a time. Emerging Explorer José Urteaga (uhr tay AH gah) is a young marine biologist who decided to find out why sea turtles were disappearing so fast that they were close to **extinction**, or dying out completely.

Urteaga discovered that **poachers**, people who hunt or fish illegally, were stealing sea turtle eggs from beach nests. Poor people in Nicaragua live on less than $1 a day, but a poacher can earn as much as $5 selling just 12 eggs a day, because of the great demand for the eggs. Sea turtle eggs are highly valued in Nicaraguan culture. Turtle meat and turtle eggs are traditional foods of the country. People also use sea turtle shells to make jewelry.

Finding a Solution

Urteaga knew that to save the turtle eggs, he had to help poor people earn money in other ways. He also had to encourage people to see their culture in a new way. Starting in 2002, he put together a team that worked with poachers, celebrities, young people, and conservation groups.

First, he convinced poachers to sell their turtle eggs to him so he could hatch them. Then he taught poachers how to make a living by farming, raising bees, guiding tourists, and making crafts. Urteaga even hired former poachers to patrol the beaches and protect the nests. He started hatcheries, places designed for hatching eggs, to make sure new generations of turtles would survive.

To inspire cultural change, Urteaga focused his message on young people. He launched a huge media campaign to reach this audience.

The campaign included sold-out rock concerts to raise awareness as well as celebrities stating, "I don't eat turtle eggs." His school programs teach children how important it is to protect the species. Children also have the opportunity to work with the hatcheries. In 2008, young people celebrated the first release of hatched baby turtles in their town's hatchery by wearing t-shirts showing a tiny turtle breaking out of its egg. Urteaga's goal is to end the demand for turtle eggs and save sea turtles from extinction by educating a new generation.

By 2010, thanks to Urteaga and his team, almost 90 percent of sea turtle nests became protected. Urteaga is committed to continue saving sea turtles by "motivat[ing] people through their brains and their hearts."

Before You Move On

Monitor Comprehension What actions are helping to change how people view Nicaragua's sea turtles?

Critical Viewing Baby sea turtles head to the sea. What difficulties might they meet along the way?

ONGOING ASSESSMENT

VIEWING LAB

GeoJournal

1. **Pose and Answer Questions** Write three questions about the efforts to save the sea turtles. Go to the **Digital Library** to watch the Explorer Video Clip. Then work with a partner to answer each other's questions.
2. **Analyze Visuals** What did you find to be the most powerful image in the video?
3. **Human-Environment Interaction** Why would it be important for Urteaga to get young people involved in his cause?

NATIONAL GEOGRAPHIC

Photo Gallery • Orange Cup Coral

For more photos from the National Geographic Photo Gallery, go to the **Digital Library** at myNGconnect.com.

Waterfall in Costa Rica

Satiny parrot snake

Vegetable market, Guatemala

Critical Viewing A cluster of orange cup coral clings to a support piling on Bonaire Island, West Indies. This type of coral catches its food with sticky tentacles, preying primarily on shrimp and small fish.

Bucket orchid

Musician with steel drum

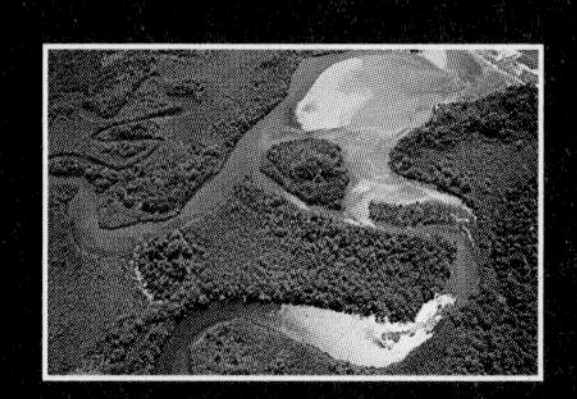

Coiba Island, Panama

Red-legged honeycreeper

2.1 Trade Across Continents

TECHTREK

myNGconnect.com For a trade map and research links on global trade

 Maps and Graphs  Connect to NG

Main Idea European exploration of the Americas led to new trade routes among the world's continents.

Shortly after Columbus landed in the Caribbean in 1492, the Spanish began to make use of the region's rich resources. Their actions were the first steps toward global trade.

Farming for Profit

Spanish settlers in the Caribbean grew **cash crops**, or crops for profit. Land on the islands was well suited for growing certain raw materials that were in short supply in Europe. For example, sugarcane grew well on the islands, and could be sold in Europe where there was a **scarcity**, or shortage. Spanish settlers **exploited**, or took advantage of, the labor of the native people for the heavy work on their farms.

The native people were also weakened by sickness. They could not fight the diseases, like smallpox and malaria, that settlers brought with them from Europe. As large numbers of native people died from disease, Spanish settlers had to look elsewhere for the workers they needed to farm their cash crops. So, after minerals and raw materials arrived in Spain, manufactured goods went to Africa to pay for slaves, beginning more than 300 years of slavery in the Americas.

New Trade Changes the World

Beginning in the 1500s, goods were exchanged in what became known as **triangular trade**—trade among three continents: the Americas, Europe, and Africa. The new trade established economic patterns that continue today.

Visual Vocabulary A **cash crop** is a crop grown for profit. Tomatoes, native to the Americas, were an important cash crop for Caribbean farmers to sell to Europe.

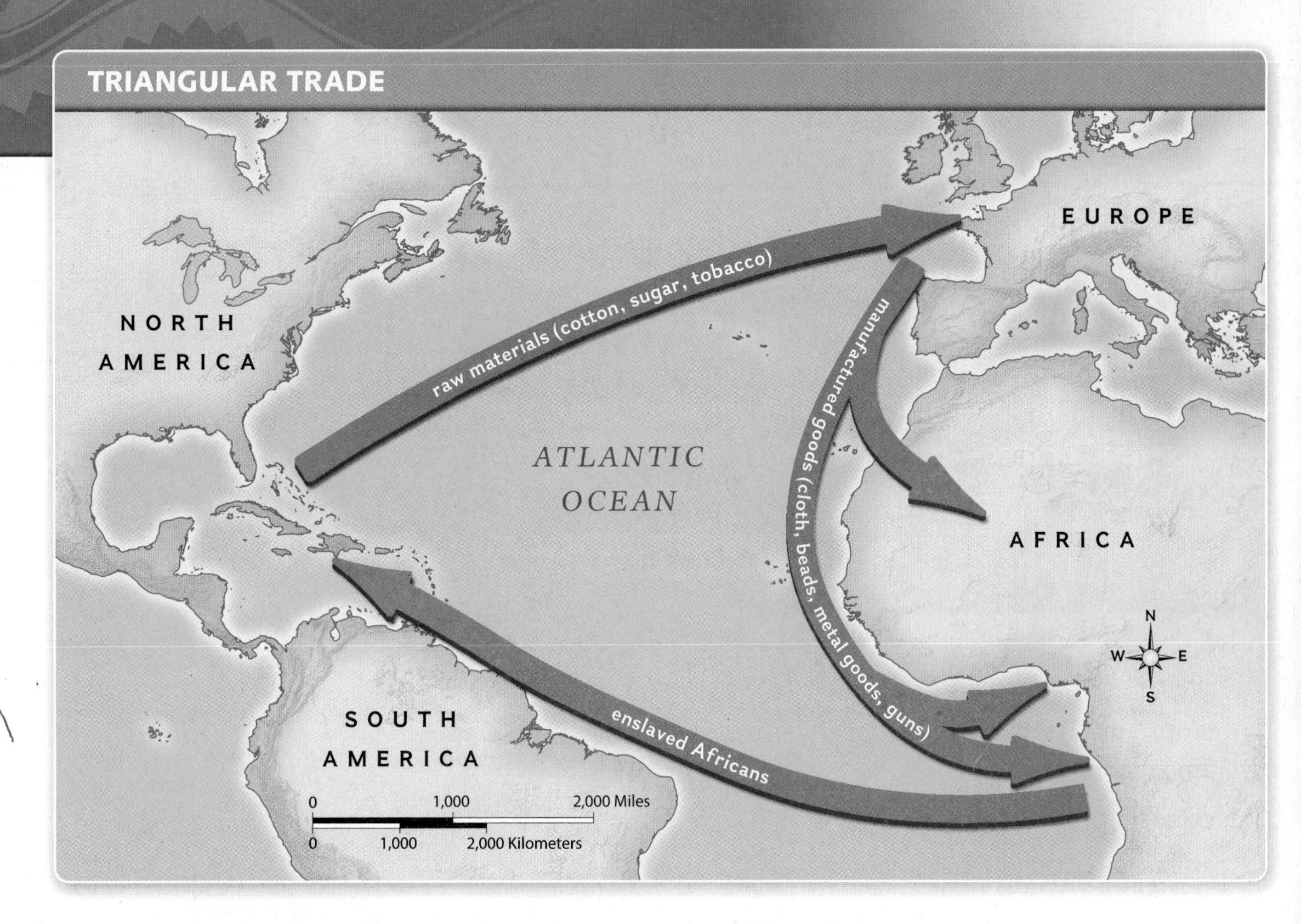

Triangular trade soon brought about competition among European countries for the resources of the region. Spain was gaining great wealth from its settlements in the Americas, causing other European countries to join the race for profit. Portugal, France, and Britain began competing with Spain for control over trade and colonization of the Americas.

Before You Move On

Monitor Comprehension In what ways did European exploration lead to new trade routes?

ONGOING ASSESSMENT

READING LAB

1. **Summarize** What effects did triangular trade have on Europeans, native groups, and Africans?
2. **Make Inferences** What role did native people play in the Caribbean economy?
3. **Movement** Trace the pattern of the triangular trade from the Caribbean to Europe to Africa. Describe the pattern of the trade using directional words. Use the scale to estimate the distance from Africa to the Caribbean.

SECTION 2 DOCUMENT-BASED QUESTION

2.2 The Columbian Exchange

TECHTREK

myNGconnect.com For photos and Guided Writing.

Student Resources Digital Library

Triangular trade between Europe, the Americas, and Africa brought about the exchange of plants, animals, and even diseases. This worldwide exchange has become known as the "Columbian Exchange," named for Christopher Columbus. The basic elements of everyday life—from what people ate to whether they lived or died—changed forever on several continents.

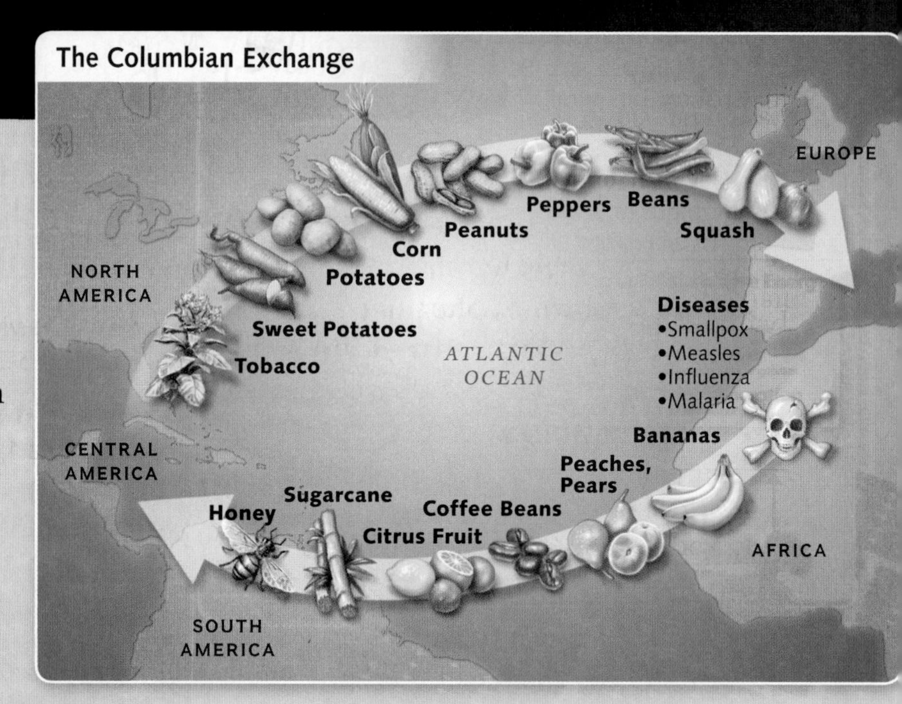

DOCUMENT 1

The diagram at right illustrates crops and diseases exchanged among continents.

CONSTRUCTED RESPONSE

1. Based on the diagram, which crops from Europe became important to the economy of the Caribbean? In which direction did disease travel in the **Columbian Exchange**?

DOCUMENT 2

Columbus's Journal

Columbus kept a journal detailing his arrival in the Caribbean in 1492. This excerpt describes the first trades made with Native Americans.

CONSTRUCTED RESPONSE

2. Explain what can be inferred from Columbus's statement "I was very attentive to them, and strove to learn if they had any gold."

Saturday, October 13, 1492

At daybreak great multitudes [numbers] of men came to the shore... loaded with balls of cotton, parrots, javelins, and other things... These they exchanged for whatever we chose to give them. I was very attentive to them, and strove [tried] to learn if they had any gold... [They] readily bartered [traded] for any article we saw fit to give them... such as broken platters and fragments of glass.

THEY CAME LOADED WITH BALLS OF COTTON, PARROTS, JAVELINS, AND OTHER THINGS. . . . THESE THEY EXCHANGED FOR WHATEVER WE CHOSE TO GIVE THEM. — CHRISTOPHER COLUMBUS

Columbus and his crew arrive in the Caribbean.

DOCUMENT 3

The Historian's Viewpoint

Alfred Crosby is the historian who invented the phrase "Columbian Exchange." Below is an excerpt from an essay he wrote on the global impact of the Columbian Exchange.

> What is the significance of the Columbian Exchange [?]. . . What is the staple [main food] of the Bantu of southern Africa? Maize [corn], an American food. What is the staple of Kansas and Argentina? Wheat, an Old World [European] food. . .
>
> How many of the six billion of us are dependent for our nourishment on crops and meat animals that didn't cross the great oceans until after 1492?

CONSTRUCTED RESPONSE

3. What point is Alfred Crosby trying to make about the main foods of southern Africa, Kansas, and Argentina?

ONGOING ASSESSMENT

WRITING LAB

DBQ Practice Look at the illustration, and reread the diary entry and the passage from Crosby's essay. What do these documents tell you about the Columbian Exchange?

Step 1. Review what you know about how trade developed between continents.

Step 2. Take notes about the main ideas in Documents 1, 2, and 3.

Document 1: Diagram of the Columbian Exchange

Main Idea(s) ________________

Document 2: Quotation from Columbus

Main Idea(s) ________________

Document 3: Excerpt from Alfred Crosby essay

Main Idea(s) ________________

Step 3. Construct a topic sentence that answers this question: How did the Columbian Exchange change the world's food supply?

Step 4. Write a detailed paragraph that explains the effect of the Columbian Exchange on the food supply of Europe and the Americas. Go to **Student Resources** for Guided Writing support.

2.3 Paths Toward Independence

Main Idea Ideas of freedom led to fights for independence in 19th-century Central America and the Caribbean.

The Columbian Exchange led to increased competition for global trade among European countries. By the 1780s, Europe had colonized nearly all of Central America and the Caribbean. Forced labor and other harsh practices used by European settlers stirred thoughts of independence among native groups, who wanted to regain control of their lands.

Haiti Leads the Way

In 1791, Haiti (then Saint-Domingue) had become the leading producer of sugarcane in the Caribbean. A small number of wealthy French colonists there used about a half million slaves. Their labor provided the backbone of this economy. As demand for sugar grew, more slaves were brought to Haiti from Africa. By 1791, slaves far outnumbered white planters.

Two years earlier, a revolution, or an overthrow of the government, had taken place in France. By August of 1791, the conflict spread to Haiti. As European planters, free people of color, and the English and French armies fought, Haitian slaves began a rebellion.

In 1794, the French government abolished slavery, but it kept control of the island. **Toussaint L'Ouverture** (too SAN loh ver CHOOR), a former slave, began a movement for independence. L'Ouverture's army struggled with the French for power until 1803, when he died in prison. Soon after his death, his army won victory over the French military. Haiti declared itself independent on January 1, 1804.

Calls for Freedom Spread

Spanish colonies in the Americas were ruled by Spanish **viceroys**—governors who represented the Spanish king and queen. Viceroys controlled the land's resources, such as gold, silver, and crops, and native labor. Most of the region fell under official control of the Spanish viceroy in Mexico City.

In 1821, Mexico seized control of much of present-day Central America. In 1823, it broke from Mexican control and became the United Provinces of Central America. **Provinces** are smaller parts of a larger nation. However, internal conflict renewed calls for independence. Over the next 20 years, each province—Guatemala, Honduras, El Salvador, Nicaragua, and Costa Rica—would declare independence.

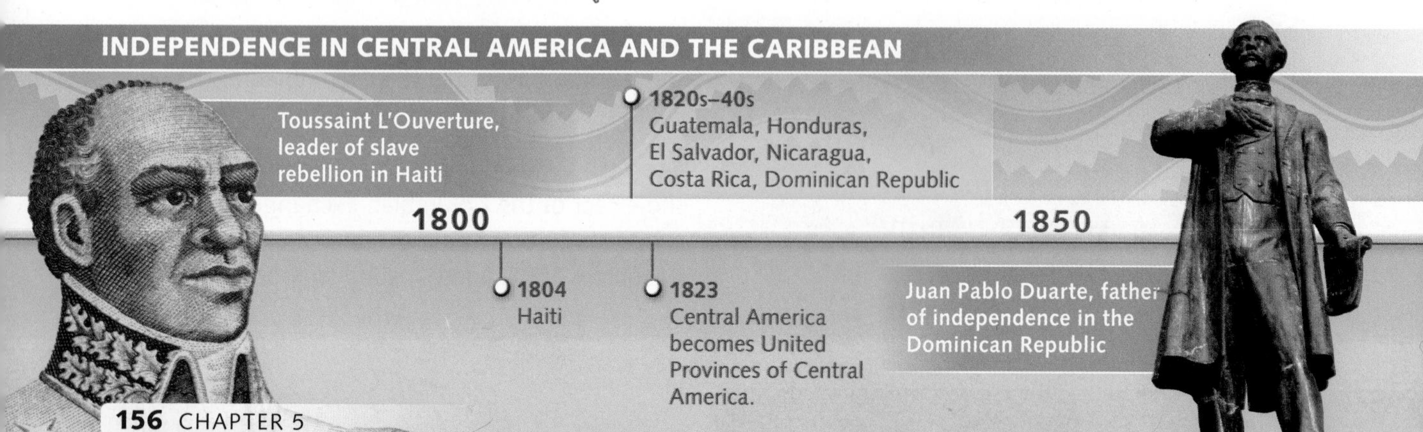

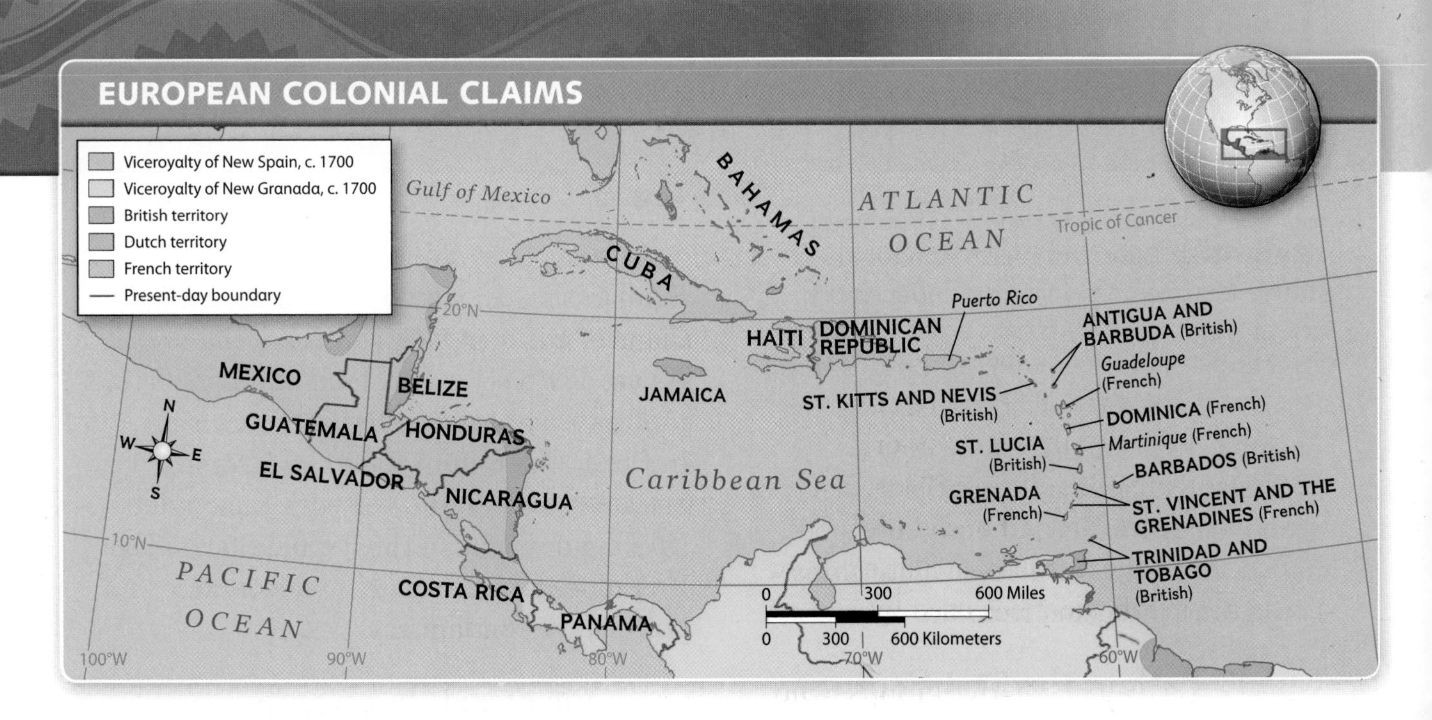

Caribbean Independence

As you've learned, Haiti was the first Caribbean island to become independent. The United States and Europe wanted to keep control of the islands' many resources. As a result, most of the islands would not be independent until the 1900s.

In fact, several islands became independent as recently as the second half of the 20th century, as the time line shows. Today, some islands remain connected to the European countries that settled them. For example, Bermuda continues to be a British territory, and Aruba is still part of the Netherlands.

Before You Move On

Make Inferences How might the Spanish desire to control resources help bring about the fights for independence?

ONGOING ASSESSMENT

MAP LAB

1. **Location** With your finger, trace the outline of the Viceroyalty of New Spain shown on the map. What present-day countries are included in this area?
2. **Interpret Maps** Find Panama on the map. Which viceroyalty did it belong to? Why do you think Panama could be considered more like South America than Central America?
3. **Make Inferences** What fact about Toussaint L'Ouverture might explain his commitment to gaining independence for slaves in Haiti?
4. **Interpret Time Lines** Between which years was there the biggest gap in Central American and Caribbean countries declaring independence?

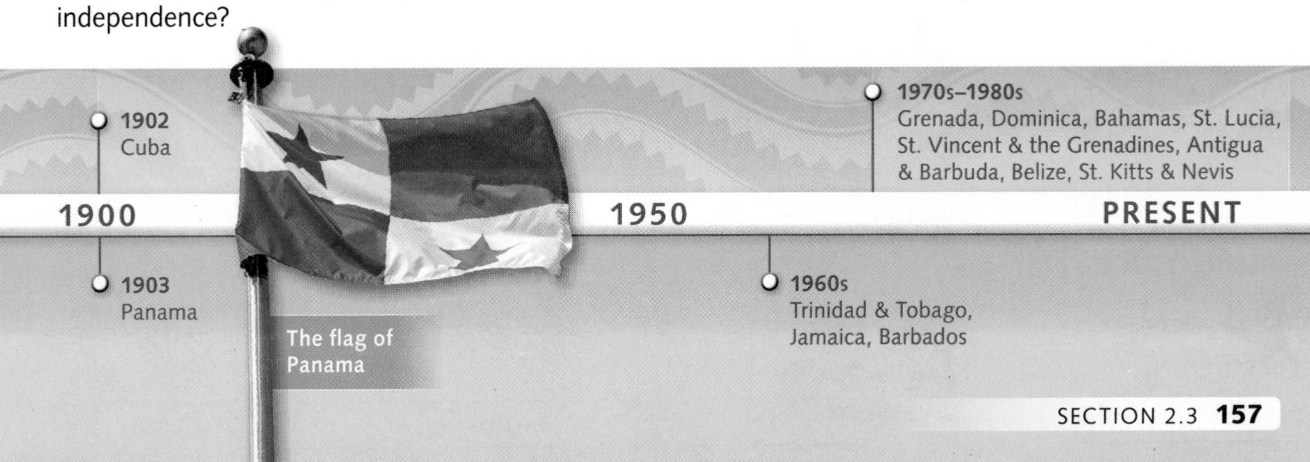

The flag of Panama

2.4 Comparing Cuba and Puerto Rico

TECHTREK
myNGconnect.com For resource maps and photos of Cuba and Puerto Rico
Maps and Graphs
Digital Library

Main Idea Cuba and Puerto Rico took very different economic paths in the 20th century.

In a quest for silver, gold, and other riches, Spain settled the Caribbean islands of Cuba and Puerto Rico as colonies. The islands had ideal conditions for growing sugarcane, and both had good natural **harbors**, or places where ships could land protected from the open sea.

Cuba's Path to the 21st Century

Spain ruled Cuba as a colony from 1511 to 1898. The Spanish built the city of Havana, where European ships carrying cargo such as silver and corn stopped before crossing the Atlantic. Spanish colonists also built sugarcane plantations.

However, native Cubans wanted to control their own resources and political destiny. During the 1800s, they staged several failed rebellions. They finally won independence from Spain in 1898, after the United States defeated Spain in the Spanish-American War. The U.S. military continued to occupy Cuba and control much of the country's economy. Over the next 50 years, Cuba's government was controlled by a series of leaders. Many of them were corrupt, or dishonest, and led with complete control as **dictators**.

In 1959, a revolutionary leader named Fidel Castro overthrew Cuba's dictator. Castro's military took control of the government, seized all land and personal property, and established communism in Cuba. Castro took over U.S.-owned businesses and built ties with the Soviet Union, an enemy of the United States. The United States eventually cut economic and political ties with Cuba.

Until the early 1990s, the government in Cuba controlled all economic activity. In 1993, to improve Cuba's economy, the government started to allow citizens to open their own businesses.

Commonwealth of Puerto Rico

Like Cuba, Puerto Rico was a Spanish colony from the 1500s until 1898. During this time the Spanish mined gold and built huge sugarcane plantations. Spanish control made most native Puerto Ricans poor, so they tried to rebel against Spain.

During the Spanish-American War, the United States sent troops to Puerto Rico. The island's location was important to U.S. military and economic interests. After Spain surrendered to the United States in 1898, the Treaty of Paris made the island a U.S. territory. In 1917, Puerto Ricans were granted U.S. citizenship.

CUBA

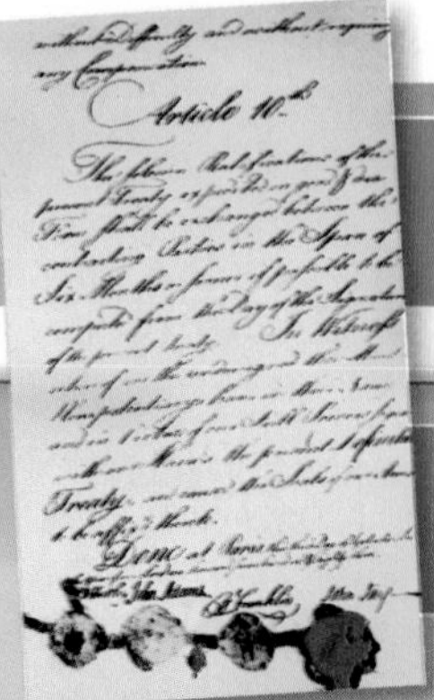

1850

1898 Treaty of Paris grants Cuban independence; U.S. continues to occupy.

1901 U.S sets up naval base at Guantánamo Bay, Cuba.

1900

PUERTO RICO

1898 Treaty of Paris grants control of Puerto Rico from Spain to U.S.

1917 Puerto Ricans are made U.S. citizens.

CUBA'S ECONOMIC RESOURCES

Economic products and resources

Bananas, Cattle, Citrus fruit, Coffee, Copper, Fish, Pineapples, Potatoes, Poultry, Rice, Sugarcane, Swine, Tobacco, Vegetables

Havana, Matanzas, Pinar del Río, Santa Clara, Cienfuegos, Camagüey, Las Tunas, Holguín, Guantánamo, Bayamo, Santiago de Cuba

Tropic of Cancer

Caribbean Sea

0 100 200 Miles
0 100 200 Kilometers

26°N, 23°N, 20°N, 17°N, 83°W, 80°W, 77°W, 74°W

PUERTO RICO'S ECONOMIC RESOURCES

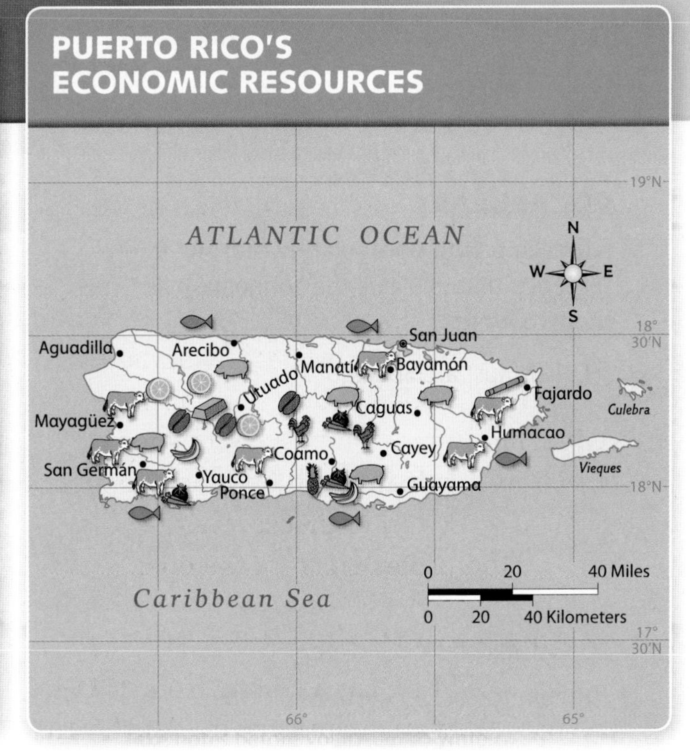

For many years, Puerto Ricans worked for more freedom from the United States. In 1952, Puerto Rico became a U.S. **commonwealth**, a nation that governs itself but is part of a larger country.

Unlike Cubans, Puerto Ricans have political freedom. Like the United States, Puerto Rico benefits from a free-enterprise economy. The government has done what it can to help Puerto Ricans start new businesses, and shift the economy away from farming toward manufacturing. Thousands of Puerto Ricans work in factories that make high-tech products. Many also work in the tourist industry.

Before You Move On

Summarize How did differences in the governments of Cuba and Puerto Rico affect economic opportunities in the two countries?

ONGOING ASSESSMENT

MAP LAB

GeoJournal

1. **Interpret Maps** Which resource is most of the land used for in Cuba? in Puerto Rico?
2. **Region** Look at the climate map in Section 1.1 and the maps above. Why would the economic resources be similar in both countries?
3. **Make Inferences** Why would the United States have wanted control over Cuba and Puerto Rico after the Spanish-American War?
4. **Interpret Time Lines** Which events show each country's connection to the United States?

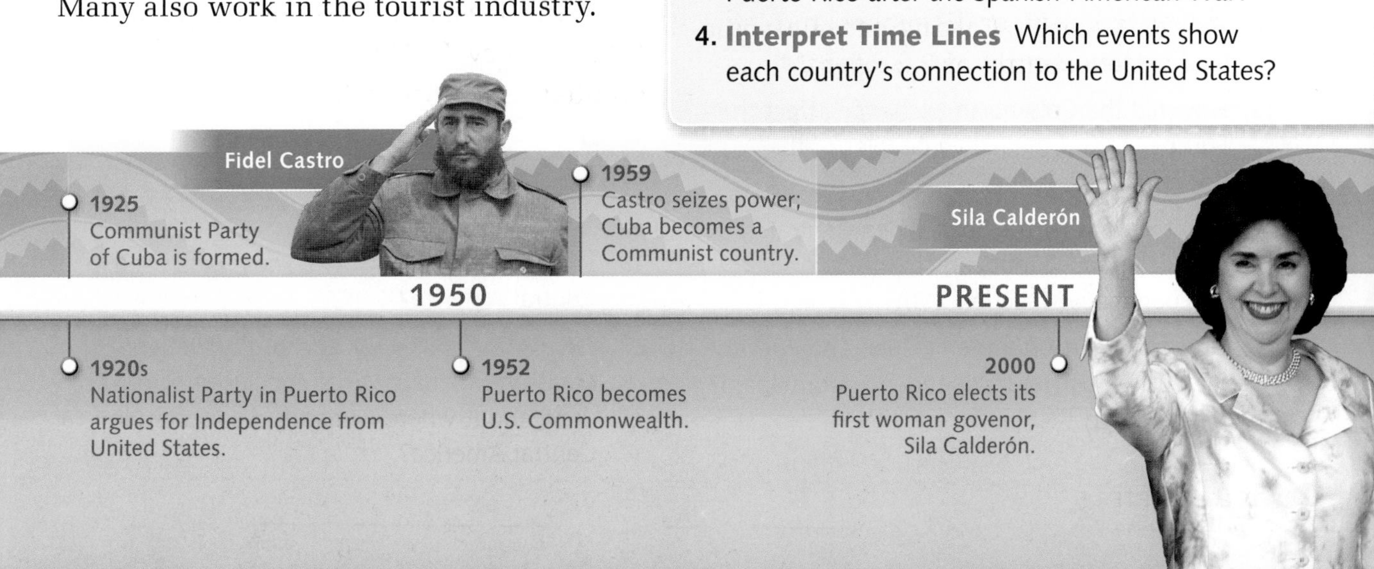

CHAPTER 5
Review

VOCABULARY

For each pair of vocabulary words, write one sentence that explains the connection between the two words.

1. isthmus; archipelago

   An isthmus is one strip of land connected to two larger land areas, but an archipelago is a group of islands.

2. seismic; tectonic plate
3. rain forest; ecosystem
4. deforestation; fertile
5. cash crops; scarcity
6. province; viceroy
7. harbor; commonwealth

MAIN IDEAS

8. What feature of Central America makes the soil fertile? (Section 1.1)
9. How do earthquakes and volcanoes affect Central America and the Caribbean? (Section 1.2)
10. What are some causes of rain forest deforestation? (Section 1.3)
11. In what way will creating new jobs help save Nicaragua's sea turtles? (Section 1.4)
12. How did Columbus's landing in the Caribbean lead to international trade? (Section 2.1)
13. How did the Columbian Exchange affect the Europeans and the native peoples of the Caribbean? (Section 2.2)
14. What actions by the Spanish led Central American and Caribbean countries to seek independence? (Section 2.3)
15. How did the economies of Cuba and Puerto Rico develop after independence from Spain? (Section 2.4)

GEOGRAPHY

ANALYZE THE ESSENTIAL QUESTION

How has physical geography been a positive or negative influence on the economy of the region?

Critical Thinking: Compare and Contrast

16. What physical features do Central America and the Caribbean have in common?
17. What is similar about how the Central American and the Caribbean land masses were created?
18. What is the difference in climate in the mountains and coastal plains? How does this difference affect farming?

INTERPRET MAPS

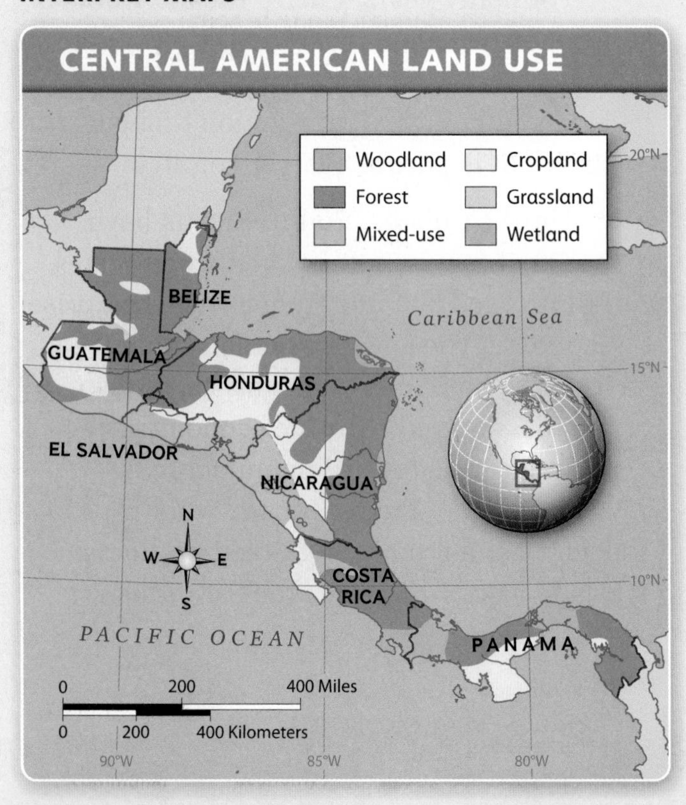

19. **Interpret Maps** How is most land used in Central America?
20. **Draw Conclusions** Locate the wetland areas on the map. What conclusion can you draw about where wetlands are located in Central America?

HISTORY

ANALYZE THE ESSENTIAL QUESTION

How have economic resources influenced the history of the region?

Critical Thinking: **Analyze Cause and Effect**

21. How did volcanoes contribute to the success of farming in the region?
22. What did the Spanish hope to gain through control of the region?
23. What effect did outside control of Caribbean islands have on their independence?
24. What political effect did Fidel Castro and his military have on Cuba?
25. Why has manufacturing become a greater part of Puerto Rico's economy than agriculture?

INTERPRET CHARTS

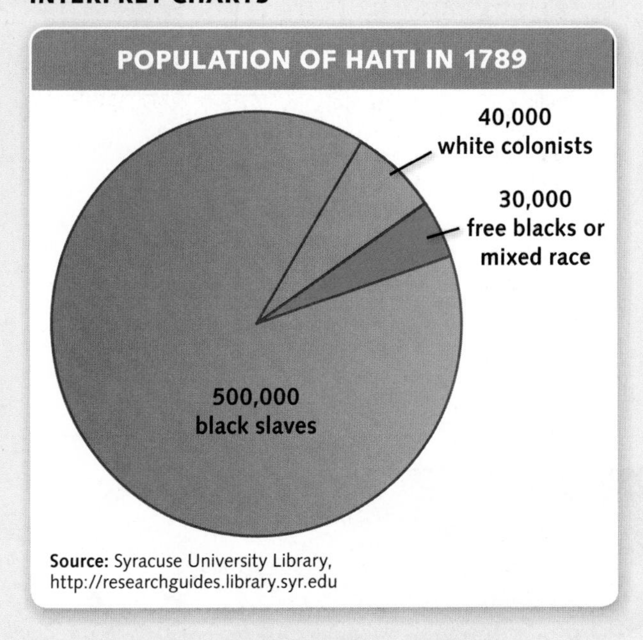

Source: Syracuse University Library, http://researchguides.library.syr.edu

26. **Analyze Data** How many more black slaves were there than white colonists in Haiti in 1789?
27. **Make Inferences** How does the racial makeup of the population in Haiti in 1789 help to explain why a rebellion started?

ACTIVE OPTIONS

Synthesize the Essential Questions by completing the activities below.

28. **Create a Poster** Persuade people to help raise funds for a Central American cause. Choose either rain forest preservation or sea turtle protection. Define the group you want to persuade—for example, your school, your friends, or the government. **Display your poster to the class. Discuss what features of the poster are persuasive.**

Poster Tips

- Think about the group you want to convince. Ask: Who might be able to help the most? How could this group help the cause?
- Take notes about what information might convince this group to join your campaign.
- Create an original slogan for your campaign that will attract attention.

TECHTREK myNGconnect.com For research links

29. **Create a Chart** Make a chart showing comparisons between three countries in the region. Be sure one of your countries is in Central America and one is in the Caribbean. Use the research links at **Connect to NG** and other online sources to gather the data. In your chart, show the following:

	(Central American country)	(Caribbean country)	(Your choice)
Population			
Square Miles of Land			
Types of Land			
Main Economic Resources			

CHAPTER 6

Central America & the Caribbean TODAY

PREVIEW THE CHAPTER

Essential Question How do trade and globalization affect the cultures of the region today?

SECTION 1 • CULTURE

KEY VOCABULARY

- tourism
- intersection
- fuse
- canal
- terrain
- lock

ACADEMIC VOCABULARY

distinct, eliminate

TERMS & NAMES

- Taino
- Calypso
- Panama Canal Zone

Essential Question How is the region trying to improve the standard of living?

SECTION 2 • GOVERNMENT & ECONOMICS

KEY VOCABULARY

- infrastructure
- reserve
- policy
- marketing
- standard of living
- food security
- global warming
- malnutrition
- surplus
- migrate
- remittance
- habitat
- ecotourism

ACADEMIC VOCABULARY

displace

TERMS & NAMES

- Human Development Index
- Port-au-Prince

Agriculture and tourism drive the economy in Souffriere, a small town on the island of St. Lucia in the Caribbean Sea.

SECTION 1 CULTURE

1.1 The Impact of Tourism

Main Idea Tourists who visit Central America and the Caribbean have a major effect on income and resources of the region.

Tourism, or the business of travel, is an important source of income for Central America and the Caribbean. However, with more than 20 million tourists each year, the environment can be damaged.

Diversity Attracts People

For centuries, this region has been an **intersection** of cultures, a meeting point for traders and settlers from many countries. Rich resources attracted European groups, including the English, French, and Dutch. The slave trade brought African culture to the region. The blend of cultures is part of the region's appeal.

Since the 1980s, global air travel and advertising have made tourism an important industry. Visitors usually stay in one of the many island resorts, or on cruise ships that sail around the islands. This "overnight tourism," though, can have long-term negative effects.

Efforts to Improve

Overnight tourists use a great deal of electricity and consume vast amounts of water and food. This causes shortages for the local people. Large resorts and cruise ships release pollution into the air and water, which threatens marine life.

Organizations such as the United Nations seek to increase environmental protections. However, because the tourism industry is such a large part of the region's economy, local governments are sometimes resistant to new limitations. Many travelers and businesses are making efforts to offset the damage.

Cruise lines are beginning to use recyclable materials and conserve fuel in an effort to continue business without further harming the environment. Some travelers even contribute to programs that plant trees in the region.

Before You Move On

Monitor Comprehension In what ways does tourism affect the region's income and resources?

Critical Viewing These pyramids in Guatemala are temples where ancient Mayans worshipped. What aspects of these ruins might make them appealing to tourists?

TOURISM IN CENTRAL AMERICA AND THE CARIBBEAN

Country	International Tourist Arrivals* (number of people)	Receipts from Tourism ** (US $billions)	Percentage of Total Receipts from Tourism ** (percent)
Bahamas	5,003,967	$2.2	64.6
Barbados	1,272,772	$1.2	56.6
Belize	1,082,268	$0.3	35.0
Dominican Republic	4,239,686	$4.0	34.0
El Salvador	966,416	$0.8	20.9
Guatemala	1,181,526	$1.1	12.1
Honduras	1,056,642	$0.6	8.5
Nicaragua	734,971	$0.3	9.4
Panama	1,004,207	$1.2	12.6
St. Lucia	802,240	$0.3	66.0

* **Source:** Association of Caribbean States
** **Source:** World Bank Online

This adventure tour in Martinique is led by local guides.

ONGOING ASSESSMENT

DATA LAB

GeoJournal

1. **Interpret Charts** Which country in the region received the most money from tourism? Which ones received the least?
2. **Analyze Data** Which country depends most on money received from tourism?
3. **Identify Solutions** What can travelers and the tourism industry do to cause less harm to the environment in this region?

1.2 Caribbean Food and Music

TECHTREK

myNGconnect.com For a music clip and photos and Guided Writing

Digital Library

Student Resources

Main Idea Caribbean food and music blend influences from indigenous cultures and other world cultures.

Worldwide trade and global communication have spread Caribbean culture traits to the world. At the same time, other cultures continue to influence the food and music in the region.

Caribbean Food

Since the Columbian Exchange, new influences continued to add to the region's diet. Foods from the native **Taino** (TY noh) and from Europe, Africa, and Asia have **fused**, or blended, into a rich cooking tradition.

The basic foods in the Greater Antilles include rice, beans, yams, peppers, plantains (similar to bananas), and avocados. Poultry and fish are also local foods. Caribbean cooks use blends of spices to flavor their foods. For example, Jamaican jerk is a **distinct**, or easily recognized, blend of strong spices used for grilling meat. Freed African slaves living in the Jamaican mountains developed the spice blend in order to preserve meat. Today, many Jamaican families follow the tradition of sharing a large lunch on Sundays. Foods served might include jerk chicken, fish, fried plantains, and a popular dish of rice and black-eyed peas.

In the past, islanders ate healthy diets based on fruits and vegetables, mixed with meat or fish. When open trade was firmly established in the 1990s, fast food restaurants arrived in the region, adding modern foods to the islanders' diet.

This pepper shrimp is a very spicy dish popular in Jamaica.

Caribbean Music

Native cultures used wind instruments and drums in their music. European colonists brought stringed instruments to the islands. Island cultures blended European and African instruments and rhythms to make their own musical styles.

Calypso began on the island of Trinidad as a type of folk music. It uses simple rhythms and local language to tell stories. *Soca*, developed in the 1970s, is a mixture of calypso and East Indian music. Popular Afro-Cuban styles made it to New York in the 1940s, where they combined with jazz to create *salsa*. Some other well-known Caribbean styles are Dominican Republic *merengue* (meh RENG gay) and Jamaican *ska* and *reggae* (REG ay).

Before You Move On

Monitor Comprehension What foods and styles of music have blended to become part of culture in the Caribbean Islands?

Dancers perform a traditional dance on National Folklore Day in the Dominican Republic.

Critical Viewing Based on the photo, how would you describe the characteristics of this folk dance?

ONGOING ASSESSMENT

WRITING LAB

GeoJournal

1. **Summarize** What are some of the influences on the musical styles of the Caribbean?
2. **Write a Report** Write a paragraph that identifies the effects of global influences on traditional culture in this region. Are they positive or negative? Make sure you think through all of the positive and negative effects and that you explain why you believe one side outweighs the other. Use evidence from the text to support your response. Go to **Student Resources** for Guided Writing support.

1.3 The Panama Canal

TECHTREK
myNGconnect.com For photos and an illustration of the Panama Canal

 Digital Library
 Student Resources

Main Idea The Panama Canal provides a water route connecting the Atlantic and Pacific oceans.

As early as the 1500s, Spanish explorers wanted to create an artificial water route, or **canal**, through the Central American isthmus. A canal would significantly reduce the time it took for ships to travel from Europe to the Pacific.

Connecting Two Oceans

Trade routes from the Atlantic to the Pacific required a difficult journey across land or a long water passage around the tip of South America. In 1855, the United States completed the first railroad across Panama. Goods and people could now travel across the isthmus by rail. However, the idea for a canal remained.

Building the Canal

Panama declared independence from Colombia in 1903. Then the government signed a treaty that gave the United States control of the **Panama Canal Zone**, the area where the canal would be built.

Many obstacles had to be overcome in building the canal. To try to keep builders healthy, doctors worked to **eliminate**, or get rid of, mosquitoes that caused serious diseases. The **terrain**, or physical land features, of Panama also posed challenges. As many as 40,000 workers at a time made plans, cleared brush, drained swamps, and drilled through rocks. The canal took 100 steam engines and 10 years to finish, from 1904 to 1914—and more than 20,000 workers died in the process.

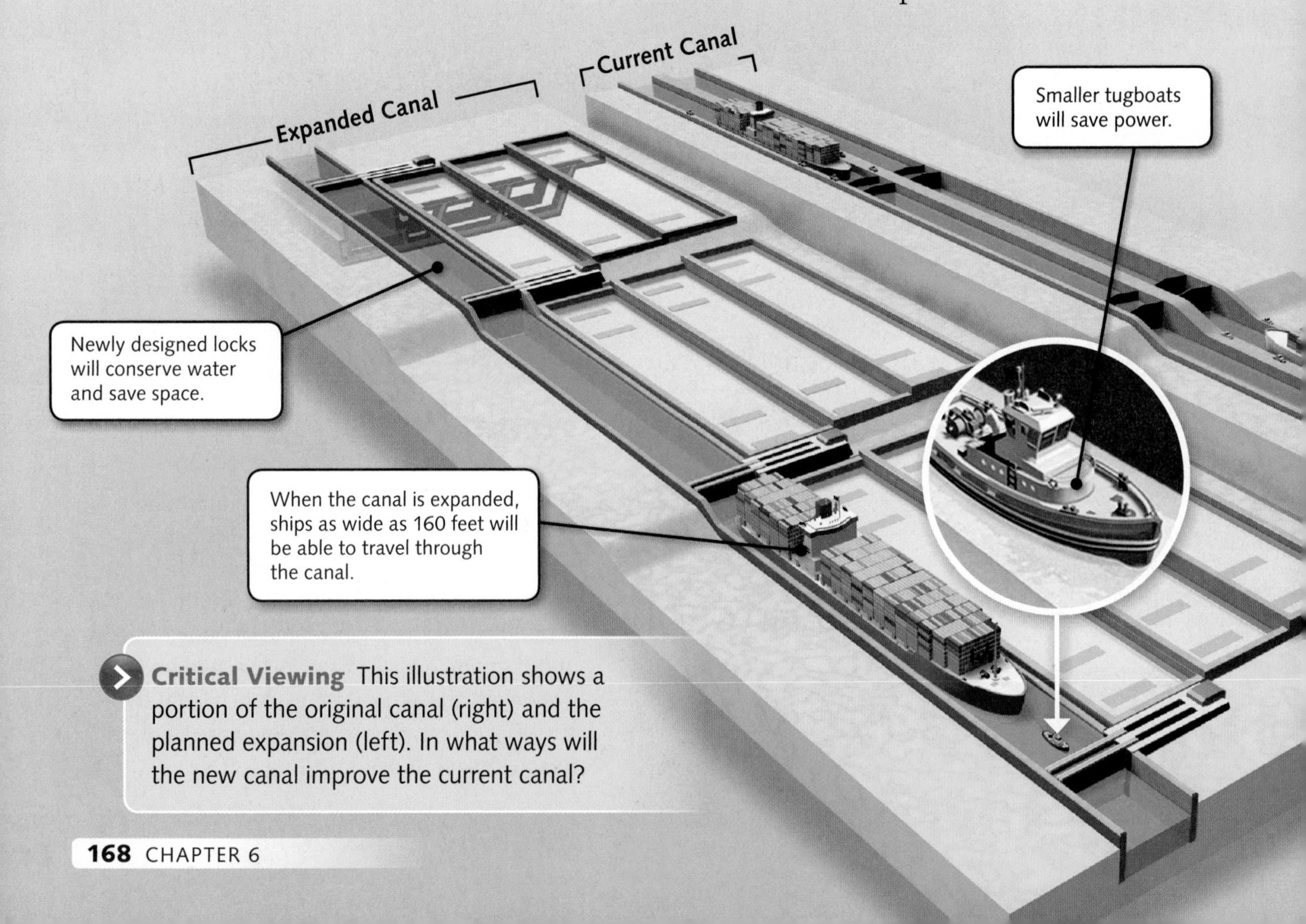

Critical Viewing This illustration shows a portion of the original canal (right) and the planned expansion (left). In what ways will the new canal improve the current canal?

PANAMA CANAL MILESTONES

1855 The United States completes a railroad across Panama.

1881 A French company attempts to build a canal across Panama.

1889 The French company's plan for a canal collapses.

1903 Panama declares independence from Colombia; canal rights are granted to the United States.

1914 The Panama Canal is opened to water traffic.

1977 Treaty with the United States grants ownership of the Canal Zone to Panama.

1999 Panama is granted complete control of the canal.

2014 Expansion of the canal is scheduled for completion.

Visual Vocabulary **Locks** are devices that help to equalize the water levels of the waterways being connected. These locks in the Panama Canal are the first step down to the Pacific Ocean.

Canal Connects the World

The Panama Canal's slogan, "The Land Divided, the World United," reflects the importance of this water passage. Before its completion, a ship traveling from New York to San Francisco had to travel 14,000 miles. The canal's 51-mile length shortened the travel distance by almost half, to approximately 6,000 miles.

As technologies advanced, larger ships were built. In 2006, Panama voted to expand the canal so it could accommodate these ships and the increased trade that caused traffic jams. The project is scheduled for completion in 2014.

Before You Move On

Make Inferences What made building a water route connecting the Atlantic to the Pacific important?

PANAMA CANAL EXPANSION PROJECT

ONGOING ASSESSMENT

VIEWING LAB

GeoJournal

1. **Location** Look at the locator map above that shows the position of the Panama Canal. Why do you think builders picked that spot instead of a different one?
2. **Analyze Visuals** Based on the illustration, how will the expansion of the Panama Canal save money?
3. **Summarize** Based on the milestones, how long did the United States control the canal?

2.1 Comparing Costa Rica and Nicaragua

TECHTREK
myNGconnect.com For photos of Costa Rica and Nicaragua
Digital Library

Main Idea Costa Rica and Nicaragua both work to build a more stable economy, despite different political conditions.

During the 20th century, Costa Rica and Nicaragua (nik uh RAHG wah) had very different histories. As a result, Costa Rica became a strong and stable democracy, while Nicaragua became the poorest country in Central America.

Different Political Paths

Costa Rica has enjoyed peace for over 60 years. In fact, since 1949 the country has never had an army. Costa Rica's capital city of San José is home to several global human rights organizations. Without political conflict to interfere, Costa Rica has had the opportunity to create a stable economy. Tourism is the country's greatest source of income. This industry provides jobs to more than 50 percent of the working population.

Unlike Costa Rica, Nicaragua has had an unstable government since its independence from Spain in 1821. This instability included dictatorships and civil wars that lasted through the 1990s. In addition, the country was hit by a major hurricane in 1998. It destroyed Nicaragua's **infrastructure**, the basic systems a society needs, such as roads, bridges, and electricity. Thousands were left without homes, jobs, or medical care.

Unstable conditions make the fight against poverty more difficult. Nicaragua entered the 21st century facing challenges. It needed to build its economy and develop social programs to help poor people, who in 2005 made up almost half the population.

Visual Vocabulary A **reserve** is land set aside for farming. Members of the Miraflor Nature Reserve in Nicaragua can set up farms on the land there.

Economic Challenges

Costa Rica's economy has grown over the past 20 years. However, the poverty rate has remained between 15 and 20 percent. Changes in government **policy**—official guidelines and procedures—might account for this lack of improvement.

In the 1980s, after years of heavy spending and borrowing, the country was running out of money. Over the next decades, various policies were enacted to address this economic crisis. The government restricted its spending and raised taxes. Funding was reduced for social programs, many of which were created to help poor people. In 2007, in an effort to bring about economic growth, Costa Rica joined other Central American countries in a free trade agreement with the United States. The agreement went into effect in Costa Rica in 2009.

COMPARING ECONOMIC DEVELOPMENT INDICATORS

	Costa Rica	Nicaragua
Life Expectancy at Birth	77.5 years	71. years
Adult Literacy Rate aged 15 and above	94.9%	67.5%
Per Capita Income (U.S. Dollars)	$10,900	$2,800
Population Below the Poverty Line	16%	48%

Source: CIA World Factbook

In 2005, Nicaragua received funds from the United States to address rural poverty. The money was used for factors of production, such as farming equipment. It was also used for **marketing**—advertising and promotion—for rural businesses, and for building roads so these new businesses could better transport goods. Efforts like these can provide long-term solutions for the problem of poverty in Nicaragua.

Before You Move On

Summarize In what ways have both countries built a more stable economy?

ONGOING ASSESSMENT

DATA LAB

GeoJournal

1. **Analyze Data** How does literacy rate relate to the percentage of people living below poverty? Answer by completing this sentence: As the literacy rate decreases, the poverty rate ______.
2. **Place** Based on the chart, which country seems to have a better standard of living, or quality of life? How can you tell?
3. **Make Inferences** Why might small farmers want to set up their own reserve in Nicaragua, rather than depend on government funding?

2.2 Challenges in Haiti

TECHTREK
myNGconnect.com For photos of the international aid effort in Haiti

Digital Library

Main Idea Haiti faces many great challenges in its efforts to build a strong economy and decrease poverty.

In the 1700s, the French colony of Saint-Domingue (san do MANG yuh), which is today's Haiti, was the richest in the Caribbean. Today, Haiti is the poorest country in the Western Hemisphere with 80 percent of its people living in poverty.

Poverty's Historical Roots

Haiti gained independence from France in 1804. Because most of Haiti's citizens had been slaves, they did not have any money or means of income. European nations feared slave revolts in their colonies in the Caribbean, and so did not support Haiti financially when it became an independent nation. In the 20th century, political conflict and outbreaks of disease kept tourists away, making it difficult for Haiti to grow economically.

Critical Viewing Members of a Chinese emergency rescue team work in Port-au-Prince two days after the 2010 earthquake. Based on the photo, how might past building practices have contributed to the extent of the devastation?

21st Century Haiti

The **Human Development Index** (HDI) is used by geographers to compare quality of life in different countries. HDI combines measures of the health, education, and **standard of living**—the level of goods, services, and material comforts—of people in a country. People in countries with a low HDI, such as Haiti, are often less healthy, less educated, and poorer than people in countries with a high HDI, such as the United States.

Politics in 21st century Haiti have been marked by instability and corruption. Due to violence between political groups, U.S. forces were sent in 2004 to Haiti's capital city, **Port-au-Prince**, to maintain security. Over the next few years, further efforts toward peace in Haiti were unsuccessful.

Adding to Haiti's challenges, a massive earthquake struck in January of 2010. Port-au-Prince and the surrounding area were nearly destroyed. Over 1.5 million people were **displaced**, or forced from their homes. Damage to the airport and to seaports made it difficult to receive immediate help from other countries.

As Haiti tried to recover, organizations from around the world began donating money and supplies such as food and medicine. Some countries also sent emergency rescue workers. Many nations and global organizations excused Haiti from paying back billions of dollars in loans, so they could rebuild at home.

Before You Move On

Summarize What factors contribute to the difficulties Haiti faces in overcoming poverty?

RESPONDING TO THE EARTHQUAKE

1.5 million
Number of Haitians displaced from their homes and living in temporary housing after the earthquake

28,000
Number of displaced Haitians that had moved into new homes six months after the earthquake

1,340
Number of tent cities and camps still being used six months after the earthquake

Source: 2010 United Press International

The day after the earthquake, Haitians set up this tent city amid the rubble. In this aerial photo, the colorful squares in the center are the roofs of tents used as temporary shelters.

ONGOING ASSESSMENT

DATA LAB

1. **Analyze Data** How many displaced Haitians were in new homes six months after the earthquake? How many were still homeless?
2. **Make Inferences** What factors might explain this difference in numbers?
3. **Draw Conclusions** What do the photos show about how Haitians and the international community responded after the earthquake?

2.3 Feeding Central America

TECHTREK
myNGconnect.com For research links and photos on food security in Central America

Digital Library

Connect to NG

Global Issues

Main Idea Food supply in Central America is affected by natural disasters and human activities.

KEY VOCABULARY

food security, n., easy access to enough food

global warming, n., an increase in temperature around the world

malnutrition, n., a lack of healthy food in the diet, which leads to physical harm

surplus, n., amount beyond what is needed

Food security, or easy access to enough food, is an important issue around the world. Understanding the causes of shortages can help local governments and global organizations take the right steps toward improving food security in Central America.

Impact on Food Supply

Natural disasters have a major impact on the food supply. Many countries struck by Hurricane Mitch in 1998—El Salvador, Guatemala, Honduras, and Nicaragua—barely had time to recover before major flooding hit Central America in 2008. Disasters can destroy a country's major crops, such as bananas in Honduras. The 2010 earthquake in Haiti made it difficult for Haitians to gain access to nutritious meals and safe water.

In some countries, such as Guatemala and Nicaragua, the rainy season doesn't always provide enough water for the crops that make up the food supply. Even one dry season can greatly reduce food production.

Human activity can also threaten the food supply. Many countries have not managed their natural resources well. Lack of water for irrigation, declining soil quality from growing the same crop year after year, and deforestation all have led to food shortages.

Today, climate change adds a new threat to the food supply. Climate scientists predict that **global warming** will cause more extreme weather patterns, leading to flooding in some areas and drought in others. Many Central Americans survive mainly on corn and beans, but drought has dried up these crops in many areas.

Before You Move On

Summarize How do nature and human activities affect the food supply in Central America?

SOME MAJOR CROPS IN SELECTED CENTRAL AMERICAN COUNTRIES

EL SALVADOR

GUATEMALA

HAITI

HONDURAS

NICARAGUA

- eggs
- beans
- maize
- bananas
- avocados
- mangoes
- plantains
- oranges
- rice
- nuts

Fresh mangoes from Haiti

Childhood Hunger

Childhood **malnutrition**, or the lack of healthy food in the diet, is one of the most serious effects of food shortages in Central America. In some countries, many pregnant women do not get the nutrition they need. As a result, some babies are very small at birth. Poor nutrition continues throughout childhood and can slow down a child's healthy growth.

In Guatemala, 23 percent of children under five are underweight and almost half are small for their age because they don't get proper nutrients in their diet. Many poor families spend their whole day trying to grow or buy enough food to get to the next day. There is no time to tend to health issues or for children to go to school. In these conditions, cycles of malnutrition are difficult to break.

Solutions For the Future

Most experts believe that the best way to improve food security in Central America is to increase each country's own crop production. Growth in agriculture means larger food supplies and lower food prices. It can also mean higher incomes for farmers both large and small.

Land quality is a determining factor in a country's ability to increase crop production. Honduras and Guatemala have large areas of good quality land for farming. In most years, much of Guatemala and the coastal areas of Nicaragua and Honduras typically experience enough rainfall for their crops to thrive.

Areas with low amounts of quality soil or inadequate rainfall, such as El Salvador and southern Honduras, benefit from programs that provide fertilizer and irrigation methods to help rural farmers.

Education is another way to improve food security in Central America. The United Nations recommends rural people be educated in productive farming methods. Programs have been designed to teach rural farmers in Central America how to keep soil healthy, and how to grow crops and sell the **surplus**, or extra.

In emergency situations such as the earthquake in Haiti, a country can benefit from the aid of foreign countries. However, when a poor country can improve its food supply by learning how to increase its own crop production, it can achieve long-term food security without outside aid.

Before You Move On

Make Inferences How does education provide a long-term solution to the problem of food security in Central America?

ONGOING ASSESSMENT

READING LAB

1. **Location** What geographic features and conditions of Central America contribute to food shortages in the region?
2. **Summarize** Why is it hard for a poor family to break out of the cycle of malnutrition?
3. **Turn and Talk** What kinds of programs would help improve food security in one country in the region? Turn to a classmate and use information from the lesson to develop some specific ideas.

2.4 Migration and the Caribbean

TECHTREK
myNGconnect.com For photos and a graphic of migration and the Caribbean
Digital Library

Main Idea Many Caribbean people migrate to other countries to find economic opportunites and help support their familes back home.

People **migrate**, or move from one place to another, because of push-pull factors. Push factors make people move away from an area because of difficulties like war or drought. Pull factors draw people to a place because it offers more security and better job opportunities.

Migration Within the Caribbean

Today, it is hard to make a living on many Caribbean islands. The collapse of major businesses, including the sugar industry, has pushed workers out of rural areas into cities to find work. As a result of this internal migration, or migration within a country or region, two-thirds of the population now lives in cities such as Santo Domingo in the Dominican Republic or San Juan, Puerto Rico.

Many of these cities have become overcrowded. Urban unemployment is high in the Caribbean. Migrants seeking better jobs and a higher standard of living have been forced to travel to other islands within the Caribbean region.

By the 1990s, the tourism industry had expanded dramatically across the region. The demand for workers pulled many people to islands with large or growing tourist industries, such as Aruba, the Bahamas, and the Virgin Islands.

Migration Out of the Caribbean

At the same time, push-pull factors played a part in workers leaving to go the United States, Canada, Europe, and other places. For example, political conflict in Cuba and Haiti pushed people to migrate to the United States.

Critical Viewing Santo Domingo in the Dominican Republic, the oldest city founded by Europeans in the Western Hemisphere, has a growing population. Based on the photo, what economic opportunities might exist in Santo Domingo?

Remittances

Most migrants who find work in another country send money back to their families in the form of **remittances**, money sent to a person in another place. Remittances have become a significant part of the economy of some Caribbean countries. For example, Jamaica receives more than $79 million each year in official aid, or money from other governments or organizations. However, the island receives 27 times that figure (see chart at right) in remittances. These remittances are sent from relatives in other countries, such as the United States, Canada, and France.

Before You Move On

Summarize How are people who migrate out of the Caribbean able to help support their families back home?

WORKERS LEAVING THE REGION
(2009/2010)

SELECTED DESTINATIONS

UNITED STATES & CANADA

EUROPE

CARIBBEAN

MEXICO

workers

remittances

SELECTED REMITTANCES

remittances

official aid

Honduras

$564 million

El Salvador

$233 million

Dominican Republic

$152 million

Jamaica

$79 million

Source: World Bank Online

ONGOING ASSESSMENT

DATA LAB

GeoJournal

1. **Analyze Visuals** Based on the graphic, what goes out of the Caribbean, and what comes back in?
2. **Analyze Data** What brings more money into countries in this region, official aid or remittances? How does the graphic show this?
3. **Draw Conclusions** How do remittances help to ease the strain of poverty in the region?

2.5 Conserving the Rain Forest

TECHTREK

myNGconnect.com For photos of ecotourism in Central America and the Caribbean

Digital Library

Main Idea Ecotourism provides a new opportunity for this region to protect rain forest habitats and fight poverty.

The rain forests and other natural resources of Central America and the Caribbean are a draw for large numbers of travelers every year. Tourism, however, can damage or even permanently destroy these valuable resources.

Rain Forest Habitats

A rain forest is an important animal **habitat**, a natural home or environment for certain species. Plants in the rain forest provide nourishment for the animal population and replenish the air with oxygen. Some animal and plant species in the rain forest are not known to exist anywhere else.

The features of the rain forest that attract visitors are the very same features that are threatened by excessive tourism. If a species loses its habitat, or the habitat is altered in any way, that species may face extinction. Conserving the resources of the rain forest is an important step towards protecting species that live there.

quetzal

three-toed sloth

Visual Vocabulary A **habitat** is a place where certain species have what they need to live. The rain forest canopy is an important animal habitat.

Rain Forest Ecotourism

Ecotourism is a way of visiting natural areas that conserves the resources of the region. The purpose of ecotourism is to allow a visitor to experience an environment in its most natural form. A rain forest ecotourist might explore from the ground on a guided hike or a birdwatching tour. Another ecotourist might explore from above by taking a tour of the rain forest canopy, home to a great variety of animal species.

This method of travel not only protects plants and animals of the rain forest, but also improves the lives of the local people. Ecotourism helps stop deforestation and soil destruction by providing alternative jobs to poor farmers who live off rain forest land. Local people can work in hotels, or as tour guides or rangers.

Ecotourism helps to preserve the environment by allowing a country to profit from and sustain its resources in their natural form. Housing for ecotourists is designed to have very little effect on the environment. For example, the construction of the Lapa Rios Resort in Costa Rica caused the loss of only one tree.

However, even ecotourists can have a negative impact. If too many ecotourists visit a single location, they can interfere with the habitat there. International conservation organizations have become involved in ecotourism. They have begun to point travelers to destinations that are working to preserve the natural resources of the region.

Before You Move On

Summarize How does ecotourism help protect against deforestation?

blue jeans frog

ONGOING ASSESSMENT

SPEAKING LAB

GeoJournal

1. **Turn and Talk** You are the owner of a resort in a Central American country. With a partner, discuss what you would include on your website to convince people that your resort preserves the environment and helps local people. Share your ideas with the class.
2. **Compare and Contrast** In what way is the focus of ecotourists different from that of other tourists? Provide an example.
3. **Analyze Visuals** From what you observe in the photos of the canopy and some of its animal inhabitants, what do you think draws ecotourists to the rain forest?

CHAPTER 6
Review

VOCABULARY

For each pair of vocabulary words, write one sentence that explains the connection between the two words.

1. tourism; intersection

> *The Caribbean was once an intersection of cultures, which makes it a popular area for tourism today.*

2. canal; terrain
3. lock; canal
4. infrastructure; policy
5. food security; global warming
6. malnutrition; food security
7. migrate; remittance
8. ecotourism; habitat

MAIN IDEAS

9. What technological advances helped bring tourists to Central America and the Caribbean? (Section 1.1)
10. What native foods are part of the diet of people in the Caribbean? (Section 1.2)
11. Why is Panama rebuilding its canal? (Section 1.3)
12. What has made it possible for Costa Rica to establish a more stable government than Nicaragua? (Section 2.1)
13. What were the most important needs of the people in Haiti after the earthquake of 2010? (Section 2.2)
14. What is one way to improve food security in Central America? (Section 2.3)
15. Why are remittances so important to many countries in the region? (Section 2.4)
16. In what ways does ecotourism protect the resources of the rain forest? (Section 2.5)

CULTURE

ANALYZE THE ESSENTIAL QUESTION

How do trade and globalization affect the cultures of the region today?

Critical Thinking: Analyze Cause and Effect

17. What characteristics of Caribbean culture attract tourists today?
18. How might global communication continue to affect cultures in the region?
19. How did Caribbean music become a blend of several different styles?
20. How will an increase in ships passing through the Panama Canal help the region?

INTERPRET MAPS

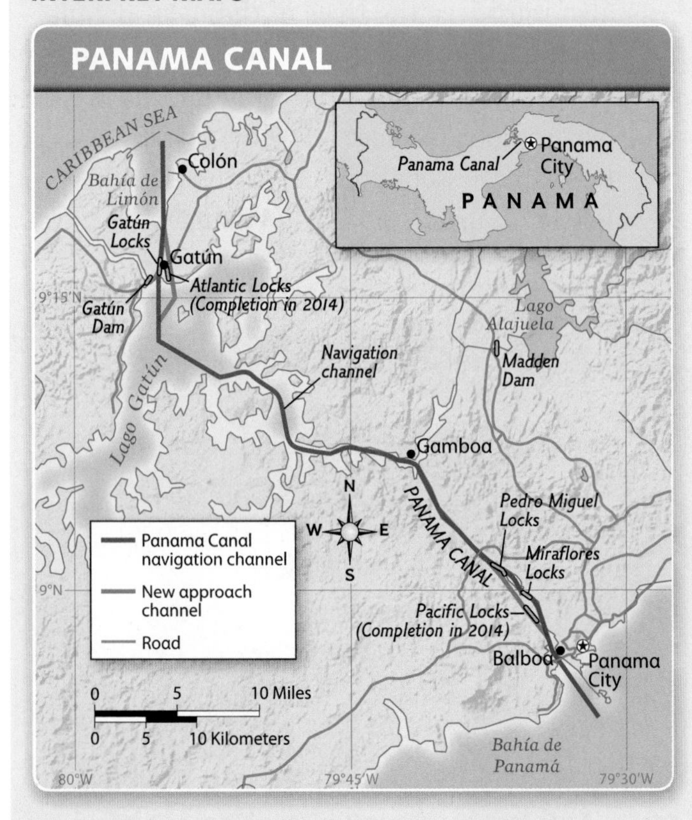

21. **Interpret Maps** What is the length of the Panama Canal in miles and in kilometers?
22. **Analyze Visuals** Will the new locks replace or add to the old ones? How can you tell?

GOVERNMENT & ECONOMICS

ANALYZE THE ESSENTIAL QUESTION

How is the region trying to improve the standard of living?

Critical Thinking: Compare and Contrast

23. What challenges does Nicaragua face in fighting poverty that Costa Rica does not?
24. Why was it helpful for Haiti to be excused from paying back some of its foreign debt?
25. Why is improving land quality and education a better way to ensure food security than providing foreign aid?

Critical Thinking: Make Inferences

26. Assume that the price of sugar and bananas has fallen and the production costs are high due to global competition. How might this affect migration in the region?
27. What economic opportunities does ecotourism provide for poor people?

INTERPRET CHARTS

POPULATION AND POVERTY IN CENTRAL AMERICA

Country	Population in Millions	Percent Living Below Poverty Line
Benchmark: United States	310.2	12.0
Guatemala	13.5	56.2
Costa Rica	4.5	16.0
Honduras	7.9	59.0
Dominican Republic	9.8	42.2

Sources: CIA World Factbook

28. **Analyze Data** Which Central American country has the lowest percentage of people living in poverty?
29. **Make Inferences** From which country or countries do you think more people might migrate? Why do you think so?

ACTIVE OPTIONS

Synthesize the Essential Questions by completing the activities below.

30. **Write and Design a Web Page** Pick a destination in Central America or the Caribbean, such as a rain forest or historic monument. Design a web page to attract people to visit your destination as ecotourists. Make sure to include how your tour preserves the environment and helps stop poverty. **Display your web page in your class.**

Writing and Designing Tips

- Describe the destination and include eye-catching visuals.
- Explain how lodging, food, and transportation help the environment.
- Show how the tour helps give jobs to local people.

31. **Create a Chart** With a partner, talk about the difference between regular tourism and ecotourism. Make a Comparison-Contrast Chart using photos from the **Digital Library** or other online sources to illustrate each type of activity. Explain why the photo represents one type of tourism and not the other. Use the example below as a model for your chart.

PHOTO	REGULAR TOURISM	ECOTOURISM	WHY?
Lots of people sun-tanning at a beach	X		People aren't concerned with how beach is affected by so many people.

Canals as Transportation

TECHTREK

myNGconnect.com For a graphic comparing canals

Student Resources

Waterways can often be effective routes for moving people and goods from one place to another. However, natural waterways such as rivers and lakes don't always connect to each other, so water traffic comes to a stop. From ancient times to today, people have built canals to connect natural waterways for transportation and to gain access to water.

Sometimes the bodies of water being joined by a canal have different water levels. In that case, the canal will require the use of locks, which are devices that help to equalize the water levels of the waterways being connected. The Dutch (Netherlands) are believed to be the first to make use of locks, as early as the 1300s.

Compare

- China
- Egypt
- France
- Netherlands
- Panama
- Scotland
- Sweden
- United States

ECONOMIC IMPACT OF CANALS

Early Middle Eastern civilizations are believed to have built canals for irrigation and drinking water, and the Romans built them to transport their military throughout Europe. Later, canals played an important role in Europe and the United States during the Industrial Revolution. These waterways provided cheaper, faster ways to move goods to new markets and to bring raw materials to factories, which helped the economy to grow.

Eventually, railroads took over much of the work of the canals. In fact, it is not uncommon to find a railroad running alongside certain canals. However, some cities today still use canals for transportation, and some of those canals have become popular tourist attractions.

IMPORTANT CANALS

In the United States, the Erie Canal provides a 363-mile water route from the Atlantic Ocean to the Great Lakes.

Like the Erie Canal, the Panama Canal (Central America) and the Suez Canal (Egypt) are important shipping routes. Use the visuals on the opposite page to compare these two major canals.

Other canals around the world serve the same important function:

- Caledonian Canal (Scotland)
- Canal du Midi (France)
- Göta Canal (Sweden)
- Grand Canal (China)
- Amsterdam-Rhine Canal (Netherlands)

COMPARE TWO IMPORTANT CANALS

PANAMA CANAL

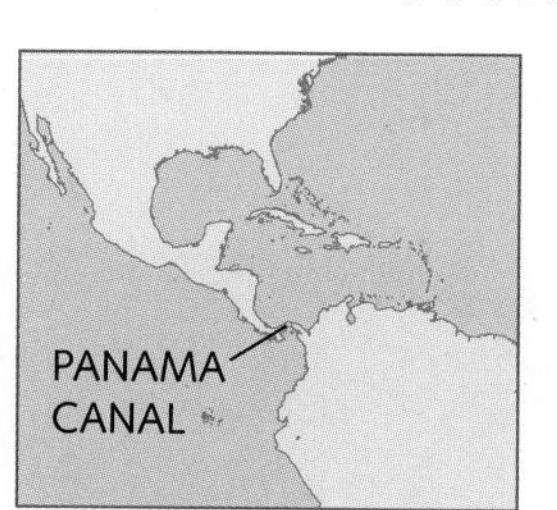

1904
year building began

1914
year the canal opened

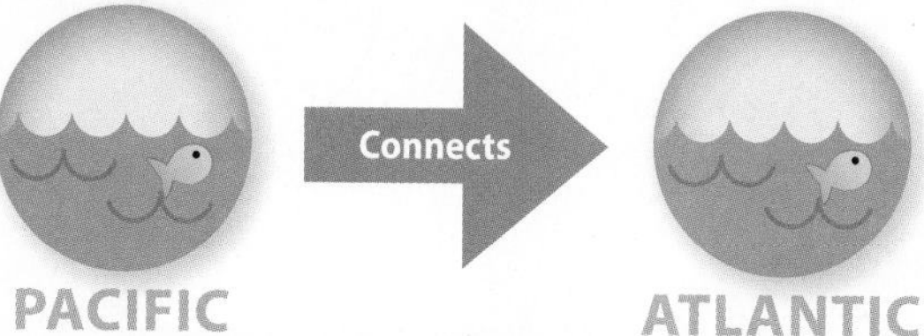

PACIFIC → ATLANTIC

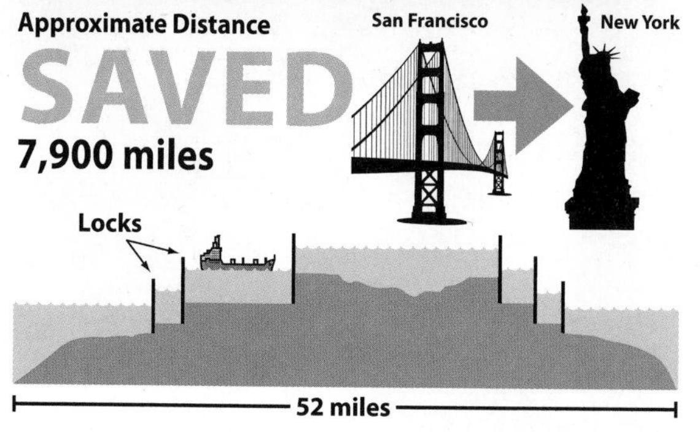

SUEZ CANAL

SUEZ
CANAL

1859
year building began

1869
year the canal opened

Connects

MEDITERRANEAN SEA → RED SEA

Source: www.britannica.com

ONGOING ASSESSMENT

RESEARCH LAB

GeoJournal

1. **Summarize** How have canals been used from ancient times to today?
2. **Compare** Based on the visual, which canal saves more miles between major cities?

Research and Create Charts Choose two of the canals from the list on the opposite page. Do research and create a chart to compare data about the two canals. Based on the information you find, you may want to change the categories in your chart from those shown on the visual above

Active Options

TECHTREK

myNGconnect.com For research links on an Explorer

Connect to NG

Magazine Maker

ACTIVITY 1

Goal: Research an animal species.

SPEAK YOUR MIND

Central American and Caribbean countries are home to many species of bats. Over 70 species of bats live on the small Panamanian island of Barro Colorado. Research bats in the region, and prepare a speech that describes the unique features of different bats that help them to function in their environment. Help others understand bats and their habitats. Indicate the role ecotourism might play in the bat's future.

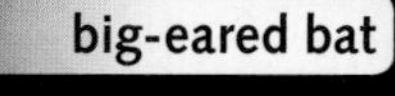
big-eared bat

ACTIVITY 2

Goal: Learn about a new technology.

WRITE A FEATURE ARTICLE

National Geographic Emerging Explorer Ken Banks created software to allow rural groups to communicate even if they don't have Internet access. Rural farmers can get up-to-date information on crop prices—without access to phone or wireless connections—and stay competitive. Use the links at **Connect to NG** to find out more about this mobile technology. Then use **Magazine Maker** to write a feature article that explains the ways this technology can be used to help small, nonprofit groups.

ACTIVITY 3

Goal: Extend your knowledge of the geography of the Caribbean.

CREATE A SKETCH MAP

The Caribbean is famous as the setting for pirate activity through the ages. Create a sketch map of the region that shows places where pirate activity took place. Write captions that describe some famous pirates and explain their routes around the islands.

NATIONAL GEOGRAPHIC
World Cultures and Geography
GFO

EXPLORE SOUTH AMERICA WITH NATIONAL GEOGRAPHIC

MEET THE EXPLORERS

NATIONAL GEOGRAPHIC

Husband-and-wife team and Emerging Explorers Cid Simoes and Paola Segura work to teach farmers in Brazil to conserve their land while growing sustainable, profitable crops.

CONNECT WITH THE CULTURE

Since 1940, this stadium, the Estádio do Pacaembu in São Paulo, Brazil, has hosted important soccer matches, including several FIFA World Cup matches in 1950. Soccer helps to unify Brazil. It has won five World Cup championships, more than any other country.

INVESTIGATE GEOGRAPHY

The Amazon River and its tributaries border eight countries and supply one-fifth of the world's river flow. The rain forests found along its winding course are the largest on Earth and home to millions of plants and animals—many found nowhere else.

Washington, D.C.
2,377 miles
Bogotá, Colombia
Go to myNGconnect.com for maps of South America.
STEP INTO HISTORY
The buildings, temples, and plazas of Machu Picchu, built at the height of the Inca Empire in the 1400s, stand high in a tropical mountain forest in Peru.

CHAPTER 7

SOUTH AMERICA GEOGRAPHY & HISTORY

PREVIEW THE CHAPTER

Essential Question How does elevation influence climate in South America?

SECTION 1 • GEOGRAPHY

KEY VOCABULARY

- vegetation
- grasslands
- adapt
- subsistence farming
- tributary
- biodiversity
- current
- rain shadow
- transpiration
- greenhouse gas

ACADEMIC VOCABULARY

saturate, acknowledge

TERMS & NAMES

- Andes Mountains
- Amazon River Basin
- Llanos
- Angel Falls
- Pampas
- Atacama Desert
- El Niño

Essential Question How did mountains, plateaus, and rivers shape the region's history?

SECTION 2 • HISTORY

KEY VOCABULARY

- descendant
- kinship
- terraced
- suspension bridge
- artifact
- geoglyph
- excavate
- lowland
- nomad
- hunter-gatherer
- slash-and-burn
- treaty
- convert
- monopoly
- rebellion
- exile
- liberate

ACADEMIC VOCABULARY

utilize, transform, transition

TERMS & NAMES

- Machu Picchu
- Guaraní
- Tupinambá
- Yanomami
- Treaty of Tordesillas
- Francisco Pizarro
- Atahualpa

110°W 0° 10°S 20°S 30°S 120°W 40°S

TECHTREK FOR THIS CHAPTER
Student eEdition
Maps and Graphs
Interactive Whiteboard GeoActivities
Digital Library
Connect to NG
Go to myNGconnect.com for more on South America.
jaguar
Caribbean Sea
Caracas
Orinoco R.
VENEZUELA
Georgetown
Paramaribo
Cayenne
GUYANA
FRENCH GUIANA (France)
Boundary claimed by Suriname
SURINAME
Equator
Bogotá
COLOMBIA
GALÁPAGOS IS. (Ecuador) (ARCHIPIÉLAGO DE COLÓN)
Quito
ECUADOR
Negro R.
AMAZON BASIN
Amazon R.
Madeira R.
BRAZIL
Tocantins R.
Araguaia R.
PERU
Lima
ANDES
Lake Titicaca
La Paz
BOLIVIA
Sucre
Brasília
PACIFIC OCEAN
Paraguay R.
PARAGUAY
Asunción
Rio de Janeiro
São Paulo
Tropic of Capricorn
Paraná R.
CHILE
ARGENTINA
Cerro Aconcagua 22,831 ft (6,959 m)
Santiago
Buenos Aires
URUGUAY
Montevideo
ATLANTIC OCEAN
0 500 1,000 Miles
0 500 1,000 Kilometers
N S E W
Stanley
FALKLAND ISLANDS (U.K.) (ISLAS MALVINAS) Administered by the United Kingdom (claimed by Argentina)
South Georgia (U.K.)
90°W 80°W 60°W 50°W 40°W
10°N 0° 10°S 20°S 30°S 40°S 50°S
100°W 90°W 80°W 70°W 60°W 50°W 40°W 30°W 20°W
A B C D E F G H
2 3 4 5 6

1.1 Physical Geography

SOUTH AMERICA PHYSICAL

Main Idea South America contains diverse physical features.

South America's physical features vary widely. The continent contains the dramatic Andes Mountains, the massive Amazon River Basin, and wide open grasslands and plains.

High Mountains, Large Basin

The **Andes Mountains** are a string of mountain ranges that stretch about 5,500 miles along the western side of the continent. Many of the mountains in the Andes rise higher than 20,000 feet above sea level. The climate in the Andes is generally cool and dry. Because of low temperatures and high elevation, few types of **vegetation**, or plants, grow here.

The **Amazon River Basin** is the largest river basin on Earth. It covers nearly 2,700,000 square miles in the north-central part of South America—nearly the width of the continent. The river drains this basin, flowing from the Andes to the Atlantic Ocean. The climate is warm and wet. Many different species of plants and animals thrive in the Amazon Basin.

Northern Grasslands, Southern Plains

The northern part of South America has a warm climate, and it contains both low and high elevations. Cattle ranching dominates the **Llanos**, or **grasslands**—wide open areas used for grazing and crops. The Guiana Highlands in the north boast unusual plants and animals. The highest waterfall in the world—**Angel Falls**—is located in Venezuela.

In much of the southern part of the continent, the climate is mild and the elevation is low. The rich soil of the grassy plain in Argentina called the **Pampas** is ideal for growing alfalfa, corn, and wheat.

Before You Move On

Monitor Comprehension What are the main physical features in South America?

ONGOING ASSESSMENT

MAP LAB

GeoJournal

1. **Location** According to the physical map, what is the elevation of Cerro (or Mount) Aconcagua, and where is it located?
2. **Interpret Maps** Look at the two maps. What is the difference in climate between the Andes and the Amazon Basin? Which climate supports more diverse vegetation?
3. **Draw Conclusions** Based on climate and elevation, which area of South America is best for growing crops and why?

1.2 Life at Different Elevations

TECHTREK
myNGconnect.com For online maps and photos of different altitude zones

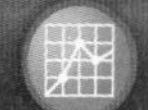 Maps and Graphs  Digital Library

Main Idea Elevation and climate affect where people live and how they use the land.

The people of South America must **adapt**, or modify, their economic activities to fit the different elevations and climates of the region. Life can vary greatly from one part of South America to another.

Elevation and Climate

People, animals, and plants adapt to a range of climates across the continent. More people live in the mild climates of the plains and grasslands than live in the extreme climates of high mountains and tropical rain forests. The low-elevation plains, such as the Llanos, the Pampas, and the Coastal Plain, are moderate, both in temperature and rainfall.

SOUTH AMERICAN ALTITUDE ZONES

Altitude Zone	Climate	Elevation Range	Crops
Tierra Caliente	hot; adequate to extreme rainfall	0 to 2,500 feet	bananas, peppers, sugarcane, cacao
Tierra Templada	warm; adequate rainfall	2,500 to 6,000 feet	corn, beans, wheat, coffee, vegetables
Tierra Fría	cool; some rainfall	6,000 to 12,000 feet	wheat, barley, potatoes
Tierra Helada	cold; little rainfall	12,000 to 15,000 feet	no substantial crops

Source: H.J. deBlij, *The World Today: Concepts and Regions in Geography*. Hoboken, NJ: John Wiley & Sons, 2009.

Critical Viewing At an elevation of 11,800 feet, La Paz, Bolivia, is the world's highest capital city. What details in this photo show how people have adapted to living in a high-elevation urban environment?

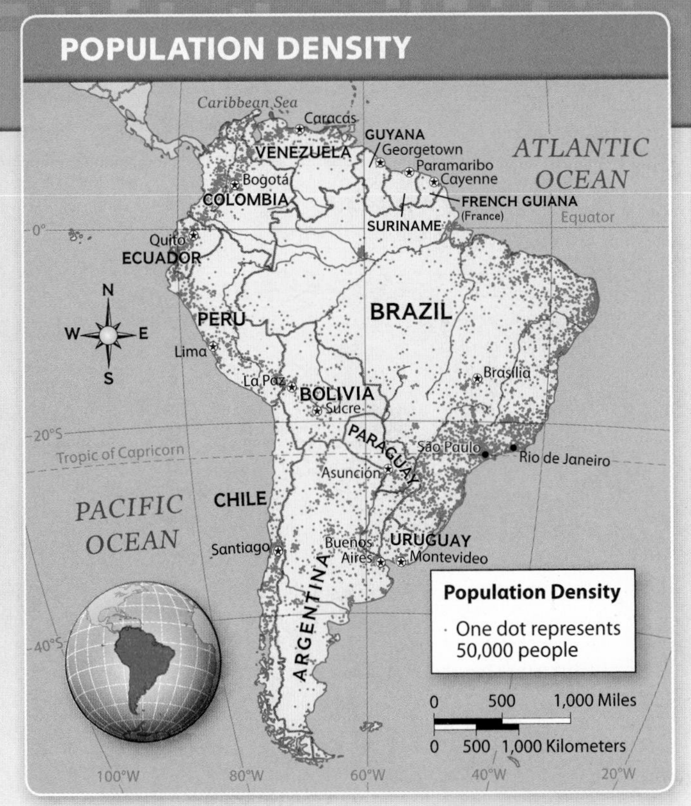

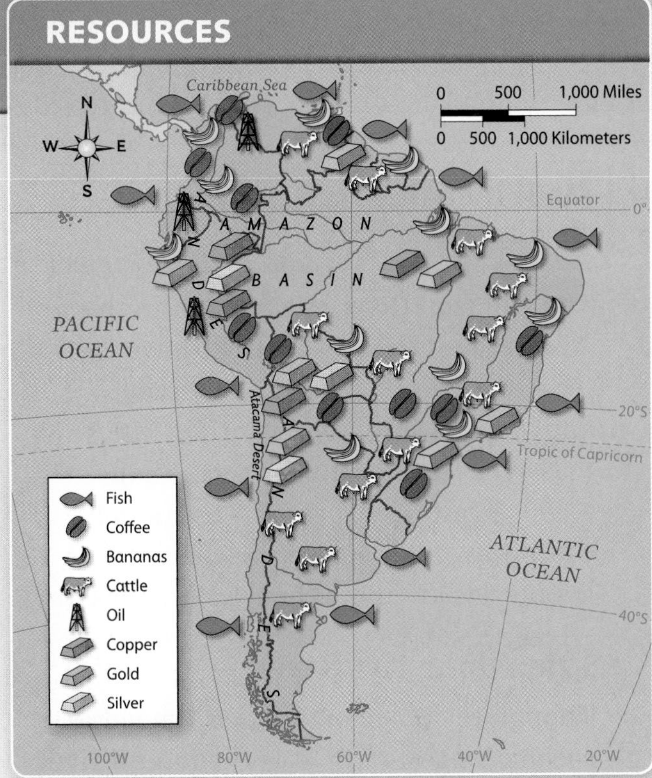

They are located in the *tierra templada* (tee EHR ah tem PLAH dah) elevation zone. The cool, dry elevations of the Andes are called *tierra fría* (FREE ah), or cold land.

Further up in the Andes, a very high and cold elevation range is called *tierra helada* (he LAH dah), or frozen land. At the other extreme, the hot and humid Amazon River Basin is *tierra caliente* (kay YEN tay), or hot land.

Using the Land

Elevation and climate determine how people use the land. Rain is scarce at high elevations in the mountains and highlands of the continent. Some farmers in the Andes herd animals and grow only enough food for their families. This is called **subsistence farming**. However, others have become part of the global economy, selling wool to European and other manufacturers.

On the plains, higher rainfall provides opportunities for ranching and profitable, large-scale farming. Crops produced at these lower elevations in the region include tropical fruits, sugarcane, coffee, corn, wheat, and soybeans.

Before You Move On

Make Inferences Why do people settle in areas with moderate climates and low elevations?

ONGOING ASSESSMENT

DATA LAB

GeoJournal

1. **Interpret Charts** According to the chart, in what altitude zone do potatoes grow? Which altitude zone experiences the most rainfall?
2. **Draw Conclusions** Based on the chart, why do no substantial crops grow in the *tierra helada* zone?
3. **Human-Environment Interaction** According to the population density map, where do most people in South America live? Use the resources map to help you explain why those areas have the most population.

1.3 The Amazon River

TECHTREK
myNGconnect.com For an online map and photos of river basin life

 Maps and Graphs

 Digital Library

Main Idea The Amazon River supports life in its vast rain forest.

The Amazon River begins high in the Andes Mountains of Peru. It flows east for 4,000 miles across the continent and empties into the Atlantic. This massive river system has more than 1,000 **tributaries**, or small rivers that drain into a larger river. Seven of these tributaries are more than 1,000 miles long.

Life Along the River

Though the Amazon River is the world's second longest river, it is the largest river in the world by volume. The Amazon creates the largest river basin—2,700,000 square miles—and includes the world's largest tropical rain forest. The Amazon rain forest is home to thousands of plant and animal species and millions of insect species. The variety of species in an ecosystem is called its **biodiversity**.

Annual flooding of the river from melting snow and rain in the Andes usually occurs between June and October. The floodwaters deposit rich nutrients, or substances that support life, in the lowland forest soils. These nutrients support the biodiversity of the rain forest.

Today, a growing population puts pressure on rain forest lands through mining, logging, farming, and land development. Conservationists suggest that setting limits on development will help protect the Amazon rain forest.

Before You Move On

Summarize In what ways does the Amazon River support life in its rain forest?

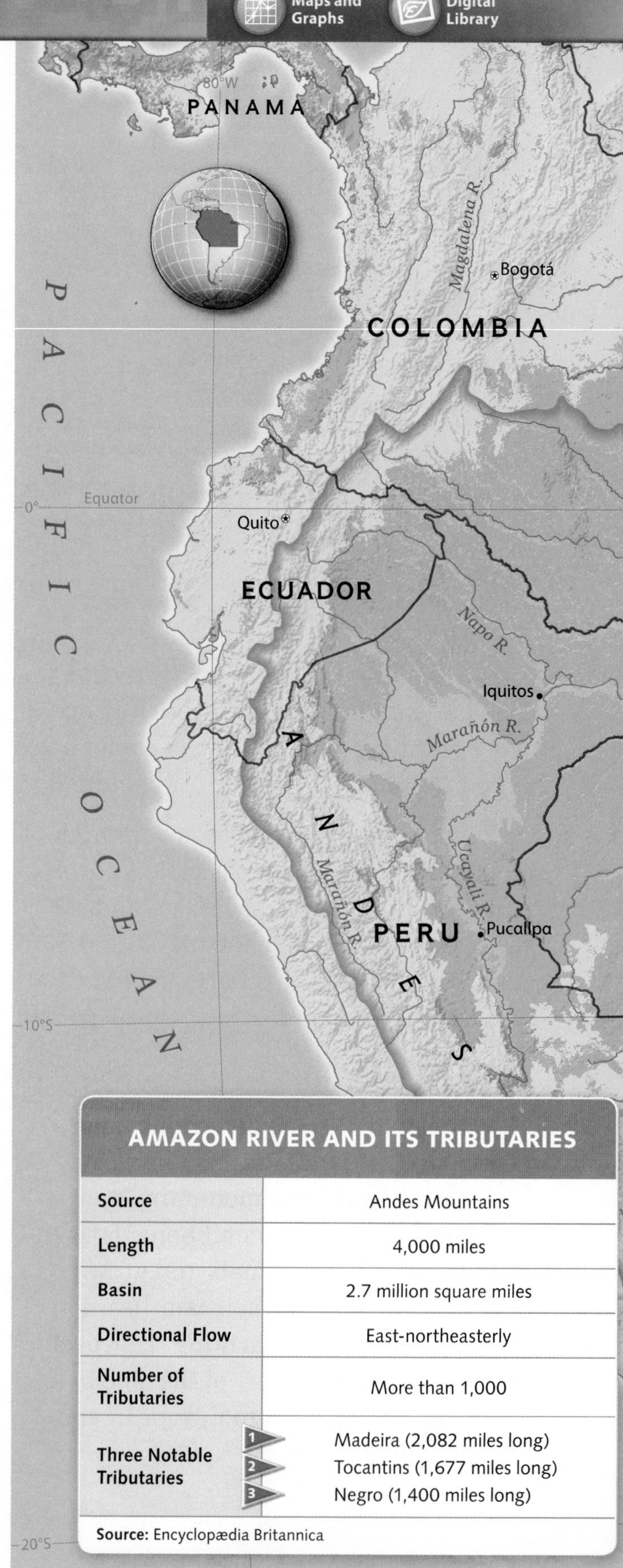

AMAZON RIVER AND ITS TRIBUTARIES	
Source	Andes Mountains
Length	4,000 miles
Basin	2.7 million square miles
Directional Flow	East-northeasterly
Number of Tributaries	More than 1,000
Three Notable Tributaries	1 Madeira (2,082 miles long) 2 Tocantins (1,677 miles long) 3 Negro (1,400 miles long)

Source: Encyclopædia Britannica

ONGOING ASSESSMENT
MAP LAB

GeoJournal

1. **Location** Locate the Purus River and Marañón rivers on the map. Use the map legend to describe the land that surrounds each river.
2. **Interpret Maps** What are three cities located on the Amazon River? Which tributary flows into Manaus?
3. **Monitor Comprehension** How is the biodiversity of the Amazon River and rain forest threatened by a growing population?

1.4 Cold and Warm Currents

TECHTREK
myNGconnect.com For current maps and photos of Peru, Chile, and Brazil

Maps and Graphs

Digital Library

Main Idea Wind currents and ocean currents influence climate across South America in powerful and unpredictable ways.

Like elevation, wind currents and ocean currents influence climate too. **Currents** are the continuous movement of air or water in the same direction. As you read about these currents, follow their patterns on the maps at right.

Currents and Climate

Cold wind and ocean currents flow from the high latitudes near the South Pole toward the equator, making the west coast of South America generally cool and dry. Warm wind and ocean currents flow in the other direction, from the equator toward the South Pole, and create a warm and humid climate on the east coast.

The Peru Current brings cold waters to the Pacific coast in the west. It flows along the southern coast of Chile and northward along the coast of Peru. The Peru Current carries nutrient-rich waters from deep in the Pacific Ocean, so fish thrive off the coasts of Chile, Peru, and Ecuador.

On the eastern side of the continent, the Brazil Current brings warm waters from the Atlantic. The coasts and inland areas of Brazil and Argentina receive warm, humid wind currents and, in some areas, plenty of rainfall. This rainfall nourishes crops and vegetation. The plains of the Pampas benefit from the Brazil Current.

These moist wind currents do not reach the **Atacama Desert**, located on the western side of the Andes. This desert lies in a **rain shadow**, a dry region on one side of a mountain range. In South America, the Andes prevent moist Atlantic winds from reaching west of the mountains. Instead, moisture condenses into rain on the eastern side of the Andes. So, even though it lies along one of the world's largest bodies of water, the Atacama Desert is one of the driest places in the world. On average, only a half inch of rain falls each year.

Critical Viewing Vicunas live in the Atacama Desert in Chile. From what you can see in the photo, how would you describe their habitat?

Critical Viewing This lush sugarcane plantation in Argentina contrasts with the dry Atacama Desert. What details in the photo tell you that this plantation does not lie in a rain shadow?

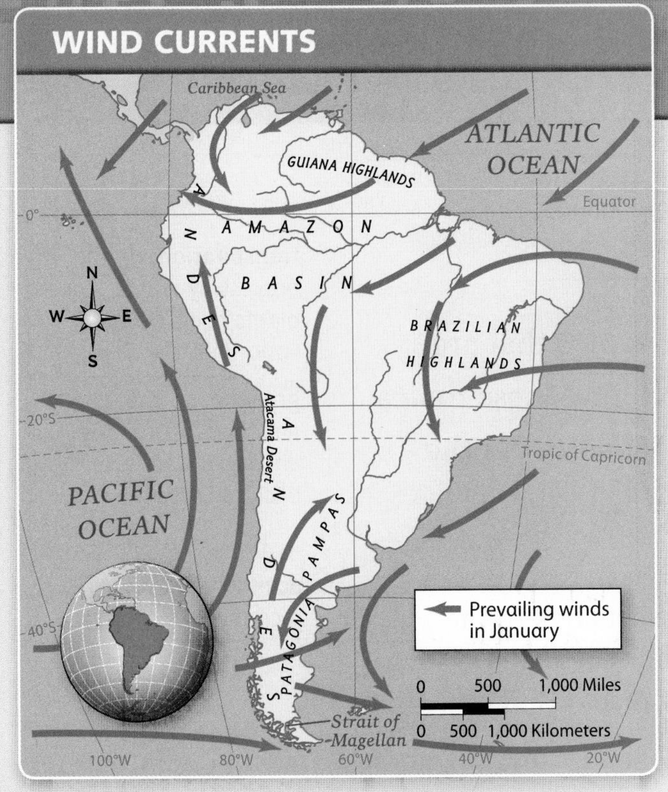

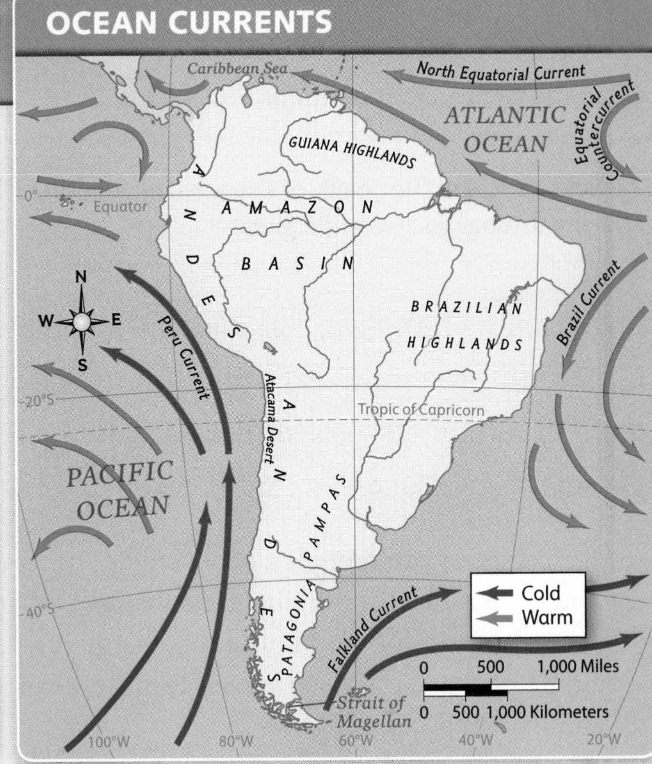

El Niños on the Pacific Coast

El Niños influence climate on the western coast of South America. An **El Niño** (ehl NEEN yoh) happens when the usual wind and ocean currents reverse. This reversal brings warm air currents and water currents that produce high rainfall in coastal areas. El Niños occur in Peru because of its location in the current system and on the Pacific coast.

El Niños do not occur every year, but they do happen somewhat regularly, at least once every 12 years. They are difficult to predict and their results can be devastating. Heavy rains **saturate**, or soak, coastal areas. These rains cause severe flooding and wipe out habitats. They can also cause damage to crops, canals, bridges, and roads. Scientists try to predict when an El Niño is likely to arrive. This helps prepare local populations to deal with the impacts of heavy rainfall.

Before You Move On

Monitor Comprehension How do wind and ocean currents influence climate in South America?

ONGOING ASSESSMENT

MAP LAB

1. **Interpret Maps** Locate the Atacama Desert on the wind currents map. Do wind currents from the Pacific flow over the desert? How might this contribute to its lack of rain?
2. **Describe Geographic Information** Scientists use complex maps like these to predict weather patterns and climate changes. In what ways is the information on these maps similar, and in what ways is it different?

1.5 Rain Forests and Climate Change

TECHTREK
myNGconnect.com For a graph on deforestation and photos of the rain forest

Maps and Graphs

Digital Library

Main Idea The health of the Amazon rain forest can influence climate change across the globe.

KEY VOCABULARY

transpiration, n., the process by which plants release water vapor into the air

greenhouse gas, n., a gas that absorbs heat energy and reflects it back to Earth

ACADEMIC VOCABULARY

acknowledge, v., to recognize

Climate change and global warming are topics that are often in the news. In order to understand how these topics are connected to rain forests, it helps to understand the scientific processes that keep rain forests healthy.

Rain forests are complex ecosystems. As you have read, thousands of different animals, birds, and insects live in the Amazon rain forest. Many unusual plants grow there, too. Some of these plants are used to fight deadly diseases like malaria and cancer.

How Rain Forests Work

Through a process called **transpiration**, plants and trees release water vapor into the air. As the water vapor rises, it cools, forming thick clouds that then produce rain. The warm air and moist soil support the growth of vegetation. In fact, trees grow so high and thick that sometimes sunlight does not reach the forest floor.

The plants in the Amazon rain forest perform an important function. They absorb greenhouse gases from Earth's atmosphere. A **greenhouse gas** is a gas that traps heat energy, warming the earth. Burning fossil fuels such as coal and oil produces greenhouse gases like carbon dioxide.

Too much carbon dioxide in the air causes the atmosphere to warm because it reflects heat energy back to Earth. Rain forest plants and trees naturally absorb this carbon dioxide from the air. In this way, rain forests help clean Earth's air.

Before You Move On

Summarize Why are rain forests important?

DEFORESTATION IN BRAZIL, 2000–2009

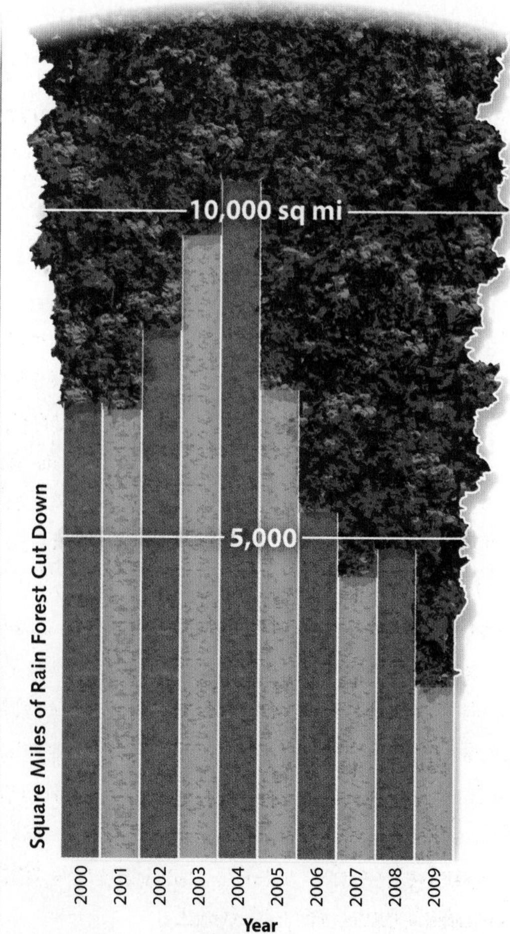

Source: National Institute for Space Research, Brazil, 2010

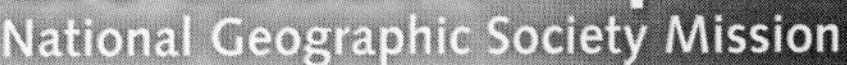

Changes in the Rain Forest

Since the mid-20th century, large sections of the Amazon rain forest have been cut down. In *National Geographic*, journalist Scott Wallace reported that 20 percent of the rain forest has been lost in the last 40 years. (January 2007)

Widespread deforestation threatens the biodiversity of the Amazon. As more of the rain forest is lost, fewer trees and plants remain to produce moisture and cloud cover and to remove greenhouse gases from the air. These changes are having a global impact.

COMPARE ACROSS REGIONS

A Global Climate Challenge

Changes in the Amazon rain forest are party responsible for changes in climate identified by scientists. Climate change is a gradual shift in Earth's climate due to natural causes. In its history, Earth has undergone many climate changes.

Global warming, on the other hand, is a term used by some scientists to describe the rapid warming of Earth's surface observed over the last century. They argue that the heavy use of fuels that release carbon dioxide contributes to rising temperatures. Since the beginning of the 20th century, the average temperature of the globe has risen 1.4°F. While that number might seem small, scientists **acknowledge**, or recognize, that a change of even one or two degrees is cause for concern. Rapid changes in average temperature can destroy habitats and change ecosystems.

Deforestation is underway in the Amazon rain forest near Rondonia State, Brazil.

Protecting the Amazon and other rain forests around the world is critical to the health of the planet. Many countries in South America are working to protect the Amazon rain forest. Across the Atlantic, countries in the African rain forest of the Congo River Basin face similar challenges. Leaders of Cameroon, Democratic Republic of Congo, Guinea, and Ghana are cooperating to manage sustainable forestry operations and to protect easily-damaged wildlife habitats.

Before You Move On

Monitor Comprehension What problems threaten the health of the Amazon rain forest?

ONGOING ASSESSMENT

READING LAB

1. **Summarize** What is the difference between climate change and global warming?
2. **Describe** In what ways do rain forests clean Earth's air?
3. **Interpret Graphs** According to the graph, what has been the general trend in deforestation since 2004? What do you think caused this trend?

2.1 The Inca

TECHTREK

myNGconnect.com For an online empire map and photos of Inca civilization

Maps and Graphs

Digital Library

Main Idea The Inca ruled a vast empire in a difficult, mountainous environment.

The Inca Empire in South America stretched along the Pacific coast. The empire included parts of present-day Colombia, Ecuador, Peru, Bolivia, Chile, and Argentina. From 1438 until the Spanish conquest in the 1530s, the Inca Empire was one of South America's largest and most advanced civilizations.

Workings of the Empire

The Inca built the capital of their empire in Cuzco, in what is now Peru. Inca government and society were highly organized. The emperor—called the Sapa Inca—was considered a **descendant**, or a relative of, the sun god, Inti. At its high point, the empire included 80 provinces and about 12 million people.

Within Inca society, families were organized into groups based on **kinship**, or blood relationship, and common land ownership. Inca married within their kinship groups. They also worked together and shared land and resources. The leader of each family worked for the empire for a few months of the year as builders, farmers, craftsmen, or foot soldiers.

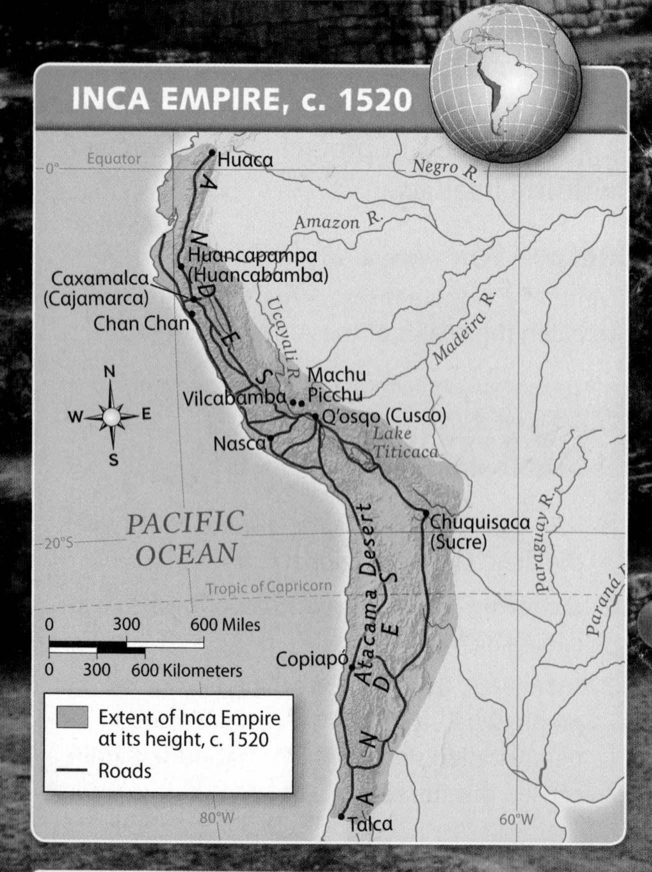

Critical Viewing The ancient ruins of Machu Picchu are located in Peru, at an elevation of 7,710 feet. Based on what is shown in the photo, what do you think might have been the challenges of building a city at high elevations?

Achievements of the Inca

The Inca **utilized**, or made practical use of, their advanced engineering skills to adapt to the mountainous environment. For example, the Inca farmed on **terraced** fields, or flat fields cut into slopes or mountainsides. They also built irrigation canals to water their crops because the climate was arid.

The Inca adapted to their mountainous surroundings in other ways. They built suspension bridges using vines and wood. A **suspension bridge** is a bridge used to cross canyons or water. The Inca also built a system of roads that helped keep the empire unified.

Another example of the Inca's engineering skills is **Machu Picchu** (MAH choo PEE choo), built in the 1400s. The Inca built this complex city on a mountain by constructing giant walls, terraces, sloping ramps, and steep stairways.

Visual Vocabulary **Suspension bridges** are bridges used to cross canyons or water. This rebuilt suspension bridge in Peru is modeled after the bridges the Inca built.

Some archaeologists believe it served as a royal estate. Today it is a UNESCO World Heritage Site because of its historical and archaeological significance.

By the 1530s, the empire faced internal problems that included a weak economy and civil war. A much smaller but better-equipped Spanish force conquered the Inca in 1532. Today, the descendants of the Inca, the Quechua, live in the Andes of Peru, Ecuador, and Bolivia.

Before You Move On

Make Inferences In what ways did the Inca maintain control of their empire?

ONGOING ASSESSMENT

PHOTO LAB

GeoJournal

1. **Analyze Visuals** Based on the photos and text, in what ways did the Inca utilize advanced engineering skills in their empire?
2. **Describe** How would you describe the Inca Empire based on the photo of Machu Picchu?
3. **Monitor Comprehension** What factors made it possible for a small force of Spanish conquerors to overcome the large Inca Empire?

TECHTREK
myNGconnect.com For photos of an Explorer at work and an Explorer Video Clip
Digital Library

Exploring Nasca Culture

with Christina Conlee

Main Idea Artifacts and other archaeological discoveries provide insight into the Nasca culture.

Archaeologists study how people lived in the past. They examine **artifacts**, or objects left behind by past cultures. NG Expeditions Council Grantee Christina Conlee is an archaeologist who studies the Nasca of Peru. The Nasca lived on a desert plateau in southern Peru nearly 2,000 years ago. They left behind beautiful ceramic artifacts as well as the Nasca lines, which are a series of **geoglyphs**, or large, geometric designs and animal shapes drawn on the ground. The Nasca were one of the first complex societies in South America.

A Surprising Find

In 2004, Conlee and her team were **excavating**, or carefully uncovering, what they thought was a house in La Tiza, Peru. She was surprised to unearth, instead, an ancient burial site. Between 2004 and 2006, Conlee and her team excavated nine burial sites. Artifacts collected at the sites included elaborate ceramics and copper objects. Conlee believes that the kinds of artifacts she found belonged to individuals with a high social rank in the community.

Some of the artifacts uncovered were not typical of traditional Nasca ceramics. The presence of copper objects, shell ornaments, and more elaborate tomb paintings suggested a different population group. Were they local Nasca or people from another culture?

Ancient Migrations

Conlee was determined to find the answer. Through chemical analysis she performed on the bones, Conlee determined that the remains did indeed belong to the Nasca. However, her analysis also proved that some remains belonged to a group of Nasca rivals called the Wari.

The Wari were more powerful than the Nasca. Conlee believes when they moved into Nasca areas, they may have caused some Nasca to move away. Ultimately, the migration of the Wari may even have led to the end of Nasca culture. Conlee's studies and discoveries continue to reveal new insights about Nasca culture and their migration patterns.

Before You Move On

Make Inferences What did archaeological discoveries reveal about ancient Nasca culture?

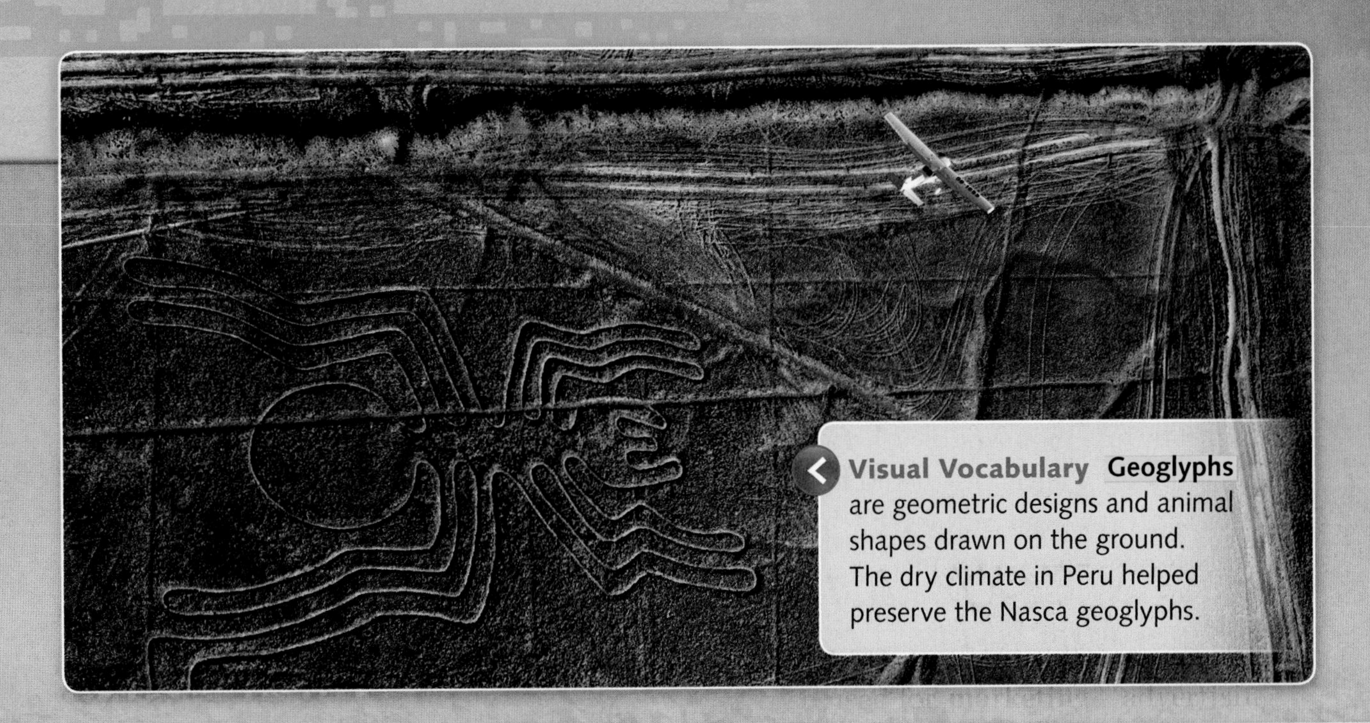

Visual Vocabulary Geoglyphs are geometric designs and animal shapes drawn on the ground. The dry climate in Peru helped preserve the Nasca geoglyphs.

Christina Conlee and fellow archaeologist Aldo Noriega excavate a site in La Tiza, Peru.

ONGOING ASSESSMENT

VIEWING LAB

GeoJournal

1. **Analyze Visuals** Go to the **Digital Library** for an Explorer Video Clip about high-altitude archaeologist Johan Reinhard. What type of discoveries has he made?
2. **Describe** How do archaeologists such as Christina Conlee and Johan Reinhard learn about ancient cultures?

National Geographic Explorer-in-Residence Johan Reinhard

2.3 People of the Lowlands

TECHTREK
myNGconnect.com For an online map and photos of the Yanomami

Maps and Graphs

Digital Library

Main Idea People have lived in the lowlands of South America for thousands of years.

People who lived in South America began to settle in low-lying areas of the continent more than five thousand years ago. The development of agriculture and a stable food supply allowed for larger groups of people to live there together.

Lowland River Basins

The **lowland**, or low-lying, areas of South America include several fertile river basins. The lowlands include the Orinoco River and surrounding grasslands, the Amazon River Basin, and the Paraguay River Basin.

These rivers and basins contained abundant animal life and vegetation to support ancient people who lived there. For example, the Amazon River Basin was an ample source of fish. The Paraguay Basin, which includes Gran Chaco, a floodplain, provided fertile soils for growing food.

Developing Agriculture

At first, lowland people lived as **nomads**, or people who move from place to place. As **hunter-gatherers**, or people who hunted animals and gathered plants and fruits for food, they moved to different locations as food became scarce.

As early as 3000 B.C., people began to farm the land and build villages in the lowlands. Establishing a stable food supply encouraged different groups to settle in villages rather than continue to move from place to place.

The **Guaraní** (GWAH rah NEE) people lived in the eastern and central lowlands on the Paraguay and Paraná Rivers. Before planting crops, the Guaraní cleared the land by cutting down and burning existing forest and vegetation. This agricultural technique is called **slash-and-burn**. Typically, Guaraní women grew corn and root vegetables such as sweet potatoes and cassava. In addition to farming, Guaraní men hunted and fished.

SLASH-AND-BURN AGRICULTURE

1 Slash Wooded areas and jungles are too thick to plant crops. Farmers slash, or cut down, trees.

2 Burn Fallen trees and foliage are burned to clear the land. Ash produced by the fires is used as fertilizer.

3 Fertilize and Plant Cleared land is fertilized with ash. Crops such as corn and sweet potatoes are planted.

Another group, the **Tupinambá** (too pee NAAM baa), settled near the mouth of the Amazon and southward along the Atlantic coast. They also cultivated crops using slash-and-burn agriculture. Because they lived in the Amazon River Basin and near the ocean, the Tupinambá also fished and hunted river mammals and turtles.

The Yanomami

Today, indigenous groups continue to live in rain forests throughout the Amazon River Basin. The **Yanomami** (yaa noh MAA mee) are still hunter-gatherers who use the slash-and-burn technique to clear the land for farming. They live in villages but migrate to different areas to meet agricultural needs. Many have moved to the north-central lowlands, where they have found more fertile land.

Before You Move On

Make Inferences How did the physical features of the lowlands shape the lives of the people who lived there?

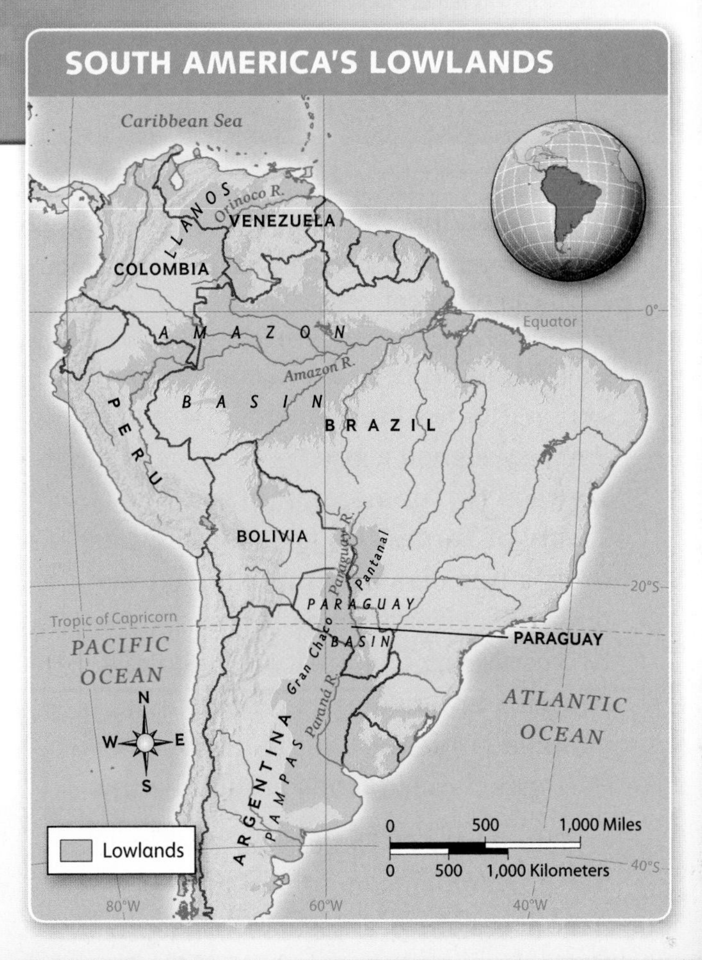

4 Migrate Groups move on to new locations after soil on cleared lands becomes less productive.

ONGOING ASSESSMENT

SPEAKING LAB

GeoJournal

1. **Turn and Talk** Turn to a partner and describe the process of slash-and-burn agriculture in the lowlands. Why did lowland groups engage in slash-and-burn agriculture?
2. **Create Charts** With a partner, make a chart comparing several aspects of nomadic and settled lifestyles. Discuss the characteristics of each lifestyle and what the advantages of each might be.

	NOMADIC	SETTLED
Movement	1. Move frequently	
Food Sources	2.	

3. **Location** In what parts of the continent did the Guaraní and Tupinambá live? Where do the Yanomami live? Use the text to help you locate these areas on the map.

2.4 The Spanish in South America

TECHTREK

myNGconnect.com For an online map of Spanish settlement and photos of artifacts

Maps and Graphs

Digital Library

Main Idea The arrival of the Spanish in the 1500s shaped the history and culture of the South American continent.

In 1494, in order to avoid conflicts over exploration and settlement, Spain and Portugal signed a **treaty**, or an agreement between two or more countries. The **Treaty of Tordesillas** (tor duh SEE uhs) drew a line on a map that divided the newly discovered lands between the two countries. The treaty divided South America into two parts. The Spanish claimed lands west of the line, and Portugal claimed lands east of the line.

The Treaty of Tordesillas set the stage for the Spanish conquest of most of South America. Four years later, Christopher Columbus, sailing on behalf of Spain, landed on South America's northern coast.

Critical Viewing *The Spaniard and the Inca Chief* by James McConnell. What stories does this painting tell about the Spanish and the Inca?

The Spanish Conquest

As they had in Mexico, Spanish conquistadors arrived in South America determined to expand Spain's empire and search for resources such as gold and silver. In 1533, **Francisco Pizarro**, with a small army of men, overthrew the Inca emperor, **Atahualpa** (AH tah WAHL pah). Pizarro founded the city of Lima, Peru, which became the center of the Spanish government and empire in South America. Other conquistadors explored and conquered Colombia, on the northern coast, and most of Chile, on the western coast. (See the map opposite.) The Spanish conquest of South America permanently **transformed**, or changed, much of the continent and its people.

This gold mask is one example of the wealth the Spanish sought in South America.

Gold statues discovered in the mountains of Argentina illustrate the extent of the Inca Empire.

1400

1438–1533
Inca Empire rules vast area of South America.

1450

1494
Spain and Portugal sign the Treaty of Tordesillas.

1498–1500
Columbus's third voyage reaches northern coast of South America.

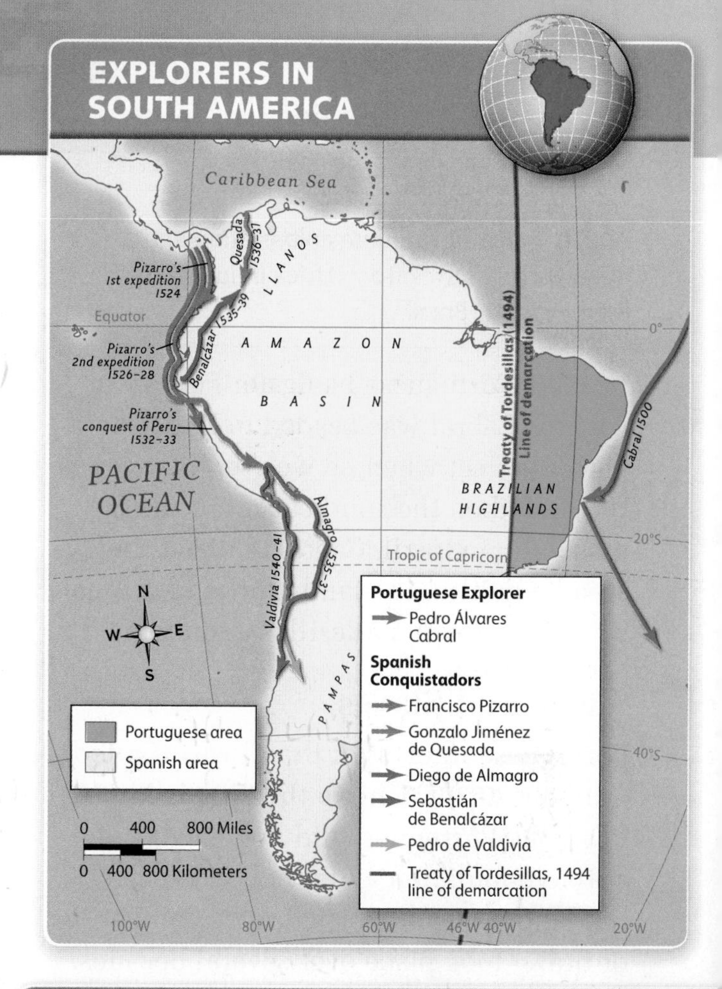

Impact on Native Populations

Deadly diseases that traveled with the Spanish to South America wiped out entire villages and native populations. Because these groups had no resistance to diseases such as smallpox, measles, and influenza, many died quickly.

The Spanish enslaved natives and forced them to work on plantations and ranches and in mines. Large numbers of enslaved native people died from the effects of harsh labor conditions.

Missionaries who arrived after the 1550s viewed South America as an opportunity to spread Christianity. The goal was to **convert**, or persuade native populations to change their religious beliefs. Some conversions were forced. Many native people began to practice the Catholic faith, and some blended aspects of Christianity with their own religious practices.

Before You Move On

Summarize What impact did the Spanish have on the history and culture of South America?

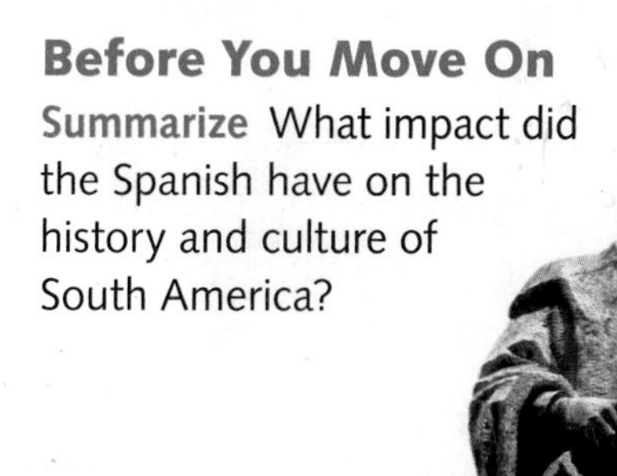

ONGOING ASSESSMENT

MAP LAB

1. **Movement** According to the map, where were Spanish conquistadors most active in the 1520s and 1530s? Who made multiple expeditions?
2. **Evaluate** Locate the line of demarcation on the map. How would you describe the impact of that line on South America?
3. **Interpret Time Lines** About how many years passed between the signing of the Treaty of Tordesillas and the overthrow of the Inca?

Pedro Álvares Cabral, Portuguese navigator, lands on eastern coast of South America in 1500.

1500 — **1550**

- **1524** Pizarro reaches northwestern coast of South America.
- **1532** Spanish conquer Inca Empire.
- **1535** Spanish establish Lima, Peru, as center of empire in South America.
- **1541** Spanish establish Santiago, Chile.

2.5 Brazil and the Slave Trade

Main Idea Portuguese colonization and the arrival of slaves from Africa influenced the history of Brazil.

In 1500, Portuguese navigator Pedro Álvares Cabral was headed to India with his fleet when he went off course. He landed on the southeastern coast of present-day Brazil. Cabral realized the area lay within the land allowed Portugal by the Treaty of Tordesillas and claimed it.

Sugar and Slaves

Portuguese interest in Brazil was limited until the 1530s. Unlike the Spanish in Peru, the Portuguese did not conquer the native population and quickly take over the land. Instead, Portuguese colonization of Brazil took place over several decades.

Portuguese settlers discovered that these new lands contained natural resources valued in European markets. First, the Portuguese exported brazilwood, which was sought for its red color used to dye fabric. Then colonists realized that sugarcane, which grew abundantly in Brazil, was a more valuable crop. They built plantations and began exporting sugarcane and sugar products to Europe. The Portuguese tried enslaving natives to work the sugar plantations, but overwork and disease killed many of them. The Portuguese then turned to another source for labor: Africa.

Critical Viewing This 1819 painting shows slaves working on a plantation in Brazil. What details do you notice?

Because of their earlier exploration, the Portuguese knew about the slave markets in Africa. By the mid-1500s, the Portuguese and other European countries were exporting African slaves across the Atlantic to South America and the Caribbean. The Portuguese were able to create a **monopoly**, or complete control, of the slave trade. A continuous trade in slaves began in Brazil in 1560 and lasted well into the 19th century.

This lighthouse still stands on the site of a colonial trading port in Salvador, Brazil.

Portuguese Wealth

Brazil's abundant natural resources, combined with the African slave labor to extract them, made the Portuguese wealthy. As valuable natural resources like gold and diamonds were discovered, Portuguese demand for slave labor increased. Slaves were brought in to work on sugar and coffee plantations and in gold and diamond mines. Portuguese slave traders imported more than five million slaves from Africa to Brazil.

The Portuguese successfully put down **rebellions**, or revolts, against their rule until the early 1800s. Brazil finally declared independence from Portugal in 1822, but not everyone was free. Even though the slave trade had ended in 1850, slavery continued in Brazil for several decades until it was abolished in 1888.

Before You Move On

Monitor Comprehension What influence did the slave trade have on Brazil's history?

1826 Jean Baptiste-Debret depicts slaves carrying sacks of coffee in Brazil.

PORTUGUESE CLAIMS IN BRAZIL

Portuguese Territory
- 1600
- 1654
- 1750
- Treaty of Tordesillas, 1494 line of demarcation
- Present-day boundary
- Administrative capital

0 300 600 Miles
0 300 600 Kilometers

ONGOING ASSESSMENT

READING LAB

GeoJournal

1. **Movement** Why did the Portuguese import slaves from Africa?
2. **Summarize** In what way did slaves contribute to Portuguese wealth?
3. **Interpret Maps** According to the map, where in South America did the Portuguese settle? What is the pattern of their settlement?
4. **Interpret Time Lines** How long did slavery exist in Brazil? What year did slavery end? Use the time line to form your answer.

1750

1725 Diamonds are discovered in Minas Gerais.

1822 Brazil declares its independence from Portugal.

1840s Coffee becomes Brazil's top export.

1850

1888 The Golden Law abolishes slavery in Brazil.

2.6 Simón Bolívar on Independence

TECHTREK

myNGconnect.com For paintings of revolutionary leaders and Guided Writing

Digital Library | Student eEdition | Student Resources

Most of South America was ruled by the Spanish for more than 300 years. Making the **transition**, or change, from colonial rule to independence was difficult. Simón Bolívar led the revolution against the Spanish. Bolívar was born in 1783 to a wealthy family in Caracas, Venezuela. Both of his parents died when he was a child. After their death, his uncle made sure that he received an education and sent Bolívar to Europe. There, Bolívar learned new ideas about freedom and government. In 1810, he joined the independence movement in Venezuela. Present-day Bolivia was named in his honor.

DOCUMENT 1

The Letter from Jamaica (1815)

Bolívar wrote this letter in Jamaica while living in **exile**, or a state of absence from one's home country, during Venezuela's struggle for independence. Here he describes the need for revolution and independence from Spain.

> The hatred we feel for the Peninsula [Spain] is greater than the sea separating us from it; it would be easier to bring the two continents together than to reconcile [unite] the spirits and minds of the two countries.

CONSTRUCTED RESPONSE

1. What sea is Simón Bolívar referring to in this passage?
2. What two countries in particular is Bolívar discussing?
3. How would you describe the overall tone of Bolívar's words in this passage?

DOCUMENT 2

The Angostura Address (1819)

After returning from exile in Jamaica, Bolívar delivered this speech in Venezuela in 1819.

> The continuation of power in the same individual has frequently led to the demise [downfall] of democratic government. . . . A just zeal [enthusiasm for a cause] is the guarantee of republican freedom, and our citizens should properly fear that the same ruler who has long ruled them will wish to rule them forever.

CONSTRUCTED RESPONSE

4. According to Bolívar, what should citizens fear most in a ruler?
5. What does Bolívar argue leads to the "demise of democratic government"?
6. How does a "just zeal" guarantee freedom, according to Bolívar?

A JUST ZEAL IS THE GUARANTEE OF REPUBLICAN FREEDOM.

— SIMÓN BOLÍVAR

Source: Tito Salas, *Retrato Ecuestre del Libertador,* 1936.

DOCUMENT 3 

Simón Bolívar, "The Liberator"

To **liberate** means to set someone or something free. Simón Bolívar earned the name "the Liberator" for his brave efforts against the Spanish during the struggle for independence. Bolívar is also often referred to as the "George Washington of South America."

CONSTRUCTED RESPONSE

7. In 1936, the Venezuelan government commissioned artist Tito Salas to paint this work for the National Pantheon, a monument built to honor national heroes. What details did Salas include to present Bolívar as a liberator?

ONGOING ASSESSMENT

WRITING LAB

GeoJournal

DBQ Practice By the early 1800s, many countries in South America began to press for independence from Spain. In what ways did Bolívar lead Venezuela's fight for independence?

Step 1. Review the colonial period of South America in Sections 2.4 and 2.5.

Step 2. On your own paper, take notes about the main ideas expressed in each document in this lesson.

Document 1: The Letter from Jamaica
Main Idea(s) ____________

Document 2: The Angostura Address
Main Idea(s) ____________

Document 3: Painting of Bolívar
Main Idea(s) ____________

Step 3. Construct a topic sentence that answers this question: Why did Bolívar want to liberate South America from Spanish rule?

Step 4. Write a detailed paragraph that explains why Bolívar wanted self-rule and freedom for South America. Go to **Student Resources** for Guided Writing support.

Review

VOCABULARY

On your paper, write the vocabulary word that completes each of the following sentences.

1. People who lived in the South American lowlands practiced __________ agriculture.
2. The Portuguese had a __________ on the slave trade.
3. Archaeologists __________ artifacts to learn about past cultures.
4. Over 1,000 __________ feed into the Amazon River.
5. A __________ is an agreement between two or more countries.

MAIN IDEAS

6. What are two extreme physical features on the South American continent? (Section 1.1)
7. What two factors help determine where people settle in South America? (Section 1.2)
8. In what ways does the Amazon River nourish the rain forest? (Section 1.3)
9. What unpredictable weather event impacts Peru and what are its effects? (Section 1.4)
10. How do rain forests protect the health of the planet? (Section 1.5)
11. How did the Inca adapt to their mountain environment? (Section 2.1)
12. In what ways have archaeologists contributed to an understanding of Nasca culture? (Section 2.2)
13. What natural resources helped the people of the lowlands thrive? (Section 2.3)
14. How did the arrival of the Spanish impact native populations? (Section 2.4)
15. In what ways did slave labor benefit the colonial Portuguese economy? (Section 2.5)
16. Why was Simón Bolívar called "the Liberator"? (Section 2.6)

GEOGRAPHY

ANALYZE THE ESSENTIAL QUESTION

How does elevation influence climate in South America?

Critical Thinking: Compare and Contrast

17. Compare and contrast climate and elevation in the Andes and the Amazon River Basin.
18. Compare and contrast these two altitude zones: *tierra templada* and *tierra fría*.
19. Compare the ways in which currents and landforms influence climate in Chile.

INTERPRET MAPS

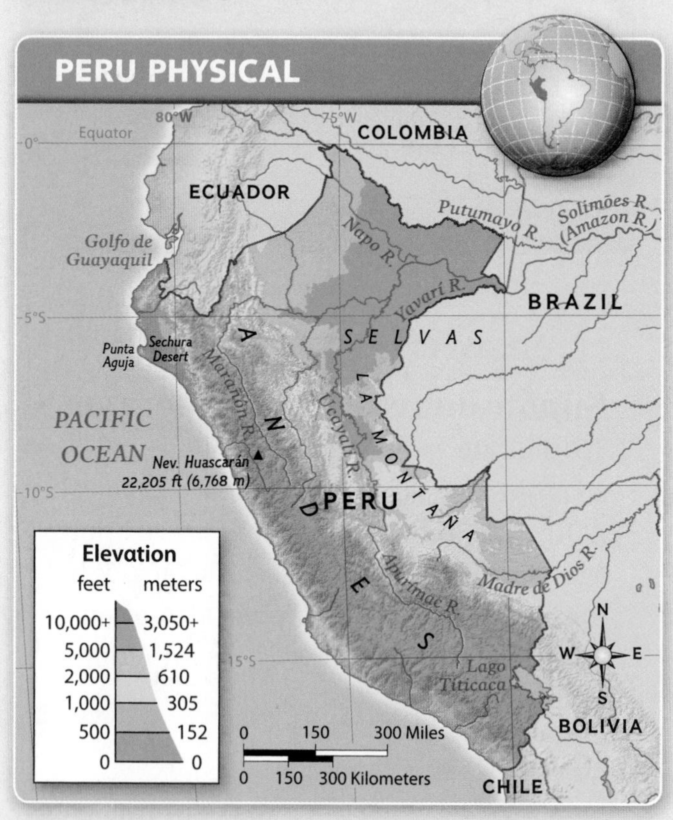

20. **Identify** What major mountain range lies in Peru, and what map element can help you determine its location?
21. **Describe Geographic Information** Where does Peru's largest area of lowlands lie in relation to the equator? Use the word "latitude" in your answer.

HISTORY

ANALYZE THE ESSENTIAL QUESTION

How did mountains, plateaus, and rivers shape the region's history?

Critical Thinking: Draw Conclusions

22. In what ways does Machu Picchu demonstrate the Inca's engineering skills?
23. Describe what archaeological evidence left behind on the desert plateau reveals about migration patterns of the Nasca culture.
24. How did people of the lowlands change from a nomadic to a settled way of life, and what role did the region's abundant rivers play in that change?
25. Which mountain resources in South America encouraged Spanish exploration, conquest, and colonization?

INTERPRET PRIMARY SOURCES

Portuguese historian Pero de Magalhães Gândavo wrote *The Histories of Brazil* in 1576. Read his description of the valuable resources found in Brazil, and then answer the questions below.

> Certain Indians arrived in the Captaincy of Porto Seguro... giving news of the existence of green stones in a mountain range many leagues inland; and they were emeralds... and there were many other mountains of blue earth in which they [Indians] assured them there was much gold.

26. **Identify** According to this passage, what important minerals were found inland?
27. **Make Inferences** Based on your reading, how did the discovery of valuable resources in Brazil influence decisions about importing slaves from Africa?

ACTIVE OPTIONS

Synthesize the Essential Questions by completing the activities below.

28. **Write Journal Entries** Describe South America from the perspective of a traveler in the region. Select two countries. In your journal, write two entries on each country: one on climate and elevation, and another on physical features, such as mountains and rivers. Use the following tips to help you write your journal entries. **Share your journal entries with a classmate or friend.**

Writing Tips
- Take notes before you begin to write.
- Write an outline to help you organize details for the two countries you select.
- Include as many details as possible.
- Write in first-person narrative.

29. **Create Charts** Work in a group to make a four-column comparison chart that compares Brazil, Argentina, Venezuela, and Ecuador. Use the research links at **Connect to NG** and other online sources to gather the data. Compare your group's data with data found by other groups. What sources did you find most useful in your research?

	Brazil	Argentina	Venezuela	Ecuador
Climate				
Elevation				
Population				
Main Crops				
Year of Independence				

CHAPTER 8

SOUTH AMERICA TODAY

PREVIEW THE CHAPTER

Essential Question In what ways is South America culturally diverse?

SECTION 1 • CULTURE

KEY VOCABULARY

- mestizo
- roots
- language family
- topography
- immigrate
- cuisine

ACADEMIC VOCABULARY

predominant

TERMS & NAMES

- Aymara
- Quechua
- Guaraní
- Mundurukú
- Creole
- Candomblé

Essential Question How is modern South America building its economies?

SECTION 2 • GOVERNMENT & ECONOMICS

KEY VOCABULARY

- prosperous
- coup
- soybean
- fertilizer
- temperate
- Mediterranean climate
- export revenue
- profitable

ACADEMIC VOCABULARY

ruthless, diversify, erratic

TERMS & NAMES

- Dry Pampas
- Wet Pampas

Essential Question How has Brazil become an economic power?

SECTION 3 • FOCUS ON BRAZIL

KEY VOCABULARY

- steel
- ethanol
- biofuel
- megacity
- slum
- infrastructure
- venue

ACADEMIC VOCABULARY

foremost, impact

TERMS & NAMES

- São Paulo
- Rio de Janeiro

TECHTREK FOR THIS CHAPTER

Student eEdition

Maps and Graphs

Interactive Whiteboard GeoActivities

Digital Library

Connect to NG

Go to myNGconnect.com for more on South America.

This cathedral in Brazil's capital city reflects the country's efforts to modernize.

1.1 Indigenous Cultures

Main Idea Indigenous cultures in South America maintain traditions in a modern world.

Indigenous groups have lived in South America for thousands of years. Their descendants still live and work in the region today.

European Contact

As you have learned, the arrival of the Spanish, the Portuguese, and other Europeans during the colonial period changed life for the indigenous people in South America. Unfamiliar diseases and warfare with Europeans killed many native people and substantially reduced their populations.

European contact also introduced a new population group: the mestizo. Many people in South America are **mestizo**, or of mixed European and native ancestry. Some mestizos also have African **roots**, or cultural origins, because of the large numbers of slaves that were imported in the colonial period.

Critical Viewing Potato farmers work near the Aymara community of San Jose, Peru. Based on what you see in the photo, what is farming like for the Aymara?

Maintaining Traditions

The Aymara, Quechua, and Guaraní are the three largest indigenous groups in South America today. The **Aymara** (eye MAHR uh) live in the Andes of Peru and Bolivia. Today, the Aymara continue some of the traditions of their ancestors, such as speaking their native language, also called Aymara. The Aymara also continue to herd llamas and alpacas and grow crops such as potatoes and quinoa (KEEN wah), a grain that grows well in the mountains.

The **Quechua** (KEHCH wah) live in the Andes of Peru, Ecuador, and Bolivia. Like the Aymara, many Quechua farmers live in isolated mountain villages, far away from modern cities and lifestyles. Their religious practices are a blend of Catholicism and native beliefs. The Quechua have maintained traditions such as weaving and speaking their native language, Quechua.

The **Guaraní** (gwah rah NEE) live in Paraguay and are the main indigenous group in that country. Most people in Paraguay trace their roots to both Guaraní and Spanish ancestors. The Guaraní culture is represented in Paraguay's folk art and the Guaraní language.

The **Mundurukú** (moon doo ROO koo), another important native group, live in Brazil. Their ancestors farmed, hunted, and fished in the rain forest. Although they live in relative isolation, one way the Mundurukú adapt to the modern world is by selling forest products such as latex, a liquid substance harvested from native rubber trees.

INDIGENOUS POPULATIONS OF SOUTH AMERICA (2007)

Country	National Population	Indigenous Population	Indigenous Percentage
Argentina	33,900,000	372,996	1.10
Bolivia	8,200,000	4,142,187	50.51
Brazil	155,300,000	254,453	0.16
Chile	14,000,000	989,745	7.07
Colombia	35,600,000	620,052	1.74
Ecuador	10,600,000	2,634,494	24.85
French Guiana	104,000	4,100	3.94
Guyana	806,000	45,500	5.65
Paraguay	4,800,000	94,456	1.97
Peru	22,900,000	8,793,295	38.40
Suriname	437,000	14,600	3.34
Venezuela	21,300,000	315,815	1.48

Source: International Union for Conservation of Nature, 2007

Before You Move On

Monitor Comprehension What traditions do indigenous groups maintain today?

ONGOING ASSESSMENT

DATA LAB

1. **Analyze Data** According to the chart, in what countries do the highest and lowest numbers of indigenous people live? In what ways might low percentages impact indigenous populations?
2. **Make Inferences** Based on the chart and text, what percentage of people in Paraguay are indigenous? What other way might they identify themselves?
3. **Explain** What does "mestizo" mean? In what way does this cultural group reflect South America's colonial past?

1.2 Language Diversity

TECHTREK
myNGconnect.com For a map of indigenous languages and research links
Maps and Graphs
Connect to NG

Main Idea South America has a rich diversity of languages.

As you know, South America is a land of varied cultures. It is also a land of diverse languages. The various languages that South Americans speak come from at least 50 **language families**, or groups of related languages.

European Influences

The **predominant**, or main, languages in South America are Spanish and Portuguese. Other European languages spoken include French and Italian. Spanish, Portuguese, French, and Italian are Romance languages, or languages that come from Latin. During the 1800s, the land extending from Mexico through South America became known as Latin America. The widespread presence of the Romance languages helped to define Latin America as a culture region.

European presence in South America also brought about the development of Creole. **Creole** is a language that results when two other languages are combined together. During the colonial period, plantation workers created a common language that was a blend of European and non-European languages. In this way, groups who spoke different first languages found a way to communicate. Some people living on the Atlantic and Caribbean coasts of South America still speak Creole today.

Although European influence on languages in South America is strong, indigenous languages continue to thrive.

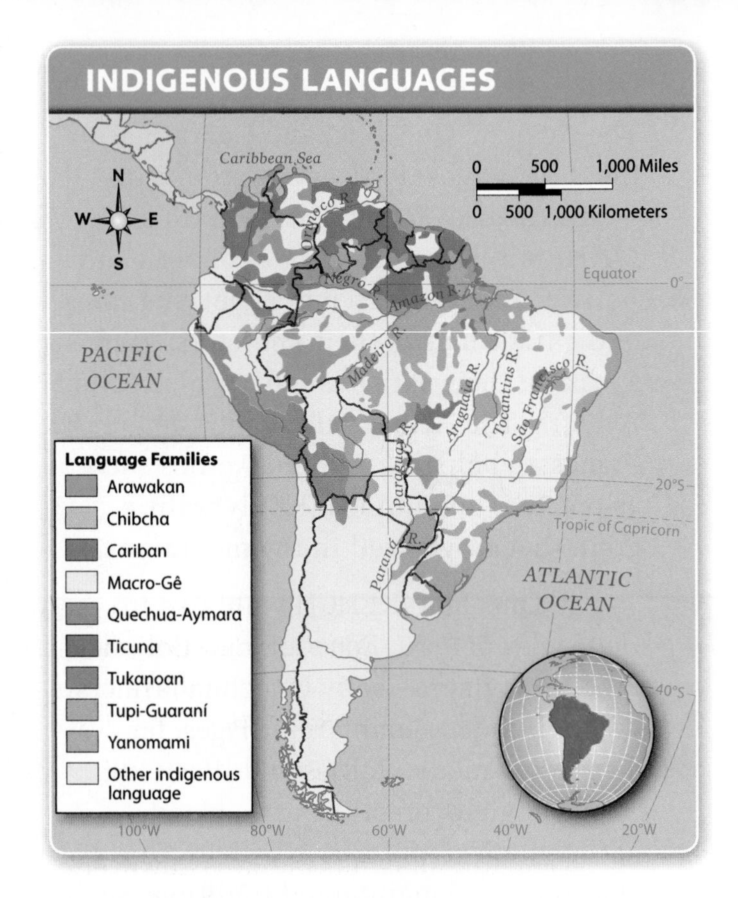

The large number of indigenous groups on the continent is one reason why more than 500 languages are spoken today. A second reason is because of the physical features, or **topography**, of the land and the density of its jungles. The mountain ranges and rain forests isolated indigenous groups and kept their languages separate from European languages.

Indigenous Languages

In some countries, indigenous languages hold official status because of their widespread use. For example, along with Spanish, Quechua is considered an official language in Peru. In Paraguay, where Guaraní and Spanish are both official languages, more people speak and understand Guaraní than Spanish.

Critical Viewing Spanish is the official language of Colombia, but more than 80 indigenous languages are also spoken there. In what ways might language be an important part of this celebration?

Unlike Quecha and Guaraní, however, some indigenous languages in South America are spoken only by small groups of people in remote parts of the continent. Several of these languages—such as Kallawaya in Bolivia, Sáliva in Colombia, and Maka in Paraguay—are endangered, or in danger of vanishing completely.

National Geographic's Enduring Voices Project is working toward preserving these and other endangered languages. Enduring Voices works in language-diverse places, such as South America, to document and record indigenous languages. As this project captures the sounds and words of these languages, it is also helping to preserve the histories, songs, and stories of the cultures that speak them.

Before You Move On

Summarize What factors have contributed to South America's language diversity?

ONGOING ASSESSMENT

LANGUAGE LAB

1. **Location** How did the topography of the continent help preserve indigenous languages?
2. **Draw Conclusions** Why do you think people who speak endangered languages such as Kallawaya and Maka would be willing to work with the Enduring Voices Project?
3. **Interpret Maps** Based on the map, which indigenous language families are represented along the Tocantins River?
4. **Monitor Comprehension** What do you think might happen to some of the indigenous languages that are spoken by a small number of South Americans? Why?

1.3 Daily Life

TECHTREK
myNGconnect.com For photos of daily life and research links
Digital Library
Connect to NG

Main Idea South Americans' daily activities reflect aspects of their varied cultures.

Like most people around the world, South Americans' daily lives revolve around how they worship and celebrate, how they learn and play, and what they eat.

Religious Practices

As you have read, Roman Catholicism became the main religion in South America during the colonial period. Spanish and Portuguese colonists were Roman Catholic, and they converted many native people to Catholicism. Roughly 80 percent of South Americans are Roman Catholic, with Brazil having the world's largest Catholic population.

After the Spanish and Portuguese, other Europeans **immigrated**, or moved permanently, to South America. Some of them were Protestants, or Christians who separated from the Catholic church. Today, most Protestants in the region live in Chile and the Guianas and in parts of Brazil, Bolivia, and Ecuador.

Other religious practices exist alongside Christianity in South America. In Brazil, a local religion called **Candomblé** combines African spiritual practices with Catholicism. The annual Carnival festival in Brazil mixes Roman Catholic practices with traditional African celebrations.

School and Sports

Every country in South America provides public education. However, some children struggle to receive adequate schooling. These children may live in rural areas with few schools and may not attend school regularly. Other children may leave school at an early age to help earn money for their families. Though some barriers to education exist, the majority of people on the continent are able to read and write.

Critical Viewing A teacher conducts class in Peru. How would you describe the students, their classroom, and what they are doing?

Playing *futbol*—known in the United States as soccer—or watching a favorite team compete is a popular pastime in South America. Brazilian soccer teams frequently compete in, and often win, the World Cup, a global soccer tournament that occurs once every four years.

Regional Food

South American **cuisine**, or the food that is characteristic of a particular place, varies in different parts of the continent. In coastal countries like Chile, seafood is often a main ingredient, such as in a stew called *paila marina* (pie yuh mah REE nah). In Uruguay and Argentina, where there is a good deal of grazing land, beef is often served. In the mountains of Peru, llamas provide a good source of meat. Stews with black beans, or *feijoadas* (fay ho AH daz), rice, and vegetables are common throughout South America.

Before You Move On

Monitor Comprehension What daily activities reflect the various cultural elements of South America?

Critical Viewing This market is located in Santiago, Chile. Based on this photo, what kind of food can you infer might be important in Chilean cuisine?

ONGOING ASSESSMENT

WRITING LAB

GeoJournal

Describe Culture Imagine you and your family are living in a South American country for a short time. Write a blog entry describing daily life during your stay.

Step 1 Outline your blog entry. Write a topic sentence for each paragraph you plan to write.

Step 2 Use the research links at **Connect to NG** to research your country of choice. Include aspects of culture mentioned in the text.

Step 3 Look for photos that might illustrate different aspects of daily life in that country.

Step 4 Share your blog entry with classmates.

2.1 Comparing Governments

myNGconnect.com For photos of government buildings in South America

Digital Library

Main Idea The governments of South America are moving toward democracy and strengthening their economies.

In the early 1800s, movements for independence in countries across South America brought an end to colonial rule. However, in many countries, real power remained in the hands of a few wealthy families. South American governments have gone through many changes since independence. Argentina, Peru, and Chile provide three examples of the challenges South American governments have struggled to overcome.

Argentina

Argentina has faced economic and political challenges. After independence in 1816, dictators held power for several decades. In the 1850s, the country adopted a new constitution and entered into a **prosperous**, or economically strong, period that lasted until the late 1920s. In 1930, military officers staged a **coup**, or takeover, of the government. After another coup in 1943, Colonel Juan Domingo Perón gained support among workers.

In 1946, Perón was elected president. He raised wages for workers and established social and economic programs. Many of Perón's reforms were expensive, and corruption weakened his administration. In the 1950s, he was overthrown by military leaders who were dissatisfied with his leadership.

For decades, Argentina sought to move forward. Since the 1980s, democratically elected presidents have faced serious economic crises. Today, though, Argentina's constitutional government is stable, and, since recovering from a serious financial crisis in the early 2000s, its economy is one of the strongest in South America.

Peru

Peru gained independence in 1821. For most of its history, Peru's government has shifted between democratically elected presidents and military rule. Many leaders—even those elected democratically—have favored wealthy landowners over ordinary citizens. Political instability made economic and social progress difficult in Peru.

In 2001, Peruvians elected their first Quechua president, Alejandro Toledo. Toledo had the support of the Quechua people and a large majority of Peru's poor population. His election demonstrated that Peru's government could represent all of its citizens.

Critical Viewing President Alejandro Toledo greets children in Peru. What does this photo lead you to think about Toledo's approach to leadership?

Chile

Chile declared independence in 1818. Since then, Chile has mostly been a representative democracy. Like many other countries in South America, though, Chile has experienced rule by a dictator. In 1973, Salvador Allende's government was overthrown by the military. General Augusto Pinochet (peen oh SHAY) acted as a dictator in Chile for nearly two decades, from 1973 to 1990. Pinochet was **ruthless**, or cruel, and no one was allowed to disagree with his policies.

Chile returned to democratic rule in 1990. In 2006, the people of Chile elected its first female president, Michelle Bachelet Jeria. Bachelet's election was especially meaningful. Her father was killed during Pinochet's rule, and she and her mother were both imprisoned and exiled because they opposed Pinochet. As president, Bachelet helped ease poverty, expanded social reforms, and used profits from copper exports to create new employment opportunities.

Before You Move On

Summarize In what ways are Argentina, Peru, and Chile alike and different in their movement toward democracy and strong economies?

ONGOING ASSESSMENT

SPEAKING LAB

GeoJournal

1. **Turn and Talk** Turn to two other classmates and compare the similar challenges facing the governments of Argentina, Peru, and Chile since independence. Take notes while you compare. Include important dates and people. Ask each of your partners a question based on your discussion.
2. **Identify** Why was Michelle Bachelet's election significant? With a partner, discuss how her election was shaped by her family history.

Michelle Bachelet Jeria arrives at the presidential palace after her inauguration in Chile in 2006.

2.2 The Pampas Economy

TECHTREK
myNGconnect.com For a map of land use and photos of agriculture in Argentina

Maps and Graphs

Digital Library

Main Idea The Pampas is a fertile region that contributes to the economic success of Argentina.

The wide, grassy Pampas is South America's major crop-growing region. The Pampas is also the grazing ground for another valuable export—cattle.

An Agricultural Heart

The Pampas is a large plain that stretches across central Argentina from the Atlantic Ocean to the foothills of the Andes Mountains. The Pampas covers about 295,000 square miles in the northern half of the country, nearly one-quarter of Argentina's land area.

During the colonial period in the 1800s, the Spanish imported horses and cattle to the region. Spanish and mestizo cowboys, called gauchos, herded cattle and sheep on the Pampas, a practice continued by modern gauchos.

Cattle grazing takes place in the region's dry zone, or the **Dry Pampas**, located in the west. The humid zone, or the **Wet Pampas**, is in the east. Agriculture abounds in the Wet Pampas, where nearly 40 inches of rain falls on average each year. Together, the cattle industry and agriculture contribute greatly to the region's economic prosperity.

In the 1980s, a new crop became profitable in Argentina: **soybeans**, a type of bean grown for food and also for industrial products such as plastics, inks, and adhesives. Soybeans grow quickly and unlike crops such as wheat and corn, do not need as much **fertilizer**, a substance added to soil to enrich it. In response to demand, Argentines increased soybean production. Low production costs and high prices in global markets make soybeans a valuable crop in Argentina.

Critical Viewing This gaucho is tending sheep on the Pampas in Argentina. Based on the photo, how would you describe the work of a gaucho?

1 DRY PAMPAS Argentines are among the world's top consumers of beef. Beef consumption per capita in Argentina is 119 pounds per year compared to 82 pounds per year in the United States.

2 SOYBEANS These soybeans are grown in the Wet Pampas. Soybeans are used to make vegetable oil and livestock feed. Soybeans can also be cooked and eaten, and are a good source of protein.

3 WET PAMPAS This patchwork of crops is in the Wet Pampas, Argentina's main agricultural area. Soybeans, wheat, corn, flax, and alfalfa are some of the crops grown here.

ARGENTINA'S LAND USE

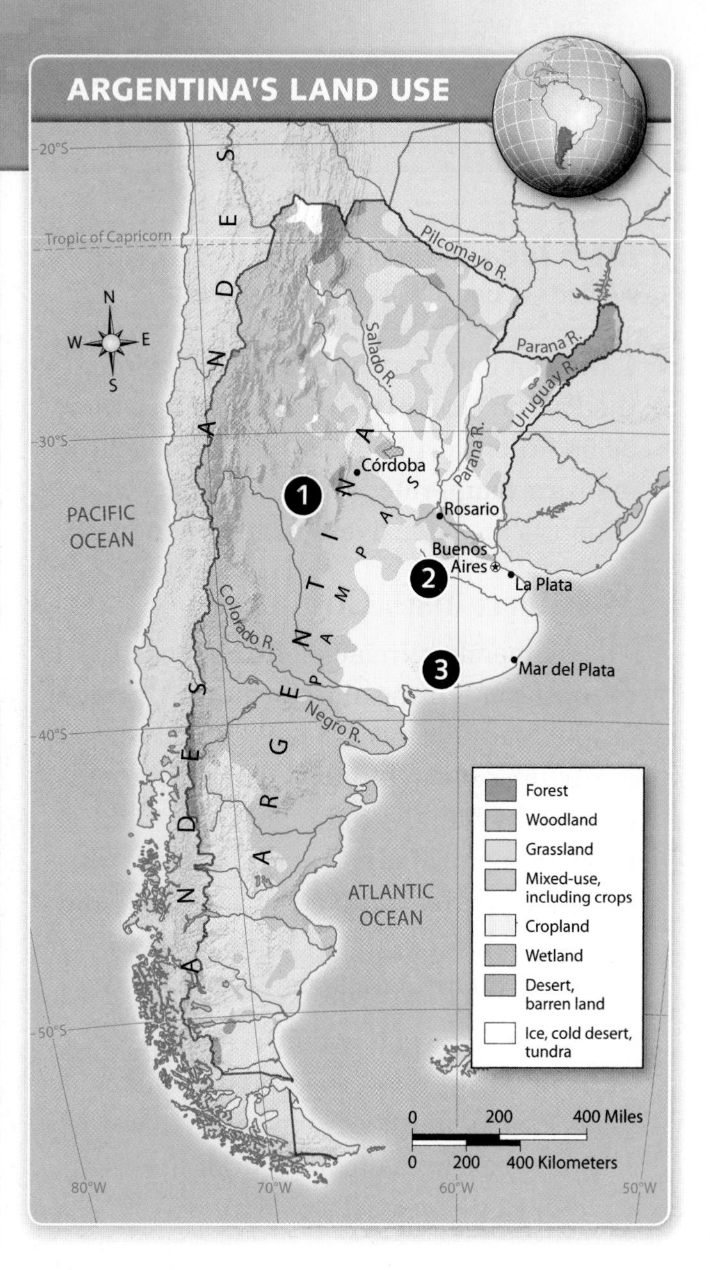

Before You Move On

Summarize How does the rich soil of the Pampas contribute to Argentina's economy?

ONGOING ASSESSMENT

READING LAB

1. **Monitor Comprehension** What crop has become an important export in Argentina, and why is it profitable?
2. **Interpret Maps** How might Argentina's coastal location benefit its agriculture industry?

2.3 Chile's Food Production

TECHTREK
myNGconnect.com For a map of Mediterranean climates and photos of agriculture

Maps and Graphs

Digital Library

Main Idea The mild climate in central Chile supports abundant agricultural exports.

Chile is a long, narrow country. It borders the Pacific Ocean and stretches more than 2,600 miles north and south. Central Chile enjoys a **temperate**, or mild, climate that supports an extensive agriculture industry.

Mediterranean Climates

The temperate climate along the coasts of southern Europe and northern Africa on the Mediterranean Sea is called a **Mediterranean climate**. This climate is defined by hot, dry summers and mild, rainy winters. Mediterranean climates are found in southern Australia, southern and central California in the United States, southern South Africa, and central Chile.

One reason such widely scattered parts of the world have similar climates is because of latitude. Remember that latitude measures distance from the equator in degrees. Places with Mediterranean climates have similar latitudes either north or south of the equator. Generally, these latitudes measure between 30°S–40°S or 30°N–40°N. In addition to similar latitudes, places with Mediterranean climates lie on western coasts and have similar precipitation patterns because of their coastal positions.

Chile's Agricultural Bounty

Because of Chile's location on the western coast of South America at nearly 30°S, farmers in central Chile can grow a rich mix of crops. Fruits such as grapes, peaches, and apples are grown for export in Chile's fertile valleys.

Two crops in particular show how Chile takes full advantage of its Mediterranean climate: grapes and olives. A thriving wine industry depends on the many varieties of grapes that grow in Chile's vineyards. Chile has also become a major exporter of olives and specialty

Critical Viewing Chile exports 40 percent of its fruit to the United States. What details in this photo tell you these peaches are imported from Chile?

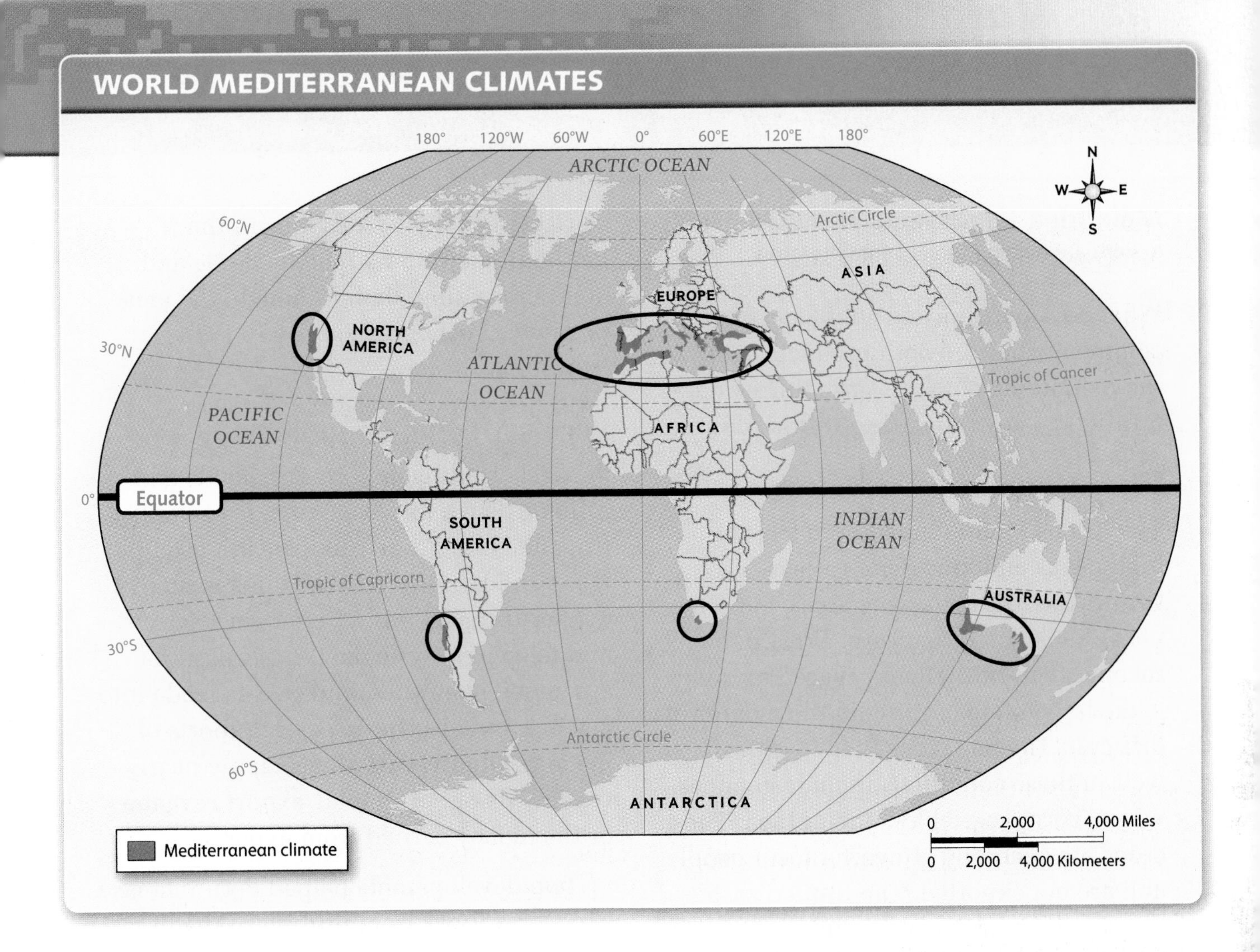

olive oils. Chilean wine and olive oil exports increased dramatically in the 1990s and early 2000s. Both wine and olive products are exports traditionally dominated by Mediterranean countries such as Italy, France, and Greece.

While agricultural production is a source of growth for Chile's economy, copper remains its most valuable export. The growth of agricultural exports since the late 1980s helped Chile **diversify**, or vary, its economy. Diverse economies depend on multiple industries—such as mining and agriculture—and tend to be stronger and more competitive.

Before You Move On

Summarize What main agricultural exports are supported by Chile's Mediterranean climate?

ONGOING ASSESSMENT

MAP LAB

1. **Interpret Maps** Look at the grid on the map of Mediterranean climates. Which Mediterranean climates are located south of the equator? At what latitude do they lie?
2. **Location** According to the map, where is the largest Mediterranean climate region located? Based on your reading, what might the weather be like during the winter months in that region?
3. **Summarize** Summarize the characteristics of a Mediterranean climate. Use the chart below to organize your summary.

MEDITERRANEAN CLIMATES	
1.	
2.	

2.4 Products of Peru

myNGconnect.com For photos of agricultural products of Peru

Digital Library

Main Idea Agriculture and mining industries in Peru are helping the economy to grow.

What do asparagus and gold have in common? They are both products of the mountains of Peru. Agriculture and mining are key industries in this country.

High Mountain Agriculture

The Andes Mountains in Peru feature high peaks and low, steep valleys. The valleys are fertile in some areas, but often not suitable for agriculture. **Erratic**, or inconsistent, rainfall and rugged terrain make large-scale farming in the mountains difficult. The few crops that do grow well in these conditions include potatoes, wheat, corn, vegetables, and quinoa. These crops are generally grown for local people and are not exported. The limited cash crops grown for export include sugarcane, wheat, coffee, and asparagus, a vegetable valued in international markets.

Other economic activities in the Peruvian mountains include cattle and alpaca ranching. By far, though, the most profitable economic activity in Peru is mining.

Peru's Mining Economy

Peru is a leading exporter of metals and minerals such as silver, zinc, lead, copper, tin, and gold. These products are used in many industries worldwide. For example, copper is made into wire for electrical and telephone systems. Lead is used for automobile batteries, and gold is made into jewelry and electronic parts. Exports of metals and minerals alone supply nearly two-thirds of Peru's total **export revenue**, or the money earned from exports.

Two developments helped Peru increase export revenues from mining so that it became more profitable than agriculture. The first was new government policies on mine ownership, which began in

Critical Viewing This Peruvian farmer uses traditional methods to plow a field. What can you infer about the challenges of this type of farming?

Critical Viewing Copper is made into electrical wire and into tubes and pipes used for plumbing. Based on the photo, what can you infer about the importance of this export?

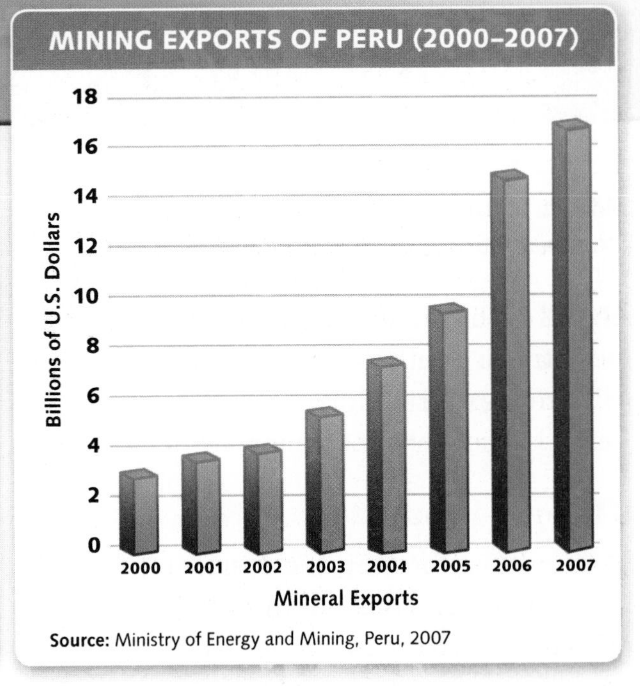

the early 1990s. These policies brought about private ownership of the mines. This shift allowed for investment in needed improvements and upgrades, which led to a rapid growth of the mining industry in Peru. Existing mines became more **profitable**, or financially successful, and new mines opened. Today the Yanacocha gold mine, which opened in Peru in 1993, is the largest gold mine in South America.

Peru's mining revenues also increased because of an upward trend in world prices for silver, gold, and other metals and minerals since the late 1990s. Because Peru is a top producer of these industrial raw materials, the country has become competitive in the global economy.

Before You Move On

Monitor Comprehension What exports have helped Peru's economy to grow?

ONGOING ASSESSMENT

DATA LAB

1. **Identify** Look at the bar graph. What happened to mining exports between the years 2000 and 2007?
2. **Interpret Graphs** What is the difference in mining revenues between 2000 and 2007, measured in dollars? What accounts for that difference?
3. **Summarize** What governmental policy change in the 1990s led to a growth of the mining industry between 2000 and 2007?

3.1 Brazil's Growing Economy

TECHTREK
myNGconnect.com For an online map of Brazil's resources and photos of economic activity

Maps and Graphs

Digital Library

Main Idea Brazil is a leading industrial country and has a strong, diverse economy.

Brazil is the largest and most populous country in South America with many different exports. Brazil is an emerging force in the global economy.

Diverse Products

Agriculture, forestry, ranching, and fishing all contribute to the country's economy. Brazil exports bananas, oranges, mangoes, cacao beans, soybeans, rice, cashew nuts, and pineapples. It is the **foremost**, or leading, global coffee producer and grows one-third of the world's total number of coffee beans. Brazil is also one of the leading exporters of sugarcane and raw sugar.

Forests in Brazil contain many raw materials, including timber, or wood prepared for use in construction. Wood from Brazilian forests is also used to make pulp for paper products. Mahogany, a rain forest hardwood, is a valuable export used to make fine furniture.

Cattle ranching is a big business in Brazil. Export products from cattle include leather and beef. Today, Brazil exports more beef than the United States, Australia, and Argentina.

Critical Viewing Sugarcane has sharp leaves that scratch and cut. Based on what you see in the photo, what protective gear is this cutter wearing?

Brazil's coastline stretches more than 4,600 miles along the Atlantic Ocean. New technology, facilities, and processes will allow Brazil to develop its commercial fishing industry.

Mining and Manufacturing

Mining contributes to Brazil's growing economy. Brazil is a top producer and exporter of iron ore, bauxite (aluminum ore), gold, copper, and diamonds. Oil wells dot Brazil's coast.

In addition to its mineral production, Brazil's manufacturing industry is central to the country's economic strength. Brazil manufactures **steel**, a strong metal made from iron and other metals. Brazilian steel is used in automobiles, transportation equipment, and aircraft. Brazil also manufactures computers and electronic equipment.

Fuel of the Future

Brazil has produced and exported sugarcane and sugar since the 1500s. For several decades, the country has been developing its sugar-based ethanol industry. **Ethanol** is a liquid removed from sugarcane or corn. It is mixed with gasoline to make an alternative fuel called a **biofuel**. Cars that run on biofuels use less gasoline. Because they are made from agricultural products, biofuels are a renewable source of energy. Today, Brazil is poised to be a global leader in the production of biofuel.

Before You Move On

Summarize What factors enable Brazil to develop a diverse economy?

RESOURCES AND INDUSTRIES

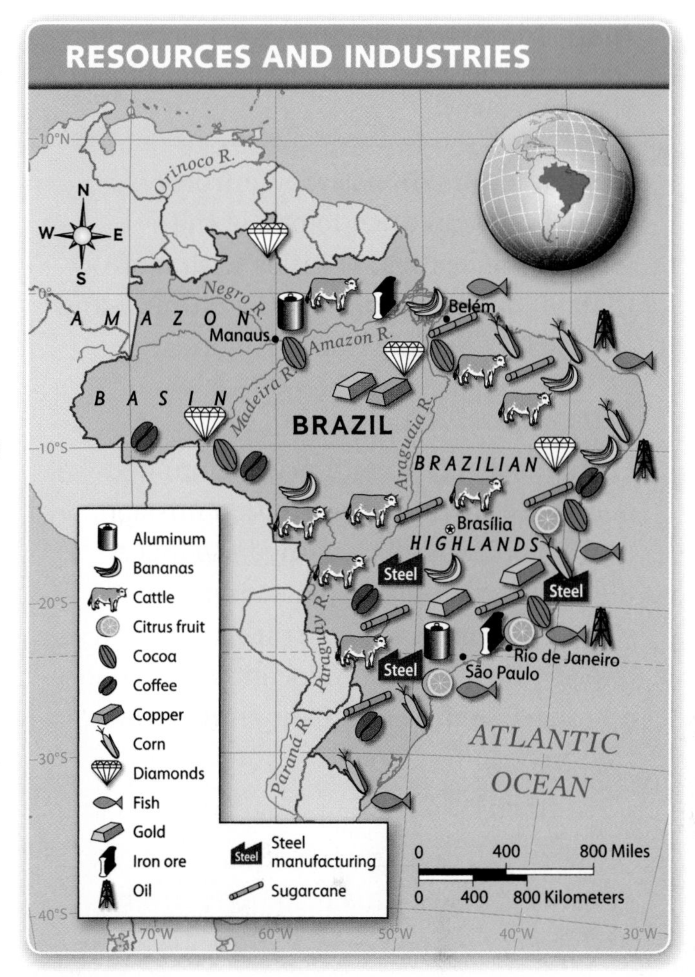

ONGOING ASSESSMENT

MAP LAB

1. **Interpret Maps** Study the map. What are two resources that provide fuel for Brazil? Where are these resources found?
2. **Draw Conclusions** Where are most cattle resources located in Brazil? Why might cattle not be concentrated in the Amazon Basin?
3. **Human-Environment Interaction** In what ways have Brazilians used natural resources to build a strong economy?
4. **Form and Support Opinions** What part of Brazil's economy do you think will be the most important in the future? Why?

3.2 São Paulo

TECHTREK
myNGconnect.com For a graph and photos of São Paulo, Brazil

Maps and Graphs

Digital Library

Main Idea São Paulo is the largest city in the Southern Hemisphere and a major contributor to Brazil's economy.

São Paulo, Brazil, has grown from a quiet mission town to one of the world's **megacities**, or cities with more than 10 million people. It is a cultural and industrial center of South America.

Early Growth

Portuguese missionary priests founded São Paulo in 1554. They built a mission and school, which remained the focus of the town for many years. São Paulo was also a point of departure for military expeditions. Its hilltop location provided a natural defense and panoramic views of the surrounding area.

In the late 1600s, gold was discovered in the nearby mountains of the state of Minas Gerais. This resource proved extremely valuable, and by the mid-1700s, Brazil was producing nearly half of the world's supply of gold. Roughly 50 years later, gold deposits were mostly depleted. However, coffee production soon replaced gold mining as the main economic activity. By the mid-1800s, coffee had become a significant export crop. Wealth gained from coffee production transformed São Paulo and contributed to its rapid growth in industry and population.

In just 20 years between 1880 and 1900, the population of São Paulo jumped from 35,000 to 240,000. Some of the population growth came from rural to urban migration. However, much of the city's growth was the result of immigration from Asia and Europe.

The Modern City

In the mid-1900s, São Paulo became the industrial center of Brazil. By the 1950s, the automotive industry was well

developed. Jobs in the industry attracted workers from other parts of Brazil and other South American countries. São Paulo continues to manufacture and export one million cars each year.

In addition to manufacturing industries, tourism has contributed to the city's economic growth. Today, São Paulo attracts people from countries all over the world. Visitors to São Paulo enjoy the city's beaches, its shopping district, its diverse collection of restaurants, and a bustling nightlife.

One result of the rapid expansion of São Paulo's economy is the explosive population growth since 1950. Many people who migrated there made their homes in **slums**, or overcrowded and poverty-stricken urban areas. Slums, or *favelas*, as they are called in Brazil, developed on the outskirts of the city. Slums are not unique to São Paulo, but are characteristic of large cities that experience rapid population growth.

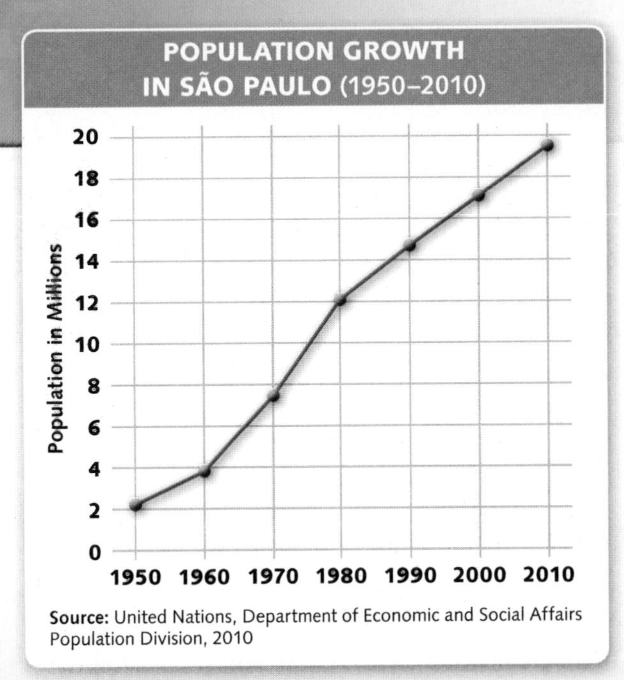

Before You Move On

Summarize What industries are most important in São Paulo today?

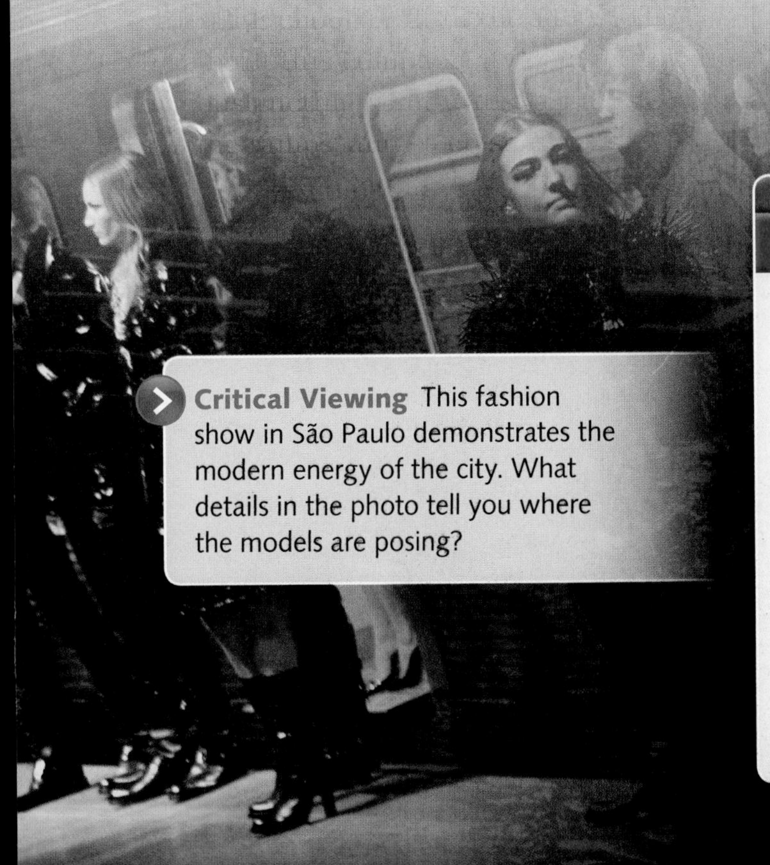

Critical Viewing This fashion show in São Paulo demonstrates the modern energy of the city. What details in the photo tell you where the models are posing?

ONGOING ASSESSMENT

DATA LAB

1. **Interpret Graphs** According to the graph, during what two decades did São Paulo experience the most rapid rate of growth? What accounts for that increase?
2. **Analyze Data** Using the data in the graph, how much larger in millions of people was São Paulo in 2010 than it was in 1950? How would you describe population growth in São Paulo since the 1990s?
3. **Make Inferences** In what ways might explosive population growth and the development of slums be related?

3.3 Impact of the Olympics

myNGconnect.com For photos of Olympic preparations and Guided Writing

Digital Library

Student Resources

Main Idea Brazil is working hard to maximize the economic and social impact of the 2016 Olympic Games.

Athens, Greece, hosted the first modern Olympics in 1896. Since that time, the International Olympic Committee (IOC) has decided which countries would host the games. Not once had the committee chosen a city in South America. That changed in 2009 when the IOC selected **Rio de Janeiro**, Brazil, as the host of the 2016 Olympics.

Preparing the City

Rio has previous experience hosting international sports events. In 2007, Brazil hosted the Pan-American Games, which are similar to the Olympics but open only to countries in the Western Hemisphere. At that time, Brazil's government made improvements to the **infrastructure**, or the basic systems that a society needs, such as roads, bridges, and sewers. However, more improvements are needed to host the Olympics. Thirty-four competition **venues**, or locations for organized events, need to be built or updated. Efficient transportation to and from the venues is also important.

The Olympic Games last only a few weeks. However, many in Rio de Janeiro hope that one **impact**, or effect, will be long-term improvements in the health of their city. Overcrowded and often dangerous favelas surround Rio. As part of its Olympic preparations, the city plans to tear down and rebuild in these locations. Areas that were once slums will become paved neighborhoods with running water, electricity, and gas—all of which are missing in the existing favelas.

Critical Viewing Brazilian soccer star Pélé (far right) celebrates winning the bid with the Rio Olympic committee. Based on this photo, what can you say about the committee's reaction?

Hosting the Games

Benefits come to countries that host the Olympic Games. First, constructing buildings such as stadiums and the Olympic Village (where athletes stay) creates jobs. Ideally, these structures will still be used long after the games are over. Businesses such as hotels and restaurants benefit, too. Athletes and tourists from around the world flood a host city and spend money on lodging, food, and other goods. Billions more watch the Olympics on television. With such wide exposure, Rio de Janeiro hopes to earn a reputation as a thriving and world-class city.

Before You Move On

Make Inferences In what ways might the 2016 Olympics change Rio?

ONGOING ASSESSMENT

PHOTO LAB

1. **Analyze Visuals** Look at the photo of Rio de Janeiro. As a tourist visiting Rio, what impression might you have of the city?
2. **Describe** In what ways might the hills and coastline shown in the photo help make Rio an appealing location for the 2016 Games?
3. **Region** What makes the IOC's choice of Rio de Janeiro as an Olympic host a departure from previous choices?
4. **Form and Support Opinions** Consider what you learned from the text and photos. Write two or three sentences that explain whether or not you consider Rio a good choice for the 2016 Games. Support your opinion with facts and details. Go to **Student Resources** for Guided Writing support.

The Christ the Redeemer statue atop Corcovado Mountain overlooks Rio de Janeiro.

CHAPTER 8

Review

VOCABULARY

For each pair of vocabulary words, write one sentence that explains the connection between the two words.

1. venue; infrastructure

> *Olympic hosts often build new infrastructure such as roads and new competition venues for the events.*

2. export revenue; profitable
3. Mediterranean climate; temperate
4. soybean; fertilizer
5. megacity; slum
6. steel; ethanol

MAIN IDEAS

7. What are the three largest indigenous groups in South America? (Section 1.1)
8. Why are so many different languages spoken in South America? (Section 1.2)
9. In what ways has religion shaped the daily lives of South Americans? (Section 1.3)
10. What governmental changes have occurred in Argentina, Peru, and Chile since independence? (Section 2.1)
11. How does the Pampas region contribute to Argentina's economy? (Section 2.2)
12. What are the main characteristics of the Mediterranean climate in Chile? (Section 2.3)
13. What are Peru's most profitable exports? Why? (Section 2.4)
14. What exports and industries shape Brazil's economy? (Section 3.1)
15. In what way has population growth put pressure on São Paulo? (Section 3.2)
16. How is Rio de Janeiro preparing to host the 2016 Olympic Games? (Section 3.3)

CULTURE

ANALYZE THE ESSENTIAL QUESTION

In what ways is South America culturally diverse?

Critical Thinking: Summarize

17. In what ways do governments in South America recognize indigenous languages?
18. How did the majority of South Americans become Roman Catholic?

GOVERNMENT & ECONOMICS

ANALYZE THE ESSENTIAL QUESTION

How is modern South America building its economies?

Critical Thinking: Find Main Ideas

19. What industry has contributed to Peru's economic growth?
20. How does Chile's Mediterranean climate influence its economy?

INTERPRET GRAPHS

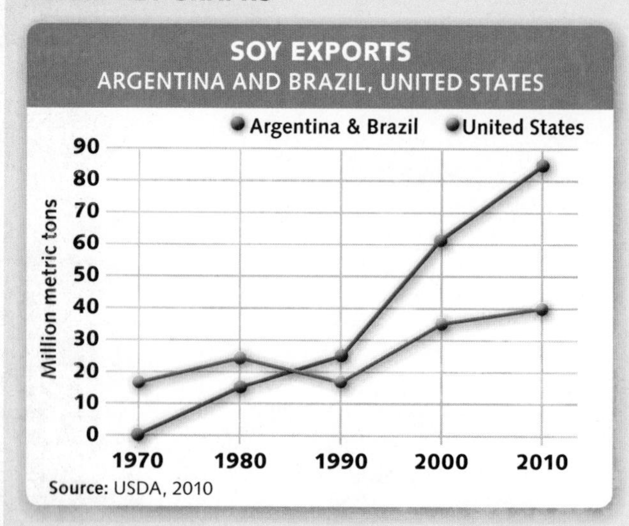

21. **Analyze Data** According to the graph, about when did Argentina and Brazil surpass the United States as soy exporters?
22. **Make Generalizations** Look at the graph. What decade represents the most rapid period of growth for soy exports overall?

FOCUS ON BRAZIL

ANALYZE THE ESSENTIAL QUESTION

How has Brazil become an economic power?

Critical Thinking: Synthesize

23. What factors point to Brazil's current and future economic strength?
24. In what ways does São Paulo reflect Brazil's economic potential?

INTERPRET MAPS

25. **Location** According to the map, where are most cities located in Brazil? How would you explain this?
26. **Make Inferences** Locate Brasília, the current capital of Brazil, and Rio de Janeiro, its former capital. What might the transfer of capital city from Rio de Janeiro to Brasília in 1960 predict about future development in Brazil?

ACTIVE OPTIONS

Synthesize the Essential Questions by completing the activities below.

27. **Create a Poster** Introduce a country in South America by creating a poster. Use the **Digital Library** to choose photos or create your own pictures that represent different aspects of that country, including its economy and cultural diversity. Include pictures of different people and languages, main exports, foods, sports, and celebrations. **Present your poster to the class.**

Poster Tips
- Create a catchy title for your poster.
- Organize your pictures in categories.
- Write labels explaining each picture on your poster.

28. **Conduct Internet Research** Work in a group to gather two new facts about the economies of Brazil, Argentina, Colombia, and Peru. Use the research links at **Connect to NG** and other online sources to search for your facts. Then create a chart to record your facts. Compare your findings with those of the other groups.

Brazil	Fact 1: ______ Fact 2: ______
Argentina	Fact 1: ______ Fact 2: ______
Colombia	Fact 1: ______ Fact 2: ______
Peru	Fact 1: ______ Fact 2: ______

Sports and the Olympics

TECHTREK
myNGconnect.com For research links about the Olympic Games

Connect to NG

Student Resources

As you have learned, the 2016 Olympics will be held in Rio de Janeiro, Brazil, which will mark the first time Brazil has hosted the games. In 2008, China hosted its first Olympic Games. Nearly 11,000 athletes from 204 countries competed in Beijing. Spectators around the world were awed by competing athletes who broke 40 world records and 130 Olympic records.

In any Olympics, the medal count draws the interest of many people. The top medal winners at the 2008 Games were the United States (110), China (100), and Russia (72). Exploring different factors that help determine high medal counts reveals patterns as well as surprises.

Compare

- Australia
- Brazil
- China
- Jamaica
- United States

POPULATION AND ECONOMICS

Countries that win medals generally have populations of at least one million. According to 2008 population statistics, the most populous country represented at the games was China, followed by India and the United States. However, high populations do not guarantee high medal counts. China and the United States won many medals at the 2008 Olympics, but India won just three. Other, less populous countries won more, including two of the least populous countries at the games, Slovenia and Jamaica.

Population is not the only factor in determining a high medal count. How strong a country is economically matters as well. The Gross Domestic Product (GDP), or the measure of the total value of goods and services provided in a country, is one indicator of the economic health of a country. Countries with higher GDPs have more resources for training and supporting athletes for international competitions. At the Beijing Games, the highest ranking GDPs, measured in U.S. billions of dollars, belonged to the United States, Japan, and China. However, countries with low GDPs were competitive with countries with higher GDPs. Zimbabwe, with one of the lowest GDPs, won four medals.

HOME ADVANTAGE

One advantage for hosting the Olympic Games has been, historically, a higher medal yield. Host countries tend to win 30 to 40 percent more medals than in the previous Olympic cycle. At the Beijing Games, China won 100 medals. Four years earlier at the Summer Games held in Athens, Greece, China had won only 63 medals.

OLYMPIC SNAPSHOT
(Beijing Olympics, Summer 2008)

Source: The Wall Street Journal, © 2011

ONGOING ASSESSMENT

RESEARCH LAB

GeoJournal

1. **Explain** Which three countries have the highest GDPs and populations? How does Australia compare to these countries?
2. **Identify** Of the countries listed above, which has the lowest GDP? How does its medal count compare with Brazil's?
3. **Make Predictions** How many medals did Brazil win in 2008? Based on historical patterns of "home advantage," how many medals can you guess Brazil might win at the 2016 Games?

Research and Draw Conclusions Select two countries, each from a different part of the world. Research their performances at one of these Summer Olympic Games: Athens (2004), Sydney (2000), or Atlanta (1996). Compare your findings to the Olympic snapshot of the 2008 Beijing games above. In what ways do your countries compare to the top medal winners? What surprises did you find?

Active Options

TECHTREK

myNGconnect.com For research links on South America and photos of Amazon rain forest life

Connect to NG

Digital Library

Magazine Maker

ACTIVITY 1

Goal: Extend your understanding of life in the Amazon rain forest.

Create a Photo Gallery

The Amazon rain forest is home to thousands of species of birds, land animals, and fish. Thousands of types of trees, plants, and insects live there, too. Do research and choose several examples of different Amazon plants, insects, and animals that interest you. Use the Magazine Maker to create a photo gallery of the plants, animals, and insects you selected. Write a description of each photo. Invite your friends to visit your virtual gallery.

Spider monkey, Amazon rain forest, Peru

ACTIVITY 2

Goal: Learn more about South American history and culture.

Prepare a Multimedia Presentation

Archaeologists have learned much about different cultures in South America. National Geographic grant recipient Christina Conlee explores artifacts left behind by the Nasca culture. Use research links at **Connect to NG** to find information about them. Then select your favorite artifact and create a multimedia presentation about it. Include at least one photo of the artifact in your presentation. Write a label that describes the artifact. Create a title for your presentation.

ACTIVITY 3

Goal: Research daily life in Latin America.

Write a Feature Article

The culture region called Latin America includes Mexico, Central America, the islands of the Caribbean, and South America. Because of years of Spanish colonial influence, Latin America is united by common cultural practices. Using research links at **Connect to NG**, write a feature article about aspects of daily life that unite Latin America as a culture region. Focus on food, sports, festivals, religion, and traditional dress.

NATIONAL GEOGRAPHIC
World Cultures and Geography

GEO

EXPLORE EUROPE WITH NATIONAL GEOGRAPHIC

MEET THE EXPLORER

NATIONAL GEOGRAPHIC

Some archaeological sites are underwater. Emerging Explorer Katy Croff Bell works with archaeologists in the Mediterranean and Black seas to help them figure out where to look for submerged secrets.

INVESTIGATE GEOGRAPHY

The Alps are the highest and most extensive mountain range in Europe. They stretch across central Europe and are concentrated in France, Germany, Italy, Switzerland, and Austria. This is the Lauterbrunnen Valley in Oberland, Switzerland.

STEP INTO HISTORY

The Colosseum in Rome is one of the Roman Empire's greatest architectural and engineering achievements. The arena, completed in A.D. 80, seated nearly 50,000 spectators who watched gladiator games, performances, and even mock naval battles.

CONNECT WITH THE CULTURE

Architect I.M. Pei's modern pyramid serves as an entrance into the Louvre in Paris, France. The museum holds some of the world's greatest art treasures.

CHAPTER 9

EUROPE GEOGRAPHY & HISTORY

PREVIEW THE CHAPTER

Essential Question How did Europe's physical geography encourage interaction with other regions?

SECTION 1 • GEOGRAPHY

KEY VOCABULARY

- peninsula
- uplands
- polder
- bay
- fjord
- canal
- waterway
- ecosystem
- marine reserve

ACADEMIC VOCABULARY

navigable, erosion

TERMS & NAMES

- Northern European Plain
- Alps
- Danube River
- Rhine River

Essential Question How did European thought shape Western civilization?

SECTION 2 • EARLY HISTORY

KEY VOCABULARY

- democracy
- city-state
- golden age
- philosopher
- republic
- patrician
- plebeian
- barbarian
- aqueduct
- feudal system
- serf
- perspective
- indulgence

ACADEMIC VOCABULARY

aristocrat, veto

TERMS & NAMES

- Acropolis
- Alexander the Great
- Julius Caesar
- Augustus
- Christianity
- Middle Ages
- Crusades
- Renaissance
- Johannes Gutenberg
- Martin Luther
- Reformation
- Counter-Reformation

Essential Question How did Europe develop and extend its influence around the world?

SECTION 3 • EMERGING EUROPE

KEY VOCABULARY

- navigation
- colony
- textile
- factory system
- radical
- guillotine
- natural rights
- apartheid
- nationalism
- trench
- reparations
- concentration camp

ACADEMIC VOCABULARY

convert, alliance

TERMS & NAMES

- Industrial Revolution
- Enlightenment
- John Locke
- Reign of Terror
- Napoleon Bonaparte
- Treaty of Versailles
- Great Depression
- Adolf Hitler
- Holocaust
- Iron Curtain
- Cold War
- Berlin Wall

TECHTREK FOR THIS CHAPTER
Student eEdition
Maps and Graphs
Interactive Whiteboard GeoActivities
Digital Library
Connect to NG
Go to myNGconnect.com for more on Europe.
Chestnut horses
ICELAND
Reykjavík
Norwegian Sea
Arctic Circle
Barents Sea
NORWAY
SWEDEN
FINLAND
Gulf of Bothnia
Faroe Islands (Denmark)
Shetland Islands (U.K.)
Orkney Islands (U.K.)
Hebrides (U.K.)
ATLANTIC OCEAN
NORTHERN IRELAND
SCOTLAND
IRELAND
UNITED KINGDOM
WALES
ENGLAND
North Sea
Baltic Sea
DENMARK
Copenhagen
Oslo
Stockholm
Helsinki
Tallinn
ESTONIA
Riga
LATVIA
LITHUANIA
Vilnius
Minsk
BELARUS
Kaliningrad
NETHERLANDS
Amsterdam
Berlin
GERMANY
POLAND
Warsaw
Kiev
UKRAINE
BELGIUM
Brussels
London
Cardiff
Dublin
Belfast
Edinburgh
English Channel
Channel Islands (U.K.)
LUXEMBOURG
Paris
FRANCE
Frankfurt
Prague
Kraków
CZECH REPUBLIC (CZECHIA)
SLOVAKIA
CARPATHIAN MOUNTAINS
MOLDOVA
Chişinău
Sea of Azov
LIECHTENSTEIN
Vienna
Bratislava
Budapest
HUNGARY
AUSTRIA
SWITZERLAND
Bern
Mt. Blanc (4,810 m) 15,781 ft
SLOVENIA
Ljubljana
Zagreb
CROATIA
ROMANIA
Bucharest
Belgrade
Black Sea
BOSNIA AND HERZEGOVINA
Sarajevo
SERBIA
BULGARIA
Sofia
MONTENEGRO
Podgorica
Prishtina
KOSOVO
Skopje
MACEDONIA
Tirana
ALBANIA
GREECE
Athens
Aegean Sea
Rhodes (Greece)
Crete (Greece)
Bay of Biscay
PYRENEES
ANDORRA
SPAIN
Madrid
PORTUGAL
Lisbon
MONACO
SAN MARINO
ITALY
Rome
VATICAN CITY
Corsica (France)
Sardinia (Italy)
Balearic Islands (Spain)
Tyrrhenian Sea
Adriatic Sea
Ionian Sea
Sicily (Italy)
MALTA
Valletta
GIBRALTAR (U.K.)
Mediterranean Sea
Volga R.
Don R.
Dnieper R.
Danube R.
Vistula R.
Oder R.
Elbe R.
Rhine R.
Rhône R.
Po R.
Loire R.
Seine R.
Thames R.
Shannon R.
Douro R.
Tagus R.
Jan Mayen (Norway)
Gotland (Sweden)
Prime Meridian
0 200 400 Miles
0 200 400 Kilometers

1.1 Physical Geography

TECHTREK

myNGconnect.com For online maps of Europe and Visual Vocabulary

Maps and Graphs

Digital Library

EUROPE PHYSICAL

Main Idea Europe is made up of several peninsulas with varied land regions and climates.

Europe is a "peninsula of peninsulas." A **peninsula** is a body of land surrounded on three sides by water.

A Peninsula of Peninsulas

Europe forms the western peninsula of Eurasia, the landmass that includes Europe and Asia. In addition, Europe contains several smaller peninsulas, including the Italian, Scandinavian, and Iberian. Europe also consists of significant islands, including Great Britain, Ireland, Greenland, Iceland, Sicily, and Corsica.

Four land regions form Europe. The Western Uplands are made up of **uplands**, or hills, mountains, and plateaus, that stretch from the Scandinavian Peninsula to Spain and Portugal. The **Northern European Plain** is made up of lowlands that reach across northern Europe. The Central Uplands are hills, mountains, and plateaus at the center of Europe. The Alpine region consists of the **Alps** and several other mountain ranges.

Varied Climates

Most of Europe lies within the humid temperate climate region. The North Atlantic Drift, an ocean current of warm water, keeps temperatures relatively mild. Winds also affect climate. The sirocco (shuh RAH koh) sometimes blows over the Mediterranean Sea and brings wet weather to southern Europe at different seasons. The mistral is a cold wind that sometimes blows through France and brings cold, dry weather to the country.

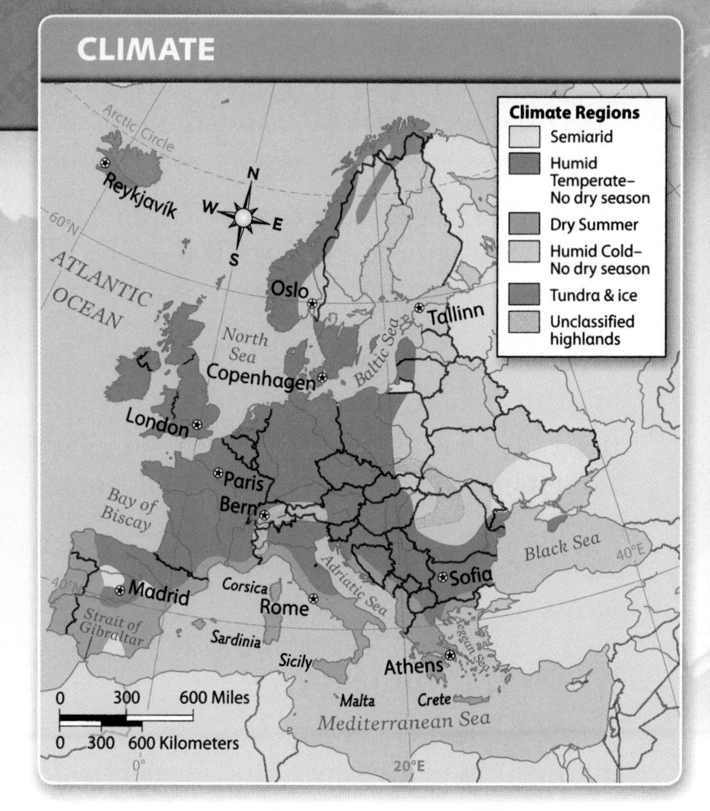

In general, a Mediterranean climate brings mild, rainy winters and hot, dry summers and supports a long growing season. Hardy plants grow best in this climate. In contrast, Eastern Europe has a humid continental climate with long, cold winters. Iceland, Greenland, and northern Scandinavia have a polar climate and a limited growing season.

Before You Move On

Monitor Comprehension What are the main land regions and climates in Europe?

ONGOING ASSESSMENT

MAP LAB

GeoJournal

1. **Interpret Maps** Study both maps in this lesson. Which climate regions are found on the Scandinavian Peninsula? Based on the climate, where do you think most of the peninsula's population is concentrated?
2. **Compare and Contrast** Use both maps to determine what places in Europe have the coldest climates. What geographic characteristics do these places have in common?

SECTION 1 GEOGRAPHY

1.2 A Long Coastline

TECHTREK

myNGconnect.com For an online map and photos of Europe's coastal features and ports

Maps and Graphs

Digital Library

Main Idea Europe's long coastline helped to promote trade, industry, exploration, and settlement on the continent.

Europe has more than 24,000 miles of coastline. If you walked 25 miles a day along the continent's coasts, it would take more than four years to walk the entire distance. These extensive coastlines provided early Europeans with great access to oceans and seas.

Trade and Industry

Europe's water access has benefited the continent in many ways. These benefits include the growth of trade and the development of industry.

Trade has been central to Europe's growth. The civilizations of ancient Greece and Rome flourished largely because of trade. Early sailors traveled to nearly every port on the roughly 2,500-mile-long Mediterranean Sea. They brought back from other lands goods and ideas, such as grains, olive oil, and new religions, that greatly influenced European culture.

Europe also developed several industries that depend on oceans and seas, including a fishing industry. In fact, Europeans have fished along their coastlines for thousands of years.

In lowland areas such as the Netherlands, the people created a way to drain water from the sea in order to increase their farming industry. They built dikes, or giant walls, to hold back the sea in order to create **polders**. Most of the low-lying land of a polder, which once was part of the seabed, was transformed into farms. Today, the Netherlands has about 3,000 polders.

Critical Viewing A boat docks at a harbor in Gdansk (guh DANTSK), Poland, on the Baltic Sea. What do you notice about the harbor?

Exploration and Settlement

The location of the continent near large bodies of water also encouraged exploration. In the 1400s, explorers helped European rulers obtain raw materials, spread religious beliefs, and build empires.

Over time, people settled around the ports where the ships docked. Towns often grew up near **bays**, which are bodies of water surrounded on three sides by land. Some of the towns, including Hamburg, Germany, became large cities as trade and industry expanded. In contrast, the deep and narrow bays of Norway, called **fjords** (fee ORDZ), did not encourage settlement.

Before You Move On

Summarize In what ways has Europe benefited from its long coastline?

ONGOING ASSESSMENT

MAP LAB

1. **Compare and Contrast** Use the map above and those in Section 1.1 to compare and contrast the Italian Peninsula with the Scandinavian Peninsula. What do the two have in common? What differences do you see?
2. **Analyze Cause and Effect** Complete the chart by writing one effect for each cause.

CAUSE	EFFECT
Trade was conducted on Europe's oceans and seas.	Goods and ideas spread.
Industry developed on the coasts.	
Explorers traveled to new lands.	
Cities grew along the coasts.	

1.3 Mountains, Rivers, and Plains

myNGconnect.com For an online map and photos of Europe's landforms and natural resources

Main Idea The landforms and resources in Europe support many economic activities.

As you have already learned, Europe consists of four main land regions. A great variety of landforms lie within these regions, including mountains and a vast plain. Many important rivers also cross the continent.

Mountain Chains

Europe's Alpine region contains several mountain chains. The Alps stretch from Austria and Italy to Switzerland, Germany, and France. The Pyrenees are located to the west of the Alps and separate Spain and France. South of the Alps lie the Apennines, which run along the Italian Peninsula. The Carpathians extend through Poland, Romania, and Ukraine.

All of these mountain chains provide natural resources for industries, including forests, which supply wood, and mineral resources, such as iron ore. The valleys between the mountains contain fertile land for growing crops.

Rivers and Plains

Europe has a wealth of rivers. Many are **navigable**, which means that boats and ships can travel easily on them. The **Danube River** is an important transportation route. The river starts in Germany and passes along or through ten countries before emptying into the Black Sea. The **Rhine River** is another vital body of water used to transport goods deep inland. The river originates in Switzerland, winds through Germany, and flows into the North Sea.

Visual Vocabulary A **canal** is a human-made water passage used for travel, shipping, and irrigation. The boats in this canal in Venice, Italy, are being used to move goods within the city.

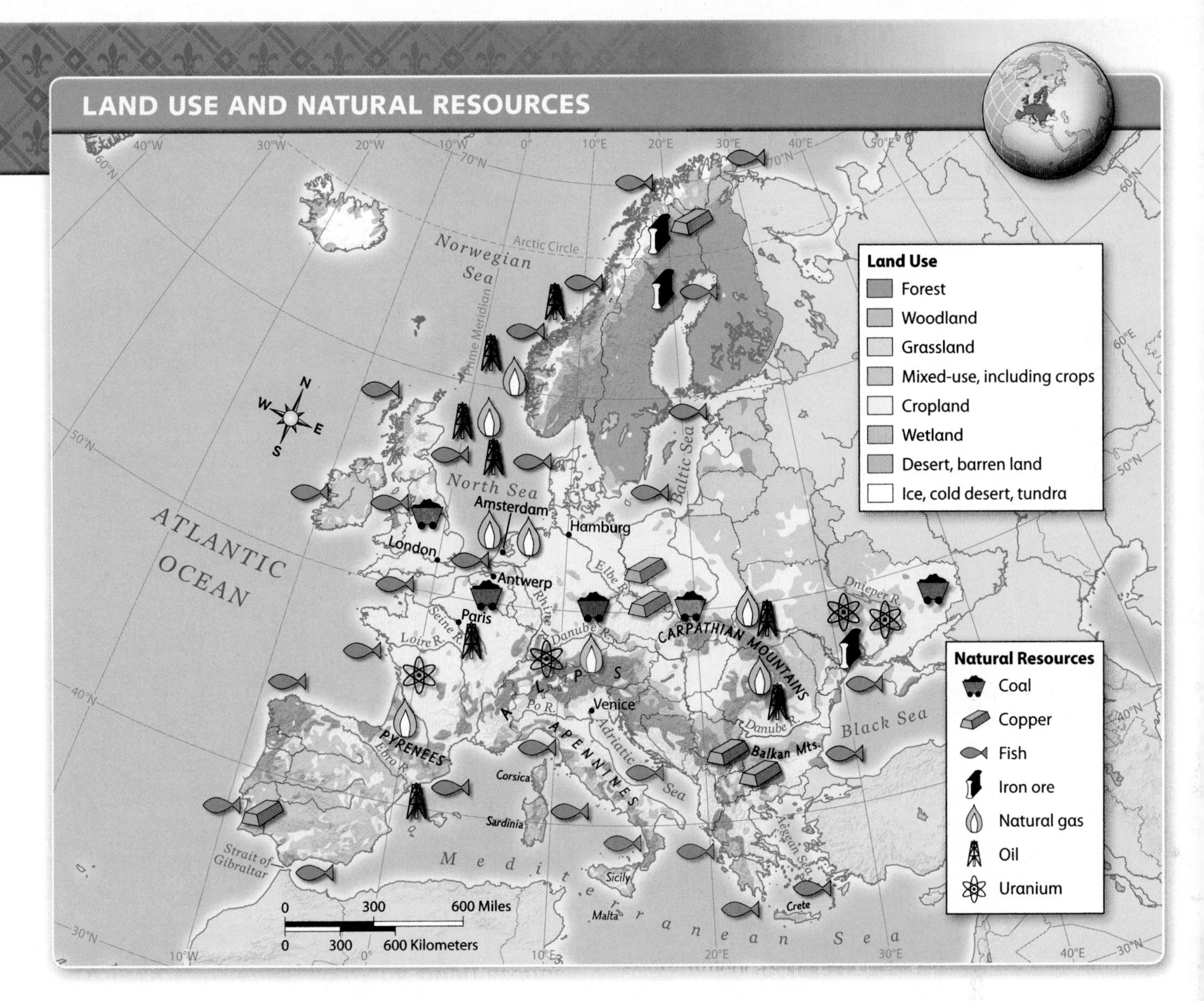

For centuries, Europeans have built canals. When linked together, canals and rivers form **waterways**, or navigable routes of travel and transport. The small country of the Netherlands has more than 3,000 miles of rivers and canals.

Many of the rivers in Europe cross the Northern European Plain. This vast lowland region stretches across France, Belgium, Germany, and Poland all the way to Russia. The fertile soil on the plain makes it ideal for growing crops, and thousands of farms are sprinkled throughout the region. The plain also contains some of the largest and most heavily populated cities, or urban centers, in Europe, including Paris, France.

Before You Move On

Summarize What economic activities are supported by Europe's landforms and resources?

ONGOING ASSESSMENT

MAP LAB

GeoJournal

1. **Interpret Maps** What natural resources are found in the Carpathian Mountains? What geographic challenges might workers deal with when they extract these resources?
2. **Draw Conclusions** Find the Danube River on the map. What natural resources are found along the river? What role might the Danube play in handling these resources?
3. **Make Inferences** What natural resources are found in the North Sea? What impact might they have on the countries bordering the sea?

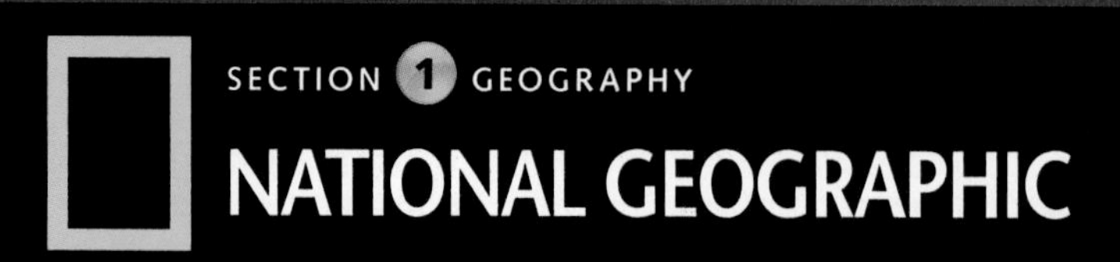

TECHTREK
myNGconnect.com For a map, photos, and an Explorer Video Clip

Maps and Graphs

Digital Library

Protecting the Mediterranean with Enric Sala

Main Idea Human activities have harmed the Mediterranean Sea's natural environment.

Under the Sea

On June 4, 2010, National Geographic Explorer-in-Residence Enric Sala began an expedition: exploring the underwater world of the Mediterranean Sea. He wanted to find out how human activities have affected the sea's ecosystem.

An **ecosystem** is a community of living organisms and their natural environment. Three human activities have had an impact on the Mediterranean's ecosystem. One is overfishing, which occurs when people catch fish at a faster rate than the fish can reproduce. Another is pollution. The third is overdevelopment, which has occurred as coastal populations have grown. (See the chart on the next page.) Growing populations have added to the Mediterranean's pollution and the **erosion**, or wearing away, of its coastline.

myNGconnect.com
For more on Enric Sala in the field today

Critical Viewing Sala swims with a sea turtle. What equipment is he using during this underwater exploration?

Sala's work was inspired by Jacques Cousteau, the French underwater explorer. When he became an explorer himself, Sala sailed with Cousteau's son, Pierre-Yves Cousteau. They compared the condition of the Mediterranean Sea now with its condition 65 years earlier and found that the sea has been damaged. Sala concluded, "We have lost most of the large fish and the red coral because of centuries of exploitation [misuse]."

POPULATION OF MEDITERRANEAN CITIES (IN MILLIONS)

City	1960	2015 (projected)
Athens, Greece	2.2	3.1
Barcelona, Spain	1.9	2.73
Istanbul, Turkey	1.74	11.72
Marseille, France	0.8	1.36
Rome, Italy	2.33	2.65

Source: UN, 2002

Marine Reserves

In spite of the harm that has been done, Sala sees signs of hope. During their Mediterranean expedition, he and Cousteau visited the Scandola Natural Reserve, near Italy. Scandola is a **marine reserve**, or protected area where people are prohibited from fishing, swimming, or anchoring their boats. As a result, marine life is thriving at the reserve. "This marine reserve," Sala said, "has restored the richness that Jacques Cousteau showed us 65 years ago."

Before You Move On

Monitor Comprehension What can be done to protect the Mediterranean Sea from the human activities that have harmed its environment?

ONGOING ASSESSMENT

DATA LAB

1. **Interpret Charts** Based on the chart, which Mediterranean city will have undergone the greatest growth by 2015? What impact will this growth have on the city's coastline?
2. **Make Inferences** According to the chart, which city will have undergone the least growth by 2015? What does this projected number mean for the city's coastline?
3. **Turn and Talk** What could be done today to preserve fish populations for the future? Get together with a partner and come up with one or two specific suggestions. Share your ideas with the rest of the class.

NATIONAL GEOGRAPHIC
Photo Gallery • Scotland's Rugged Coast

For more photos from the National Geographic Photo Gallery, go to the **Digital Library** at myNGconnect.com.

The Carnival in Venice

Ancient Roman aqueduct

Musicians in Krakow, Poland

Critical Viewing These rocky pinnacles, or pointed formations, took shape thousands of years ago on a hill in Scotland called the Storr. The largest of the formations shown here is known as the Old Man of Storr.

Italy's Amalfi Coast

French pastries on display

The Parthenon in Athens

Armor from the 1600s

2.1 Roots of Democracy

TECHTREK
myNGconnect.com For an online map of ancient Greek city-states

Maps and Graphs

Main Idea The ideas out of which democracy grew first took root in ancient Athens.

The rulers of Athens laid the groundwork for **democracy**, a government in which people can influence law and vote for representatives. Democracy was one of the great achievements of Greek civilization.

The Greek City-States

Greece lies on both the Balkan and Peloponnesus (pehl uh puh NEE suhs) peninsulas. The two are connected by an isthmus, or narrow strip of land. People first arrived in Greece around 50,000 B.C. Early civilizations developed between 1900 B.C. and 1400 B.C.

Around 800 B.C., several Greek city-states started to thrive. A **city-state** is an independent community that includes a city and its surrounding territory. The mountains on the peninsulas made transportation and communication difficult. As a result, each city-state developed independently. The two largest and most important Greek city-states were Athens and Sparta.

Each city-state established its own community and government. The earliest form of government in the city-states was a monarchy, in which a king or queen rules. Over time, a group of upper-class noblemen called **aristocrats** began to act as advisors to the king. In some city-states, the aristocrats set up a ruling council that served as the government. This council was a form of oligarchy (AHL ih gahr kee), in which a small group rules.

Around 650 B.C., tyrants in many of the city-states seized power away from the councils, took control of the government, and re-established one-person rule. Today, any harsh ruler may be called a tyrant. However, not all tyrants in ancient Greece were bad leaders. Some were fair and had the support of the Greek people.

Democracy in Athens

Around 600 B.C. in Athens, a statesman named Solon controlled the government of the city-state. He established assemblies in which all the wealthy people of Athens—not just the aristocrats—made the laws.

Then in 508 B.C., a leader named Cleisthenes (KLIHS thuh neez) increased the people's power even more. He established a direct democracy. Under this government, all citizens voted directly for laws. However, only Athenian adult males were citizens and had the right to vote.

Athens and Sparta

Democracy developed in Athens but not in all city-states. Sparta, Athens' rival, had an oligarchy ruled by a small group of warriors. They supervised a military training system for Spartan boys.

In 490 B.C., Athens and Sparta joined together to defeat the invading army of the Persian Empire under King Darius I. After that, however, the two city-states became fierce enemies.

Before You Move On

Summarize What ancient Greek ideas served as the roots of modern democracy?

Visual Vocabulary The **Acropolis** of Athens is a rocky hill that once served as the city's fortress and contained its most important temples.

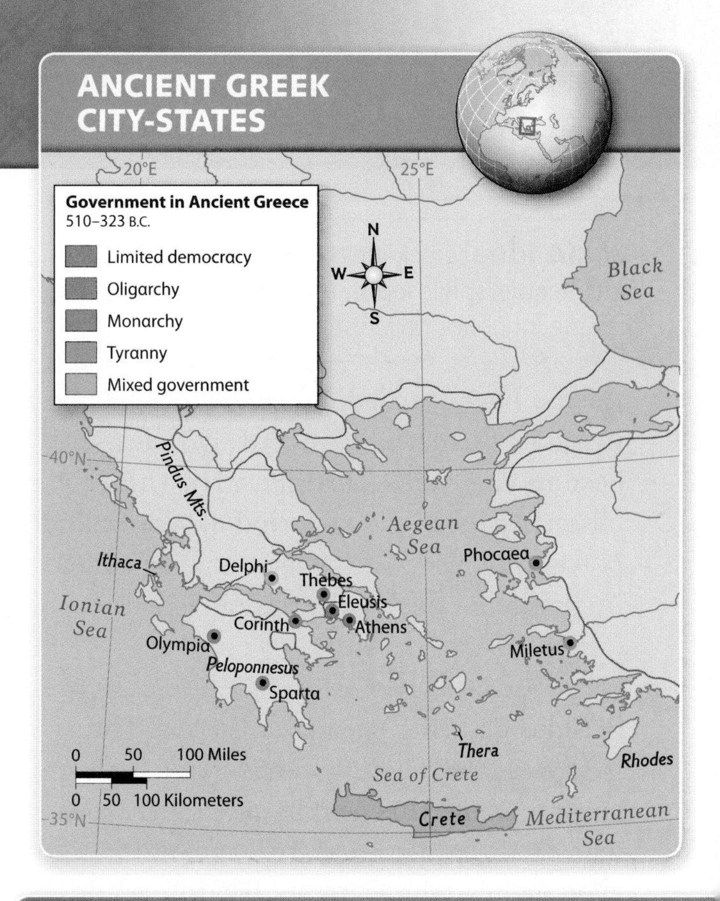

ONGOING ASSESSMENT

READING LAB

Synthesize Use the Greek roots in the chart to form a word in English that completes each of the following sentences:

a. The form of government that represents the people is called ______(use your own paper)______.

b. People empowered to enforce a city's laws are the ______(use your own paper)______.

c. The ruler of a kingdom is also called a ______(use your own paper)______.

SELECTED ENGLISH WORDS FORMED FROM GREEK ROOTS

Greek Root	Meaning	English Word
demos	people	democracy
polis	city-state	policy
aristo	best	aristocracy
monos	one	monarchy
oligo	few	oligarchy

2.2 Classical Greece

myNGconnect.com For an online map and photos of Classical Greece

Main Idea Greek ideas about democracy, architecture, philosophy, and science have had a lasting influence on Western culture.

As you have learned, democracy began in ancient Greece. In 461 B.C., Pericles became the leader of Athens. His rule began a **golden age**, a period of wealth and power during which democracy developed further and Greek culture flourished.

Golden Age of Greece

Pericles had three goals for Greece. The first was to strengthen democracy. He accomplished this goal by paying citizens who held public office. This meant that even people who were not wealthy could afford to serve in government.

The leader's second goal was to expand the empire. Pericles built a strong navy and used it to increase Athens' power over the other Greek city-states.

Pericles' third goal was to make Athens more beautiful. He began rebuilding the city, including the Acropolis. Many of Athens' temples had been destroyed during the war with Persia. Pericles constructed a new temple called the Parthenon, dedicated to the goddess Athena for whom Athens was named.

Greek Achievements

The golden age of Greece was a period of extraordinary achievements. Greek architects designed temples and theaters with graceful columns. **Philosophers**, people who examine questions about the universe, searched for the truth. Socrates (SAHK ruh teez) and his student, Plato (PLAY toh), were leading philosophers.

In the sciences, the mathematician Euclid (YOO klihd) developed the principles of geometry. The physician Hippocrates (heh PAH kruh teez) changed the practice of medicine by insisting that illnesses originated in the human body and were not caused by evil spirits.

Greek Culture Spreads

Greece's golden age ended around 431 B.C., when war broke out between Athens and Sparta. The conflict, known as the Peloponnesian War, lasted 27 years and weakened both Athens and Sparta.

Statue of the goddess Athena

1900–1400 B.C. Early Greek civilizations develop.

800 B.C. Greek city-states begin to thrive.

1900 B.C. | 1000 B.C. | 750 B.C.

Early Greek gold lion's head

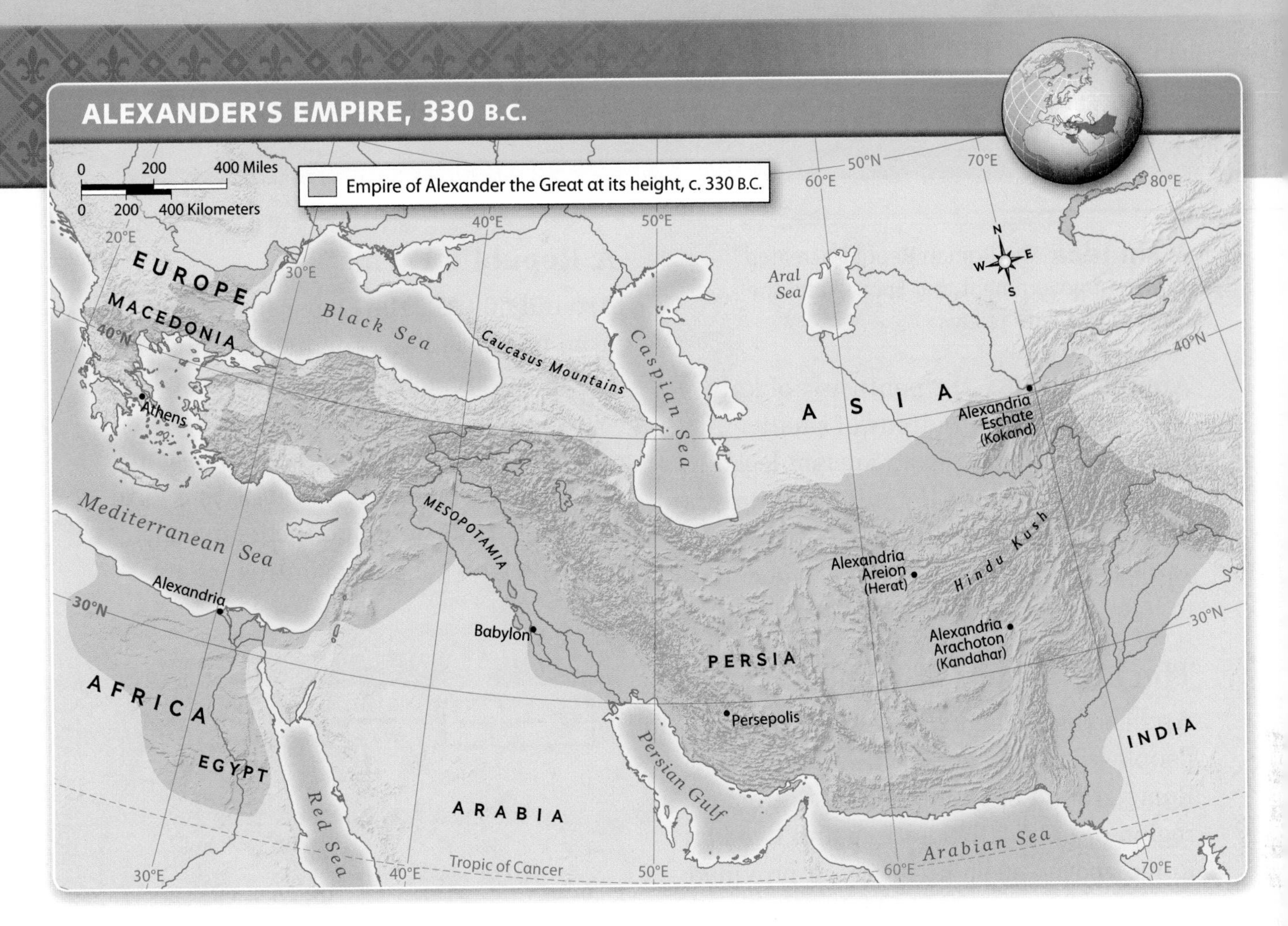

Around 340 B.C., King Philip II of Macedonia took advantage of the weakened city-states and conquered Greece. In 334 B.C., Philip's son, **Alexander the Great**, became king and began to extend his father's empire. Alexander loved Greek culture and spread its ideas throughout the lands he conquered. Alexander died in 323 B.C. at the age of 33. The Greek ideas about democracy, science, and philosophy that he helped spread shaped the modern world.

Before You Move On

Monitor Comprehension What Greek ideas have had a lasting influence on Western culture?

ONGOING ASSESSMENT

MAP LAB

1. **Draw Conclusions** Study the map. Across which continents did Alexander's empire spread? What helped him unite his vast empire?
2. **Make Inferences** Based on the map, what great empires did Alexander conquer? What do these conquests suggest about Alexander?

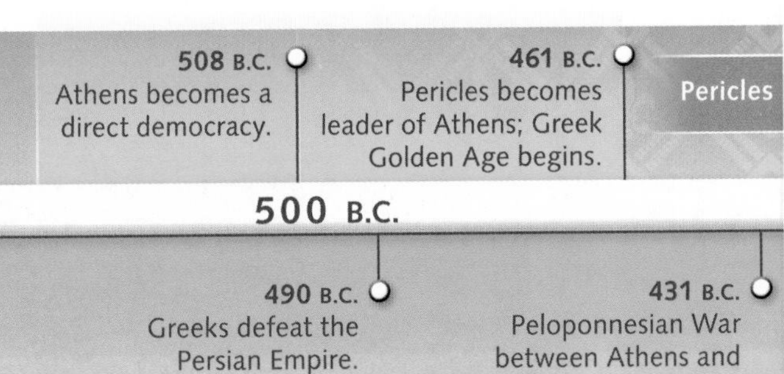

340 B.C. Philip II of Macedonia conquers Greece.

250 B.C.

334 B.C. Alexander the Great begins to extend his father's empire.

2.3 The Republic of Rome

myNGconnect.com For an online map of ancient Rome and photos of Roman ruins

Maps and Graphs

Digital Library

Main Idea The Roman Republic created a form of government that Europe and the West would later follow.

Around 1000 B.C., the peninsula of Italy was dotted with hundreds of small villages. According to an ancient legend, two brothers named Romulus and Remus founded Rome in 753 B.C. The brothers were said to be the children of a god and to have been raised by a wolf.

The Beginnings of Rome

Archaeologists actually believe that people known as the Latins founded Rome around 800 B.C. They came from a region of Italy called Latium and lived on Rome's seven steep hills, which provided protection from enemy attack. The Tiber River, which flows through Rome, provided water for farming and a route for trade. Over time, Rome developed into a wealthy city-state.

A Republic Forms

Around 600 B.C., the Etruscans, a people from northern Italy, conquered Rome. One Etruscan king named Tarquin was a brutal tyrant. In 509 B.C., the Romans rebelled against him, and Roman leaders began to create a republic. A **republic** is a form of government in which the people elect officials who govern according to law.

THE HILLS OF ROME

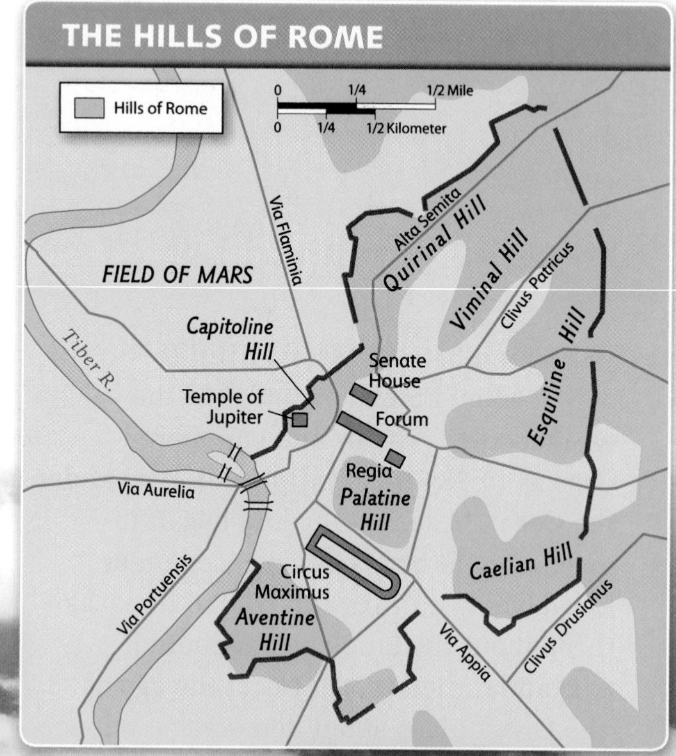

Critical Viewing The Roman Forum contained the ancient city's most important buildings, including the Senate. In what ways does this photo reflect Rome's former glory?

GOVERNMENT OF THE ROMAN REPUBLIC

EXECUTIVE
- Two consuls
- Elected to one-year term
- Led the government and controlled the army

LEGISLATIVE

Senate
- 300 members made up of patricians
- Not elected; selected by the consuls to serve for life
- Made the laws and advised consuls

Assembly
- Made up of plebeians
- Elected tribunes as representatives
- Made the laws and selected consuls

JUDICIAL
- Eight judges
- Governed provinces
- Served for one year

Legal Code
- Twelve Tables
- Established rights and responsibilities of Roman citizens

Two main classes of people lived in Rome at this time. The **patricians** were mostly wealthy landowners. The **plebeians** were mostly farmers. At first, only the patricians could take part in government. They controlled the Senate and made laws.

In 490 B.C., plebeians gained the right to form an assembly and elect legislative representatives called tribunes. The assembly made laws and elected the consuls, the two executive officials who led the government for a year at a time. One consul could **veto**, or reject, a decision made by the other consul.

The judicial branch was made up of eight judges who served for one year. These judges oversaw the lower courts and governed the provinces.

Around 450 B.C., the government published the Twelve Tables. These were bronze tablets that set down the rights and responsibilities of Roman citizens. At this time, only adult male landowners born in Rome were citizens. Roman women were citizens but could not vote or hold office.

The Roman Way

Citizens of Rome believed in values that were known as the Roman Way. These values included showing self-control, working hard, doing one's duty, and pledging loyalty to Rome. The Roman Way helped to unite all Roman citizens.

The Romans applied these values as the Republic began to conquer new lands and expand. During the second century B.C., Rome defeated the empire of Carthage in northern Africa. By 100 B.C., Rome controlled most of the lands around the Mediterranean. About this time, tensions began to grow between patricians and plebeians. These tensions triggered a war between the two groups. The war set the stage for the end of the Roman Republic—and the birth of the Roman Empire.

Before You Move On

Summarize What structures, laws, and values made up the government of the Roman Republic?

ONGOING ASSESSMENT

DATA LAB

1. **Interpret Charts** According to the chart, members of the Senate were selected by the consuls. However, the assembly elected the consuls. In what way did this arrangement help to control the Senate's power?
2. **Turn and Talk** Study the chart and consider what you know about the U.S. government. Then, with a partner, compare and contrast the two systems of government.

2.4 The Roman Empire

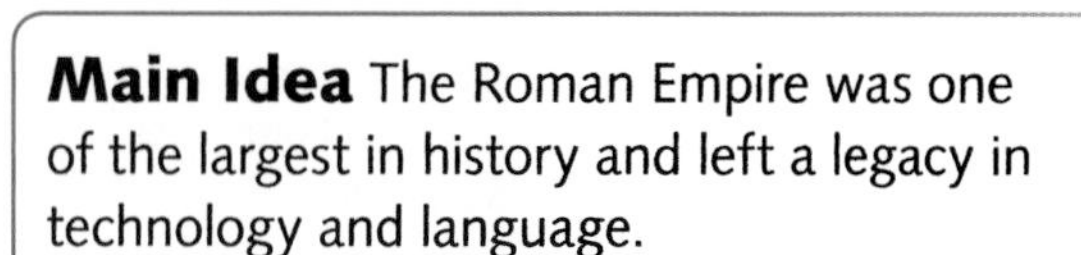

Main Idea The Roman Empire was one of the largest in history and left a legacy in technology and language.

As you have read, tensions grew between the patricians and plebeians not long after Rome defeated Carthage. The Roman soldiers had brought back great wealth from the conquered territory. They used their wealth to buy large plots of farmland. Small farmers could not compete with them, and the gap between rich and poor widened. In 88 B.C., war erupted between the patricians and plebeians.

Creation of the Empire

After years of war, a general named **Julius Caesar** rose to power and became the sole ruler of Rome in 46 B.C. Caesar started projects to help the poor and tried to re-establish order in Rome, but he developed powerful enemies. In 44 B.C., a small group of senators stabbed him to death.

Octavian, Caesar's nephew, fought in the long, internal war for power that followed Caesar's death. Octavian won, but the war put an end to the Roman Republic. Calling himself **Augustus**, which means "honored one," Octavian became the ruler of the Roman Empire in 27 B.C. His rule began a period called the *Pax Romana*, which is Latin for "Roman peace."

Rome's Decline

For about 500 years, the Roman Empire was the most powerful in the world and extended over three continents. Around A.D. 235, however, Rome had a series of poor rulers. In addition, German tribes, whom the Romans called **barbarians**, began invading the empire from the north.

In 330, Emperor Constantine moved the capital of the weakened empire from Rome to Byzantium, in present-day Turkey, and renamed the city Constantinople. He also made Christianity lawful throughout the empire. **Christianity**, which is based on Jesus' life and teachings as described in the Bible, began in the Roman Empire.

In 395, the empire was divided into an Eastern Empire and a Western Empire with two different emperors. In 476, invaders overthrew the last Roman emperor and ended the Western Empire.

Rome's Legacy

The Western Empire fell, but Rome left the world a great legacy, or heritage. For example, Roman engineers built a network of roads that connected the empire. Many of the roads are still in use.

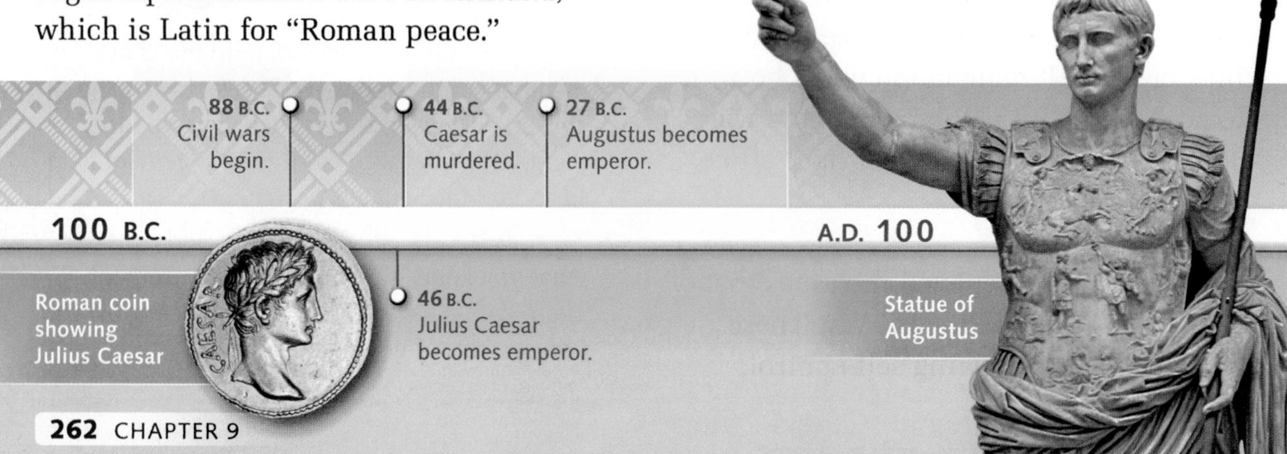

Roman coin showing Julius Caesar

Statue of Augustus

THE ROMAN EMPIRE AT ITS HEIGHT, A.D. 200

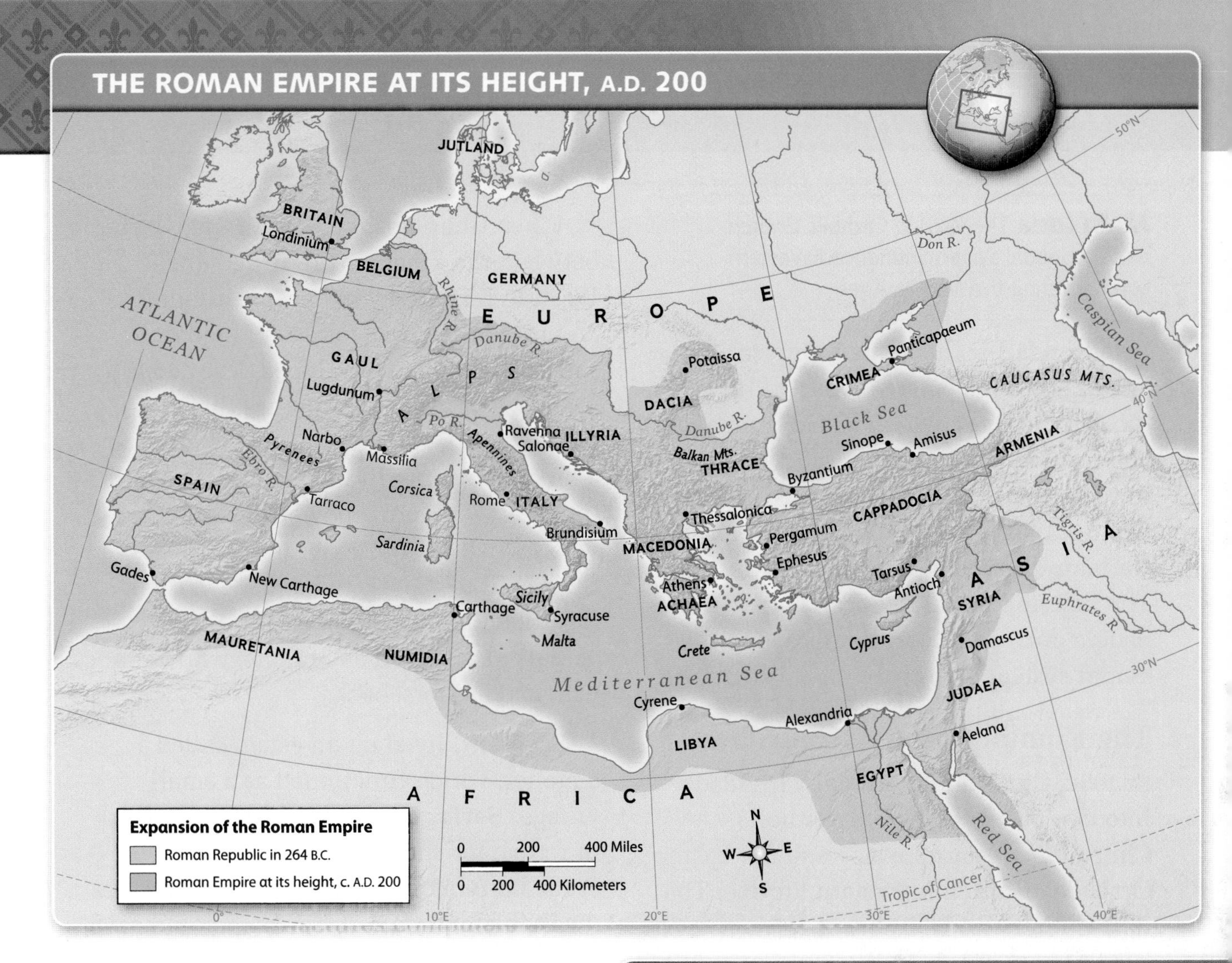

The engineers also developed the arch and used it to construct buildings and **aqueducts**, which carried water to parts of the empire. Latin, the language of Rome, became the basis for Romance languages, such as Spanish and Italian. Many English words have Latin roots.

Before You Move On

Summarize Describe the Roman Empire's rise, fall, and legacy.

ONGOING ASSESSMENT

MAP LAB

GeoJournal

1. **Interpret Maps** According to the map, over which continents did the Roman Empire extend? What challenges might the size of the empire have presented to its rulers?
2. **Location** Find Byzantium on the map. Why do you think Constantine chose this location to become the capital of the Eastern Empire?

A.D. 395 Empire is divided into an Eastern Empire and a Western Empire.

A.D. 300

A.D. 500

A.D. 330 Constantine moves the capital from Rome to Byzantium.

A.D. 476 Rome falls to invaders.

Painting of German invader

2.5 Middle Ages and Christianity

TECHTREK
myNGconnect.com For portrayals of the Crusades in fine art
Digital Library

Main Idea The Roman Catholic Church and the feudal system influenced Western Europe during the Middle Ages.

After the fall of the Roman Empire, Western Europe entered a period known as the **Middle Ages**, which lasted from about 500 to 1500. During this period, Western Europe consisted of numerous kingdoms. Castles, like those on the Rhine River in Germany, served as defensive fortresses. The Roman Catholic Church helped unite people during the Middle Ages, and the feudal system provided a social structure.

The Roman Catholic Church

In 1054, Christianity officially divided into two parts: the Roman Catholic Church in Western Europe and the Eastern Orthodox Church in Eastern Europe. The Roman Catholic Church was the center of life for most people in Western Europe. It cared for the sick, provided education, and helped preserve books and learning.

The Church also played a leading role in government. It collected taxes, made its own laws, and waged wars. In 1096, the Church began a series of **Crusades**. These were military expeditions undertaken to take back holy lands in Southwest Asia from Muslim control. The Crusades cost many lives and ended in 1291.

The Feudal System

The many kingdoms of Western Europe were often at war. From about 400 to 800, a German group called the Franks stopped the fighting and unified most of Western Europe. Their most important leader was Charlemagne (SHAHR luh mayn).

When Charlemagne died in 814, warfare between the kingdoms returned and Western Europe again became divided. To provide security for each kingdom, the feudal system developed. The **feudal system** was a social structure that was organized like a pyramid. At the top was a king who owned vast territory. Beneath the king were lords, powerful noblemen who owned land. The lords gave pieces of their land to vassals, who pledged their loyalty and service to the lord. Some vassals also served as knights, who were warriors on horseback.

Each lord lived on an estate called a manor, which functioned as a small village. **Serfs**, who farmed the lord's land in return for shelter and protection, were at the bottom of the pyramid. Serf families dwelt in small huts on the manor and gave most of the crops they grew to their lord.

The Growth of Towns

In time, the growth of towns helped end the feudal system. Trade and businesses developed, and people began to leave the manors. A deadly disease called the bubonic plague, which swept through Europe in 1347, also weakened the feudal system. The plague killed millions and greatly reduced the workforce in the towns. Desperate for workers, employers offered higher wages. As a result, farmers left the country to seek the higher-paying jobs in the towns.

Before You Move On

Summarize In what ways did the Roman Catholic Church and the feudal system influence Western Europe during the Middle Ages?

ONGOING ASSESSMENT

VIEWING LAB

GeoJournal

1. **Interpret Visuals** What details in the illustration suggest the measures taken to protect those who lived in the manor?
2. **Make Inferences** Notice the position of the church in the manor. Why might it have been positioned near the lord's castle?
3. **Compare and Contrast** Based on the illustration and what you have read, in what ways did life probably differ for lords and serfs?

2.6 Renaissance and Reformation

myNGconnect.com For examples of Renaissance art

Digital Library

Main Idea Both the Renaissance and the Reformation brought great change to Europe.

You have learned that the growth of towns in Western Europe helped put an end to the feudal system. The Roman Catholic Church also began to lose some of its power at this time. As these key structures of the Middle Ages weakened, the **Renaissance** began to take hold. The Renaissance was a rebirth of art and learning that started in Italy in the 1300s and had spread through Europe by 1500.

The Renaissance

Several other factors led to the Renaissance. Increased trade in the growing towns brought some Italian merchants great wealth. This wealth allowed them to buy the artists' work.

As you have learned, the Western part of the Roman Empire fell in 476. The Eastern part continued and became known as the Byzantine Empire. The fall of this empire in 1453 also advanced the Renaissance. Scholars from the empire came to Italy, bringing with them ancient writings of the Greeks and Romans. Studies of these works encouraged humanism, which focuses on human rather than religious values. In addition, a German inventor named **Johannes Gutenberg** developed a printing press in 1450 that printed many books in a short amount of time. Soon, more people in Europe had access to knowledge—including the new humanist ideas.

The result was an explosion in art, architecture, and literature. Renaissance artists such as Leonardo da Vinci, Michelangelo (my kuhl AN juh loh), and Raphael (rah fee ELL) used **perspective** to make a painting look as if it had three dimensions. Architects used elements of ancient Greek and Roman design to create churches and buildings. Writers wrote in the vernacular, the language spoken in a particular region. For example, Dante Alighieri (ah lah GYER ee), who bridged the Middle Ages and Renaissance, wrote his work, *The Divine Comedy*, in Italian, not Latin.

The Reformation

Meanwhile, some people began looking more critically at the Church. **Martin Luther**, a monk in Germany, was shocked by the corrupt practices of some priests. To raise funds, they often sold **indulgences**, which relaxed the penalty for sin.

Gutenberg's printing press

Portrait of Dante

1300s Renaissance begins in Italy.

1300

1308 Dante starts writing *The Divine Comedy* in Italian.

1400

1450s Johannes Gutenberg develops the printing press.

Critical Viewing *The School of Athens* by Italian artist Raphael portrays important ancient Greek philosophers and scientists. Why do you think the artist chose to celebrate ancient Greece?

In 1517, Luther wrote the 95 Theses, in which he objected to such practices, and nailed them to a church door. Luther's actions started the **Reformation**, the movement to reform Christianity. Over time, people founded Protestant churches. The term comes from the word *protest*.

In response, the Church began a reform movement called the **Counter-Reformation** and placed more emphasis on faith and religious behavior. Nonetheless, the conflict between Catholics and Protestants would continue for the next 300 years.

Before You Move On

Monitor Comprehension What changes did the Renaissance and Reformation bring to European culture and society?

ONGOING ASSESSMENT

READING LAB

GeoJournal

1. **Summarize** What was the Renaissance?
2. **Analyze Cause and Effect** In what ways did the development of Gutenberg's printing press help spread humanist ideas?
3. **Draw Conclusions** What humanist ideas might have led people to look at the Roman Catholic Church more critically?

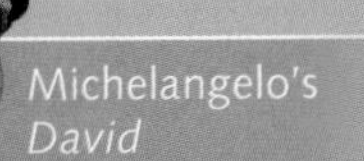

Michelangelo's *David*

1497 Leonardo da Vinci finishes painting *The Last Supper.*

1500

1504 Michelangelo completes the sculpture *David.*

1517 Martin Luther nails his 95 Theses to the door of a church in Wittenberg, Germany.

1600

In this illustration, Martin Luther posts the 95 Theses.

3.1 Exploration and Colonization

TECHTREK

myNGconnect.com For an online map of European colonization in Africa, Asia, and the Americas

Maps and Graphs

Main Idea To expand trade, Europeans explored Africa, Asia, and the Americas and established colonies on all three continents.

Around 1415, Prince Henry of Portugal decided that he would send explorers to Africa to establish new trade routes. Henry, who became known as Prince Henry the Navigator, founded a **navigation** school. The school taught sailors about mapmaking and shipbuilding and marked the beginning of the Age of Exploration.

European Exploration

Portugal was the first of many European countries to sponsor voyages of exploration. Europeans wanted to find gold and establish trade with Asia to obtain spices, silk, and gems. They also wanted people in other lands to **convert**, or change their religion, to Christianity.

The voyages were filled with danger. Explorers often sailed for months in ships that were small and not always able to withstand strong storms at sea. The men also faced disease and attacks by native peoples. Furthermore, the explorers were traveling to unknown lands. Mapmakers often marked unexplored places with the phrase "Here be dragons."

Nonetheless, Portuguese explorers such as Bartolomeu Dias and Vasco da Gama sailed along the coast of Africa in the late 1400s to open up trade with Asia. Italian explorer Christopher Columbus uncovered a "new world"—the continents of North America and South America—in 1492. In the 1530s, Jacques Cartier (kahr TYAY) explored parts of North America for France. An Englishman, Sir Francis Drake, sailed around the world in 1577.

Critical Viewing In this painting, Columbus and his crew land in North America as Native Americans arrive to meet the explorers in their canoes. What qualities must explorers have had to undertake their voyages?

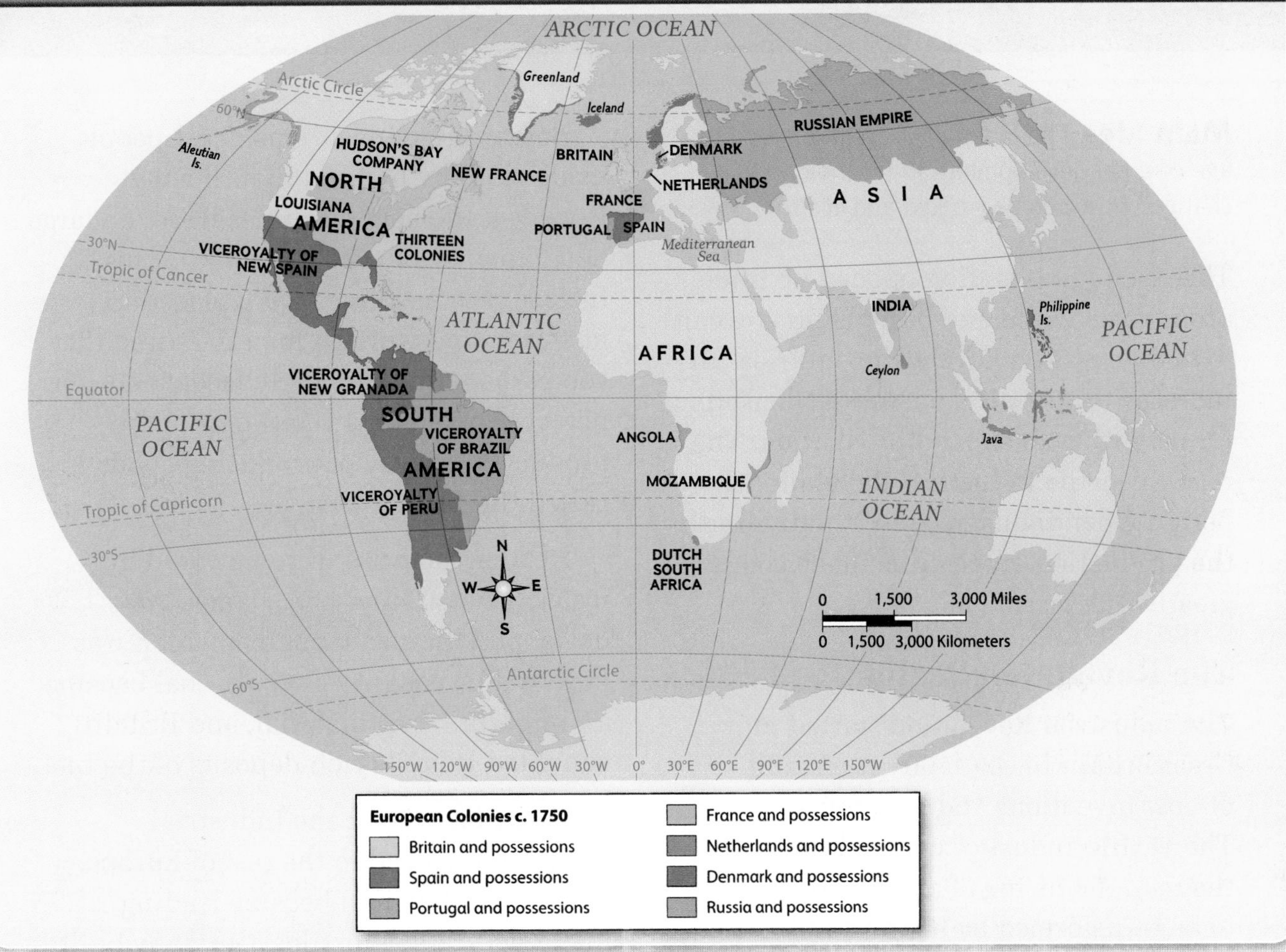

Establishing Colonies

In addition to trade, Europeans used the voyages of exploration to claim lands for their own countries. When explorers landed in a new place, they declared it a colony. A **colony** is an area controlled by a distant country. As you have learned, Spanish explorers claimed colonies in Mexico and South America. The French and the English also established colonies in North America. By 1650, European countries controlled parts of Africa and Asia as well.

European exploration and colonization resulted in a sharing of goods and ideas known as the Columbian Exchange. From the Americas, Europeans obtained new foods, such as potatoes, corn, and tomatoes. Europeans introduced wheat and barley to the Americas. They also introduced diseases like smallpox. The diseases killed millions of native peoples.

Before You Move On

Monitor Comprehension What inspired Europeans to undertake voyages of exploration, and what did they gain as a result?

ONGOING ASSESSMENT

MAP LAB

1. **Interpret Maps** According to the map, where in Asia did France establish a large colony? Why was this location beneficial geographically?
2. **Identify Problems and Solutions** Study the map. Who was Spain's main rival for colonies in South America? What problems might have arisen from their rivalry?

3.2 The Industrial Revolution

TECHTREK

myNGconnect.com For an online map and images of the Industrial Revolution

Maps and Graphs

Digital Library

Main Idea The Industrial Revolution was an age of great developments in technology that changed how people worked and lived.

The Age of Exploration opened up trade around the world and brought great wealth to many western European countries. To increase this wealth, businesses looked for new ways to expand production. The result was the **Industrial Revolution**, a period when industry grew rapidly, and the production of machine-made goods greatly increased.

The Revolution Begins

The Industrial Revolution started in Great Britain in the 1700s as a result of new inventions and technologies. The **textile** industry, which deals with the manufacturing of cloth, was the first to be transformed by the revolution. In 1769, textile manufacturers began using machines that were run using water from a stream. Then, around 1770, James Hargreaves invented the spinning jenny. This machine allowed workers to make cotton and wool yarn at a much faster rate.

Before these inventions, most people made cloth by hand in their homes. However, the new machines were too large and expensive to use in small houses. Instead, the machines were placed in factories, and workers manufactured the goods there. In these early factories, each person worked on a small part of the product. This way of producing goods is called the **factory system**.

At first, factories were powered by water. Then around 1776, James Watt developed the steam engine, which was powered by coal. As a result, coal became an important raw material, and Britain benefited from its rich deposits of the fuel.

In the late 1700s, the Industrial Revolution spread to the rest of Europe. France and Belgium became leading manufacturers of textiles. Germany built factories for processing iron. Railroad systems developed in the 1800s. In 1825, George Stephenson built the first railroad in England. By 1850, thousands of miles of tracks crossed Europe.

Critical Viewing England's Iron Bridge, built in 1779, was the world's first arch bridge made of iron. Based on what you have read, what made this bridge possible?

Impact of the Revolution

The Industrial Revolution had a tremendous impact on how people worked and lived. Cities grew rapidly because people migrated there for factory jobs. Standards of living rose, and a prosperous middle class grew.

However, factory workers often faced harsh conditions. Laborers worked as many as 16 hours a day. Child labor was common. Boys and girls as young as five years of age worked in factories and mines. Some were chained to their machines.

Many workers lived in small, crowded houses in neighborhoods where open sewers were common. Diseases spread quickly in these cramped buildings. Over time, the workers' quality of life improved as sewer systems were created and other public health acts were passed.

Before You Move On

Summarize In what ways did the Industrial Revolution change how people lived and worked?

ONGOING ASSESSMENT

MAP LAB

GeoJournal

1. **Interpret Maps** Where in Europe were most of the industries concentrated? What does this suggest about the economies of countries in other parts of Europe?
2. **Human-Environment Interaction** Which industry was the most widespread in Europe? Why was this such an important industry?
3. **Evaluate** Based on the map, which countries probably imported the fewest raw materials?

3.3 The French Revolution

myNGconnect.com For images of the French Revolution

Digital Library

Main Idea The late 1700s in France was a period of economic and political unrest, which led to the French Revolution and the rise of Napoleon.

By the summer of 1789, the French people had not yet benefited from the Industrial Revolution. Harvests were poor, and prices skyrocketed. On July 14, mobs attacked the Bastille, Paris's ancient prison. This action sparked the French Revolution.

Roots of the Revolution

For years, France's lower and middle classes had suffered injustices. French society was composed of three large groups, called the Three Estates. The First Estate was made up of clergy. The Second Estate was made up of the nobility, or aristocrats. The Third Estate included everyone else, from merchants to peasants. The Third Estate paid most of the taxes but had no voice in government.

The people of the Third Estate began to call for change. Many of them were influenced by the **Enlightenment**. This movement stressed the rights of the individual. The ideas of Enlightenment thinkers like Voltaire and **John Locke** had helped inspire the American Revolution in 1776. The American Revolution, in part, inspired the revolution in France.

The Revolution Begins

In May 1789, the Third Estate demanded reforms, but the king of France, Louis XVI, refused. In response, the Third Estate formed the National Assembly. On August 26, 1789, the assembly issued the *Declaration of the Rights of Man and of the Citizen.* This document guaranteed liberty, equality, and property to citizens. The assembly tried to form a new government in which Louis would share power with an elected legislature. However, he again refused to cooperate.

French citizens stormed the Bastille because they thought it held guns and gunpowder.

The guillotine was considered an efficient and painless method of execution.

1785 — 1790 — 1795

- 1789 Mobs attack the Bastille.
- 1792 Jacobins seize power.
- 1793 King Louis XVI and Marie Antoinette are executed; Reign of Terror begins.
- 1794 Robespierre is executed, and the Reign of Terror ends.

The Radicals Take Over

Finally, in 1792, the Jacobins, a group of **radicals**, or extremists, seized power and formed the National Convention. The following year, the group executed Louis XVI and Marie Antoinette, his queen.

The violence soon got worse. Jacobin leader Maximilien Robespierre led a **Reign of Terror**. The Jacobins used a machine called the **guillotine** (GHEE uh teen) to cut off the heads of an estimated 40,000 people. In July 1794, the French finally turned on Robespierre and executed him.

Napoleon's Rise

After five years of violence, the French were exhausted. France was at war with Prussia, Austria, and Britain, and the government was not ruling effectively.

A young general, **Napoleon Bonaparte**, saw his chance and overthrew the government. Over the next five years, Napoleon increased his powers. He then declared himself Emperor Napoleon I and set about conquering other European powers and building an empire. Britain and Prussia finally defeated him in 1815.

Before You Move On

Summarize What led to the French Revolution and the rise of Napoleon?

Critical Viewing Marie Antoinette, shown here, was often accused of reckless spending. What details in this painting support this accusation?

ONGOING ASSESSMENT

SPEAKING LAB

Express Ideas Through Speech Get together in a group and do research to prepare a panel discussion in which you will present the viewpoints of various figures from this section.

Step 1 Decide who each person in your group will be. You might choose from King Louis XVI, Marie Antoinette, Maximilien Robespierre, and Napoleon or be a member of the Third Estate.

Step 2 Come up with a few questions that your panel will discuss. The questions should focus on the French Revolution, the Reign of Terror, and Napoleon's rise.

Step 3 Present the panel discussion before the class. At its conclusion, invite questions and answer them in character.

1800 — 1805 — 1810

1799 Napoleon overthrows the French government.

1804 Napoleon names himself Emperor.

1815 Napoleon is defeated.

Statue of Napoleon on horseback

SECTION 3 DOCUMENT-BASED QUESTION

3.4 Declarations of Rights

TECHTREK

myNGconnect.com For photos of the documents and Guided Writing

Digital Library

Student Resources

As you have learned, thinkers like John Locke and Voltaire led the Enlightenment. They asserted that people have **natural rights**, or rights that people possess at birth, such as life, liberty, and property. Two key documents describe these rights: the American Declaration of Independence and the French *Declaration of the Rights of Man and of the Citizen*. In 1993, Nelson Mandela of South Africa received the Nobel Peace Prize. In his speech at the ceremony, he explained that the rights detailed in the declarations are still important.

DOCUMENT 1

from the Declaration of Independence (July 4, 1776)

We hold these truths to be self-evident, that all men are created equal, that they are endowed [provided] by their Creator with certain unalienable [guaranteed] Rights, that among these are Life, Liberty, and the pursuit of Happiness; that, to secure these rights, Governments are instituted among Men, deriving their just powers from the consent of the governed.

This painting illustrates the signing of the Declaration of Independence.

CONSTRUCTED RESPONSE

1. What rights are citizens guaranteed?

DOCUMENT 2

from the Declaration of the Rights of Man and of the Citizen (August 26, 1789)

The representatives of the French people, organized as a National Assembly, . . . have determined to set forth in a solemn declaration the natural, unalienable, and sacred rights of man. Articles:

1. Men are born and remain free and equal in rights. Social distinctions [classes] may be founded only upon the general good.
2. The aim of all political association is the preservation of the natural . . . rights of man. These rights are liberty, property, security, and resistance to oppression.

CONSTRUCTED RESPONSE

2. Think about what you learned in Section 3.3 about the roots of the French Revolution. In what ways might the ideas in this document have inspired the French people to revolt?

MEN ARE BORN AND REMAIN FREE AND EQUAL IN RIGHTS.

— DECLARATION OF THE RIGHTS OF MAN AND OF THE CITIZEN

Mandela and fellow Nobel recipient, F. W. de Klerk, were elected co-presidents of South Africa in 1994.

DOCUMENT 3

from Nobel Lecture by Nelson Mandela (December 10, 1993)

Nelson Mandela helped lead the struggle to end **apartheid** (uh PAHRT hyt) in South Africa. This system had denied black South Africans their rights. In recognition of his efforts, Mandela received the Nobel Peace Prize. The following excerpt is from his acceptance speech.

> The value of our shared reward will and must be measured by the joyful peace which will triumph, because [of] the humanity that bonds both black and white into one human race. . . .
>
> Thus shall we live, because we will have created a society which recognizes that all people are born equal, with each entitled in equal measure to life, liberty, prosperity, human rights, and good governance.

CONSTRUCTED RESPONSE

3. How do the rights Mandela discusses reflect those described in Documents 1 and 2?

ONGOING ASSESSMENT

WRITING LAB

GeoJournal

DBQ Practice Think about the ideas in the Declaration of Independence and the *Declaration of the Rights of Man and of the Citizen*. How did these ideas influence Nelson Mandela?

Step 1. Review your answers to Constructed Response questions 1, 2, and 3.

Step 2. On your own paper, jot down notes about the main ideas expressed in each document.

Document 1: Declaration of Independence

Main Idea(s) ______________________

Document 2: Declaration of the Rights of Man and of the Citizen

Main Idea(s) ______________________

Document 3: Nobel Lecture

Main Idea(s) ______________________

Step 3. Use your notes to construct a topic sentence that answers this question: How did the Declaration of Independence and the *Declaration of the Rights of Man and of the Citizen* influence Nelson Mandela?

Step 4. Write a paragraph that explains specific phrases and ideas in the Declaration of Independence and the *Declaration of the Rights of Man and of the Citizen.* Go to **Student Resources** for Guided Writing support.

3.5 Nationalism and World War I

myNGconnect.com For an online map of Europe before World War I

Main Idea Nationalism, new alliances, and growing tensions in Europe led to World War I.

After the French Revolution, the French people developed powerful feelings of nationalism. **Nationalism** is a strong sense of loyalty to one's country. During the 1800s, nationalism swept through Europe.

Italy and Germany Unify

Nationalism led to unification efforts in Italy and Germany. In 1800, the Italian Peninsula was made up of separate city-states. In 1870, the states came together to form a unified Italy. Germany was also composed of many different states in the early 1800s. Beginning in 1865, Prussia, the most powerful German state, led the way to unification. Driven by nationalist feelings, Prussia fought to take control of other German states away from their non-German rulers. In 1871, the states came together as a united German Empire.

Growing Tensions in Europe

By 1900, tensions had begun to grow among European powers. Nationalism had united some countries from within. However, nationalism also created fierce competition among rival countries. Mainly, the countries competed for raw materials and colonies in Africa and Asia. To strengthen their position, Britain, France, and Russia formed an **alliance**, or agreement to work toward a common goal, called the Triple Entente. The German Empire and Austria-Hungary formed an alliance known as the Central Powers.

These alliances were tested in June 1914, when Archduke Franz Ferdinand of Austria-Hungary was assassinated in Serbia by a nationalist from Bosnia-Herzegovina. The assassin belonged to a group that was unhappy with Austrian rule of Bosnia-Herzegovina and wanted to unite with Serbia. Immediately after the assassination, Austria-Hungary declared war on Serbia. Then, because Serbia was a Russian ally, Russia declared war on Austria-Hungary. Within weeks, much of Europe had been drawn into war.

A Brutal War

The Great War, as it was called, dragged on for four brutal years. Both sides fought from **trenches**, or long ditches that protected soldiers from the enemy's gunfire. Both sides also used deadly technology, including machine guns, airplanes, tanks, and poison gas. German U-boats, or submarines, sank British ships.

1870 Italy unifies.

1870

1871 German states unite to form the German Empire.

Prussian prime minister Otto von Bismarck oversaw German unification.

1885

Illustration of Archduke Ferdinand's assassination

1900

1914 Archduke Ferdinand is assassinated; World War I begins.

In 1917, Germany seemed to gain an advantage when the Communist Party seized control of Russia's government and economy and made peace with Germany. That same year, the United States entered the war on the side of France and Britain. The fresh American troops helped turn the tide against Germany. In 1918, Germany surrendered to France, Britain, and the United States. By the time the war ended, ten million soldiers had died. About seven million civilians also lost their lives.

Impact of the War

In 1919, Germany signed the **Treaty of Versailles**. Under this peace treaty, Germany was forced to pay several billion dollars in damages and accept full blame for the war. Many of Germany's territories were taken away, and new countries were formed, including Austria, Hungary, Czechoslovakia, Yugoslavia, and Turkey. The treaty angered and humiliated the German people and did little to ease tensions in Europe. These tensions would help lead the way to another world war in a little more than 20 years.

EUROPE BEFORE WORLD WAR I 1914

Triple Entente
Neutral countries that joined the Triple Entente
Central Powers
Neutral countries that joined the Central Powers
Countries that remained neutral

Arctic Circle, Iceland (Denmark), 20°W, 60°N, Prime Meridian, ATLANTIC OCEAN, NORWAY, SWEDEN, North Sea, DENMARK, Copenhagen, Baltic Sea, UNITED KINGDOM, London, NETH., GERMAN EMPIRE, BELGIUM, LUX., Paris, FRANCE, SWITZ., AUSTRIA-HUNGARY, RUSSIA, Moscow, Sarejevo, ROMANIA, SERBIA, Sofia, BULGARIA, Black Sea, 40°E, PORTUGAL, SPAIN, 40°N, Corsica, Rome, MONT., ALBANIA, ITALY, Sardinia, Sicily, GREECE, Athens, OTTOMAN EMPIRE, Mediterranean Sea, 0°, 20°E, N, E, S, W

0 300 600 Miles
0 300 600 Kilometers

Before You Move On

Monitor Comprehension In what ways did nationalism, new alliances, and growing tensions in Europe lead to World War I?

ONGOING ASSESSMENT

MAP LAB

GeoJournal

1. **Region** According to the map, which empires ruled much of Europe in 1914?
2. **Make Inferences** Note the countries that remained neutral in the war. What geographic factors might have encouraged their neutrality?

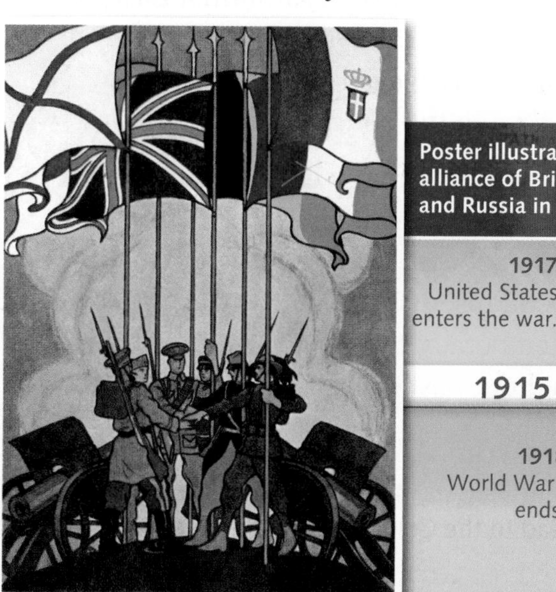

Poster illustrating the alliance of Britain, France, and Russia in 1915

Leaders who signed the treaty

French prime minister Georges Clemenceau

American president Woodrow Wilson

British prime minister David Lloyd George

3.6 World War II and the Cold War

TECHTREK
myNGconnect.com For an online map of post-war Europe

Maps and Graphs

Main Idea After World War II was fought to defeat the Axis Powers, the Cold War developed between the democratic United States and the Communist Soviet Union.

At the end of World War I, Germany lost its military power. As you have read, the Treaty of Versailles also placed full blame for the war on Germany and forced it to pay **reparations**, or money to cover the losses suffered by the victors. The Great Depression, which began in 1929, further damaged Germany's economy. The **Great Depression** was a severe downturn in the world's economy. During this crisis, **Adolf Hitler** rose to power in Germany.

World War II

Hitler became the leader of the National Socialist German Workers' Party, or the Nazis. In 1936, Hitler made an alliance with Italy. Germany also formed an alliance with Japan, where the military had seized power. Germany, Italy, and Japan formed the Axis Powers.

Germany's invasion of Poland in 1939 started World War II. Two of Poland's allies, Great Britain and France, declared war on Germany soon after the invasion. Germany responded by conquering Poland and then quickly took over most of Europe, including France.

In 1941, Japan attacked the United States at Pearl Harbor, Hawaii. As a result, the United States abandoned its neutrality and entered the war on the side of Britain and the Soviet Union. Together, they were known as the Allies. Over time, many other countries took sides and joined either the Allies or the Axis Powers.

After more than five years of war, Germany surrendered on May 8, 1945. Allied troops were stunned to find the Nazi **concentration camps** where six million Jews and other victims had been murdered. This mass slaughter was called the **Holocaust**. Japan continued to fight until the United States dropped atomic bombs on Hiroshima and Nagasaki. Japan surrendered on September 2, 1945.

The Cold War

After World War II, the Soviet Union established Communist governments in Eastern Europe. Germany was divided into Communist East Germany and democratic West Germany. The imaginary boundary that separated Eastern and Western Europe was called the **Iron Curtain**. The division marked the beginning of the **Cold War**, a period of great tension between the United States and the Soviet Union.

To defend against possible attack, both sides forged military alliances. Western Europe and the United States formed NATO (North Atlantic Treaty Organization), while Communist Eastern Europe formed the Warsaw Pact. The two never directly waged war against each other during the course of the Cold War.

In the 1980s, many eastern European countries overthrew their Communist governments. In 1991, the Soviet Union itself collapsed. The Cold War ended, and democracy replaced communism throughout Eastern Europe.

Before You Move On

Make Inferences In what ways did World War II help lead to the Cold War?

Visual Vocabulary
The **Berlin Wall** divided Communist East Berlin from democratic West Berlin. It was torn down in 1989.

ONGOING ASSESSMENT

PHOTO LAB

GeoJournal

1. **Analyze Visuals** Study the photo. In what way was the Berlin Wall a solid symbol of the Iron Curtain?
2. **Evaluate** Soldiers patrolled the wall on East Berlin's side. How can you tell that this photo was taken on the West Berlin side?

Review

VOCABULARY

Match each word in the first column with its definition in the second column.

WORD	DEFINITION
1. ecosystem	a. sold to relax penalty for sin
2. democracy	b. strong sense of loyalty to one's country
3. plebeians	c. community of living organisms and their environment
4. indulgence	d. area controlled by a distant country
5. colony	e. common people
6. nationalism	f. government of the people

MAIN IDEAS

7. What are some of the significant islands in Europe? (Section 1.2)
8. Where is much of Europe's farming industry located? (Section 1.3)
9. What events led to the development of democracy in ancient Greece? (Section 2.1)
10. How were plebeians represented in the Roman Republic? (Section 2.3)
11. In what way did the Roman Empire influence language? Why? (Section 2.4)
12. In what ways did the Roman Catholic Church serve as a unifying force in Western Europe? (Section 2.5)
13. What achievements in arts and literature did the Renaissance inspire? (Section 2.6)
14. How did the Industrial Revolution change Europe? (Section 3.2)
15. Why did the French people welcome Napoleon's rise to power? (Section 3.3)
16. Why did Russia declare war on Austria-Hungary in 1914? (Section 3.5)
17. What was the Cold War? (Section 3.6)

GEOGRAPHY

ANALYZE THE ESSENTIAL QUESTION

How did Europe's physical geography encourage interaction with other regions?

Critical Thinking: Evaluate

18. In what ways do rivers like the Danube make trade easier within Europe?
19. Why has trade been central to Europe's growth throughout its history?

EARLY HISTORY

ANALYZE THE ESSENTIAL QUESTION

How did European thought shape Western civilization?

Critical Thinking: Draw Conclusions

20. What elements of the democracy practiced in ancient Greece did the United States adopt?

INTERPRET MAPS

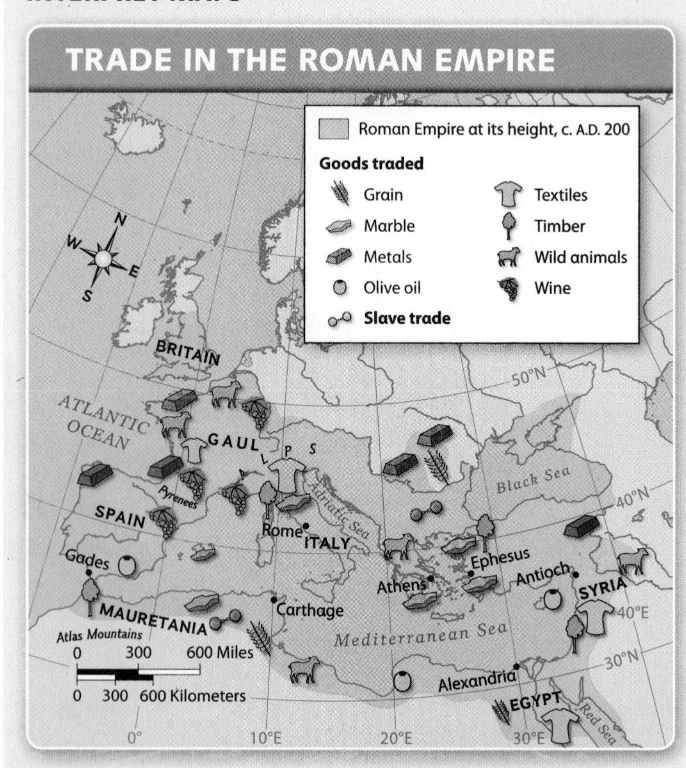

21. **Movement** From what part of the Empire did Rome obtain its grains? its textiles?

EMERGING EUROPE

ANALYZE THE ESSENTIAL QUESTION

How did Europe develop and extend its influence around the world?

Critical Thinking: Make Inferences

22. In what way did improvements in navigation and shipbuilding lead Europe to establish colonies in other parts of the world?

23. Why did the Industrial Revolution make European leaders eager to establish colonies in the Americas and Asia?

24. What are some of the positive effects of nationalism? What are some negative effects?

INTERPRET TABLES

MILES OF RAILWAY TRACK IN SELECTED EUROPEAN COUNTRIES (1840–1880)

	1840	1860	1880
Austria-Hungary	144	4,543	18,507
Belgium	334	1,730	4,112
France	496	9,167	23,089
Germany	469	11,089	33,838
Great Britain	2,390	14,603	25,060
Italy	20	2,404	9,290
Netherlands	17	335	1,846
Spain	0	1,917	7,490

Source: Modern History Sourcebook

25. Analyze Data How much railway track did the Germans build between 1840 and 1880? What might account for this increase?

26. Draw Conclusions France and Spain are almost the same size. Note the difference in the extent of the railway system in each one in 1880. What does this suggest about the level of industrialization in each country?

ACTIVE OPTIONS

Synthesize the Essential Questions by completing the activities below.

27. Write Tour Notes Suppose that you are going to lead a group of tourists on a trip on one of Europe's rivers. Select and research the river. You might choose the Danube, Rhine, Tiber, Rhone, Thames, or Seine. Then write notes for a guided tour of the river. Start by describing its location, size, and appearance. Next, point out important sites along the river and explain their historical significance. Finally, discuss the river's uses today. **Gather photos of the river and conduct the tour with your group of "tourists."**

Writing Tips
- Use language that appeals to the senses to help your tourists see and experience the river.
- Include stories about the sites and historical events to hold your audience's interest.
- Involve your audience by comparing the river to one that is familiar to them.

myNGconnect.com For research links on European history

28. Create a Slide Show Prepare a slide show of famous European buildings, using **Connect to NG** or other online sources. Research and identify five buildings, such as the Pantheon in Rome, Italy. Write two sentences that explain the importance of each building to European history. Copy the chart below to help you organize your information.

BUILDING	IMPORTANCE TO EUROPE
1.	
2.	
3.	

CHAPTER 10

EUROPE TODAY

PREVIEW THE CHAPTER

Essential Question How is the diversity of Europe reflected in its cultural achievements?

SECTION 1 • CULTURE

KEY VOCABULARY

- dialect
- heritage
- perspective
- abstract
- troubadour
- opera
- genre
- epic poem
- novel
- staple
- cuisine

ACADEMIC VOCABULARY

cosmopolitan

TERMS & NAMES

- Romantic period
- impressionism
- Baroque period
- Classical period

Essential Question What are the costs and benefits of European unification?

SECTION 2 • GOVERNMENT & ECONOMICS

KEY VOCABULARY

- tariff
- currency
- euro
- sovereignty
- eurozone
- consumer
- democratization
- privatization
- demographics
- aging population

ACADEMIC VOCABULARY

exchange, assimilate

TERMS & NAMES

- Common Market
- European Union (EU)
- Orange Revolution

Buildings with traditional red-tiled roofs line this square in Prague in the Czech Republic.

1.1 Languages and Cultures

TECHTREK
myNGconnect.com For photos reflecting European culture

Digital Library

Main Idea Europe has a great variety of languages, cultures, and cities.

Europe has more than a half billion people, yet they live in an area that is one-half the size of the United States. In addition, the continent of Europe contains more than 40 countries. The result is a great diversity, or wide variety, of languages and cultures.

European Languages

Many of the languages spoken in Europe today fall into three language groups: Romance, Germanic, and Slavic. The Romance languages include French, Spanish, and Italian. The Germanic languages are spoken mostly in northern Europe and include German, Dutch, and English. Most people in Eastern Europe speak Slavic languages, such as Russian, Polish, and Bulgarian.

Some countries in Europe have more than one official language. Belgium, for example, has three: Dutch, French, and German. Even in countries with only one official language, people may speak different dialects. A **dialect** is a regional variety of a language. In Italy, for instance, people in Rome speak a dialect of Italian that differs from that in other cities.

Cultural Traditions

Because Europe is composed of many different countries and ethnic groups, it has a rich cultural **heritage**, or tradition. Europe's cultural diversity is reflected in its religions and celebrations.

Christianity is the dominant religion in Europe. Today, about 45 percent of the continent's total population is Catholic. Protestantism is most common in Northern Europe.

Critical Viewing In this photo of the Palio, a centuries-old cultural tradition is honored as these horses race in Italy. What details in the photo convey the excitement of the race?

In recent years, Islam has become the fastest-growing religion in Europe. Immigrants from Turkey, North Africa, and Southwest Asia move to Europe and bring their Muslim faith with them.

Many of the holidays celebrated in Europe are rooted in religion. However, Europeans enjoy other kinds of festivals as well. One of the most colorful is the Palio, a horse race held each summer in Siena, Italy. In this race, which dates back to the Middle Ages, ten riders from ten of the city's neighborhoods compete.

City Life

More than 70 percent of Europeans live in urban areas. In Belgium, over 95 percent of the people live in or near its cities. Most of Europe's cities are **cosmopolitan**, which means that they bring together many different cultures and influences. London is an example of a cosmopolitan city. Its restaurants and shops reflect the South Asian, Caribbean, and East Asian origins of some of its newer citizens.

Many European cities date back hundreds of years. As a result, they developed in ways very different from American cities. These cities are often smaller in area than those in the United States and have narrow, winding streets. Most people live in apartments rather than individual houses. For recreation, city dwellers visit their many parks. They also tend to use public transportation more often than most Americans.

Before You Move On

Make Inferences What might be some of the advantages and disadvantages of Europe's great variety of languages and cultures?

ONGOING ASSESSMENT

PHOTO LAB

GeoJournal

1. **Analyze Visuals** Note the way the riders are dressed in the photo. What do you suppose their clothes represent?
2. **Draw Conclusions** Study the photo and recall what you have read about the Palio. How is this race an example of a cultural tradition?
3. **Turn and Talk** Get together with a partner and discuss the holidays and festivals that you celebrate. Take notes on your discussion and be prepared to share with the rest of the class.

SECTION 1 CULTURE

1.2 Art and Music

TECHTREK
myNGconnect.com For photos of European art and samples of European music

Digital Library

Main Idea European art and music have developed over thousands of years.

Throughout the centuries, European art and music have changed to reflect different styles and beliefs.

European Art

European art grew out of the artistic achievements of ancient Greece and Rome. Greek and Roman gods and goddesses were frequent subjects of the artists from these cultures, but they were portrayed to represent realistic human forms.

Much of the art of the Middle Ages reflected the influence of Christianity. The religious subjects were often presented as two-dimensional figures. During the Renaissance, artists used **perspective** to give their work greater depth. Although religious subjects were common, artists also painted portraits of people.

In the **Romantic period** of the early 1800s, artists moved away from religious themes to paint landscapes and other natural scenes that would convey emotion. **Impressionism** emerged in the late 1800s. Impressionist artists, such as Claude Monet, used light and color to capture a moment. By 1900, artists wanted to create a new form of art. These modern artists often worked in an **abstract** style, which emphasized form and color over realism.

Critical Viewing The *Mona Lisa*, by Italian Renaissance artist Leonardo da Vinci, is probably the most famous portrait of all time. What about this painting might account for its popularity?

Critical Viewing This painting, *Impression, Sunrise*, by French artist Claude Monet, gave the impressionist movement its name. What kind of mood is conveyed in this painting?

Most opera houses contain a stage, an orchestra pit, and several levels of balconies. The ceiling of the Paris Opéra, shown here, was painted by Russian-born artist Marc Chagall.

European Music

Like art, European music began in ancient Greece and Rome. Musicians played on a few simple instruments and were often accompanied by singers.

During the Middle Ages, music was used in religious ceremonies. Singers called **troubadours** performed songs about knights and love. These songs influenced Renaissance music, when instruments such as the violin were introduced.

The new instruments helped inspire the complex rhythms in the music of the **Baroque period**, which lasted from about 1600 to 1750. **Opera**, which tells a story through words and music, was born then.

The **Classical** and Romantic periods followed the Baroque period and continued until about 1910. Composers from these two periods, such as Ludwig van Beethoven (BAY toh vuhn) of Germany, wrote works using instruments and techniques that are still used today.

Before You Move On

Monitor Comprehension What styles and beliefs have influenced European art and music?

ONGOING ASSESSMENT

LISTENING LAB

GeoJournal

1. **Analyze Audios** Listen to the music clip of Beethoven's Fifth Symphony in the **Digital Library**. Describe the music's mood. What instruments help to convey this mood?
2. **Form and Support Opinions** What do you think of the opening? Support your opinion by referring to specific details in the music.

1.3 Europe's Literary Heritage

TECHTREK
myNGconnect.com For photos of European writers and Guided Writing

Digital Library

Student Resources

Main Idea European literature has reflected new ways of thinking over the centuries.

Plays by the English playwright William Shakespeare (1564–1616) are performed almost every day. European writers such as Shakespeare have influenced literature for centuries. They wrote in many different **genres**, or forms of literature, including poetry, plays, and novels.

Literary Origins

European literature began with the ancient Greeks and Romans. Around 800 B.C., the Greek poet Homer wrote the epic poems *The Iliad* and *The Odyssey*. An **epic poem** is a long poem that tells the adventures of a hero who is important to a particular nation or culture. Around 20 B.C., the Roman poet Virgil wrote *The Aeneid*, an epic poem about the founding of Rome.

One of the greatest writers of the late Middle Ages and early Renaissance was the Italian poet Dante (1265–1321). As you have learned, Dante wrote *The Divine Comedy* in Italian, not in Latin. The epic poem deals with the religious beliefs and politics of his time.

Many later works of the Renaissance focused on human behavior. Shakespeare explored this theme in plays such as *Hamlet*. Spanish writer Miguel de Cervantes (1547–1616) wrote what is considered the first modern novel, *Don Quixote* (kee HO tee). A **novel** is a long work of fiction, containing characters and a plot. The printing press, which Johannes Gutenberg developed in the 1450s, helped spread the popularity of these books.

The 1700s and 1800s

In the mid-1700s, Enlightenment ideas about reason and government inspired the movement toward democracy. These ideas, in turn, led French and English writers of the time, such as Voltaire and John Locke, to explore the rights of the individual.

In the 1800s, writers of the Romantic period continued this exploration, with an emphasis on emotion and nature. For example, German author Johann Wolfgang von Goethe (GHER tuh) (1749–1832) wrote *The Sorrows of Young Werther*, a novel about a sensitive young artist.

Other writers of the 1800s took a much more realistic look at life. In novels such as *Sense and Sensibility*, British writer Jane Austen (1775–1817) used humor to examine women's role in society.

Critical Viewing Inspired by tales of knights, Don Quixote (right) goes to battle evil with his servant Sancho Panza (left). What details in this painting suggest that Cervantes' novel is a comedy?

Critical Viewing Austen's novels typically end with marriage, such as the one shown in this scene from a film adaptation of *Sense and Sensibility*. Based on the photo, how would you describe an English wedding in the 1800s?

Another British writer, Charles Dickens (1812–1870), commented on social issues, including poverty, in such novels as *Oliver Twist*. Norwegian playwright Henrik Ibsen (1828–1906) wrote plays, such as *A Doll's House*, which criticized the traditional role of husbands and wives at that time.

Modern Literature

The two world wars had a great impact on the modern literature of the 20th century. Writers at this time reflected the sense that life was uncertain and unpredictable. Some rejected traditional genres and experimented with writing new forms of plays, poems, and novels.

Many modern writers examined the inner workings of the mind. In the novel *Ulysses*, for example, Irish writer James Joyce (1882–1941) focused on the thought processes of the main character over the course of a single day. Romanian playwright Eugene Ionesco (1909–1994) used ridiculous situations to comment on what he saw as the emptiness of life. Many European and other writers today have been influenced by these authors.

Before You Move On

Summarize What new ways of thinking has European literature reflected over the centuries?

ONGOING ASSESSMENT

WRITING LAB

Write Reports Think about what you have learned about Europe's literary heritage. Then consider the following question: In what ways do beliefs and events influence literature? Select a writer mentioned in this lesson and write a report in which you answer this question.

Step 1 Research to learn more about the writer and the period in which he or she lived.

Step 2 Find out how the beliefs and events of the time influenced the writer.

Step 3 Write a brief report in which you explain these influences. Support your ideas with specific references to one or two of the writer's works. Go to **Student Resources** for Guided Writing support.

1.4 Cuisines of Europe

TECHTREK 
myNGconnect.com For photos of traditional European foods
Digital Library

Main Idea Landforms and climate have influenced the cooking traditions of Europe.

Throughout most of Europe, foods such as meat, bread, and cheese are **staples**, or basic parts of people's diets. However, the **cuisine**, or cooking traditions, of most European countries is largely determined by the landforms and climate of a particular region.

Foods of Western Europe

The hot, dry climate in the Mediterranean countries of Spain, France, Italy, and Greece is ideal for growing olives, tomatoes, and garlic. As a result, these are key ingredients in their cuisines. Fish from the bodies of water surrounding the countries is also an important menu item.

The cuisines of France and Italy have influenced cooking throughout the world and are especially known for their sauces. French sauces are typically made of milk or cheese, while those of Italy are often tomato based.

People in western European countries with cooler climates often eat more filling fare. In Germany, Great Britain, and Ireland, potatoes grow well and are popular side dishes. In contrast with the light sauces of France and Italy, German cooks often serve heavier gravies.

In Scandinavian countries such as Sweden and Norway, the people often eat herring and other fish. Scandinavians also enjoy deer meat provided by the herding culture of these northern countries.

Critical Viewing The French often enjoy long, relaxed meals with friends and family. What attitude toward life and food does this traditional way of eating suggest?

Foods of Eastern Europe

Eastern Europe's cold climate results in a shorter growing season than that of Western Europe. In Russia, root vegetables such as turnips and beets are well adapted to the country's climate. A soup called *borscht*, which is made from beets, is a traditional offering on cold winter nights.

The fertile soil of Hungary allows Hungarian farmers to grow grains and potatoes. These crops are used to make a variety of breads and dumplings. A meat stew called *goulash* is Hungary's national dish. It is made with beef, potatoes, and vegetables, and seasoned with paprika. Paprika is a red spice that was brought to Hungary by the Turks in the 1500s.

Bread like this round loaf is the traditional centerpiece at a Ukrainian wedding.

Like Hungary, Ukraine has fertile soil and fields of wheat and other grains. The country is known for its bread. On special occasions, cooks prepare breads decorated with ornaments made of dough.

Before You Move On

Summarize How have landforms and climate influenced European cooking traditions?

ONGOING ASSESSMENT

SPEAKING LAB

GeoJournal

Turn and Talk What is the traditional cuisine of the United States? What impact have dishes brought by immigrants from other countries had on how and what Americans eat? Get together in a small group and discuss these questions. Be prepared to share your ideas with other groups.

2.1 The European Union

TECHTREK
myNGconnect.com For an online map and news about the European Union

Maps and Graphs

Connect to NG

Main Idea The European Union was formed to unite Europe and benefit it economically.

In 1948, the United States established a program called the Marshall Plan to help Europe rebuild after World War II. To manage the U.S. aid money, European countries formed the Organization for European Economic Cooperation in 1948. As a result, European countries discovered that they could rebuild their countries faster when they worked together.

The Common Market

In 1957, some European countries sought even closer economic ties. They formed the European Economic Community (EEC), which became known as the **Common Market**. The first countries to join were Belgium, France, Italy, Luxembourg, the Netherlands, and West Germany. Several more countries joined during the 1980s.

The Common Market pledged to create "an ever closer union among the European peoples." However, it was primarily formed to create a single market among the member nations. A single market is one in which a group of countries trades across its borders without restrictions or tariffs. A **tariff** is a tax paid on imports and exports.

A United Europe

In 1992, the countries of the Common Market sought to extend their economic organization throughout Europe. They met in Maastricht in the Netherlands and signed the Treaty of Maastricht, which created the **European Union (EU)**. By 2010, the EU had 27 member nations with a total of more than 500 million people. When considered as a single economy, the EU is the largest in the world.

European Union flag

Critical Viewing The EU flag, seen here in front of the organization's Parliament building in Strasbourg, France, is also the flag of Europe. The flag's circle of stars represents European unity. Why is this a fitting symbol for the EU?

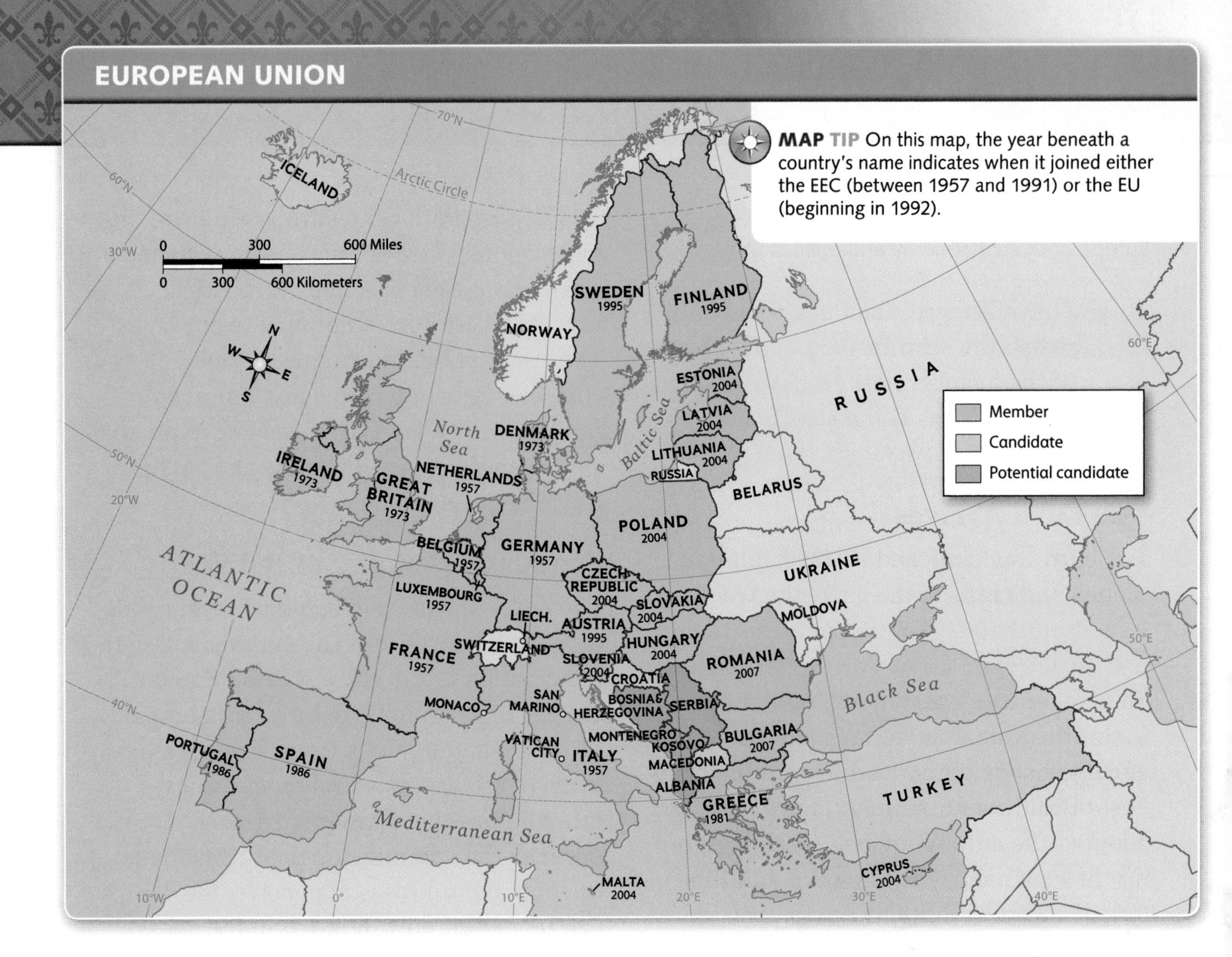

The EU has a government with an executive, a legislative, and a judicial branch. These branches propose, pass, and enforce the organization's policies and legislation. The EU also has agencies that direct economic policies. Through these agencies, the EU has eliminated tariffs among most member nations and founded a European bank. In 1999, the EU created a common **currency**, or form of money, called the **euro**. By 2011, 17 member nations had adopted this currency.

One of the requirements for joining the EU is having a stable democracy that respects human rights. Some countries that have applied for membership, such as Turkey, are still under review. However, other European countries, including Norway, have chosen not to join. Norway does not want to give up its **sovereignty**, or control over its own affairs.

Before You Move On

Monitor Comprehension How has the European Union helped unite Europe?

ONGOING ASSESSMENT

MAP LAB

GeoJournal

1. **Interpret Maps** Study the map and recall what you have learned about Europe's geography. In what ways does Europe's geography encourage unity?
2. **Draw Conclusions** Based on the map, in what area are most of the members who joined in later years located? What political situation might have prevented them from joining earlier?

2.2 The Impact of the Euro

TECHTREK
myNGconnect.com For photos of the euro

Digital Library

Main Idea The euro has helped to unify Europe both economically and politically.

As you have learned, the European Union (EU) created the euro in 1999. Since then, many of the member nations have adopted the euro, which has had a significant impact on Europe.

The Euro Arrives

The euro was launched in 1999, but paper money and coins of the currency were not issued until 2002. The 17 countries that use the euro are known as the **eurozone**. Some countries that belong to the EU, including Romania, hope to join the eurozone soon. Other countries, including Great Britain and Denmark, have not adopted the euro. They believe that giving up their own currency might result in a loss of control over their economies.

The symbol for the euro is €. Euro paper money, or banknotes, is the same throughout the eurozone. Euro coins, however, differ from country to country. The front, or common, side of each coin has the same image and a number indicating its value. The back, or national, side shows a design that was chosen by the member nation.

Economic Benefits of the Euro

The euro allows people, money, and goods to move freely within the eurozone. Before the creation of the euro, when a French citizen traveled to Germany, for example, he or she had to pay a fee to **exchange**, or convert, francs—the French currency—into marks—the German currency. Because the common currency has made travel easier and less expensive, tourism has increased within much of Europe.

The prices of fruit in this Italian market are in euros.

A single currency means lower fees for conducting business. As a result, trade has increased among European nations by an estimated 10 percent since 2002. The currency has also made costs easier to compare for companies within the eurozone. As a result, companies can import the least expensive products and then pass along the savings to **consumers**, the people who buy the goods.

Political Benefits of the Euro

In 2010, the unity of the eurozone was tested. Greece and Ireland—countries in the eurozone—were deeply in debt. To help them manage their debt, the other eurozone nations loaned the two countries money. In return, Greece and Ireland had to raise taxes and reduce spending. By cooperating, the eurozone was able to help two of its member nations.

1-EURO COIN OF THREE EUROZONE COUNTRIES

Front	Back	
		AUSTRIA Mozart, Austrian composer
		GERMANY Eagle, German symbol
		IRELAND Harp, Irish symbol

Before You Move On

Summarize In what ways has the euro helped Europe unify economically and politically?

ONGOING ASSESSMENT

VIEWING LAB

GeoJournal

1. **Compare and Contrast** Study the euros in the diagram. Note that the front of the coin is the same for all three countries shown. What elements appear on the front of each euro? What does each element represent?
2. **Make Inferences** Why might the countries have wanted their own design on the euro?
3. **Conduct Internet Research** Go online to find out what impact helping Greece and Ireland has had on the euro and the eurozone. Share your findings with the class.

2.3 Democracy in Eastern Europe

TECHTREK
myNGconnect.com For photos reflecting democratic progress in Eastern Europe

Digital Library

Main Idea Eastern European countries have faced many challenges in their transition to democracy.

After World War II, many eastern European countries came under the control of the Soviet Union. The citizens of these countries lacked democratic freedoms and had a low standard of living. In 1981, Poland rebelled peacefully against its Communist government. By the late 1980s, similar rebellions had spread throughout Eastern Europe. Finally, in 1991, Russia and several other republics declared their independence, and the Soviet Union collapsed.

The Road to Democracy

Since gaining their independence, Poland, Hungary, and the Czech Republic developed stable democratic governments In other countries, **democratization**, or the process of becoming a democracy, has been more difficult to achieve. In 1991, civil war broke out among ethnic groups in Yugoslavia. Over time, the country divided into several new democratic countries, including Serbia and Croatia.

Ukraine has also had setbacks. In 2004, the Ukrainian people staged the **Orange Revolution** and peacefully removed their prime minister, Viktor Yanukovych.

Critical Viewing Polish citizens shop and relax in a spacious mall in Warsaw. What does the shopping complex in this photo suggest about Poland's economy?

Many believed that he was corrupt and was being controlled by Russia. However, their new leader, Viktor Yushchenko, disappointed the Ukrainians. Some believed he had become anti-democratic and blamed him for their weakened economy. In 2010, the voters brought Yanukovych back to power.

Rebuilding Economies

The former Communist countries of Eastern Europe also began to rebuild their economies. They changed from government-controlled economies to market economies. They accomplished this goal through **privatization**. That means that government-owned businesses became privately owned.

Eastern European countries have had mixed results since making the adjustment to a market economy. Poland has had the greatest success. It has a fast-growing economy and exports goods throughout Europe. Other countries have been slower to establish new businesses and become competitive. They have also experienced rises in prices and unemployment.

The leaders of many eastern European countries wish to integrate with the rest of Europe. They want to join the European Union and NATO, a military alliance of democratic states in Europe and North America. While some citizens of eastern European countries believe that they were more secure under Communist leaders, others—particularly young people—disagree. They favor democracy and feel that this form of government can better help them solve their countries' problems.

Before You Move On

Summarize What challenges have eastern European countries faced in their transition to democracy and a market economy?

ONGOING ASSESSMENT

READING LAB

GeoJournal

1. **Identify Problems and Solutions** What problem did Ukraine face in its transition to democracy? In what way did the solution to the problem reflect democratization?
2. **Make Inferences** Why do you think younger eastern Europeans might be more willing than older people to support the democratic movements in their countries?

2.4 Changing Demographics

Main Idea New immigrants are changing Europe.

Every May, people from Germany, Denmark, Hungary, Bulgaria, and other European countries come together to celebrate Europe's diversity on Europe Day. The celebrations reflect Europe's changing **demographics**, the characteristics or the profile of a human population. The population has become more diverse as people from Africa and Asia have immigrated to Europe.

KEY VOCABULARY

demographics, n., the characteristics of a human population, such as age, income, and education

aging population, n., a trend that occurs as the average age of a population rises

ACADEMIC VOCABULARY

assimilate, v., to be absorbed into a society's culture

An Aging Population

For years, Europe has had an **aging population**. In other words, the average age of people on the continent has been rising. This trend has had a number of causes. For one thing, Europeans have been living longer because of better medical care. For another, most families are having fewer children. So today, senior citizens form a higher percentage of Europe's total population.

The trend created a need for more workers to replace the many senior citizens who were retiring. Workers were also needed to keep the economy strong and pay taxes to support such social services as education and health care. As a result, immigrants came to Europe to take the newly available jobs. People immigrated to Great Britain from former colonies, such as India, Pakistan, and Bangladesh. France also attracted immigrants from former colonies, especially Algeria and Morocco. In the 1970s, people from Turkey began coming to Germany for jobs. Germany now has approximately 2 million people of Turkish heritage. The fall of communism in Eastern Europe in the 1980s and 1990s also resulted in increased migration within Europe.

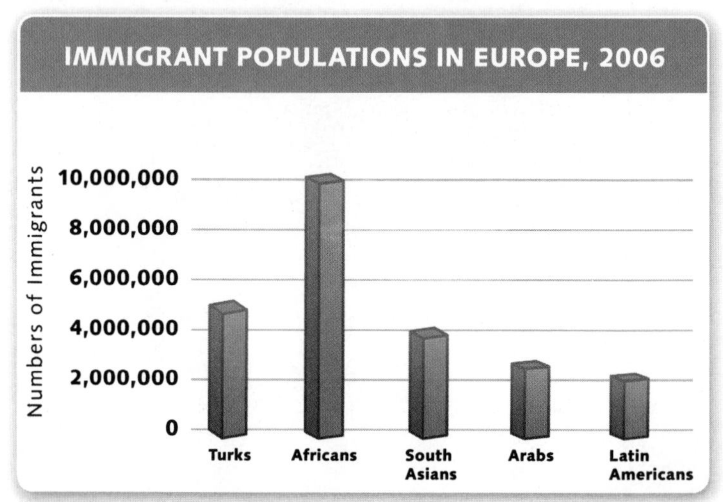

Source: Council of Europe, 2006

Before You Move On

Summarize In what way did Europe's aging population create a need for immigrants?

Reasons for Immigration

Most immigrants have come to Europe to seek a better life. Some left their home countries for economic reasons—usually to find jobs. Others left for political reasons—to escape conflict in their countries or an unjust government. Once they have found work in Europe, many immigrants send money to relatives back home. Eastern Europeans have migrated to Western Europe for similar reasons. Membership within the European Union (EU) has made their migration easier.

Challenges of Immigration

Tensions have sometimes developed between immigrants and native citizens. These tensions often arise when the two groups compete for the same jobs. Also, some immigrants enter European countries illegally—according to some EU estimates, about a half million each year. Housing, educating, and caring for these illegal immigrants can create economic strains on the host countries.

The mix of different cultures has also created problems. Many immigrants are Muslims, with cultural and religious practices that differ from those of their Christian neighbors. Some Europeans would like to see the Muslim immigrants **assimilate**, or be absorbed into their society's culture. They believe the immigrants should adopt European traditions and values. Others believe a more multicultural approach is better. This approach encourages tolerance and embraces all cultures.

COMPARE ACROSS REGIONS

Australia's Skilled Immigrants

Like Europe, Australia has an aging population. Many in the Australian government believe that immigration can help change that trend. As a result, each year the government identifies gaps in the country's workforce. It then determines the number of skilled immigrants that can come to Australia. Between 2008 and 2009, more than 110,000 immigrants arrived on this skilled migration program. Most of these immigrants came from Great Britain, India, and China.

Of course, Europe receives many more immigrants than Australia each year. Italy alone took in more than 400,000 immigrants in 2008. Many Europeans appreciate the cultural enrichment that immigrants bring to their countries. However, some believe that Europe, like Australia, should set immigration limits.

Before You Move On

Monitor Comprehension What are the ways in which new immigrants are changing Europe?

ONGOING ASSESSMENT

READING LAB

1. **Interpret Graphs** According to the bar graph, from which two continents has the immigrant population in Europe primarily come?
2. **Make Inferences** Why might some immigrants resist assimilation?
3. **Identify Problems and Solutions** What has Australia done to solve some of the problems posed by immigration?

CHAPTER 10
Review

VOCABULARY

Match each word in the first column with its meaning in the second column.

WORD	DEFINITION
1. abstract	a. tax on imports and exports
2. dialect	b. control over one's own affairs
3. sovereignty	c. art style emphasizing form and color
4. tariff	d. absorb into another culture
5. privatization	e. privately owned businesses
6. assimilate	f. regional language

MAIN IDEAS

7. Why are there so many different dialects in Europe? (Section 1.1)
8. What is the fastest-growing religion in Europe? Why might that be so? (Section 1.1)
9. During the Romantic period, what themes did artists use in their paintings? (Section 1.2)
10. Why do you think the ancient Greeks and Romans celebrated historic events in epic poems? (Section 1.3)
11. How does the soup called *borscht* reflect Russia's geography? (Section 1.4)
12. What is one of the requirements for joining the European Union? (Section 2.1)
13. In what way has the euro helped increase trade among European nations? (Section 2.2)
14. Why did the Ukrainian people stage the Orange Revolution? (Section 2.3)
15. What factors have led people from other parts of the world to immigrate to Europe? (Section 2.4)

CULTURE

ANALYZE THE ESSENTIAL QUESTION

How is the diversity of Europe reflected in its cultural achievements?

Critical Thinking: Draw Conclusions

16. In what ways do the many dialects in Europe show its great diversity?
17. What aspects of ancient Greek and Roman art inspired Renaissance artists?
18. Why did pasta with tomato sauce develop in Italy rather than Russia?

INTERPRET MAPS

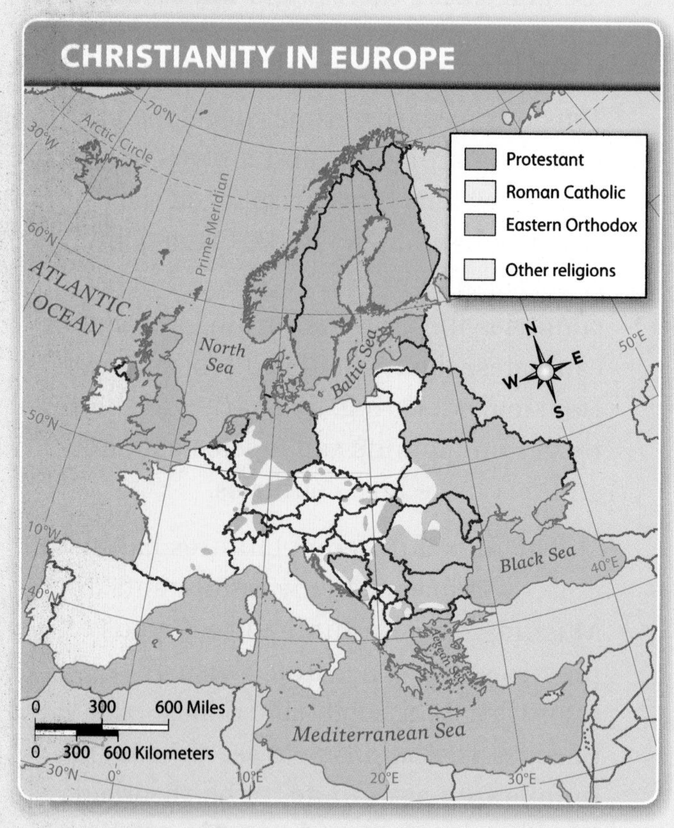

19. **Region** Where do most Protestants live in Europe? Where do most Catholics live?
20. **Make Inferences** Find the countries on the map in which a relatively small part of the population belongs to a different Christian denomination. What challenges might the people in the minority religion face?

GOVERNMENT & ECONOMICS

ANALYZE THE ESSENTIAL QUESTION

What are the costs and benefits of European unification?

Critical Thinking: Analyze Cause and Effect

21. In what way did the Marshall Plan help bring about the formation of the European Union?
22. What was the response of the eurozone countries when Greece and Ireland became deeply in debt in 2010?
23. What was the impact of the fall of communism on Eastern Europe? What was the impact on Western Europe?
24. What problems are caused by illegal immigration to Europe?

INTERPRET CHARTS

COST OF A TEN-MINUTE PHONE CALL TO THE U.S. IN EUROS (€)*

Country	1997	2006
Belgium	7.50	1.98
Czech Republic	3.09	2.02
Denmark	7.41	2.38
Ireland	4.61	1.91
Spain	6.17	1.53
France	6.78	2.32
United Kingdom	3.50	2.23

* 1997 prices have been converted to euros
Source: Eurostat

25. **Analyze Data** According to the chart, how did the cost of making a telephone call change between 1997 and 2006?
26. **Analyze Cause and Effect** What move made by the European Union might have brought about the change in the cost of a telephone call?

ACTIVE OPTIONS

Synthesize the Essential Questions by completing the activities below.

27. **Write a Speech** Suppose that you are the leader of a European country that has been invited to join the European Union. Write a speech that will persuade the citizens of your country to vote in favor of joining. **Deliver your speech to the class and ask the members of your audience to vote on whether they are in favor of joining the European Union.**

Writing Tips
- Take notes on three benefits that would result from joining the European Union.
- Support each benefit using facts and statistics.
- Address any concerns your audience might have about joining and explain why the advantages outweigh any disadvantages.

TECHTREK myNGconnect.com For photos of European art

28. **Create an Art Chart** Select three European works of art. You can choose Raphael's *School of Athens* or Leonardo da Vinci's *Mona Lisa* from the **Digital Library**, or you can search for other works online. Then research each piece to find out what period it is from and what theme, or subject matter, it represents. Copy the chart below to help you organize your information. Display the artwork and your findings on a poster. Be prepared to explain the relationship between each work's period and theme.

WORK OF ART	PERIOD	THEME

World Languages

TECHTREK

myNGconnect.com For an online graph and research links on world languages

Maps and Graphs

Connect to NG

In this unit, you learned about the diversity of languages in Europe. Every culture in the world uses language to communicate. Scholars estimate that there are about 7,000 languages spoken today.

Most countries name one or more languages as official languages. An official language is the one used by a country's government. For example, French is the official language of France, and English and French are the official languages of Canada. Most countries have groups of people whose first language differs from the official language. It is estimated that at least half of the people in the world speak one or more languages in addition to their first language.

Compare

- Africa
- Americas
- Asia
- Europe
- Oceania

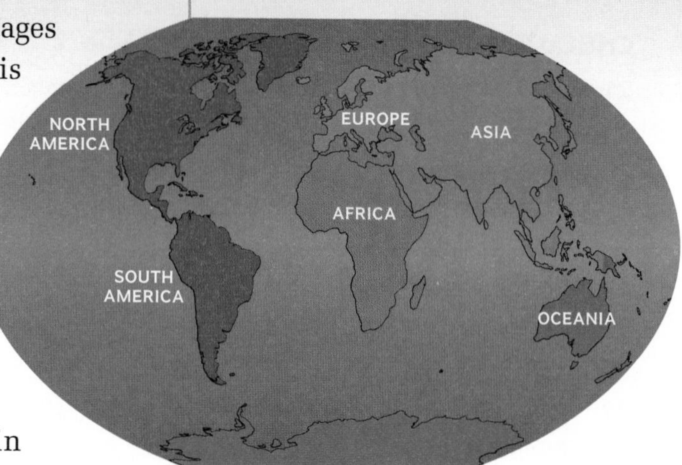

LIVING LANGUAGES

Although there are thousands of world languages, many are not widely spoken. In addition, sometimes the distinction, or difference, between a language and a dialect is not clear.

The following are the ten most widely spoken languages in the world in order of their ranking. Each is spoken as a native language by at least 100 million people. Some are official languages in widely different regions of the world. Some of the languages appear in only one region of the world.

1. Mandarin Chinese
2. Spanish
3. English
4. Hindi/Urdu
5. Arabic
6. Bengali
7. Portuguese
8. Russian
9. Japanese
10. German

DYING LANGUAGES

Some languages are spoken by so few people that they are in danger of dying out. In fact, linguists, or people who study languages, estimate that one language dies every two weeks.

Languages die for different reasons. Some simply disappear with the death of the last speaker. Others fade more slowly as a dominant language replaces it. A few linguists are trying to document some of these languages to preserve the history and culture of the people who spoke them.

The graph at right shows the number of languages spoken on the world's continents, as well the names of a few of the languages spoken. Note that "Americas" consists of the number of languages spoken on the North American and South American continents. Compare the data in the graph and use it to answer the questions.

NUMBER OF LANGUAGES SPOKEN BY CONTINENT

Source: *Ethnologue*, 16th Edition, 2009

- EUROPE
- AFRICA
- ASIA
- AMERICAS
- OCEANIA

234 (Europe)

SARDINIAN Norwegian BOSNIAN greek Scottish Gaelic Irish russian DUTCH BELARUSIAN ITALIAN Latvian french Danish Czech LATIN POLISH

2,110 (Africa)

SWAHILI Tswana Arabic Lingala BERBER kongo somali Kirundi setswana Kanuri Rwanda-Rundi BAMBARA Tsonga LUO Shona OROMO Tigrinya Zulu VENDA HAUSA PEDI FRENCH SOTHO FULA CHICHEWA Sango Ibibio SWAZI UMBUNDU Afrikaans Xhosa Spanish GBE Malagasy TSHILUBA yoruba PORTUGUESE Gikuyu Twi igbo Ndebele Sesotho Kiswahili Malinke English LUHYA

2,322 (Asia)

ARABIC Lao TAGALOG hindi PERSIAN xiang Bengali PASHTU tibetan Thai VIETNAMESE NYAW Zhuang URDU cantonese HLAI Korean Jingpho Kashmiri DOGRI MONGOLIAN WU NICOBARESE INDONESIAN KUY javanese TAMIL MIN punjabi CHAM GAN TURKMEN TETUM Saraiki burmese MON TATAR ARAMAIC bodo Kurdish KAREN Gondi Hmong Filipino azeri Dzongkha HEBREW Mandarin Mizo Japanese Ainu Sundanese CEBUANO KAZAKH

993 (Americas)

french English xinca KEKCHI CREE PAPIAMENTO CREOLE ojibwe ALEUT navajo quiche Oneida DANISH tagish HOPI Spanish ZUNI QUECHUA Aymara nahua HINDI Sranan Tongo MAM INUIT mayan Haida PORTUGUESE DUTCH

1,250 (Oceania)

ULITHIAN Takuu NAURUAN Aranda Nukuoro ANUTA PILENI Mae POHNPEIAN Rapa pitcairnese TRUKESE palauan Kapingamarangi bislama yapese CAROLINIAN SAMOAN HINDUSTANI Chamorro French Tahitian English TUVALUAN niuean Tok Pisin Sonsoralese Marshallese Maori I-KIRIBATI Tongan Hiri Motu FIJIAN

▷Hola
▷Guten tag
Hello
▷Kia ora
▷Hujambo
▷Ohayou

ONGOING ASSESSMENT

RESEARCH LAB

GeoJournal

1. **Compare and Contrast** On which continent are the most languages spoken? the fewest? What do these numbers suggest about the cultural unity of each continent?
2. **Analyze Data** Study the graph and the list of languages with the most native speakers. How many of the total languages in Europe are among those spoken most in the world?

Research and Create Charts Research to find out more about the people who speak Hindi and Portuguese. Create a chart for each language showing approximately how many people speak it and where it is spoken. Which language has more native speakers? Which language is an official language in more places? What might account for this?

Active Options

TECHTREK

myNGconnect.com For photos of Renaissance art, nuclear power plants in Europe, and European cuisine

Digital Library

Connect to NG

Magazine Maker

ACTIVITY 1

Goal: Extend your understanding of Renaissance art.

Write a Renaissance Arts Magazine

The Renaissance was a period of great artistic activity in Europe. Choose a city in Europe that was influenced by the Renaissance between 1400 and 1600. With a group, plan and publish a magazine showcasing that city's artistic achievements. Use the Magazine Maker to find photos and information on the following:

- art
- architecture
- literature
- fashion

Brunelleschi's dome atop the Cathedral of Florence in Italy

ACTIVITY 2

Goal: Research the use of nuclear power in Europe.

Create a Pro-and-Con Chart

Some European countries are planning to build new plants, while others have chosen to close existing plants. Use the research links at **Connect to NG** to create a pro-and-con chart that explains some of the advantages and disadvantages of nuclear power. Be prepared to present your chart and explain the issues.

ACTIVITY 3

Goal: Learn about European culture through its food.

Plan a Dinner Menu

Get together in a group and plan a dinner menu featuring typical European dishes. Discuss what the courses for the meal will be. Each group member should be in charge of one course, each of which should come from a different European country. Design a poster presentation of the menu.

NATIONAL GEOGRAPHIC
World Cultures and Geography
GEO

EXPLORE RUSSIA & THE EURASIAN REPUBLICS WITH NATIONAL GEOGRAPHIC

MEET THE EXPLORER

NATIONAL GEOGRAPHIC

Tracing ancient trade routes, NG Fellow Fredrik Hiebert excavated a 4,000-year-old Silk Road city in Turkmenistan. He also searches for underwater settlements in the Black Sea.

INVESTIGATE GEOGRAPHY

North of the Arctic Circle in Russia, reindeer herded by a nomadic clan charge across the tundra. Many groups in the Arctic rely heavily on the reindeer. The animals, also known as caribou, provide food, clothing, and shelter, as well as transportation.

CONNECT WITH THE CULTURE

A young Mongolian Kazak family stands in front of a yurt, a shelter used by nomads. Kazaks are a nomadic, animal-herding people. Yurts allow them to live in harsh climates and give them the freedom to move and graze their animals in traditional tribal communities.

Go to myNGconnect.com for maps of Russia and the Eurasian Republics.

STEP INTO HISTORY

Saint Basil's Cathedral in Red Square, Moscow, Russia, was built between 1554 and 1560. The cathedral has nine chapels, each topped with an onion-shaped dome.

CHAPTER 11

Russia & THE EURASIAN REPUBLICS

GEOGRAPHY & HISTORY

PREVIEW THE CHAPTER

Essential Question How have size and extreme climates shaped Russia and the Eurasian republics?

SECTION 1 • GEOGRAPHY

KEY VOCABULARY

- landmass
- steppe
- permafrost
- tundra
- taiga
- nonrenewable fossil fuel
- peat
- hydroelectric power
- methane
- greenhouse gas
- semiarid
- arid
- pesticide

ACADEMIC VOCABULARY

isolate, remote

TERMS & NAMES

- Ural Mountains
- North Atlantic Drift
- Siberia
- Black Sea
- Caspian Sea
- Aral Sea

Essential Question How has geographic isolation influenced the region's history?

SECTION 2 • HISTORY

KEY VOCABULARY

- state
- tribute
- czar
- reign
- secular
- invader
- scorched earth policy
- serf
- strike
- communism
- socialism
- collective farm
- propaganda

ACADEMIC VOCABULARY

expand, promote

TERMS & NAMES

- Slav
- Kievan Rus
- Genghis Khan
- Mongol Empire
- Silk Roads
- Peter the Great
- Catherine the Great
- Nazi Germany
- Industrial Revolution
- V. I. Lenin
- Bolshevik
- Russian Revolution
- Soviet Union
- Cold War
- Mikhail Gorbachev

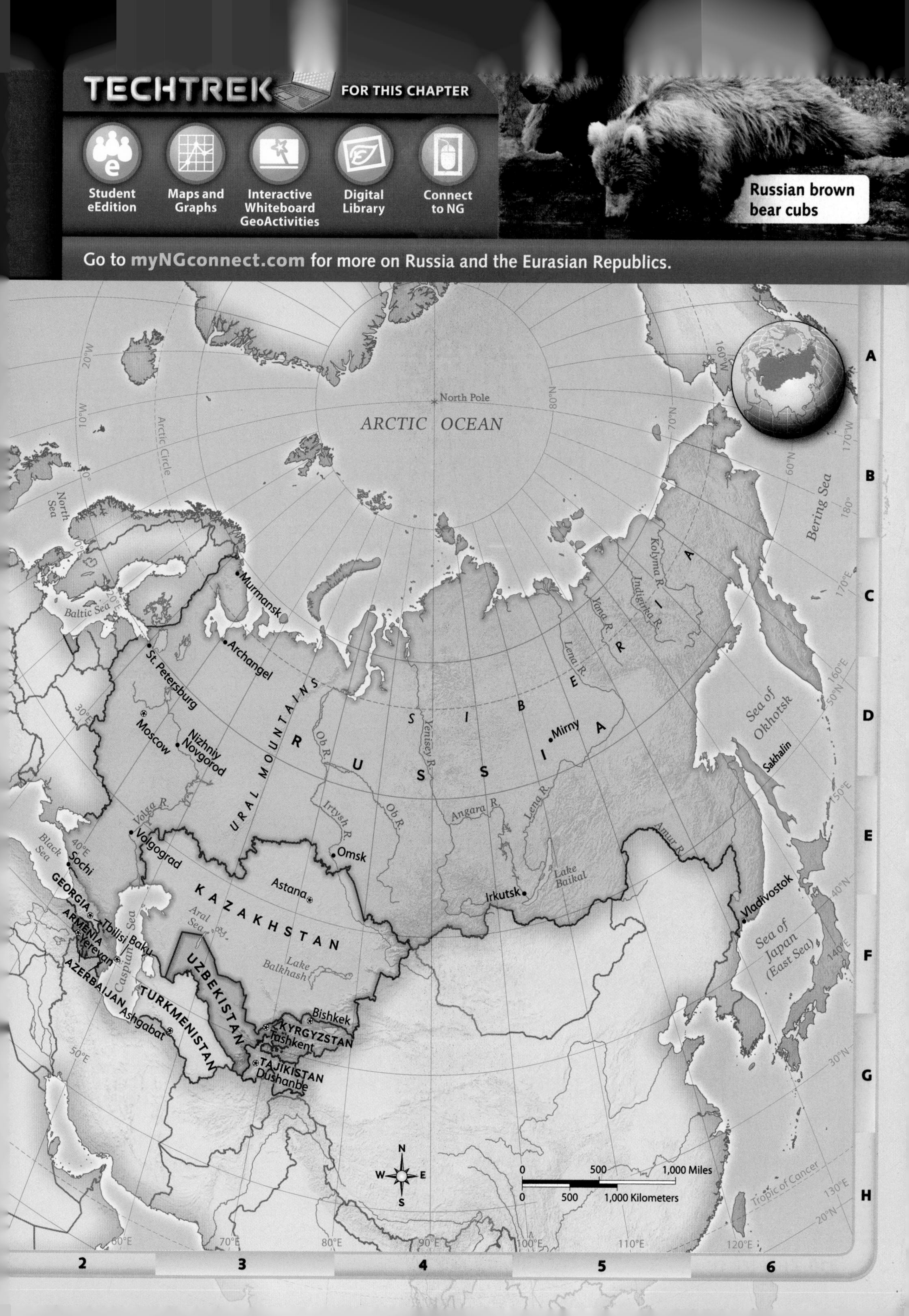

TECHTREK FOR THIS CHAPTER
Student eEdition
Maps and Graphs
Interactive Whiteboard GeoActivities
Digital Library
Connect to NG
Russian brown bear cubs
Go to myNGconnect.com for more on Russia and the Eurasian Republics.
North Pole
ARCTIC OCEAN
Arctic Circle
North Sea
Baltic Sea
Bering Sea
Sea of Okhotsk
Sakhalin
Sea of Japan (East Sea)
Black Sea
Caspian Sea
Aral Sea
Lake Balkhash
Lake Baikal
Tropic of Cancer
Murmansk
Archangel
St. Petersburg
Moscow
Nizhniy Novgorod
Volgograd
Sochi
Omsk
Astana
Irkutsk
Mirny
Vladivostok
Tbilisi
Baku
Yerevan
Ashgabat
Tashkent
Bishkek
Dushanbe
RUSSIA
SIBERIA
URAL MOUNTAINS
KAZAKHSTAN
UZBEKISTAN
TURKMENISTAN
KYRGYZSTAN
TAJIKISTAN
GEORGIA
ARMENIA
AZERBAIJAN
Volga R.
Ob R.
Irtysh R.
Yenisey R.
Angara R.
Lena R.
Yana R.
Indigirka R.
Kolyma R.
Amur R.
0 500 1,000 Miles
0 500 1,000 Kilometers
N S E W
A B C D E F G H
2 3 4 5 6

SECTION 1 GEOGRAPHY

1.1 Physical Geography

TECHTREK

myNGconnect.com For maps of Russia and the Eurasian republics and Visual Vocabulary

 Maps and Graphs

 Digital Library

RUSSIA AND THE EURASIAN REPUBLICS PHYSICAL

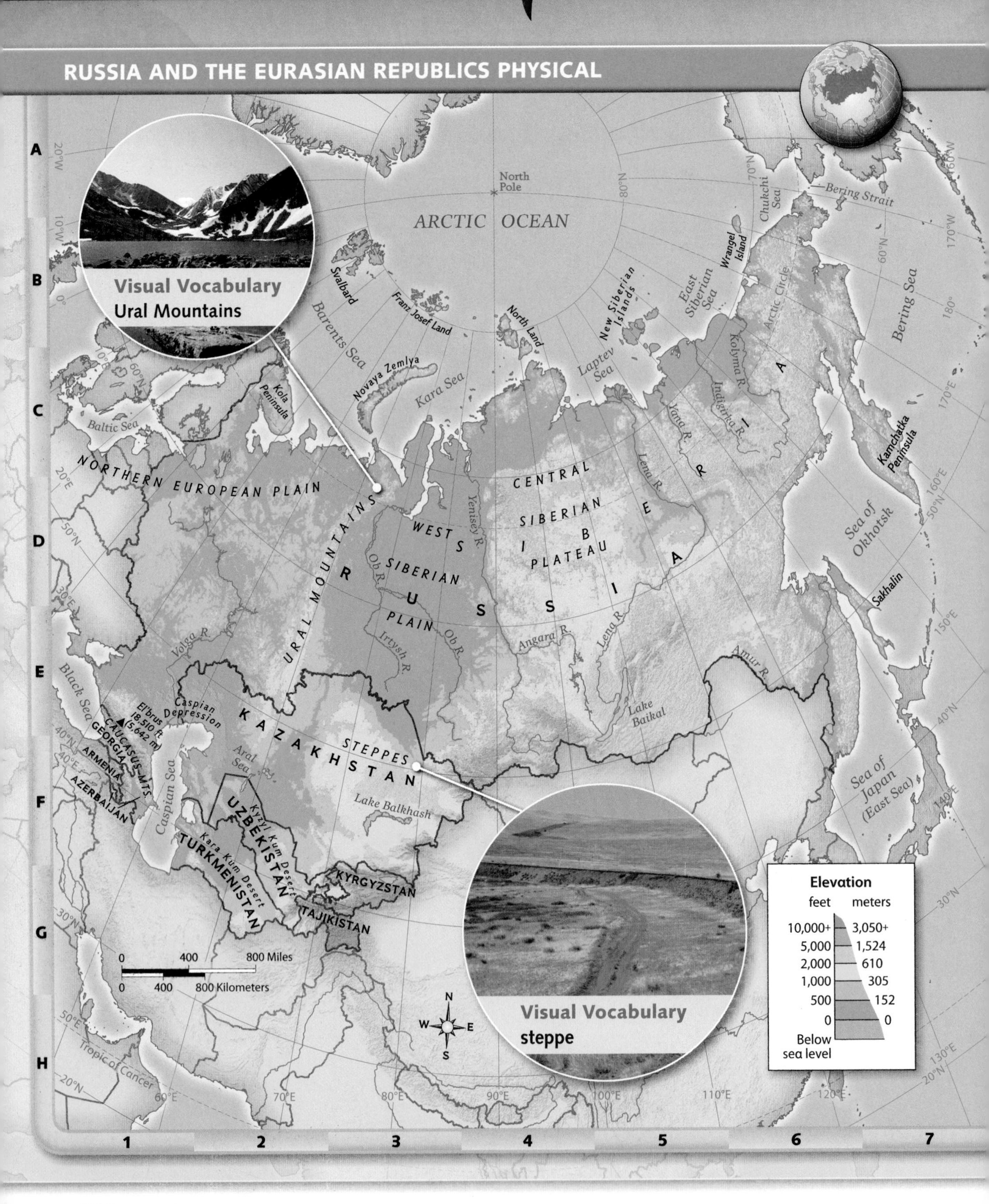

Main Idea Russia and the Eurasian republics cover a huge area and contain a variety of geographic features.

Russia and the Eurasian republics take up about one-sixth of the land surface of the entire earth. The region's geographic features have limited its population.

Huge Landmass

Russia is the largest country in the world in total area. Its **landmass**, or continuous extent of land, stretches almost 6,000 miles from east to west. The Eurasian republics lie south of Russia. The republics in the Caucasus Mountains are Armenia, Azerbaijan (ahz ur by JAHN), and Georgia. Those in Central Asia include Kazakhstan (kah zahk STAHN), Kyrgyzstan (kihr gih STAN), Tajikistan (tah jik ih STAN), Turkmenistan (turk MEN ih stan), and Uzbekistan (ooz BEK ih stan).

Plains, or large areas of level ground, cover much of the region. The relatively low **Ural Mountains** separate the Northern European Plain from the West Siberian Plain. In much of southwestern Russia and northern Kazakhstan, the very large plains are called **steppes**. Much of this land is good for agriculture and grazing.

Natural Barriers

Russia has oceans on its northern and eastern borders and mountainous areas along much of its southern border. These natural barriers separate Russia from its neighbors. The deserts and mountains of Central Asia keep republics such as Kyrgyzstan and Tajikistan **isolated**, or cut off from other countries.

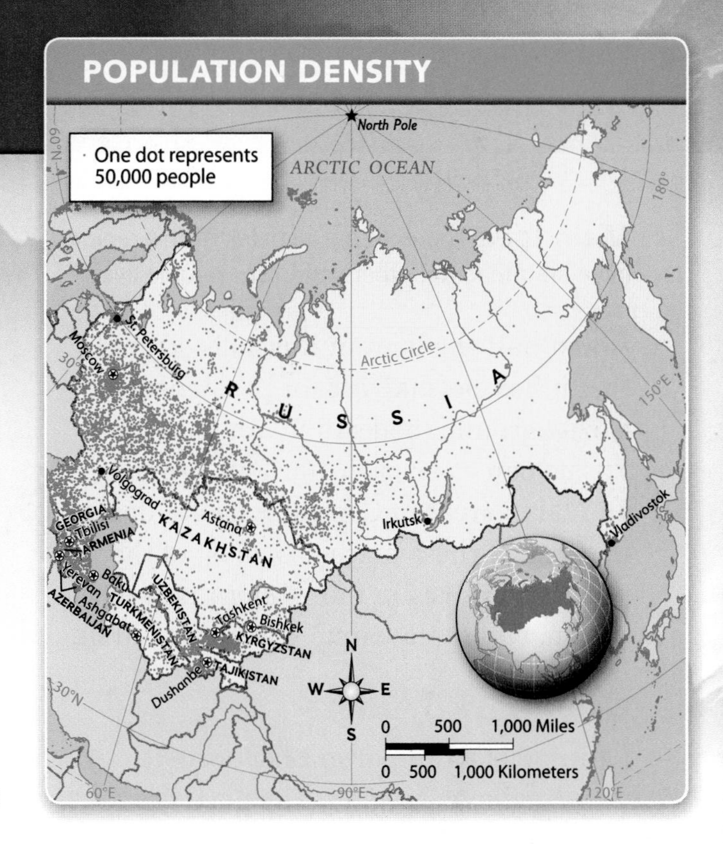

Because so much of Russia's coastline lies to the north of the Arctic Circle, few ports stay open to ships and trade all year long. Murmansk, located far to the north on Kola Peninsula, is one such port. An ocean current called the **North Atlantic Drift** warms the waters around Murmansk and keeps them ice-free most of the time.

Before You Move On

Monitor Comprehension What are some of the key geographic features of this region?

ONGOING ASSESSMENT

MAP LAB

1. **Place** Find the Ural Mountains on the physical map. What is the elevation of the mountains? Locate the steppes. How does its elevation compare with that of the Ural Mountains?
2. **Interpret Maps** Use the population density map to determine how density differs west and far east of the Urals. Then study the physical map. What physical features might contribute to the population distribution?

1.2 Land of Extreme Climates

TECHTREK
myNGconnect.com For a climate map and photos of extreme climates

Maps and Graphs

Digital Library

Main Idea The extreme climates of this region have an impact on where and how people live.

About half the land in Russia is so cold that it has **permafrost**, or permanently frozen ground, beneath it. Yet parts of Russia can also reach 100°F in summer, and large areas of Central Asia are desert. Because of these extremes, most of the population lives in the western part of the region, where the climate is not as harsh.

Cold, Dark Winters

Latitude is an important factor in the climate of a region. The northern boundary of Russia is a coastal plain along the Arctic Ocean, with no natural barriers to keep out arctic winds. The high northern latitudes of Moscow and areas to its north help bring this region long, dark, snowy winters. St. Petersburg, for example, has a latitude of almost 60° N. For about one month each winter, there is has hardly any daylight in the city.

Climate and Vegetation

Climate affects the types of vegetation that grow in different areas. **Tundra**, or flat land found in arctic and subarctic regions, exists in **Siberia**, which lies in central and eastern Russia. Here, only small plants can grow. Permafrost prevents most trees from growing because their roots can't spread deep under the ground.

Critical Viewing Snow and ice cover the ground throughout much of the long winter in the arctic city of Noril'sk. Based on the photo, what challenges might people encounter in the city's wintry weather?

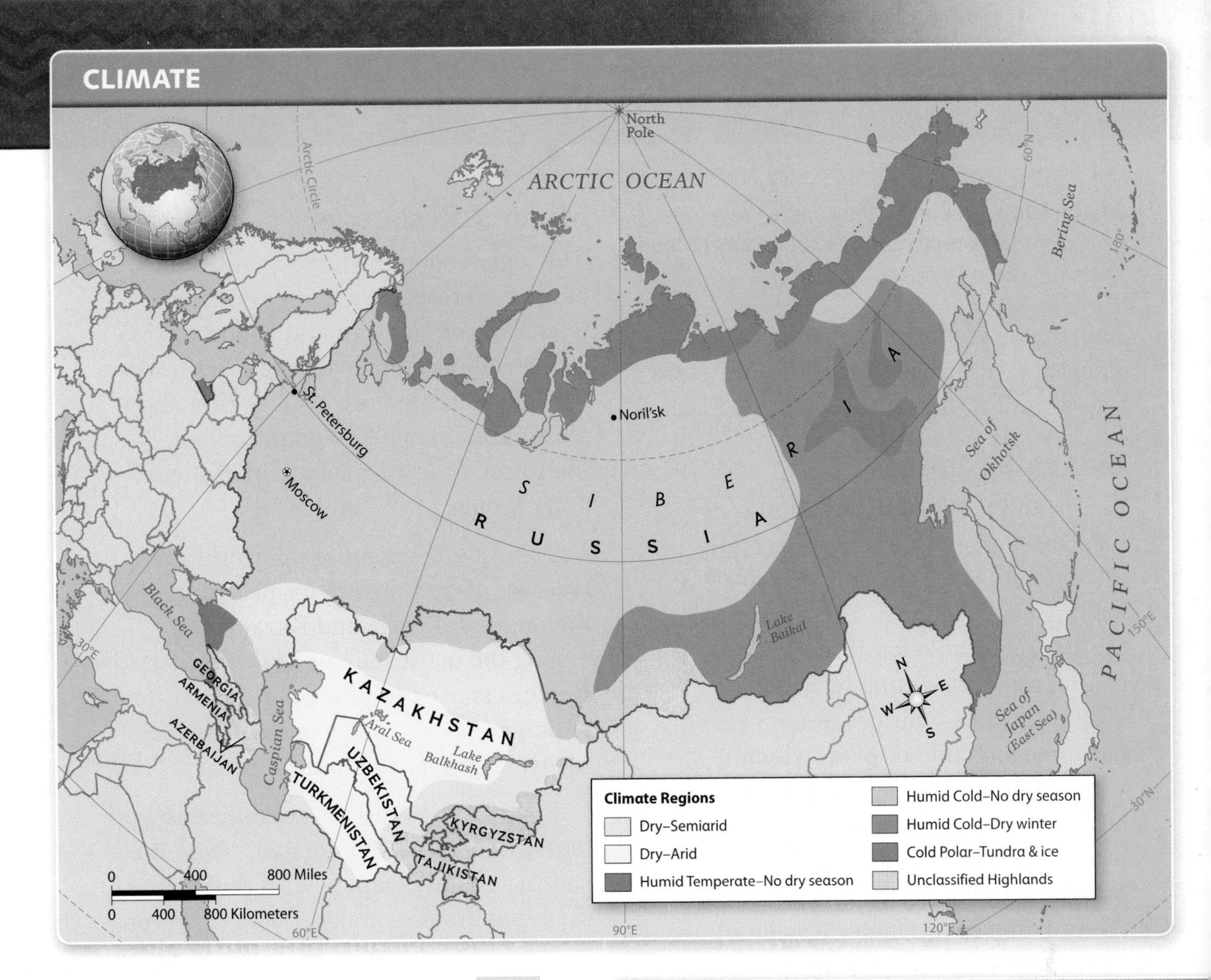

Just south of the tundra is the **taiga** (TY guh), or forest area. Mostly small evergreens like pines grow here. This area provides valuable timber resources.

Extremes in temperature and moisture make it hard for some areas to be used for agriculture. Much of the northern territory has short summers and, as a result, short growing seasons. The semiarid and desert areas are limited to herding and grazing. Farming is concentrated in the fertile soils of the western plains and steppes, along the **Black Sea**, the **Caspian Sea**, and in some river valleys.

Before You Move On

Summarize In what ways do this region's extreme climates affect the people who live there?

ONGOING ASSESSMENT

MAP LAB

1. **Location** What is the climate like in the northeastern part of Russia? How does latitude affect the climate in this region?
2. **Interpret Maps** What areas in the region have arid or semiarid climates? What areas have a humid cold climate with dry winters?
3. **Pose and Answer Questions** Study the map and then create another climate question to ask a partner.
4. **Human-Environment Interaction** Look at the population map in Section 1.1 and then compare it with the climate map on this page. In what ways might climate determine where people live in the region?

1.3 Natural Resources

TECHTREK
myNGconnect.com For a resource map and photos of resources

Maps and Graphs

Digital Library

Main Idea Russia and the Eurasian republics have plentiful natural resources, but many of them are in remote locations.

This region is among the world's richest in natural resources. These resources are important for the countries' economies.

Energy Resources

Russia and the Eurasian republics have plentiful energy resources, especially oil and natural gas. Russia is also a leading coal producer. These resources are **nonrenewable fossil fuels**. They cannot reproduce quickly enough to keep pace with their use. Russia also has large amounts of **peat**, which is very old decayed plant material. Peat is burned like coal. In addition, some rivers provide **hydroelectric power**. Power production plants use the force of the rivers' water to generate electricity.

Mineral Resources

The region contains large quantities of mineral resources that provide raw materials for factories and support industrial development. These resources include metallic ores, such as iron and aluminum, along with gold, copper, platinum, uranium, cobalt, manganese, and chrome.

Almost 20 percent of the world's reserves of iron ore are located in the region, with Russia and Kazakhstan among the main sources of this mineral. Iron ore is used to produce iron and steel, which are used in the construction of roads, railways, and buildings.

In 2010, huge reserves of minerals were found in nearby Afghanistan. Soon, this country might compete with Russia and the republics as a major producer of iron, copper, and other metals.

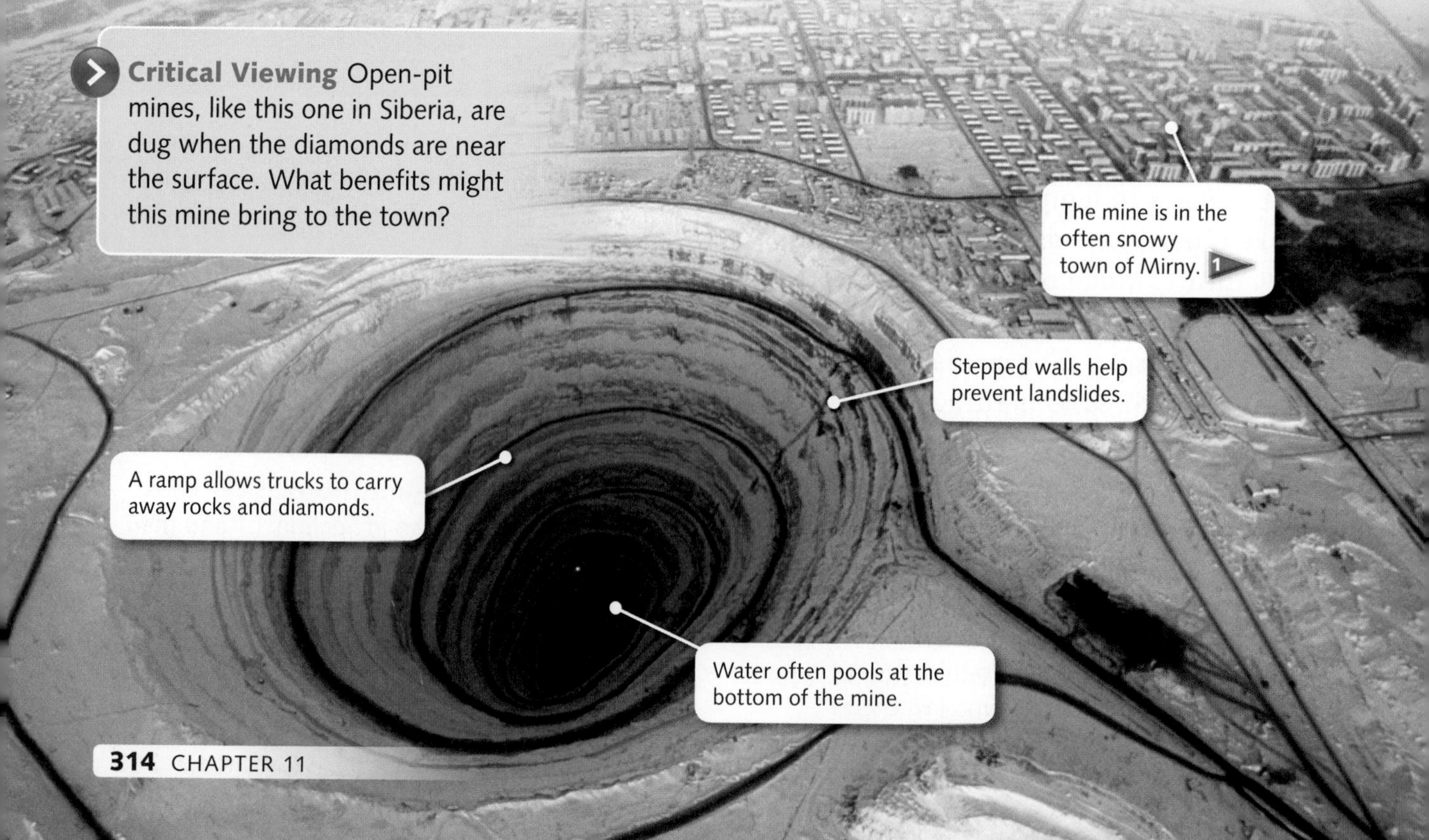

Critical Viewing Open-pit mines, like this one in Siberia, are dug when the diamonds are near the surface. What benefits might this mine bring to the town?

The Challenge of Location

Much of the region's resources are in **remote**, or hard to reach, locations. For example, Siberia contains oil fields, hydroelectric power sources, and minerals, such as nickel and gold. Many of these resources are located in the far-eastern, and coldest, parts of Siberia. The permafrost there makes it difficult to drill or mine for the natural resources and transport them to market. As a result, the resources in these areas of Siberia remain largely untouched.

Before You Move On

Summarize What are some important natural resources of Russia and the Eurasian republics, and why are some of them hard to reach?

ONGOING ASSESSMENT

DATA LAB

Create Charts You have learned that Russia and the Eurasian republics contain plentiful mineral resources that provide raw materials for factories. Identify the region's mineral resources in the map above. Then research to find out what consumer goods are made from these minerals. Create a chart like the one below to record your findings.

MINERALS	GOODS
Iron	steel, medicine, magnets, auto parts, paper clips
Aluminum	kitchen utensils, drink cans, foil, airplane and car parts

TECHTREK
myNGconnect.com For photos of the explorer's work and an Explorer Video Clip
Digital Library

Exploring Siberian Lakes with Katey Walter Anthony

Main Idea As permafrost thaws in Siberia, it releases methane gas into the atmosphere.

Expedition to Siberia

Emerging Explorer Katey Walter Anthony first went to Siberia as a high school exchange student. Now she works with other scientists from Alaska and Russia at the Northeast Science Station in Cherskiy, Siberia, 1 to study how climate change is affecting the area—and possibly the entire world. Siberia's frigid climate makes the work difficult.

Anthony and some other scientists are concerned about Siberia's permafrost. They believe that because of global warming, the permafrost below its lakes is thawing, releasing carbon that was locked inside the frozen ground. Carbon is formed from dead prehistoric animals and the plants they ate. The carbon is then turned into **methane**, a colorless, odorless natural gas that can have a negative impact on the environment.

myNGconnect.com
For more on Katey Walter Anthony in the field today

Anthony checks for methane on a frozen lake.

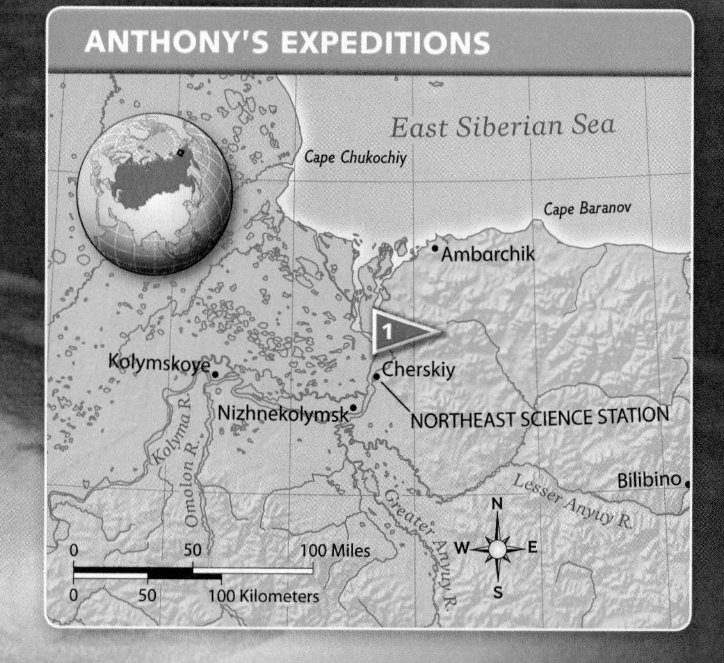

PROCESS OF METHANE RELEASE

3 Methane is released into the atmosphere.

2 Methane is released into the lake as the permafrost thaws.

1 Organic matter is trapped in the permafrost.

soil

frozen lake

sediments

permafrost

METHANE RELEASE Warming temperatures thaw the permafrost and release methane gas. The methane heats the air as it is released into the atmosphere as a greenhouse gas.

Methane and Climate Change

Methane is a **greenhouse gas**, a gas that traps the sun's heat over the earth. Anthony explains, "It's 25 times more powerful than carbon dioxide on 100-year time scales" and could have the most powerful effect of all on global warming. Siberian lakes could release about ten times as much methane as is in the atmosphere now. Some experts believe this could cause temperatures across the world to rise higher and faster.

To check for methane, Anthony chops holes into the ice on the lakes to collect gas samples that she brings back to the lab. Sometimes she wants to know right away what gases might be present, so she lights a match. If a giant flame shoots up, she knows she's found methane.

Before You Move On

Monitor Comprehension What impact does methane have on the atmosphere?

ONGOING ASSESSMENT

VIEWING LAB

GeoJournal

1. **Analyze Visuals** Go to the **Digital Library** to view the video clip on Emerging Explorer Katey Walter Anthony. How might her work in Siberia be applied to other places in the world?
2. **Interpret Models** Study the model. What does the methane have to pass through to enter the atmosphere? Why does the gas penetrate this material so easily?
3. **Analyze Cause and Effect** List some causes and effects of methane release using a chart like the one below. Why do some scientists believe that increasing methane in Siberian lakes is both a cause and an effect of global warming?

CAUSES	EFFECTS

1.5 Central Asian Landscapes

TECHTREK
myNGconnect.com For online maps and photos of Central Asia

Maps and Graphs

Digital Library

Main Idea Human activities have led to the shrinking of the Aral Sea, which has damaged the surrounding landscape.

As you have learned, methane gas is threatening Siberia's environment. Central Asia's landscape has also been damaged. Human activities have nearly destroyed one of its most important bodies of water.

Adapting to Dry Conditions

Central Asia includes landforms such as deserts, mountains, forests, and steppes. Though the area doesn't experience the extreme cold of northern Russia, there are places in northern Kazakhstan that can reach 0°F in winter. In general, summers are hot and longer than in the northern parts of the region. Central Asia's temperatures vary so widely in part because it is not protected by a large body of water, which would help to keep temperatures moderate.

Large parts of Central Asia are **semiarid** or **arid**, meaning there is little or no rainfall. These dry lands are best suited for livestock grazing. Because of irrigation in some river valleys, however, farmers have also been able to grow crops such as cotton.

The Shrinking Aral Sea

Major efforts to grow cotton led to the shrinking of the **Aral Sea** in Kazakhstan and Uzbekistan. The body of water is actually a salt-water lake. The rivers that fed the lake were redirected into canals for irrigation. The Aral Sea was once the fourth largest lake in the world, but now it is only a fraction of the size it was in 1960.

Pollution contributed to problems in the Aral Sea. Fertilizer and **pesticides**, which are chemicals that kill harmful insects and weeds, ran into it. As the lake shrank, salt and pesticides destroyed the habitat of many plants and animals and threatened human health. The lake's once-thriving fishing industry was also damaged. One resident of the area said, "My father and grandfather were fishermen in this town, but as you can see, the boats are now sitting in the middle of a desert."

In 2005, Kazakhstan, with the help of the World Bank, built a dam to save the North Aral Sea. That part has increased in size, and fishing has returned to the area. However, the southern part of the lake, in Uzbekistan, is almost completely gone.

Before You Move On

Monitor Comprehension What human activities caused the Aral Sea to shrink, and what damage to the landscape has occurred as a result?

Critical Viewing Camels walk past a stranded ship on land where the Aral Sea used to be. What does this picture suggest about the Aral Sea?

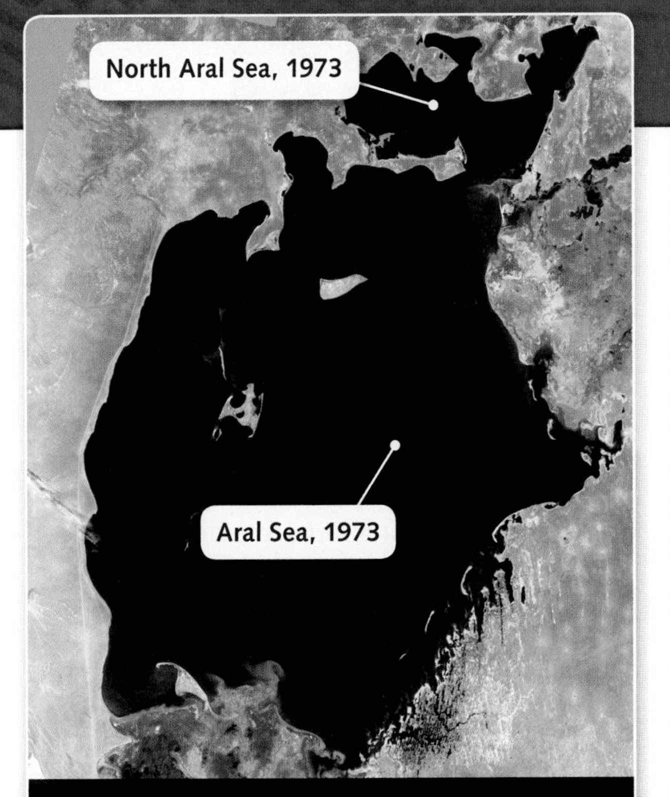

This satellite image shows what the Aral Sea looked like in 1973. At this time, the North and South Aral seas were full of water.

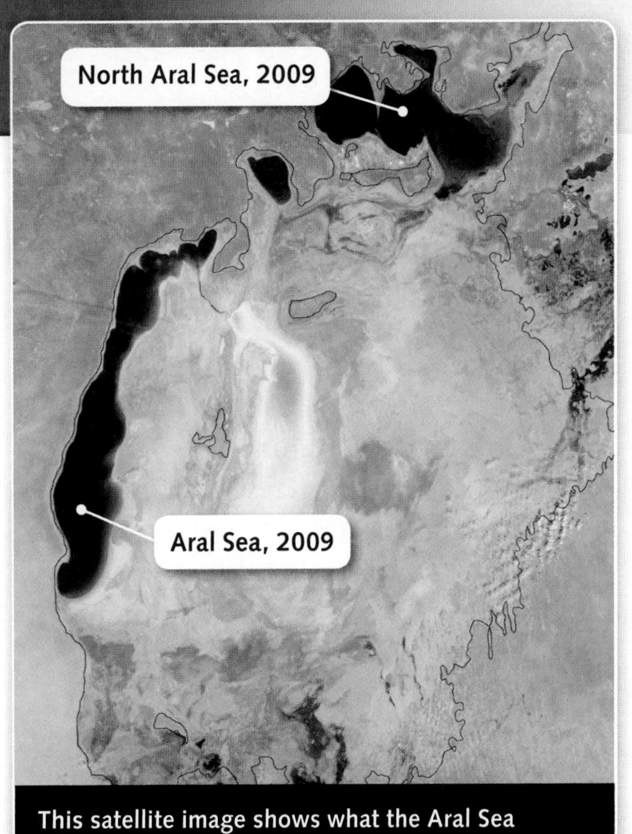

This satellite image shows what the Aral Sea looked like in 2009. The photo reveals that efforts to save the North Aral Sea have worked.

ONGOING ASSESSMENT

PHOTO LAB

1. **Analyze Visuals** Study the photos above of the Aral Sea. How much time passed between the photo on the left and the photo on the right? What happened during that time period?
2. **Identify** What details in the photo below tell you that this part of the Aral Sea has been devastated for a long time?
3. **Human-Environment Interaction** Why did people redirect the Aral Sea? What steps have been taken to fix what happened as a result?

2.1 Early History

Main Idea Settlers and conquerors from Europe and Asia shaped the early history of Russia and the Eurasian republics.

Settlement of Russia by different groups dates back to around 1200 B.C. At that time, a people called the Cimmerians lived north of the Black Sea in what is now southern Ukraine. Over hundreds of years, many other groups ruled this region.

Kievan Rus

The people whose culture had the most lasting influence on early Russia were the **Slavs** (slahvz). Some historians think they were farmers near the Black Sea around 700 B.C. or earlier. Others think they came from Poland around the A.D. 400s. By the A.D. 800s, the Slavs had built towns near rivers in Ukraine and western Russia.

In 862, Vikings from Scandinavia called the Varangian Russes (vahr ANG ee ehn ROOS ehz) took control of the Slavic town of Novgorod. Russia's name may have come from this tribe. About 20 years later, a Varangian prince captured Kiev (KEE ehf). He established a **state**, or defined territory with its own government, that came to be known as **Kievan Rus**.

Mongol Rule and Trade

By the late 1100s, Kiev's power had declined. Struggles for control within the ruling family weakened the state.

In the early 1200s, **Genghis Khan** (JEHNG gihs KAHN) established the **Mongol Empire** in Central Asia. In 1240, his grandson, Batu Khan, extended the empire by taking over Kievan Rus and much of Russia.

Russian princes had to declare their loyalty and pay **tribute**, or taxes, to the Mongol ruler, the khan. Mongols selected one of the princes to serve as grand prince and represent Russian interests.

The empire isolated Russia from European influence for more than 200 years. However, Mongol rule kept Russia and Central Asia open to the East. The **Silk Roads**, ancient trade routes, carried goods and new ideas throughout the Mongol Empire. As you can see on the map on the next page, the Silk Roads connected Southwest Asia and Central Asia with China. Goods traded included gold, jade, and silk. Merv, in Turkmenistan, and Samarkand, in Uzbekistan, were important stops on the Silk Roads.

Gold horse and rider made by early settlers of Ukraine

1000 B.C. — A.D. 800 — 1000

1200 B.C. Cimmerians settle in southern Ukraine.

A.D. 800s Slavs build towns on the Dnieper River.

c. 882 Varangian Russes found Kievan Rus.

Ceiling of St. Sophia, Kievan Rus cathedral

SILK ROADS

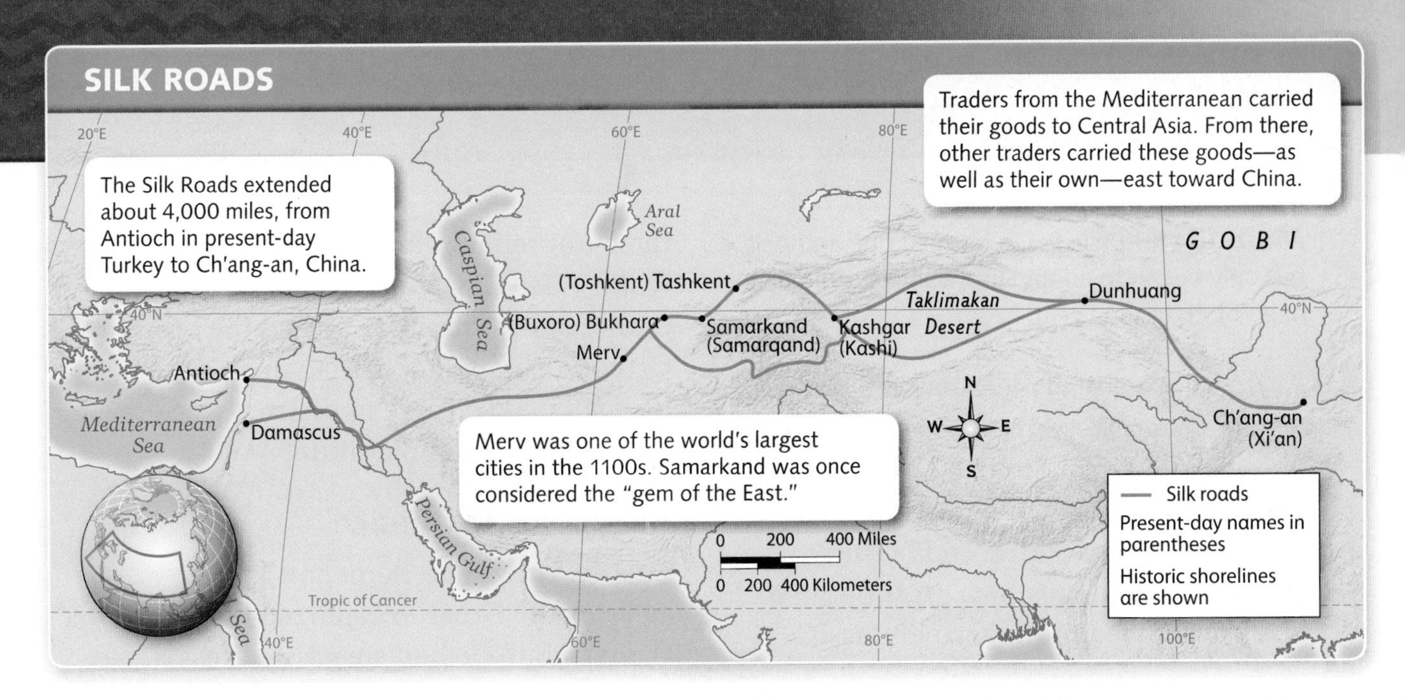

The Rise of Moscow

Around 1330, the Mongols allowed Grand Prince Ivan I of Moscow to collect tribute for them. Ivan kept some of the money for himself and used it to buy land and **expand** his territory. Moscow became stronger as Mongol rule weakened.

In 1380, Grand Prince Dmitry defeated the Mongols in battle. In 1480, Ivan III—also called Ivan the Great—refused to pay tribute to the Mongols, and they never demanded it again. This action ended Mongol rule in Russia. Ivan the Great's grandson, Ivan IV, expanded Russia and made it into an empire. He became Russia's first **czar** (zahr), or emperor.

Before You Move On

Summarize How did settlers and conquerors shape the early history of Russia?

ONGOING ASSESSMENT

MAP LAB

GeoJournal

1. **Place** Why do you think the main route of the Silk Roads split in two between the cities of Kashgar and Dunhuang?
2. **Interpret Maps** Study the map. Why was Samarkand an important stop on the routes?
3. **Analyze Cause and Effect** What actions strengthened Moscow and what resulted from its rise? Use a chart like the one below to list the causes and effects.

CAUSES	EFFECTS

1237–1240
Mongols conquer Russia.

1380
Prince Dmitry defeats Mongols.

1480
Ivan III ends Mongol rule.

1200

1400

Mongol emperor Genghis Khan

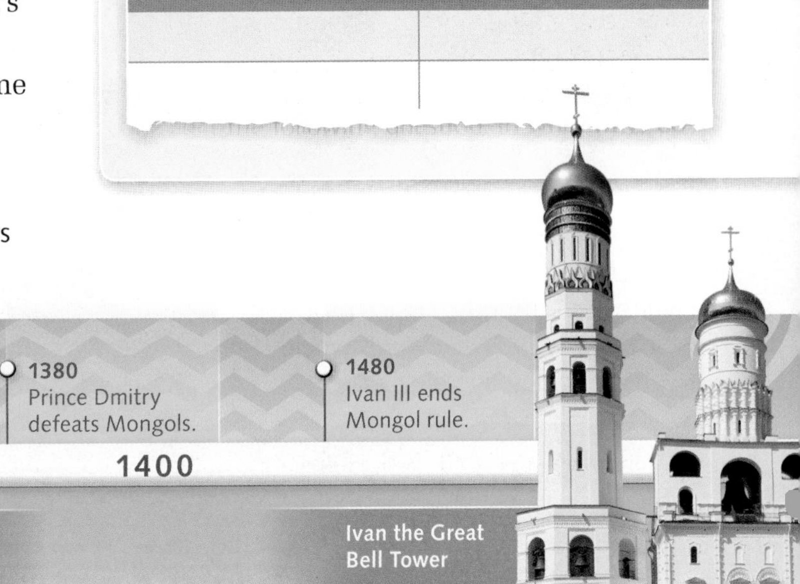

Ivan the Great Bell Tower

2.2 European or Asian?

TECHTREK
myNGconnect.com For an online map and photos of historic buildings

Maps and Graphs

Digital Library

Main Idea Russia lies in both Europe and Asia but only began to adopt European ideas under the reign of Peter the Great.

As you have already learned, Russia was isolated from Western Europe for a couple of centuries under Mongol rule. This isolation continued under the Russian czars, beginning with Ivan IV. Then, starting in the 1600s, two czars began to bring a European influence to Russia.

Spanning Europe and Asia

Geographically, Russia extends across the landmass of Europe and Asia, which is often called Eurasia. The Ural Mountains separate the continents. Many of the Russian people live in the European area west of the mountains.

In spite of lying partly in Europe, Russia did not develop a European culture. In the late 900s, Christianity became the main religion in Russia. However, while Western Europe adopted Roman Catholicism, Russia embraced the Eastern Orthodox branch of Christianity. The Russian people came to distrust western European ideas and culture.

European Influence

Finally, Czar Peter Romanov, known as **Peter the Great**, recognized that Western Europe had surpassed Russia both economically and militarily. Peter, who ruled from 1682 to 1725, decided to modernize Russia. As a result, he became the first Russian czar to travel to Western Europe. From his travels, he brought back European ideas about how government and business should be run. He also introduced a Western style of architecture. Many buildings in St. Petersburg, including the Peterhof Palace shown below, reflect this style.

Critical Viewing The Peterhof Palace is often called "the Russian Versailles," after the French palace of King Louis XIV. What words would you use to describe the palace?

Catherine the Great, empress from 1762 to 1796, also introduced European ideas to Russia, focusing on the arts and education. Western forms of entertainment, such as opera, became popular during her **reign**, or rule. In addition, Catherine founded many new schools and supported the idea of educating women. She also built hospitals and **promoted**, or encouraged, vaccinating against smallpox. Planners redesigned cities, and European architecture replaced older Russian styles.

The empress attempted to create a more **secular** country, one in which the church was less powerful. In fact, Catherine wanted to pass laws that would allow citizens to practice the religion of their choice. However, her attempt at reform was unsuccessful.

Before You Move On

Summarize What European ideas did Russia adopt under Peter the Great?

EUROPEAN AND ASIAN RUSSIA

MAP TIP Map projections sometimes distort, or bend, the area shown. The Mollweide projection here makes Europe look somewhat mashed together but it is helpful in showing vast expanses of polar regions in small spaces.

ONGOING ASSESSMENT

READING LAB

GeoJournal

1. **Make Inferences** Based on the map, in what ways do you think Russia's location encouraged its isolationism?
2. **Synthesize** How was Russia under Peter and Catherine different than under Mongol rule?
3. **Interpret Maps** In which continent does most of Russia lie?
4. **Analyze Visuals** What impression of Peter the Great is this statue meant to convey? Explain your answer with specific references to the statue.

2.3 Defending Against Invaders

Main Idea For centuries, Russia used its geographic strengths—location, size, isolation, and climate—to defend against invaders.

Throughout history, Russians adapted to the challenges of distance and climate in their country. **Invaders**, or enemies entering by force, however, often had a hard time overcoming these challenges.

Russia Builds an Empire

You have learned that Ivan IV became the first czar of Russia in 1547. Ivan began building Russia's empire by conquering territory from the Mongols. Eventually, Russia expanded across Siberia to the Pacific Ocean.

As Russia continued to grow under Peter the Great and Catherine the Great, the country was often forced to defend its empire against invaders. Enemies from Europe had to cope with Russia's vast expanse and harsh climate. They were rarely successful at invading the country. Two famous examples are the failed invasions of Napoleon and Hitler.

Napoleon Invades Russia

Napoleon I became emperor of France in 1804, and his empire included much of Europe. Angered that Britain had defeated him in battle, he came up with a plan to limit Britain's trade with European countries. Czar Alexander I refused to go along with the plan. Alexander thought reducing trade with Britain would be bad for Russia's economy.

In revenge, Napoleon invaded Russia in the summer of 1812. The Russians followed a **scorched earth policy**, where troops retreat in front of the advancing army and destroy crops and other resources that might supply the enemy. When Napoleon arrived in Moscow, the city was nearly deserted. The Russians had burned it to destroy any supplies that might have helped the French troops.

By mid-October, Napoleon started his retreat. He knew the harsh Russian winter was coming. Soon, snow and extreme cold began to weaken the French army. Russian troops attacked the army as it fled. Of the roughly 420,000 troops that set out for Moscow, only about 10,000 survived.

Ivan IV stands before St. Basil's, which he had constructed.

1500

1547
Ivan IV is crowned first czar.

1600

1604–1613
Civil war and invasions

1613
Romanov rule begins. The family, which included Peter the Great and Catherine the Great, ruled until 1917.

1700

1703
St. Petersburg is founded.

Picture of Nicholas II, the last of the Romanovs

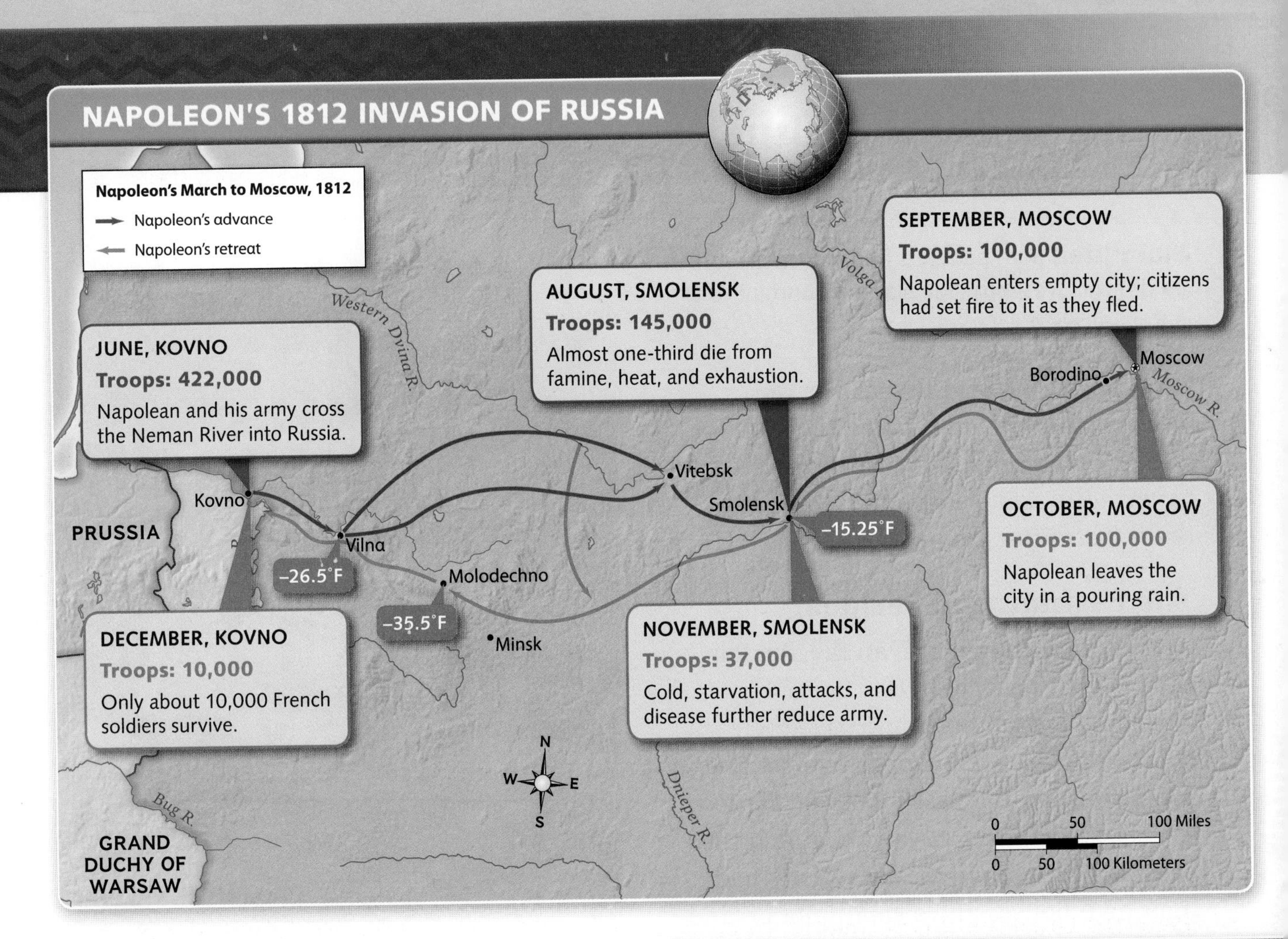

Hitler Invades Russia

History repeated itself when Adolf Hitler sent troops from **Nazi Germany** to invade Russia during World War II. Frigid winter weather resulted in the deaths of many Nazi soldiers. Hitler was unable to take Moscow in 1941, or Stalingrad in 1943. Of the 300,000 Nazi troops that fought at Stalingrad, only about 5,000 came home.

Before You Move On

Monitor Comprehension What role did Russia's geography play in defeating Napoleon and Hitler?

ONGOING ASSESSMENT

MAP LAB

1. **Movement** Using the map, figure out how far Napoleon's army had to travel to get to Moscow from Kovno.
2. **Interpret Maps** What does the map tell you about the weather as the French troops retreated from Moscow?
3. **Compare and Contrast** In what ways were Nazi Germany's experiences in Russia similar to those of Napoleon?

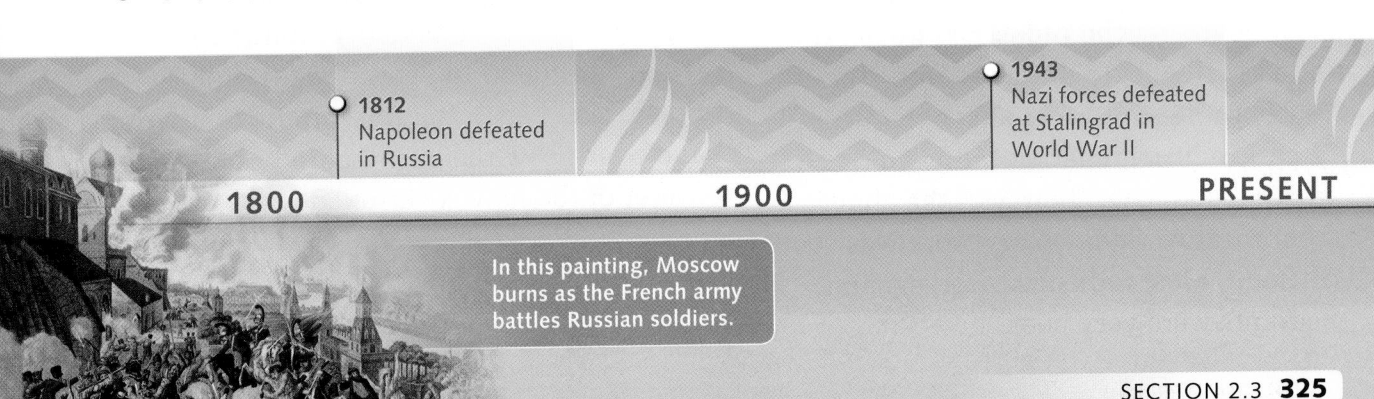

In this painting, Moscow burns as the French army battles Russian soldiers.

SECTION 2 HISTORY

2.4 Serfdom to Industrialization

TECHTREK

myNGconnect.com For an online map of Russia's resources and industries

Maps and Graphs

Main Idea Both peasants and industrial workers led difficult lives, but they played important roles in Russia's history.

For centuries, most Russian workers were peasants. They worked the land for wealthy landlords. The peasants were free to leave the estate as long as they paid their debts after the harvest.

The Beginning of Serfdom

Ivan IV, also known as Ivan the Terrible, changed this system in the 1500s. He murdered many nobles and gave their lands to people who supported him. These people needed the peasants to farm their estates, so Ivan passed laws that tied the peasants to the land as **serfs**. Serfs had their own houses and a small plot of land to farm, but they had to pay the landlord rent. Serfs could not leave the estate without permission and had few rights.

In 1861, Czar Alexander II freed the serfs. He wanted these workers to be free to work in industry and help modernize Russia. As a result, he ended their legal ties to their landlords. Serfs could then become industrial workers for wages.

Industrial Revolution in Russia

Under Peter the Great, early industries, such as shipbuilding and metalworking, had begun. Yet the **Industrial Revolution** didn't really begin in Russia until the 1890s. By 1913, it had become the fifth largest industrial nation in the world. As in other countries, however, the change from rural work to factory work was difficult. Most industrial workers and peasants were very poor.

Critical Viewing Serfs clear the land of large stones in this illustration. Based on the illustration, what can you conclude about the type of work serfs performed?

Many peasants starved because of poor harvests, and city factory workers were unhappy with their working conditions. Frequent worker **strikes**, or work stoppages, and protests gave way to political unrest. Revolutionary activist and politician **V. I. Lenin** led a political group called the **Bolsheviks**. They wanted workers to take over industry and the government. In February 1917 the Bolsheviks began the **Russian Revolution** and overthrew the czar. Lenin became leader of the new government.

Before You Move On

Summarize What roles did serfs and industrial workers play in Russian history?

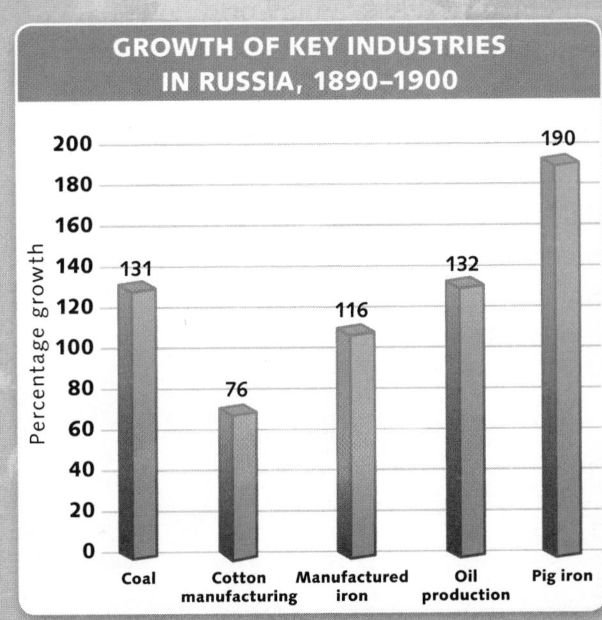

Source: Peter Stearns, *The Industrial Revolution in World History*, 1998

ONGOING ASSESSMENT

DATA LAB

GeoJournal

1. **Interpret Graphs** According to the graph, which industry grew the least between 1890 and 1900? The most? Why do you think these industries grew at these rates?
2. **Synthesize** Based on the graph, the map, and the text, what conclusions can you draw about the Industrial Revolution in Russia?
3. **Movement** Examine the map shown above. Why do you think several railroad lines ran through Moscow?

2.5 The Soviet Union

Main Idea The Communist Soviet Union had a powerful central government that controlled the region from 1922 until 1991.

The Bolsheviks, who led the Russian Revolution, believed that a Communist form of government and a socialist economic system were the answers to the problems of the Industrial Revolution. Under **communism**, a single political party controls the government and economy. **Socialism** is a system in which the government controls economic resources. The Bolsheviks wanted to end private ownership of land and resources and establish a classless society.

A Communist State

In 1922, Russia, Ukraine, Belarus, and the Transcaucasian republics of Armenia, Azerbaijan, and Georgia formed the Union of Soviet Socialist Republics (U.S.S.R.), also known as the **Soviet Union**. Later, other republics came under the control of the Soviet Union and the central Communist government in Moscow.

From 1927 to 1953, the Soviet people lived under the total command of Josef Stalin. Stalin's government isolated its citizens from contact with the West.

The Cold War

After World War II, the Soviet Union and the United States were the two most powerful countries in the world. Tension and conflict arose between the two because of their very different political and economic systems. The conflict came to be known as the **Cold War** because the countries did not fight each other directly.

The Cold War led the United States and Soviet Union to develop nuclear weapons. The United States went to war in Korea and Vietnam during the 1950s and 1960s to try to prevent communism from spreading to these countries. The Cold War also resulted in a "space race." The Soviet Union won the race in 1957 when it launched its Sputnik satellite into space.

A Controlled Economy

The Soviet Union also became an industrial leader and a world power, second only to the United States. The government owned most businesses and agriculture. On **collective farms**, workers produced a certain amount of food—determined by the government—and received a share of surplus crops. Still, the Soviet Union had trouble feeding all of its people.

A banner of Lenin overlooks Red Square as Soviet Russia celebrates the anniversary of the 1917 Russian Revolution.

1922 Soviet Union forms.

1924 Stalin becomes Soviet leader.

1925

1941–1945 Soviet Union fights Germany in World War II.

1945 Cold War begins.

1950

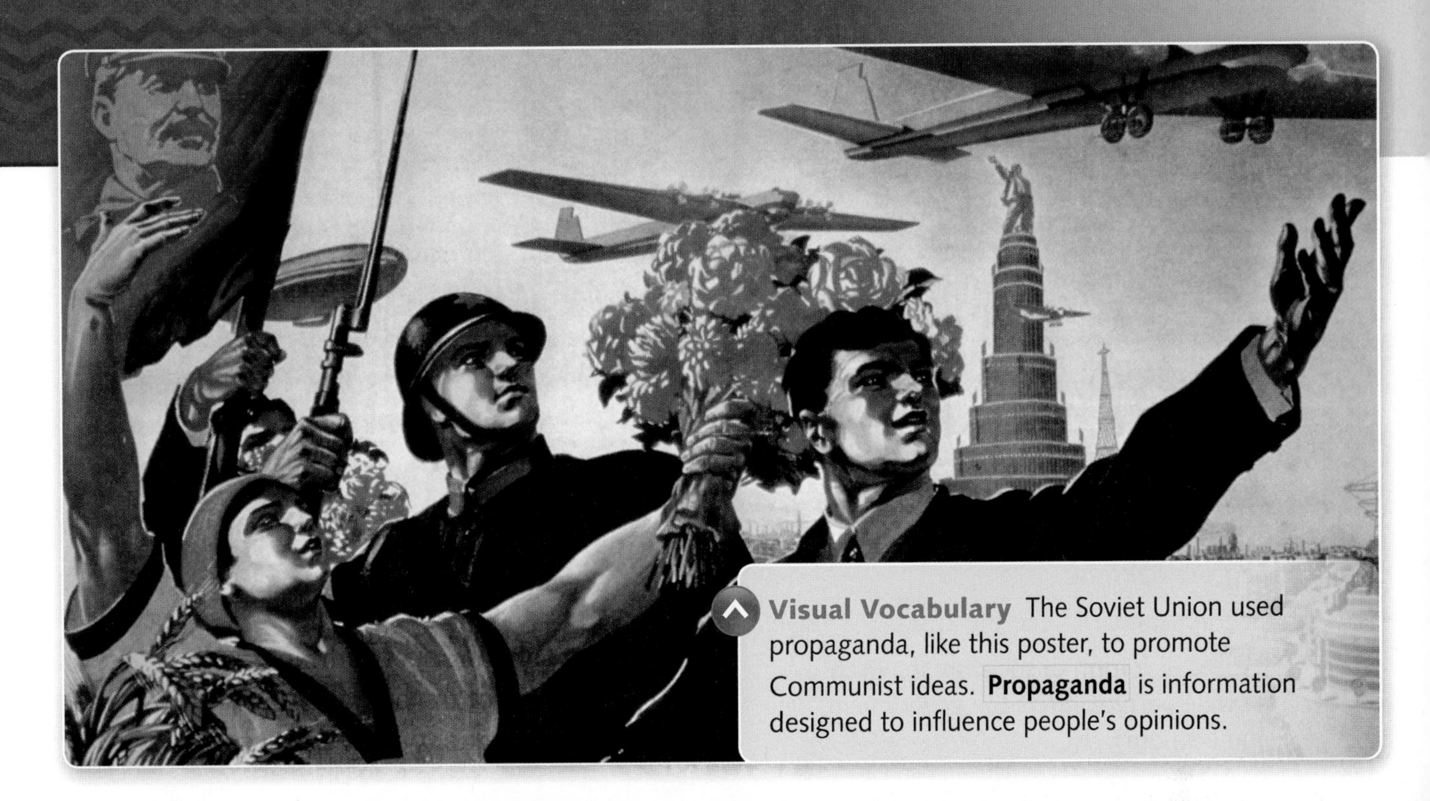

Visual Vocabulary The Soviet Union used propaganda, like this poster, to promote Communist ideas. **Propaganda** is information designed to influence people's opinions.

Quality of life was poor, too. People had guaranteed jobs, but their standard of living was much lower than that in Western countries. For example, they had little access to consumer goods. President **Mikhail Gorbachev** (mih KYL GAWR buh chawf) tried to reform and improve the economy. However, a movement toward adopting democratic forms of government was spreading across Eastern Europe. In 1991, the Soviet Union collapsed, and its republics gained their independence.

Before You Move On

Monitor Comprehension In what ways did the Communist government of the Soviet Union control its republics?

ONGOING ASSESSMENT

VIEWING LAB

GeoJournal

1. **Analyze Visuals** In the propaganda poster, the figure in the back is holding a flag with Stalin's picture on it. What are the other figures holding? What might each figure represent?
2. **Make Inferences** Notice the attitude and gestures of the figures in the poster. How do you think the poster was supposed to make the Russian people feel?
3. **Draw Conclusions** How do you think the American people felt when the Soviet Union sent the first satellite into space? How might the space race have intensified the Cold War?

1957 Soviets launch Sputnik satellite.

1961 Soviets send first person into space.

1975

U.S. president Ronald Reagan and Gorbachev

1985 Gorbachev begins reform programs and works with President Reagan to end Cold War.

1991 Soviet Union collapses.

PRESENT

CHAPTER 11

Review

VOCABULARY

On your paper, write the vocabulary word that completes each of the following sentences.

1. _______ under the tundra keeps trees from growing there.
2. Some rivers provide _______, using the power of water to generate electricity.
3. _______ is information designed to influence people's opinions.
4. Czar Alexander I freed the _______ in 1861.
5. _______ is a form of socialism in which a single party controls the government and economy.

MAIN IDEAS

6. What natural barriers separate Russia and the republics from their neighbors? What impact do the barriers have on these countries? (Section 1.1)
7. Why are Russian winters cold and dark in some places? (Section 1.2)
8. What are the two main types of natural resources in this region? (Section 1.3)
9. What does Katey Walter Anthony hope to learn by studying the presence of methane in Siberian lakes? (Section 1.4)
10. What type of agriculture is suited to the dry climate of Central Asia? (Section 1.5)
11. How was Kievan Rus founded? (Section 2.1)
12. What western European influences did Peter the Great and Catherine the Great introduce to Russia? (Section 2.2)
13. What factors helped defeat the Nazis in Russia? (Section 2.3)
14. In what ways were serfs different from peasants in Russia? (Section 2.4)
15. What methods did the government of the Soviet Union use to control the country's economy? (Section 2.5)

GEOGRAPHY

ANALYZE THE ESSENTIAL QUESTION

How have size and extreme climates shaped Russia and the Eurasian republics?

Critical Thinking: Analyze Cause and Effect

16. Why does Russia have few ice-free ports?
17. What impact do Russia's size and extreme climate have on its use of natural resources?
18. In what way did the climate in Central Asia contribute to the shrinking of the Aral Sea?

INTERPRET MAPS

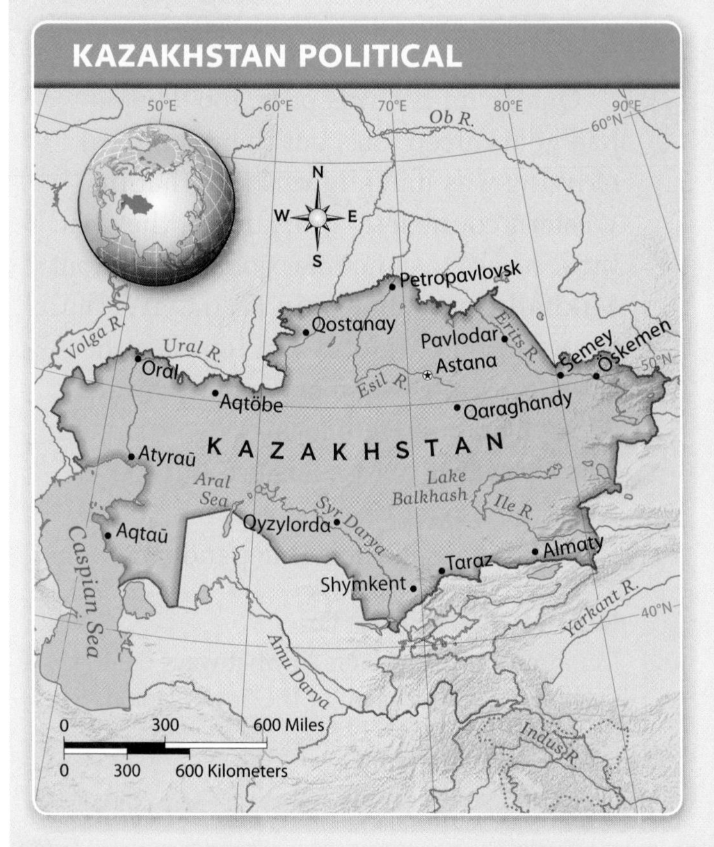

19. **Location** Which two cities are best located to take advantage of energy resources in the Caspian Sea?
20. **Make Inferences** Why do you think Kazakhstan moved its capital from Almaty to Astana in 1997?

HISTORY

ANALYZE THE ESSENTIAL QUESTION

How has geographic isolation influenced the region's history?

Critical Thinking: Make Inferences

21. What effect did the Mongols have on Russia's relationship with Europe?
22. During Napoleon's invasion of Russia, what impact might distance have had on the army's horses and supplies?
23. In what way might physical geography have prevented the Industrial Revolution from starting in Russia as early as it started in western European countries?

INTERPRET TABLES

TIME INDUSTRIAL LABORERS NEEDED TO WORK TO BUY SELECTED GOODS, 1986

	Moscow	United States
Loaf of bread	11 min*	18 min
Liter of milk	20 min	4 min
Grapefruit	112 min	6 min
Chicken	189 min	18 min
Bus fare (2 miles)	3 min*	7 min
Postage stamp	3 min	2 min
Pair of jeans	56 hrs	4 hrs
Washing machine	177 hrs	48 hrs

* price supports by government

Source: Radio Free Europe, published in the *New York Times, June 28, 1987*

24. **Compare and Contrast** What type of food must have had the biggest price difference? What food was probably more expensive in the United States than in Moscow?
25. **Make Generalizations** Based on the chart, what do you think life was like for industrial laborers in the Soviet Union in 1986?

ACTIVE OPTIONS

Synthesize the Essential Questions by completing the activities below.

26. **Write an Email** Schools sometimes conduct exchange programs with schools in other countries. Given what you know about the geography, history, and culture of Russia, decide which area of Russia you would like to visit. Write an email to a classmate recommending a particular location in Russia using the tips below. **Send the email to a classmate and ask for feedback on the information provided.**

Writing Tips
- Organize your ideas under two or three main headings before you begin writing.
- Use a clear, straightforward style to present specific, useful, and interesting details in your email.
- Make sure you explain why you offer the advice that you do.

Go to **Student Resources** for Guided Writing support.

27. **Gather and Share Information** Work in a group to gather two new facts about Russia and each of the republics. Use the research links at **Connect to NG** and other online sources to help you find the facts. Then create a chart to record your facts and share them in an oral presentation. Compare your findings with those of other groups.

Russia	Fact 1: ______ Fact 2: ______
Armenia	Fact 1: ______ Fact 2: ______
Azerbaijan	Fact 1: ______ Fact 2: ______

CHAPTER 12

Russia & THE EURASIAN REPUBLICS TODAY

PREVIEW THE CHAPTER

Essential Question What features, such as size and climate, have influenced Russian culture?

SECTION 1 • CULTURE

KEY VOCABULARY

- culture
- nomad
- yurt
- terrain
- gauge
- port
- diplomacy

ACADEMIC VOCABULARY

enlist

TERMS & NAMES

- Trans-Siberian Railroad
- Hermitage Museum

Essential Question How have Russia and the Eurasian republics dealt with recent political, economic, and environmental challenges?

SECTION 2 • GOVERNMENT & ECONOMICS

KEY VOCABULARY

- perestroika
- glasnost
- coup
- federal system
- proportional representation
- revenue
- pipeline
- radioactive
- fallout
- half-life
- contaminate

ACADEMIC VOCABULARY

autonomy, vulnerable

TERMS & NAMES

- Russification
- Kremlin
- Chernobyl

Student eEdition

Maps and Graphs

Interactive Whiteboard GeoActivities

Digital Library

Connect to NG

Go to myNGconnect.com for more on Russia and the Eurasian Republics.

Uzbek women and girls from a mountain village wear a *rumol*, a traditional head scarf.

SECTION 1 CULTURE

1.1 Climate and Culture

TECHTREK 

myNGconnect.com For photos reflecting climate and culture in Russia and the republics

Digital Library

Main Idea The variety of climates in Russia and the Eurasian republics has a major impact on the region's cultures.

Culture refers to a group's unique way of life. Cultural traits include food—what people eat and how they obtain it—shelter, clothing, religion, and language. The first three traits are influenced by climate.

Enduring the Cold in Siberia

About 30 different native peoples live in the tundra and taiga of Siberia in Russia. These include the Yakut, Nenets, Evenki, Chukchi, and Inuit (IHN yoo iht). All of these people have similar cultures but different languages.

People living in these cold climates, with their long winters and short, hot summers, traditionally dress in clothes made from reindeer fur. Some people, such as the Nenets, still live as **nomads**, moving from place to place according to the seasons in search of food. This way of life is changing, however, and today many native people live in towns or cities.

Herding in Central Asia

Like the Nenets in Siberia, some herders in Kazakhstan and other parts of Central Asia also continue to lead nomadic lives. In the region's dry climate, this traditional culture focuses on herding sheep, camels, cattle, and horses. Animals provide food and milk as well as hides and wool for clothing and tents. Although herding continues to be important in this region, many herders now live in rural villages in houses made of mud bricks that have been baked by the sun.

Farming in the Steppes

On the steppes of Russia and parts of Central Asia, the rich soil and moderate climate—cold winters and warm summers—are suited to agriculture. A peasant farming culture developed in this climate. In this culture, people grew grain crops, especially wheat, and raised livestock. They lived in settled villages. Later they worked for landlords on large estates and became serfs.

Critical Viewing A Nenet boy studies for school in the frozen tundra of northwest Siberia. What impact might the intense cold reflected in this photo have on everyday activities?

ADAPTING TO CLIMATE

SIBERIA Many native people of Siberia center their culture around reindeer. The warm, durable clothing worn by this family is made from reindeer skin. Reindeer meat and fat also provide nutritious food.

CENTRAL ASIA The animals the Kazakhs of Central Asia raise provide almost everything the herders need to live. Felt, or wool fibers pressed together, is used to make hats and traditional tents called **yurts**.

STEPPES Life for farmers on the steppes revolves around the seasonal growing cycle. Farmers often use simple tools to grow their crops. Animals help with the work and provide meat and milk.

Today, large corporations or individuals own the farms. Farmers still grow wheat as in the past, and they grow other grain crops as well, including maize, or corn, and barley. Many of these grain products are exported to other countries. Beets and potatoes are also common crops, along with sunflowers, which are grown to make oil used in cooking.

Before You Move On

Summarize What impact have the variety of climates in Russia and the Eurasian republics had on the region's cultures?

ONGOING ASSESSMENT

PHOTO LAB

1. **Analyze Visuals** The Central Asian yurt shown above is movable. Why might this be an important consideration for nomadic herders?
2. **Compare and Contrast** Study all three photos above. In what ways does the women's clothing reflect the climate in each region?
3. **Human-Environment Interaction** In what ways has climate influenced traditional culture in Siberia, Central Asia, and the steppes?

1.2 Trans-Siberian Railroad

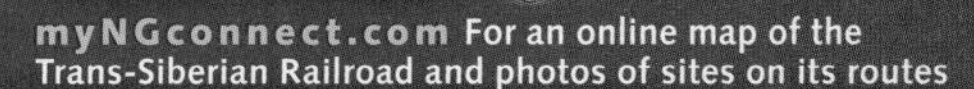

myNGconnect.com For an online map of the Trans-Siberian Railroad and photos of sites on its routes

Main Idea The Trans-Siberian Railroad links the western and eastern parts of Russia.

The **Trans-Siberian Railroad** is the world's longest continuous railroad, spanning about 6,000 miles and crossing eight time zones. *Trans-Siberian* means "across Siberia." The railroad links Moscow with eastern Russia. It also carries people and goods to East Asia and Europe.

Building the Railroad

Russia began to build the railroad in 1891. It was designed to connect Moscow with Vladivostok (vlah duh VAWH stock), a busy port on the Pacific Ocean. At that time, no reliable form of transportation linked this far eastern port with the European part of Russia.

Laborers began at both ends of the route and worked their way toward the center, but the climate and **terrain**, or physical features of the land, made progress very difficult. Workers had to lay tracks across long stretches of permafrost, and the route went through mountains, forests, rivers, and lakes. Materials—including explosives used to blast through rocks and cliffs—had to be brought thousands of miles to the work sites.

The construction project required many workers. As a result, thousands of Russian peasants, convicts, and soldiers were **enlisted**, or selected, to labor on the railroad. They were given only picks and shovels to do their difficult work. Human workers and horses hauled the heavy materials. The laborers worked long hours in extreme heat in the summer and extreme cold in the winter. They also had to deal with attacks by thieves and, occasionally, tigers. The main route, running from Moscow to Vladivostok, was finally completed in 1916.

Critical Viewing A Trans-Siberian Railroad train runs along the banks of Lake Baikal in Siberia. In this photo, the conductor looks out of the train as it curves around the lake. What challenges did the workers probably face while building this part of the railroad?

Effects of the Railroad

The railroad transformed Siberia and its traditional culture. Between 1891 and 1914, more than five million people immigrated to Siberia. New towns and cities grew up along the train's route. Soviet leaders began to industrialize Siberia and mine its plentiful raw materials. During World War II, the railroad moved Soviet troops and materials across Russia.

The Railroad Today

Today, the railroad operates several more routes and has replaced all of the old steam engines with electric trains that carry passengers and freight. The railroad also plays an important role in the world economy. Container cargo, or goods packed in large steel boxes, travels from China and other parts of East Asia to Europe. One challenge is that the **gauge**, or width of the tracks, in Russia is wider than in Europe or China. As a result, containers need to be transferred to different trains at the borders. Still, shipping containers by land across Russia is much faster—and cheaper—than shipping by sea.

Before You Move On

Monitor Comprehension In what ways has the Trans-Siberian Railroad helped to link Russia?

ONGOING ASSESSMENT

MAP LAB

GeoJournal

1. **Interpret Maps** Use the map scale to determine the distance from Novosibirsk to Irkutsk on the Trans-Siberian Railroad.
2. **Movement** Trace the Trans-Siberian Railroad route to China. What benefit does it offer to manufacturers of consumer goods in China?
3. **Make Inferences** Why are there many tunnels and bridges along the route of the Trans-Siberian Railroad?

1.3 St. Petersburg Today

Main Idea St. Petersburg is Russia's second largest city and a center of culture, industry, and trade.

St. Petersburg is located in northwest Russia on the Neva River. The river flows into the Gulf of Finland, which is at the most eastern part of the Baltic Sea. Because of its far northern location—at a latitude of about 60°N—St. Petersburg has long winter nights. In fact, for about one month a year, there's barely any daylight. For about three weeks in summer, the sky never gets completely dark. Special music and dance events are held during these "White Nights" to take advantage of the long days.

Window to the West

St. Petersburg originally sat on isolated swampland, making it a poor place on which to build. Nevertheless, Peter the Great chose the site in 1703 to gain a **port**, or harbor, on the Baltic Sea for trade. He also wanted to create a modern city—for the times—that would resemble those in Western Europe and become Russia's "window to the West." Peter got his wish. He had St. Petersburg filled with islands, canals, and bridges like those in the western European cities of Amsterdam and Venice. He also built wide boulevards like those in Paris and London.

FAST FACTS ON ST. PETERSBURG	
Population	4.6 million people
Land Area	550 square miles
Location	Average temperature in January: 21°F Average temperature in July: 65°F Latitude: 59° 57'N; longitude: 30° 19'E
Date Founded	1703

St. Petersburg's historic architecture also reflects western European influences. Many of the buildings from the 1700s still survive, including the Winter Palace, **Hermitage Museum**, and summer palaces of several czars. St. Petersburg is also filled with many beautiful and historic Russian Orthodox churches and cathedrals in Western and Eastern style.

The city is Russia's cultural center. St. Petersburg boasts world-famous museums and ballet companies, such as the Kirov. It also contains many universities and theaters and the country's oldest music academy. In addition, St. Petersburg offers a variety of contemporary music, including jazz and rock.

A Vibrant Economy

Manufacturing and construction are important industries in St. Petersburg. The city is also a center for trade and distribution of goods to and from Europe.

Above all, the local economy depends on tourism. More than three million visitors came to see the city's attractions in 2003 on its 300th anniversary. Many historic buildings that had been damaged during World War II were restored in time for the celebration. Vladimir Putin, Russia's president in 2003, is from St. Petersburg. He wanted the city to become a center of **diplomacy**, a place where international affairs could be conducted. Like Peter the Great, Putin wanted Russia to be more connected to the West.

Before You Move On

Summarize In what ways is St. Petersburg a center of culture, industry, and trade?

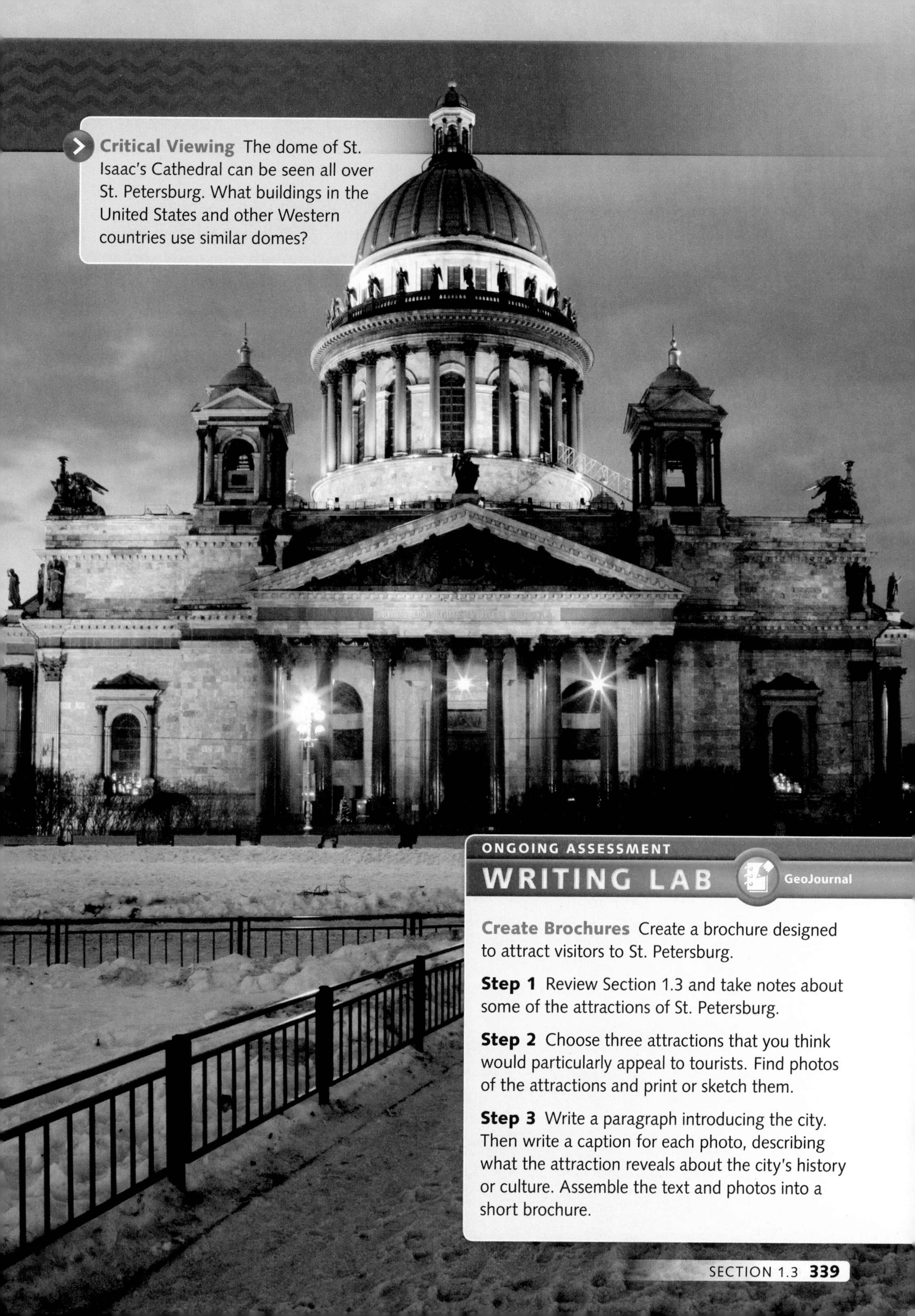

Critical Viewing The dome of St. Isaac's Cathedral can be seen all over St. Petersburg. What buildings in the United States and other Western countries use similar domes?

ONGOING ASSESSMENT

WRITING LAB

GeoJournal

Create Brochures Create a brochure designed to attract visitors to St. Petersburg.

Step 1 Review Section 1.3 and take notes about some of the attractions of St. Petersburg.

Step 2 Choose three attractions that you think would particularly appeal to tourists. Find photos of the attractions and print or sketch them.

Step 3 Write a paragraph introducing the city. Then write a caption for each photo, describing what the attraction reveals about the city's history or culture. Assemble the text and photos into a short brochure.

2.1 The Soviet Collapse

TECHTREK
myNGconnect.com For photos of the Soviet Union

Digital Library

Main Idea Economic problems and people's desire for independence caused the Soviet Union to collapse in 1991.

As you have learned, the Soviet Union formed in 1922. During the 1970s and 1980s, the economy of the huge region slowed. Soviet money had almost no value outside the country. Stores had little food on their shelves and few goods to buy. In addition, people in the Eurasian republics had long resented the Soviet policy of **Russification**. Under this policy, the Soviet Union moved Russians into the republics and put them in charge. They also forced local people to learn the Russian language.

Gorbachev Brings Reform

In 1985, Mikhail Gorbachev became head of the Soviet Communist Party and began to introduce reforms. He promoted a movement known as **perestroika** (pehr ih STROY kuh), which means "restructuring." Gorbachev wanted to restructure the economy. Less government control, he believed, would make the economy more effective.

After the economy had failed to improve by 1990, the government presented a new plan to change the economy in 500 days. According to this plan, the republics would have greater control over their economies, and state ownership of businesses would decline.

Gorbachev also introduced the policy of **glasnost** (GLAHS nuhst), or "openness," which encouraged people to speak openly about the government. This freedom of expression went beyond what Gorbachev had intended, however. People started to criticize and protest against the central government, and they began to demand even more freedom.

Critical Viewing Food shortages in the Soviet Union resulted in empty shelves, as you can see in this 1990 photo of a Moscow grocery store. What impact do you think seeing stores like this had on the Soviet people?

The Soviet Union Falls

After the collapse of the Berlin Wall in Germany in November 1989, the Communists no longer had control over Eastern Europe. People in many of the Soviet republics also wanted freedom from the central Communist government. Weakened by the economy, the government could no longer manage its vast empire. By the fall of 1990, all of the republics had declared their **autonomy**, or determination to govern themselves.

In August 1991, conservative Communists who opposed Gorbachev's political and economic reforms attempted a coup against him. A **coup** (KOO) is a sudden overthrow of the government by force. The coup failed, but Gorbachev knew he had lost power. On December 25, 1991, he resigned as president, and the Soviet Union dissolved. All of the republics became independent countries.

Life After Independence

Some groups within the newly independent republics and Russia's own borders sought independence for themselves. In Azerbaijan, Armenians continue to fight over a part of the country where most of the people are ethnic Armenians. Within Russia, the Muslim republic of Chechnya (CHEHCH nee uh) has struggled to gain full independence.

The economic transition has been difficult as well. Less government control in Russia has led to a large gap between rich and poor, higher prices, unemployment, and an increase in corruption and organized crime. Some of the Eurasian republics still have centrally controlled economies and governments.

Before You Move On

Make Inferences In what ways did attempts to reform the Soviet economy and political system contribute to the collapse of the Soviet Union?

ONGOING ASSESSMENT

SPEAKING LAB

Turn and Talk With a partner, discuss the causes and effects of the Soviet Union's collapse. Use a cause-and-effect chart like the one below to list and organize your ideas.

CAUSES	EFFECTS
1.	
2.	

2.2 Russia's Government

TECHTREK
myNGconnect.com For current events on Russia and photos of its leaders

 Connect to NG

 Digital Library

Main Idea Russia's central government consists of three branches and, together with the president, holds most of the power.

Following the collapse of the Soviet Union, Russia adopted a new constitution. This set up a **federal system**, with a strong central government and local government units. The United States also has a federal system. Russia's government is more democratic than that of the former Soviet Union. Everyone 18 years and older can vote, and there are several political parties. However, the central government and the president hold most of the power.

Presidential Power

In Russia, the president is the head of the executive branch and the most powerful government leader. In 2008, the presidential term was extended from four to six years. That year, Dmitry Medvedev (med VYED if) was elected president.

The president appoints the prime minister. Medvedev appointed Vladimir Putin, who had been president from 2000 to 2008 (two four-year terms). Unlike earlier Russian prime ministers, Putin exercises a great deal of power. Some observers believe that he is the real leader of Russia, not Medvedev.

Legislators and Judges

There are two houses in the legislative branch. The lower house is the State Duma. Members are elected based on **proportional representation**. Under this system, a political party gets the same percentage of seats as the percentage of votes it received. A party must receive at least seven percent of the vote to get seats. Putin leads the United Russia Party, which holds about 64 percent of the seats.

The upper house of the legislative branch is the Federation Council. Executive and legislative leaders in each of the local government units appoint the members of this house. The State Duma is the more powerful legislative house. All bills must first be considered in the State Duma, even those proposed by the upper house.

Critical Viewing The **Kremlin** is a historic complex of palaces and churches in the heart of Moscow. What feature in the photo suggests that the Kremlin once served as the city's fortress?

COMPARE RUSSIAN AND U.S. GOVERNMENTS		
Russia	**Government Branches**	**United States**
• President, elected to a six-year term Prime Minister, appointed • Government Ministers, appointed	**Executive**	• President and Vice President, both elected to a four-year term • Cabinet, appointed
• Federal Assembly: – Federation Council (166 members), appointed to a four-year term – State Duma (450 members), elected to a five-year term	**Legislative**	• Congress: – Senate (100 members), elected to a six-year term – House of Representatives (435 members), elected to a two-year term
• Constitutional Court Judges, appointed for life	**Judicial**	• Supreme Court Judges, appointed for life

The highest court in Russia is the Constitutional Court. The Federation Council appoints these judges based on the president's recommendations. The judges are appointed for life. In general, the judicial branch in Russia is more **vulnerable**, or open, to political pressure than in most Western democracies. That means that officials in the executive and legislative branches are sometimes able to influence the judges.

Central Control

The central government in Moscow still tries to control most levels of government. Those in power generally choose the people they want to be elected.

In 2000, then-president Putin reduced the number of local government units in Russia from 89 to 7 to increase central control of them. The president already exercised some power over these units because he nominated their governors.

Before You Move On

Summarize What branches make up the central government, and in what ways do the government and the president exercise their power?

ONGOING ASSESSMENT

DATA LAB

GeoJournal

1. **Identify** Based on the chart, which officials in the Russian executive branch probably serve a role similar to that of the Cabinet members in the U.S. government?
2. **Compare and Contrast** Study the chart. In what way is the legislative branch in Russia similar to that in the United States? In what way is it different?
3. **Express Ideas Through Speech** Is the government in Russia more or less democratic than that in the United States? Think about which officials are elected or appointed and how power is handled. Discuss with a partner.

2.3 Unlocking Energy Riches

TECHTREK
myNGconnect.com For an online map and photos of the region's energy pipelines

Main Idea Oil and natural gas enrich the economies of Russia and several countries around the Caspian Sea.

In addition to reforming its government, Russia has worked to develop its economy. Oil and natural gas are its greatest sources of wealth. In fact, the country is the largest exporter of oil and natural gas in the world. In 2008, these resources accounted for about two-thirds of the value of all Russia's exports and one-third of its **revenue**, or income.

Siberian Boom Towns

About 70 percent of Russian oil comes from western Siberia. Since the end of the Soviet Union, the region has been booming, or growing rapidly. Workers from western Russia and Central Asia come to Siberia for good-paying jobs. They earn enough to buy apartments in cities such as Surgut. New suburbs have developed, and many people enjoy a higher standard of living. Oil wealth has also helped fund new airports, museums, and schools.

Caspian Sea Riches

Some experts believe that the Caspian Sea may actually have more energy reserves than the Persian Gulf. Russia would like to control the energy from the Caspian. In 2007, Russia, Turkmenistan, and Kazakhstan signed an agreement to build a pipeline that would transport natural gas from the Caspian through Kazakhstan into Russia. From there, Russia would export the gas to Europe and make a large profit.

In 2009 and 2010, new gas pipelines were opened to carry gas from Turkmenistan to China and Iran. Pipelines in Kazakhstan also bring oil from the Caspian Sea to China.

Visual Vocabulary Oil workers seal a pipeline in Kazakhstan. A **pipeline** is a series of connected pipes used to transport liquids or gases.

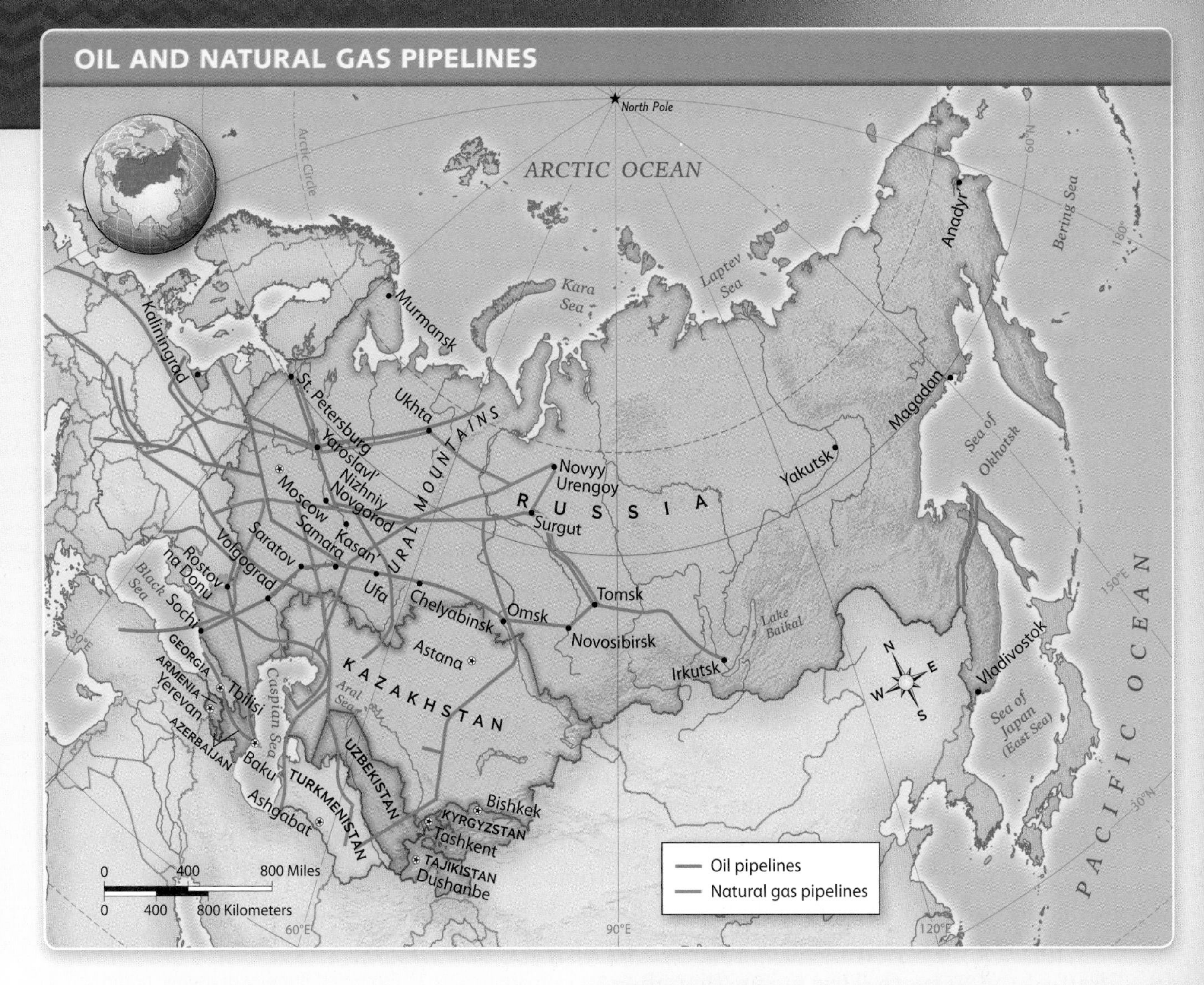

Azerbaijan has large reserves of both oil and natural gas. It used to send the fuel through pipelines to Russia. Now it makes more money by trading mainly with the United States and by piping the fuel directly through Georgia and Turkey. By so doing, Azerbaijan does not have to share export profits with Russia.

For now, Russia gains a great deal of wealth from its exports of natural gas and oil. However, by relying too heavily on energy exports, the country risks using up these natural resources.

Before You Move On

Monitor Comprehension In what ways have oil and natural gas enriched Russia and countries around the Caspian Sea?

ONGOING ASSESSMENT

MAP LAB

1. **Movement** About how far does natural gas travel by pipeline from Surgut in western Siberia to St. Petersburg in Russia? Why do you think pipelines might be a good way to transport the gas over long distances?
2. **Pose and Answer Questions** Come up with your own question about the information shown on the map. Then challenge a partner to answer your question.
3. **Draw Conclusions** What would Russia gain by controlling the energy from the Caspian Sea?
4. **Make Inferences** Why are countries in Central Asia developing pipelines to transport oil and natural gas directly to China and Iran?

2.4 After Chernobyl

TECHTREK  
myNGconnect.com For photos of the Chernobyl disaster and Guided Writing
Digital Library
Student Resources

Main Idea The Chernobyl nuclear disaster severely damaged the environment in parts of Belarus, Ukraine, and western Russia.

On April 26, 1986, a nuclear reactor at a power plant in Chernobyl exploded and caught fire, resulting in the worst nuclear disaster in history. **Chernobyl** is in the country of Ukraine. At the time of the disaster, Ukraine was part of the Soviet Union. A **radioactive** cloud about 3,280 feet high spread over parts of Ukraine, Belarus, and Russia. Winds carried the **fallout** into parts of northern and central Europe. These radioactive materials have caused harm to humans, animals, and plants.

KEY VOCABULARY

radioactive, adj., giving off energy caused by the breakdown of atoms

fallout, n., radioactive particles from a nuclear explosion that fall through the atmosphere

half-life, n., the time needed for half the atoms in a radioactive substance to decay and decrease

contaminated, adj., unfit for use because of the presence of unsafe elements

Health Effects

Scientists know that it takes some radioactive materials a long time to break down and disappear. Each material has a particular **half-life**, the time needed for half of its atoms to decay and decrease. For example, the highly radioactive element cesium has a half-life of 30 years. That means that after 30 years, half of its atoms will still be radioactive.

Thirty people died within three months of the accident, most from being exposed to huge amounts of radioactive material. Over time, thousands of the people who helped clean up the disaster have developed health problems. Some of the children born to exposed parents carry the effects in their genes. In addition, radioactive iodine got into the milk of cows that grazed on **contaminated**, or infected, grass after the accident. This caused an increase in thyroid cancer, especially in children. Millions of people are still living on contaminated land. Officials cannot yet predict the long-term effects on people's health.

Before You Move On

Summarize What are some of the health problems that occurred as a result of the Chernobyl nuclear disaster?

CHERNOBYL DISASTER BY THE NUMBERS

400
Estimated number of atomic bombs it would take to equal the accident

4,000
Estimated number of people who may die from cancers caused by radiation exposure from the accident

600,000
Estimated number of people who received significant radiation exposure, including evacuees, residents, and those who helped clean up after the accident

5 million
Estimated number of people still living on contaminated areas in Ukraine, Belarus, and Russia

Source: UN Chernobyl Forum, 2006

Environmental Damage

Belarus was hardest hit because of the direction the wind blew just after the accident. About 23 percent of the country was contaminated, including agricultural and forest land. In Ukraine, 7 percent of the land and 40 percent of its forests were contaminated. In Russia, the area bordering with Belarus was most affected.

Radiation is expected to remain in the soil for many years. Plants that grow in the region's forests are contaminated and so are the animals that feed on them. Many reindeer herds had to be killed shortly after the accident because they were so infected by the radiation.

The fallout has especially contaminated the fish in the rivers and lakes of Ukraine because these waters flow down from the site of the disaster. Some radiation that has gotten into groundwater sources will last for hundreds of years.

COMPARE ACROSS REGIONS

France's Nuclear Program

In spite of what happened at Chernobyl, many countries have continued to develop nuclear energy. France, for example, has more than 55 nuclear plants that generate more than 75 percent of its electricity. Many of the French people welcome a new nuclear plant in their town because it brings jobs and prosperity to the area.

The French are aware of the dangers of nuclear energy plants, but they have little fear. Many of the plants offer tours, and advertisements help reinforce the idea that nuclear energy is a fact of life in France.

Few humans live in the 18-mile fenced area around the Chernobyl reactor now.

As a result, the people know that the plants are technologically more advanced and safer than the one in Ukraine. The French scientists and engineers who built them learned the lesson of Chernobyl.

Before You Move On

Monitor Comprehension What has been learned from the Chernobyl disaster?

ONGOING ASSESSMENT

READING LAB

1. **Summarize** What damage did the Chernobyl disaster cause to the environment?
2. **Make Predictions** In what ways do you think the disaster will continue to affect people in Ukraine, Belarus, and Russia?
3. **Write Comparisons** Imagine that two new nuclear plants were being proposed: one in Russia and one in France. What do you think would be the reactions of the townspeople in each country? Write two sentences comparing their reactions. Make sure that you explain what led you to believe that the people in each country would feel that way. Go to **Student Resources** for Guided Writing support.

CHAPTER 12
Review

VOCABULARY

For each pair of vocabulary words, write one sentence that explains the connection between the two words.

1. nomad; yurt

   For centuries, nomads in Central Asia have traditionally lived in felt tents called yurts.

2. perestroika; glasnost
3. federal system; proportional representation
4. radioactive; contaminated
5. fallout; half-life

MAIN IDEAS

6. Describe some of the different types of climate in Russia and the Eurasian republics. (Section 1.1)
7. In what ways has Russia benefited from the Trans-Siberian Railroad? (Section 1.2)
8. Why is St. Petersburg called Russia's "window to the West"? (Section 1.3)
9. How did installing Russian officials in the republics help bring about the collapse of the Soviet Union? (Section 2.1)
10. In what way did Gorbachev's changes to the government contribute to the fall of the Soviet Union? (Section 2.1)
11. Which branch of Russia's central government has the most power? Which branch has the least power? (Section 2.2)
12. Why is western Siberia said to be "booming"? (Section 2.3)
13. What was the impact of the Chernobyl nuclear disaster on Ukraine, Belarus, and Russia? (Section 2.4)

CULTURE

ANALYZE THE ESSENTIAL QUESTION

What factors, such as size and climate, have influenced Russian culture?

Critical Thinking: Make Inferences

14. What impact does climate have on nomadic herders in Siberia and Central Asia?
15. Why was it difficult to build the Trans-Siberian Railroad across Russia?
16. At what time of year might tourists prefer to visit St. Petersburg? Why?

INTERPRET MAPS

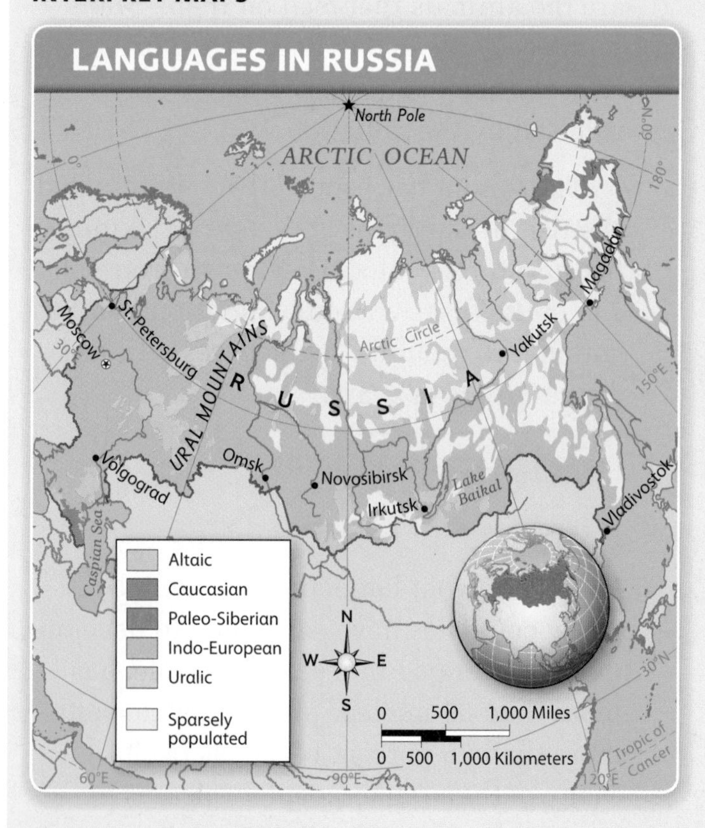

17. **Region** What are the two major language families in Russia?
18. **Compare and Contrast** What do the languages spoken west and east of the Ural Mountains have in common? In what ways do they differ?

GOVERNMENT & ECONOMICS

ANALYZE THE ESSENTIAL QUESTION

How have Russia and the Eurasian republics dealt with recent political, economic, and environmental challenges?

Critical Thinking: Make Generalizations

19. What impact did the collapse of the Soviet Union have on Russia's role in the region?
20. Why did Russia's central government reduce the number of local government units?
21. In what ways do Russia's location and power help the country control energy resources in the region?
22. Why was Russia less affected by the Chernobyl nuclear disaster than Ukraine and Belarus?

INTERPRET TABLES

NATURAL GAS RESERVES (2009)

Country	Amount (cu m)*	World Rank
Russia	47.6 trillion	1
Turkmenistan	7.5 trillion	4
United States	6.9 trillion	6
Kazakhstan	2.4 trillion	15
Uzbekistan	1.8 trillion	19
Ukraine	1.1 trillion	25
Azerbaijan	850.0 billion	27

Source: CIA Factbook / *cu m = cubic meters

23. **Analyze Data** According to the table, about how much more natural gas reserves does Russia have than the United States?
24. **Evaluate** What does this table suggest about where most of the world's natural gas reserves can be found?
25. **Draw Conclusions** Why do you think Ukraine imports natural gas from Russia?

ACTIVE OPTIONS

Synthesize the Essential Questions by completing the activities below.

26. **Write Journal Entries** Write several journal entries from the point of view of a worker on the Trans-Siberian Railroad. Describe the difficulties of the work and the obstacles you encounter. Use the writing tips below to help you write your entries. **When you have finished, trade your entries with a partner and compare them.**

Writing Tips
- Include details on where you worked, the tools you used, and the conditions you endured.
- Describe any unexpected experiences, such as an encounter with thieves or wild animals along the tracks.
- Express your feelings about having been enlisted to carry out the work.

myNGconnect.com For research links on Russia and the Eurasian republics

27. **Create Charts** Make a chart in which you compare the size, population, and major religions of Russia and the Eurasian republics. Include Kazakhstan, Kyrgyzstan, Tajikistan, Turkmenistan, and Uzbekistan in your chart. Use the research links at **Connect to NG** and other sources to gather the data. With a partner, discuss what the biggest similarities and differences are between these countries.

COMPARE COUNTRIES

Country	Size	Population	Major Religions
Russia			
Armenia			
Azerbaijan			
Georgia			

Natural Resources and Energy

TECHTREK

myNGconnect.com For an online graph and research links on energy sources

Maps and Graphs

Connect to NG

In this unit, you learned that Russia and many of the Eurasian republics have plentiful supplies of natural gas and oil. In fact, Russia has the world's largest supply of natural gas. Russia's Asian neighbor, China, is the world's biggest producer of coal. These energy sources are in great demand from countries all over the world.

Compare

- Brazil
- China
- Mexico
- Russia
- United States

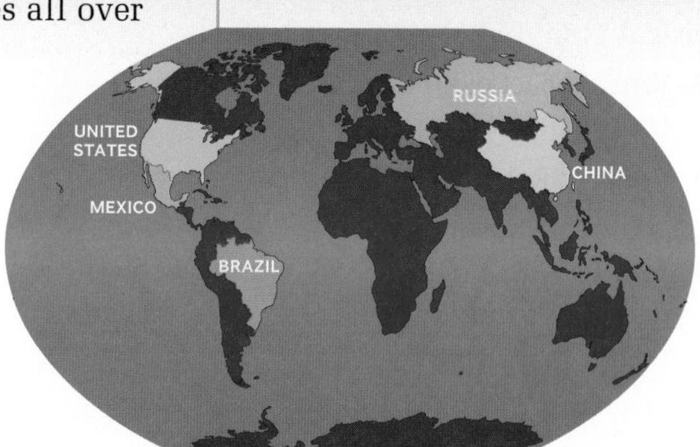

The choices people make about energy are critical to a healthy planet. As you have learned, some energy sources pollute the environment as they burn. Many scientists believe that the carbon dioxide those sources give off contributes to an overall global temperature increase. To ease these problems, they would like to reduce the use of fossil fuels and develop nonfossil energy sources.

FOSSIL FUELS

Fossil fuels are formed by buried plants and animals that have been dead for millions of years. This type of fuel is found in deposits beneath the earth's surface and must be burned to release its energy. These fuels supply about 85 percent of the world's energy. Fossil fuels include coal, oil, and natural gas.

These energy sources are classified as nonrenewable because they take millions of years to form and supplies are being used up faster than new ones can redevelop. Although natural gas is relatively clean—meaning that it doesn't create much pollution—fossil fuels, especially coal, tend to be dirty and release a great deal of harmful carbon dioxide into the air.

NONFOSSIL FUELS

Nonfossil fuels are alternative sources of energy and include the following: hydroelectric power from water, nuclear power, wind power, solar energy from the sun, and biofuels made from vegetable oil.

Many of these energy sources are renewable because they can be quickly replenished. Some countries use significant amounts of nuclear energy and hydroelectric power. However, the other nonfossil fuels make up only two percent of the world's energy.

Russia and China use both fossil and nonfossil fuels. The graphs on the next page show the energy consumption in both countries. Compare the data in the graphs and use it to answer the questions.

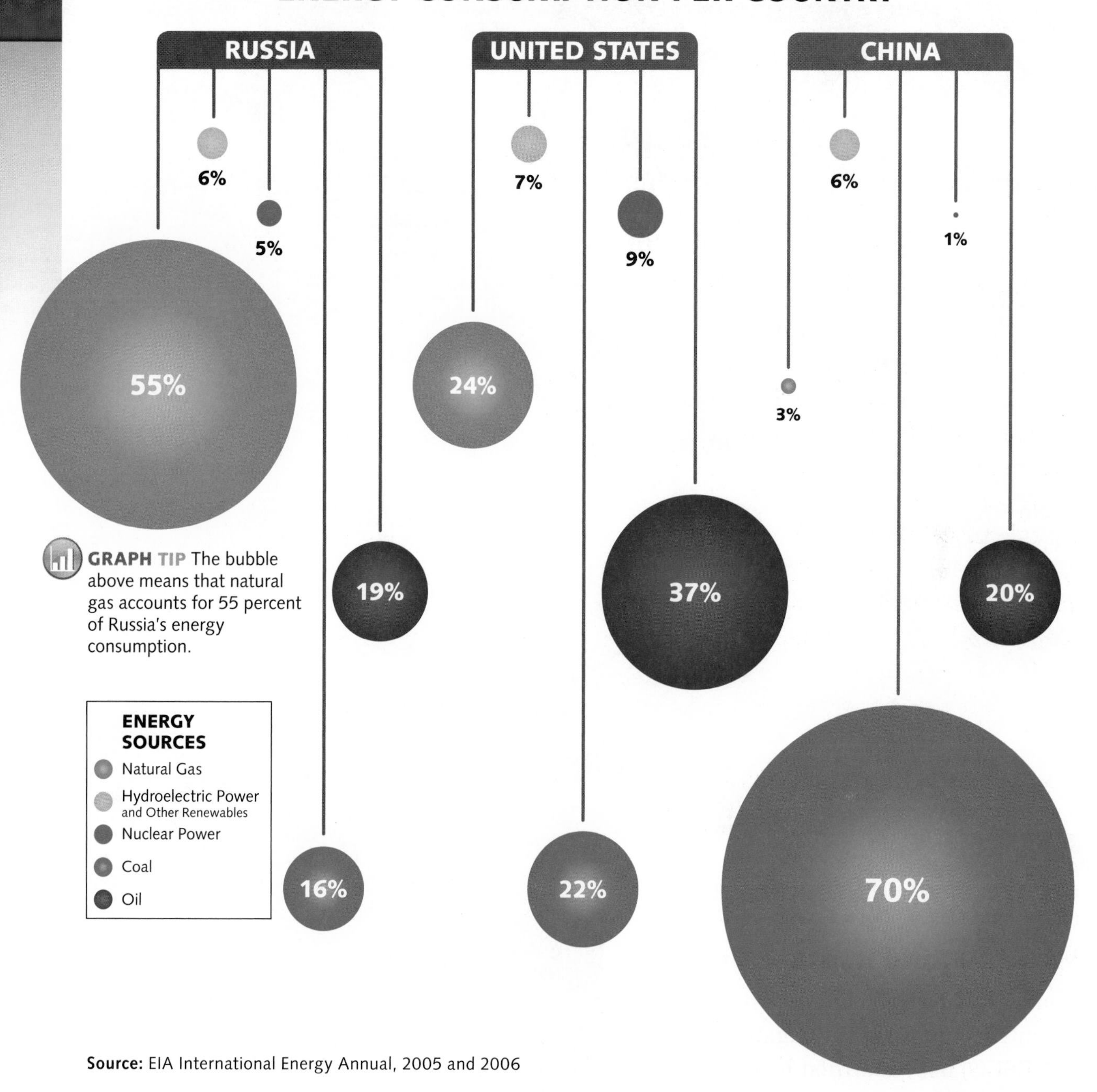

Source: EIA International Energy Annual, 2005 and 2006

ONGOING ASSESSMENT

RESEARCH LAB

GeoJournal

1. **Explain** What is the main source of energy in each country? Why might that be the case?
2. **Analyze Data** What is the energy consumption of nonfossil fuels in each country? What does this suggest about the development of alternative energy in Russia and China?

Research and Make Comparisons Research energy consumption in Brazil and Mexico and then compare with that in Russia and China. Which country appears to be investing more in nonfossil and renewable energy sources? What might this statistic suggest about its pollution levels?

Active Options

TECHTREK

myNGconnect.com For photos of natural and cultural sites in Russia and the Eurasian republics and other UNESCO World Heritage sites

Digital Library

Magazine Maker

ACTIVITY 1

Goal: Extend your understanding of animals native to Russia and the Eurasian republics.

Design a Poster for Native Animals

Russia is home to both land and sea animals. Some of these animals, including those listed on the right, have been over-hunted or are considered undesirable neighbors. Research one of these animals and use the information to design a poster that will help viewers appreciate it. Consider the animal's habitat, surviving population, and efforts made to protect it.

- Blue whale
- Brown bear
- Gray wolf
- Polar bear
- Siberian tiger

Polar bears

ACTIVITY 2

Goal: Research the natural and cultural heritage of Russia and the Eurasian republics.

Write a World Heritage Guide

The United Nations Educational, Scientific and Cultural Organization (UNESCO) has identified hundreds of sites that form part of the world's natural and cultural heritage. A variety of these World Heritage sites are located in the Russia Federation and the Eurasian republics. Prepare a guide for students visiting the area. Choose five to ten UNESCO World Heritage sites in the region. Use the **Magazine Maker** to create the guide and provide brief background information about each site.

ACTIVITY 3

Goal: Learn about Russian culture through the country's folk tales.

Hold a Russian Folk Tale Festival

Folk tales, such as "The Old Man and the Bear" and "Baba Yaga," are an age-old part of the Russian culture. With a group, find some Russian folk tales online and read them. Decide which ones you will present in the festival. Then decide how you will present the tales. You may each choose to retell a story, or you may decide as a group to act out one or more of the tales.

REFERENCE HANDBOOK

Skills Handbook

Find Main Idea and Details

Every book, paragraph, or passage has a **main idea**, which is the sentence or sentences that state the subject of the text. A main idea can sometimes be unstated, or *implied*. In that case, the **details** in the text provide clues about the main idea. To find a main idea and details, follow the steps at right.

Step 1 Look for a stated main idea in the first and last sentences of the paragraph. If no main idea is clear, look for details that offer clues about what the implied main idea is.

Step 2 Next, find details that support and clarify the main idea. If the main idea is in the first sentence, supporting details follow it. If the main idea is in the last sentence, the details come before it.

GUIDED MODEL

Habitat Loss and Restoration

A The loss of habitats can destroy an entire ecosystem. An ecosystem is a community of plants and animals and their habitat. **B** Earth has many different ecosystems that interact with each other. **B** The destruction of one affects all the other ecosystems. **B** For example, many scientists believe the destruction of rain forest habitats has led to global climate change.

TIP When an author doesn't directly state the main idea of a passage, it's up to you to figure out the **implied main idea**. To do this, ask yourself, "What do the details of the passage have in common?" Find the connection between the details, put it into your own words, and you have the implied main idea.

Step 1 Look for a stated main idea.

Read the first and last sentences in the paragraph. Do they state the main idea? If the main idea is not obvious, look for details that offer clues about what the main idea is. Some details explain the main idea. Other details give examples of the main idea.

MAIN IDEA A The loss of habitats can destroy an entire ecosystem. In this example, the stated main idea is the first sentence of the paragraph.

Step 2 Once you have figured out the main idea, find details that support it.

If the main idea is in the first sentence, it is usually followed by details that support and explain it. If the main idea is in the last sentence, the supporting details come before it. In the example at left, the details follow the main idea.

DETAIL B Earth has ecosystems that interact with each other.

DETAIL B The destruction of one affects all the other ecosystems.

DETAIL B Many scientists believe the destruction of rain forest habitats has led to global climate change.

APPLY THE SKILL

Turn to *Europe Geography & History,* Section 2.2, "Classical Greece." Read the "Golden Age of Greece" passage. Identify the main idea and details. Record them in a web like the one at right.

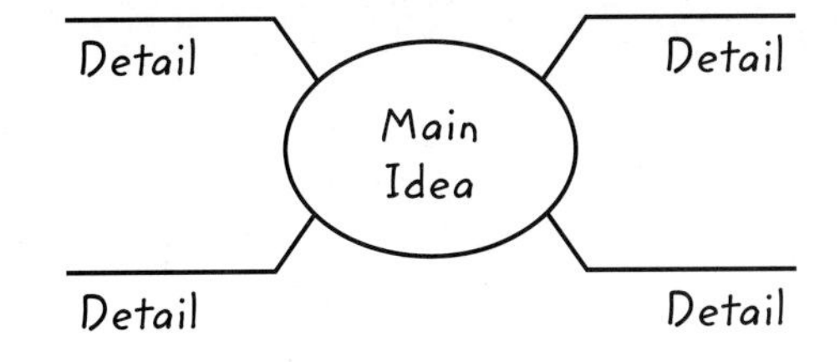

Take Notes and Outline

Taking notes while you read helps you understand and remember important information, including facts, ideas, and details. Many writers organize their notes in an **outline**. To take notes, follow the steps shown at right.

Step 1 Read the title to learn the topic of the passage.

Step 2 Write down the main ideas and important details of the passage.

Step 3 Summarize the main ideas and details in your own words.

Step 4 Search for key words and write down the words and their definitions.

Step 5 To save time and space, use abbreviations.

GUIDED MODEL

Ⓐ Economic Indicators

Ⓑ A country's economic strength can be measured by several indicators. One is Ⓒ **gross domestic product** (GDP), or the total value of the goods and services that a country produces. Ⓒ Other indicators include **GDP** per capita (or per person), income, basic literacy rate, and life expectancy.

Ⓑ Economies are also placed in one of three categories. Countries with high GDPs, such as the United States, are Ⓒ **more developed countries**. Countries with lower GDPs are Ⓒ **less developed countries**.

TIP Outlines list the main ideas and supporting details in the order in which they best make sense. You can use time order, order of importance, or the order in which they appear in the text. Once the outline is complete, you can use it as a guide for your writing.

Step 1 Read the title to learn the topic of the passage.

TITLE Ⓐ Economic Indicators

Step 2 Write down the main ideas and important details of the passage.

MAIN IDEAS Ⓑ A country's economic strength can be measured by several indicators. Economies are also placed in one of three categories.

Step 3 Summarize the main ideas and details in your own words.

You can measure a country's economic strength using indicators, such as GDP, GDP per capita, income, basic literacy rate, and life expectancy. Countries can be placed in categories, such as as more developed, less developed, and newly industrialized.

Step 4 Search for key words and write down the words and their definitions.

SAMPLE KEY WORD: Ⓒ **Gross domestic product** is the total value of the goods and services a country produces.

Step 5 To save time and space, use abbreviations. Here, for example, use "N" for *north* and GDP for *gross domestic product*.

APPLY THE SKILL

Turn to *Europe Today*, Section 1.1, "Languages and Cultures." Read the "European Languages" passage. Take notes about what you read. Then organize your notes using an outline like the example shown at right.

I. Main Language Groups in Europe
  A. Romance
    1. French
    2. Spanish
    3.
  B. Germanic

Summarize

When you **summarize**, you restate text in your own words and shorten it. A summary includes only the most important information and details. You can summarize a paragraph, a chapter, or a whole book. To summarize, follow the steps shown at right.

Step 1 Read the text looking for the most important information. Watch for topic sentences that provide the main ideas.

Step 2 Restate each main idea in your own words.

Step 3 Write your summary of the text using your own words and including only the most important information.

GUIDED MODEL

Foods of Eastern Europe

Ⓐ Eastern Europe's cold climate causes a shorter growing season than that in Western Europe. In Russia, root vegetables such as turnips and beets are well adapted to the country's climate. A soup called borscht made from beets is a traditional dish on winter nights.

Ⓑ The fertile soil of Hungary allows Hungarian farmers to grow grains and potatoes. These crops are used to make a variety of breads. A meat stew called goulash is Hungary's national dish. It is made with beef, potatoes, and vegetables and seasoned with a spice called paprika.

TIP Try using a chart to record and organize the main idea and details you want to summarize. You can then use your notes to write your summary.

Step 1 Read the text looking for the most important information. Watch for topic sentences that provide the main ideas.

TOPIC SENTENCE Ⓐ Eastern Europe's cold climate causes a shorter growing season than that in Western Europe.

TOPIC SENTENCE Ⓑ The fertile soil of Hungary allows Hungarian farmers to grow grains and potatoes.

Step 2 Restate each main idea in your own words.

RESTATED Ⓐ The climate of Eastern Europe is cold, so it has a shorter growing season than Western Europe.

RESTATED Ⓑ Hungarian farmers grow grains and potatoes.

Step 3 Write your summary of the text using your own words and including only the most important information.

SUMMARY: Eastern Europe has a cold climate, so it has a shorter growing season than Western Europe. Certain crops grow well there, such as potatoes and grains in Hungary.

APPLY THE SKILL

Turn to *Europe Today,* Section 1.1, "Languages and Cultures." Read the "Cultural Traditions" passage. Identify the main ideas and important information, and restate them in your own words. Then write a summary of the passage using the topic sentence shown here.

Cultural Traditions

Europe's cultural traditions reflect the region's ethnic diversity. __________

____

____

Sequence Events

When you **sequence events**, you put them in order based on when they occurred in time. Thinking about events in time order helps you understand how they relate to each other. To sequence events, follow the steps shown at right.

Step 1 Look for time clue words and phrases such as names of months and days, or words such as *before, after, finally, a year later,* or *lasted* that help you sequence events.

Step 2 Look for dates in the text and match them to events.

GUIDED MODEL

Greek Culture Spreads

Greece's golden age ended around **B** 431 B.C., when war broke out between Athens and Sparta. The conflict, known as the Peloponnesian War, **A** lasted 27 years. The war weakened both Athens and Sparta.

Around **B** 340 B.C., King Philip II of Macedonia took advantage of their weakness and conquered Greece. In **B** 334 B.C., Philip's son, Alexander the Great, began to extend the Macedonian Empire. Alexander's love of Greece led him to spread its culture throughout his empire. Alexander died in **B** 323 B.C. at the age of 33.

Step 1 Look for time clue words and phrases.

TIME CLUES **A** lasted 27 years

Step 2 Look for specific dates in the text.

Be sure to read the text carefully because the dates in a paragraph may not always be listed in time order. Always match an event with its date. A chart like the one below can be a useful way to sequence dates and events.

SAMPLE DATE **B** 431 B.C., when Greece's golden age ends and the Peloponnesian War begins

DATE IN TEXT	EVENT
431 B.C.	Greece's golden age ends; Peloponnesian War breaks out between Athens and Sparta.
340 B.C.	King Philip of Macedonia conquers Greece.
334 B.C.	Alexander the Great starts extending the Macedonian Empire.
323 B.C.	Alexander the Great dies.

TIP A **time line** is a visual tool that can be useful for sequencing events. Time lines often move from left to right, listing events from the earliest to the latest.

APPLY THE SKILL

Turn to *Europe Geography & History,* Section 2.3, "The Republic of Rome." Read the "A Republic Forms" passage. Identify the dates and events that occurred on each date. Use a time line like the one below to sequence the events.

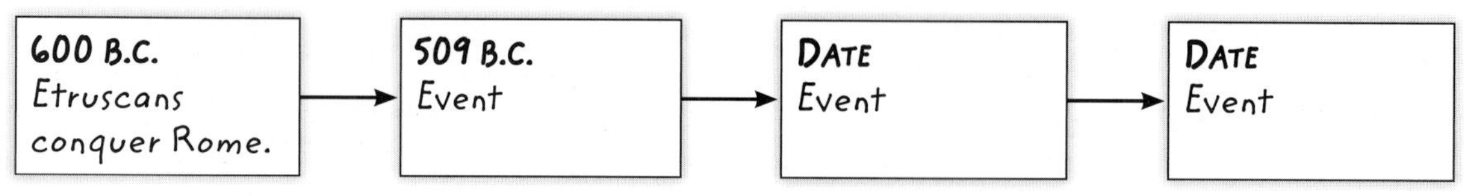

Categorize

When you **categorize**, you sort things into groups, or categories. Almost everything can be categorized, including objects, ideas, people, and information. Categorizing is important because it helps you recognize data patterns and trends. To categorize, follow the steps shown at right.

Step 1 Read the title and text and ask yourself what the passage is about to determine how the information can be categorized.

Step 2 Look for clue words to help you categorize information.

Step 3 Decide what the categories will be.

Step 4 Sort the information from the passage into the categories.

GUIDED MODEL

Ⓐ Political and Physical Maps

Cartographers, or mapmakers, create Ⓑ different kinds of maps for many Ⓑ different purposes.

Ⓐ A political map shows features that humans have created, such as countries, states, and cities. These features are labeled, and lines show boundaries, such as country borders.

Ⓐ A physical map shows features of physical geography. It includes landforms, such as mountains, plains, deserts, and bodies of water. A physical map also shows elevation and relief. Elevation is the height of a physical feature above sea level. Relief is the change in elevation from one place to another.

TIP Every day, we put information into categories, such as healthy food versus junk food or fiction versus nonfiction. Knowing how information is categorized helps you understand how things are related and organize ideas.

Step 1 Read the title and text and ask yourself what the passage is about to determine how the information can be categorized.

THE PASSAGE IS ABOUT Ⓐ the features of two types of maps.

Step 2 Look for clue words to help you categorize information.

CLUE WORDS Ⓑ *different kinds of maps, different purposes*

Step 3 Decide what the categories will be.

THE CATEGORIES political maps and physical maps

Step 4 Sort the information into the categories.

Use a chart like the one below to help you organize the information.

TYPE OF MAP	PURPOSES
Political	Shows man-made features, such as countries, states, cities, and boundaries
Physical	Shows features of physical geography, such as mountains, plains, deserts, and bodies of water; also shows elevation and relief

APPLY THE SKILL

Turn to *Europe Today,* Section 1.2, "Art and Music." Read the section carefully. Determine how categorizing can help you understand and organize the information about European art. Then create a chart like the one at right and categorize the information.

EUROPEAN ART	
PERIOD	**CHARACTERISTICS**
Middle Ages	religious subjects; two-dimensional
Romantic Period	landscapes, nature, conveys emotion

Describe Geographic Information

When you read text or study a chart or graph about a geographic subject, you take in information. One way to enhance your understanding of **geographic information** is to describe it. You can describe geographic information by following the steps shown at right.

Step 1 Read the title of a passage, chart, or other visual to find out what geographic information it contains.

Step 2 Read the passage or study the visual and identify its main ideas or topics.

Step 3 Describe the information by summarizing it or responding to questions.

GUIDED MODEL

Ⓐ The Geography of Europe

Ⓑ Europe forms the western peninsula of Eurasia, the landmass that includes Europe and Asia. In addition, Europe contains several peninsulas and significant islands.

Four land regions form Europe. The Western Uplands are hills, mountains, and plateaus that stretch from the Scandinavian Peninsula to Spain and Portugal. The Northern European Plain is made up of lowlands that reach across northern Europe. The Central Uplands are at the center of Europe. The Alpine region consists of the Alps and several other mountain ranges.

Step 1 Read the title of a passage, chart, or other visual to find out what geographic information it contains.

Ⓐ *From its title, I see that this passage is about the geography of Europe.*

Step 2 Read the passage or study the visual and identify its main ideas or topics.

MAIN IDEAS Ⓑ *Europe is a peninsula on Eurasia. It is formed by four land regions.*

Step 3 Describe the information by summarizing it or responding to questions.

SUMMARY *Europe is a peninsula and contains several peninsulas and islands. It is formed by four land regions: the Western Uplands, the Northern European Plain, the Central Uplands, and the Alpine region.*

QUESTIONS

- *What landforms does Europe contain?*
- *What land regions make up Europe?*

TIP You can use the steps above to describe the geographic information in a chart. Just imagine that the column headings are the main ideas and the information in the rows are the supporting details.

APPLY THE SKILL

Turn to *Europe Geography & History*, Section 1.3, "Mountain Chains." Use the information in this section to describe the location and natural resources of Europe's mountains. You might use questions like those shown at right to help you describe the information.

- What mountain chains are part of the Alpine region?
- What natural resources do Europe's mountain chains provide?

Express Ideas Through Speech

When you **express ideas through speech**, you say what you are thinking using effective language. The purpose of your speech may be to narrate events, explain information, or convince your audience of your point of view. You can express ideas through speech by following the steps at right.

Step 1 Determine the topic you will be presenting and the purpose of your speech. Do research to gather information using a wide variety of reliable sources.

Step 2 Organize your ideas and create speaking notes and any visual aids you may need.

Step 3 Practice speaking honestly, respectfully, clearly, and appropriately.

GUIDED MODEL

A European Literature

B Ancient Greeks and Romans (focus on journeys, heroes, and adventures)
- Iliad and Odyssey (Homer)
- Aeneid (Virgil)

Middle Ages (focus on religious beliefs and politics of the time)
- Divine Comedy (Dante)

Renaissance (focus on human experiences)
- Hamlet (Shakespeare)
- Don Quixote (Cervantes)

—first modern novel

TIP When you express ideas through speech, it is important to hold your audience's attention. Vary your volume, speaking rate, and sentence structure to add interest to your presentation and emphasize meaning. Your body language is also important. Gestures, posture, facial expressions, and eye contact will engage your audience and enhance your speech.

Step 1 Determine the topic and purpose of your speech and do research to gather the information you need.

A The topic of this student's speech is the history of European literature. The purpose is to explain the main periods of European literature.

Step 2 Organize your ideas and create speaking notes and any visual aids you may need.

Notes and visual aids will help you stay on topic, remember your points, and emphasize your message.

B Well-organized notes will help you if you lose your place while speaking or forget a point you wanted to make. Keep your notes simple—just list headings and key points and important people and events.

Step 3 Practice speaking honestly, respectfully, clearly, and appropriately.

Consider your audience and subject matter. Speak honestly and respectfully and use language that is appropriate to your audience.

APPLY THE SKILL

Select a topic from the book, research it, and prepare speaking notes for a presentation about it like those shown at right. Then express your ideas through speech to an audience.

World War I

Began 1914 (explain causes)
- nationalism
- alliances
- Archduke Ferdinand

Write Outlines for Reports and Comparisons

Writing an outline is a useful way to synthesize, organize, and summarize information before you write a report or comparison. A good outline lists the most important ideas and supporting details in order, either chronologically or by level of importance. To write an outline, follow the steps shown at right.

Step 1 Give your outline a title.

Step 2 Determine and record the main ideas in logical order using Roman numerals.

Step 3 Determine and record the ideas that support and explain the main ideas.

Step 4 Determine and record specific details below the appropriate supporting ideas.

GUIDED MODEL

(A) Earth's Water

(B) I. Fresh Water

- (C) A. Sources
  - (D) 1. Rivers, water flowing downward
  - 2. Lakes, large inland bodies of water
  - 3. Streams, brooks, and creeks
- B. Uses
  - 1. Drinking and cooking
  - 2. Irrigating crops

II. Salt Water

- A. Oceans
  - 1. Large bodies of salt water
  - 2. Atlantic, Pacific, Indian, Arctic
  - 3. Ocean currents and their effects
- B. Seas
  - 1. Small bodies of salt water enclosed partly or completely by land
  - 2. Red, Caspian, Arabian

Step 1 Give your outline a title.

(A) If you are writing an outline to organize information for a report or comparison, the outline title will probably become the title of your paper.

Step 2 Determine and record the topics in logical order using Roman numerals.

(B) Each topic will probably make up at least one paragraph in your paper.

Step 3 Determine and record the main ideas that support and explain each topic.

(C) Label these using indented capital letters below the topics. Not all topics need the same number of main ideas.

Step 4 Determine and record specific details below the appropriate main ideas.

(D) Label these details using more deeply indented numerals below the main ideas. Not every main idea will have the same number of details.

TIP When you write an outline for a comparison, you can list all of the features of one subject under one topic and all of the features of the other subject under another topic. You can also use the features as topics and list facts about both subjects under each one.

APPLY THE SKILL

Turn to *Europe Geography and History,* Section 3.1, "Exploration and Colonization." Then organize the information from the passage in an outline like the one shown at right. Use your outline to write a report.

European Exploration

I. Portuguese explorations

- A. Reasons
  - 1. Find gold
  - 2. Establish trade

Write Journal Entries

When you **write journal entries** in response to something you have read, you record your reactions, make notes, ask and answer questions, and respond to new information in an informal written format. To use written journal entries to respond to a question or writing prompt, follow these steps.

Step 1 Use what you have learned and read about the topic to answer the question.

Step 2 Respond and react in a personal way to the question.

Step 3 Record any related questions, comments, or notes you might have.

GUIDED MODEL

QUESTION:
What challenges might immigrants moving to Europe face today?

JOURNAL ENTRY:
A Some immigrants may feel pressure to assimilate into European society and fear they will lose their identity and culture. Those belonging to non-Christian religions may feel discriminated against simply for having different religious beliefs.

B When my family moved, I hated that students at my new school had to wear a uniform. I refused to wear the jacket for the first week and got in trouble. I felt angry that I wasn't allowed to dress the way I wanted. I think I know a little bit about how the immigrants living in Europe feel.

C NOTE: Look up countries in Europe with large immigrant populations on the CIA World Factbook Web site.

Step 1 Use what you have learned and read about the topic to answer the question.
A The journal writer answers the question using facts from the book and his or her prior knowledge.

Step 2 Respond and react in a personal way to the question.
B The journal writer relates to the topic on a personal level and provides insight into the question based on personal experience.

Step 3 Record any related questions, comments, or notes you might have.
C The journal writer writes himself or herself a note to look up more information about the topic online.

TIP Journals can be used to record the steps in a research project (such as your observations during an experiment). You can also use a journal to write your thoughts and questions as you read a book or carry out a project.

APPLY THE SKILL

Turn to *Europe Today,* Section 1.1, "Languages and Cultures." Then write a journal entry to respond to the following question: What is life like for Europeans who live in cities today? The beginning of a sample entry is shown at right.

Most Europeans today live in cities that reflect the cultures and ethnic origins of immigrant populations.

SKILLS HANDBOOK

Pose and Answer Questions

Pose questions about what you have read to check your understanding, keep track of details, and identify points you may have missed. To **answer questions**, review what you have read, search for answers in the text, and use your prior knowledge. To pose and answer questions, follow these steps.

Step 1 Read the passage and identify information that is important or confusing.

Step 2 Pose questions about key concepts or confusing information using question words, such as *who, what, when, where, why,* and *how*.

Step 3 Answer the questions using the text and what you already know.

GUIDED MODEL

Establishing Colonies

Ⓐ European exploration and colonization resulted in an exchange of goods and ideas known as the Columbian Exchange. From the Americas, Europeans obtained new foods, such as potatoes, corn, and tomatoes. Europeans introduced wheat and barley to the Americas. They also introduced diseases like smallpox. The diseases killed millions of native peoples.

TIP A two-column chart is a helpful way to record your questions and answers. Be sure to include the page number for each answer you found in the text in case you need to review or reread a passage or section.

Step 1 Read the passage and determine which information is important or confusing.

Ⓐ The most important topic seems to be the result of European exploration in the colonies. I'm confused about what the Columbian Exchange is.

Step 2 Pose questions about key concepts or confusing information.

My Questions

1. **What** is the Columbian Exchange?
2. **Who** were killed by European diseases?
3. **Which** foods did Europeans obtain from the Americas? **Which** foods did Europeans introduce to the Americas?

Step 3 Answer your questions using the text and what you already know.

Don't forget to examine charts, graphs, maps, and photographs as you answer.

Answers

1. The Columbian Exchange is the exchange of goods and ideas between Europeans and the areas they explored and colonized.
2. European diseases killed native peoples in the Americas.
3. Europeans obtained potatoes, corn, and tomatoes; the Americas obtained wheat and barley.

APPLY THE SKILL

Turn to *Human & Physical Geography,* Section 1.1, "Earth's Rotation and Revolution." Read "Revolution and Rotation" and pose questions about information in the passage that confuses you. Record your questions in a chart like the one at right. Use the text and diagram to answer your questions.

MY QUESTIONS	ANSWERS
What process creates the four seasons?	the rotation and tilt of Earth

Make Predictions

When you **make predictions,** you think about the events described in a passage or selection, use your prior knowledge, and guess or predict what will happen next. Making predictions as you read can help you understand and remember what you have read. To make predictions, follow these steps.

Step 1 Preview the passage or selection to anticipate what it is about.

Step 2 Use your personal knowledge. Ask yourself what you know about the topic.

Step 3 As you read, make predictions about what will happen next.

Step 4 Confirm or revise your predictions as you continue to read.

GUIDED MODEL

(A) European Music

European music began in ancient Greece and Rome. Musicians played on (C) simple instruments. During the (B) Middle Ages, music was used in religious ceremonies. Also, singers called troubadours sang about knights and love. These songs influenced (B) Renaissance music, when the (C) violin was introduced.

The new instruments helped inspire the complex music of the (B) Baroque period (1600–1750). Opera, which tells a story through words and music, was born then. The (B) Classical and Romantic periods followed the Baroque and continued until about 1910.

Step 1 Preview the passage or selection to anticipate what it is about.

(A) The title tells me that this passage is about European music. By skimming ahead, I see the names of many different types of music, (B) musical periods, and (C) words relating to instruments.

Step 2 Use your personal knowledge.

I know that the Romantic period in art took place in the early 1800s. I have listened to Baroque music before.

Step 3 As you read, make predictions about what will happen next.

I predict that this text will describe how music evolved in Europe.

Step 4 Confirm or revise your predictions as you continue to read.

My prediction was correct. The passage describes how music evolved from the Middle Ages to the present.

TIP Use a prediction chart to take notes as you read. A prediction chart allows you to record predictions, state whether they were correct, and explain why.

PREDICTION	CORRECT?	EVIDENCE
This passage will explain music I am familiar with.	Yes	It describes what opera is and when it developed.

APPLY THE SKILL

Turn to *Human & Physical Geography,* Section 1.2, "Earth's Complex Structure." Read the "Earth's Layers" text. Then use a prediction chart like the one shown at right to record and analyze your predictions.

PREDICTION	CORRECT?	EVIDENCE
This passage will describe the high temperatures deep within the earth.		

SKILLS HANDBOOK

Compare and Contrast

When you **compare** two or more things, you examine the similarities and differences between them. When you **contrast** things, you focus on only their differences. To compare and contrast, follow the steps shown at right.

Step 1 Determine what the subject of the passage or paragraph is.

Step 2 Identify two or more ideas, examples, or features relating to the subject that can be compared and contrasted.

Step 3 Search for clue words that indicate similarities (comparing) and differences (contrasting).

GUIDED MODEL

Ⓐ Varied Climates

Most of Europe lies within the humid temperate climate region. The North Atlantic Drift, an ocean current of warm water, keeps temperatures relatively warm. Winds also affect climate. The sirocco sometimes blows over the Mediterranean Sea and brings humid weather to southern Europe at different seasons. The mistral is a cold wind in winter that blows through France, bringing cold, dry weather.

Ⓒ In general, Ⓑ a Mediterranean climate brings cool, wet winters and hot, dry summers. This climate supports a long growing season. Ⓒ In contrast, Ⓑ Eastern Europe has long, cold winters. Greenland, northern Scandinavia, and Iceland have polar climates and a limited growing season.

Step 1 Determine what the subject of the passage or paragraph is.

SUBJECT Ⓐ the climates of Europe

Step 2 Identify features of the subjects that can be compared and contrasted.

FEATURE Ⓑ climates of southern and eastern European countries

Step 3 Search the passage for clue words that indicate similarities and differences.

Those sentences will help you compare. Then look for clue words that indicate how the two aspects are different. Those sentences will help you contrast.

CLUE WORDS Ⓒ in general (comparing); in contrast (contrasting)

TIP A Y-Chart and a Venn diagram, shown below, are useful graphic organizers for comparing and contrasting two topics. In a Y-Chart, list unique information on the branches and shared characteristics in the straight section. In a Venn diagram, list unique characteristics in the left and right sides and common characteristics in the overlapping area.

APPLY THE SKILL

Turn to *Human & Physical Geography*, Section 2.4, "Natural Resources." Read the "Categories of Resources" passage. Compare and contrast renewable resources and nonrenewable resources using a Y-Chart or Venn diagram.

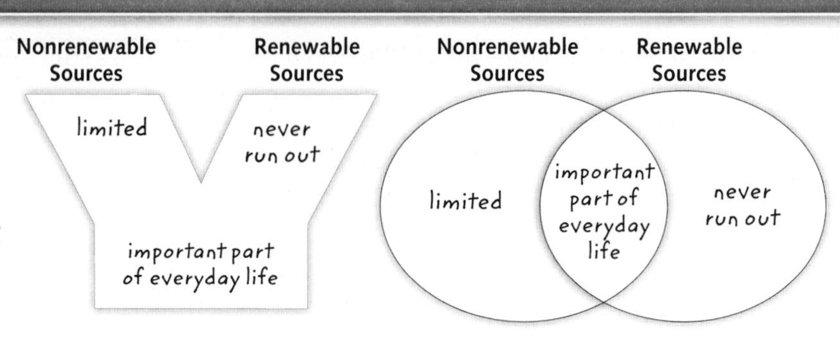

Analyze Cause and Effect

A **cause** is an event or action that makes something else happen. The **effect** is an event that happens as a result of the cause. Analyzing cause-and-effect relationships can help you understand how events are related. To analyze cause and effect, follow the steps shown at right.

Step 1 Determine the cause of an event. Look for signal words that show cause, such as *because, due to, since,* and *therefore*.

Step 2 Determine the effect that results from the cause. Look for signal words such as *led to, consequently,* and *as a result*.

Step 3 Look for a chain of causes and effects. An effect may be the cause of another action or event.

GUIDED MODEL

Rome's Decline

For about 500 years, the Roman Empire was the most powerful in the world and extended over three continents. Ⓐ Beginning around A.D. 235, however, Rome had a series of poor rulers. Ⓐ In addition, German tribes began invading the empire.

Ⓑ As a result, in 312, Emperor Constantine moved the capital from Rome to Byzantium, in present-day Turkey, and renamed the city Constantinople. In 395, the empire was divided into an Eastern Empire and a Western Empire with two different emperors. Ⓒ Because Rome had been severely weakened, invaders overthrew the last Roman emperor and ended the Western Empire in 476.

Step 1 Determine the cause.

Ask yourself why the event took place. Notice that an event may have more than one cause.

CAUSE(S)

Ⓐ Beginning around A.D. 235, Rome had a series of poor rulers.

Ⓐ German tribes began invading the empire.

Step 2 Determine the effect.

Ask yourself what happened as a result of the event or events.

EFFECT Ⓑ As a result, in 312, Emperor Constantine moved the capital from Rome to Byzantium.

Step 3 Look for a chain of causes and effects.

An effect may cause another event.

CAUSE/EFFECT Ⓒ Because Rome had been severely weakened, invaders overthrew the last Roman emperor and ended the Western Empire in 476.

TIP Test whether events have a cause-and-effect relationship by using this construction: "Because [insert cause], [insert effect] happened." If the construction does not work, one event did not lead to the other.

APPLY THE SKILL

Turn to *Human & Physical Geography*, Section 3.5, "Protecting Human Rights." Read "The Impact of Human Rights" passage. Identify a cause-and effect-relationship in the passage. Record the cause(s) and effect(s) in a graphic organizer like the one at right.

CAUSE	EFFECT
Declaration of Human Rights was developed	World pressured South Africa to grant human rights to non-whites

Make Inferences

Inferences are conclusions or interpretations a reader makes from information that a writer does not state directly. When you make inferences, you use common sense and your own experiences to figure out what the writer means. To make inferences, follow the steps shown at right.

Step 1 Read the text looking for facts and ideas.

Step 2 Think about what the writer does not say but wants you to understand.

Step 3 Reread the text and use what you know to make an inference.

GUIDED MODEL

Aryan Migration

Many historians believe a group of nomadic herders called the Aryans migrated from Central Asia into the Indus Valley around 2000 B.C. From there, they moved into northern India. Ⓐ Their language, Sanskrit, became the basis of many modern languages in South Asia. The Aryans recorded religious teachings in Sanskrit in sacred texts called the Vedas. Ⓑ The Aryans established the beginnings of Hinduism, the major religion of India today.

TIP Use a two-column chart to keep track of inferences. In the left column, write details, quotations, examples, statistics, and other facts. In the right column, write the inference that you draw from each fact. Note that an inference can be based on one fact or several facts.

Step 1 Read the text looking for facts and ideas.

You must know the facts before you can make inferences. Ask yourself, "What facts does the writer state directly in this text?"

FACT Ⓐ The Aryan language, Sanskrit, became the basis of many modern languages in South Asia.

FACT Ⓑ The early religion of the Aryans established the beginnings of Hinduism, the major religion of India today.

Step 2 Think about what the writer does not say but wants you to understand.

Ask yourself:

- *How do these facts connect with what I already know?*
- *How does this information help me better understand Hinduism?*

Step 3 Reread the text and use what you know to make an inference.

INFERENCE Ⓐ Many modern languages have ancient origins.

INFERENCE Ⓑ Modern religions are influenced by ancient religions.

APPLY THE SKILL

Turn to *Europe Geography & History,* Section 1.2, "A Long Coastline." Read the "Exploration and Settlement" text. Then make two inferences about how Europe's location near large bodies of water affected its people. Use a two-column chart like the one at right to record the facts you found and the inferences you made.

FACTS	INFERENCES

Draw Conclusions

When you **draw conclusions**, you make a judgment based on what you have read. You analyze the facts, make inferences, and use your own experiences to form your judgment. To draw a conclusion, follow the steps shown at right.

Step 1 Read the passage to identify the facts.

Step 2 Make inferences based on the facts.

Step 3 Use the inferences you have made and your own experiences and common sense to draw a conclusion.

GUIDED MODEL

Ⓐ The euro was launched in 1999, but paper money and coins of the currency were not issued until 2002. Ⓐ As of early 2011, 17 of the 27 European Union (EU) nations had adopted the euro. The countries using the euro are known as the eurozone. Ⓐ Some EU countries, including Romania and Bulgaria, hope to join the eurozone soon. Other countries, including Great Britain and Sweden, have not adopted the euro. They believe that giving up their own currency might result in a loss of control over their economies.

TIP Use a diagram to organize the facts you have identified, the inferences you have made, and your conclusion. A diagram can help you clarify your thinking.

Step 1 Read the passage to identify the facts.

FACTS Ⓐ The euro was introduced to Europe in 1999. By 2011, 17 out of 27 EU nations used it. Some countries hope to join the eurozone. Others, including Great Britain and Sweden, have not adopted the euro because they think it would result in a loss of control over their economies.

Step 2 Make inferences based on the facts.

INFERENCES *I see that most of the EU countries have adopted the euro, and at least two more want to adopt it. However, some countries have chosen not to adopt the currency. That must mean that there are both advantages and disadvantages to using the euro.*

Step 3 Use the inferences you have made and your own experience and common sense to draw a conclusion.

CONCLUSION: *In the future, more European countries will probably adopt the euro and become part of the eurozone. However, other countries will not adopt the euro because there are disadvantages to doing so.*

APPLY THE SKILL

Turn to *Human & Physical Geography,* Section 2.3, "Extreme Weather." Read the "Scientific Solutions" passage and draw conclusions about the solutions scientists have come up with to combat the dangers of extreme weather. Write a few sentences summarizing your conclusion. A sample sentence is shown at right.

Scientists are making great progress in predicting extreme weather ______________

Make Generalizations

Generalizations are broad statements that apply to a set of ideas, a group of people, or a series of events. You make generalizations based on information you have read or heard. You can also draw on personal experiences. To make generalizations, follow the steps shown at right.

Step 1 Look for the overall theme or message of the selection.

Step 2 Find information in the passage that supports the theme.

Step 3 Draw on your personal knowledge.

Step 4 Make a generalization about the topic and put it into sentence form.

GUIDED MODEL

Life in the Ocean

Dr. Sylvia Earle has led more than 60 diving expeditions to explore marine life. During these dives, she has seen the incredible variety of ocean life—**B** "More than 30 major divisions of animals are known, from sponges and jellies to many kinds of beautiful worms and mollusks."

A Dr. Earle has also seen how people have harmed the oceans. **B** "Taking too much wildlife out of the sea is one way," she claims. "Putting garbage, toxic chemicals, and other wastes in is another." Dr. Earle has witnessed a huge drop in the number of fish in the sea. She has also noted pollution's impact on coral reefs.

TIP You can also make generalizations about information from multiple sources. Determine what information the sources have in common. Then form a generalization that all of the sources would support.

Step 1 Look for the overall theme.

COMMON THEME **A** Humans harm the diverse marine life in the oceans.

Step 2 Find information that supports the theme.

SUPPORTING INFORMATION **B** Dr. Sylvia Earle, an ocean expert, states that people harm ocean life by taking too much wildlife out of the sea and putting garbage, toxic chemicals, and other wastes into it. Dr. Earle has witnessed a huge drop in the number of fish in the sea and pollution's impact on coral reefs.

Step 3 Draw on your personal knowledge.

I know that living in polluted water is bad for living things. I watched a television program on how toxic waste in the oceans has caused the extinction of some species.

Step 4 Make a generalization about the topic and put it into sentence form.

GENERALIZATION *If we don't stop polluting the waters and taking too much wildlife from the oceans, many species could become extinct.*

APPLY THE SKILL

Turn to *Europe Geography & History*, Section 2.5, "Middle Ages and Christianity." Read "The Feudal System" passage. Then make generalizations about the feudal system's role in Europe, like the one at right, using the information in the passage and what you may already know.

The feudal system provided security for Europeans during the Middle Ages.

Form and Support Opinions

When you **form an opinion**, you determine and assess the importance and significance of something. While an opinion is a person's own judgment, not a fact, sound opinions must be supported with examples and facts.

Step 1 Read the passage. Look for reliable information about the subject, including facts, statistics, and quotations.

Step 2 Form your own opinion about the subject.

Step 3 Find facts to support your opinion.

GUIDED MODEL

Germany in World War II

Ⓐ Hitler became the leader of the National Socialist German Workers' Party, or the Nazis. He forged alliances with Italy and Japan.

Ⓐ Germany's invasion of Poland in 1939 started World War II. Great Britain and France, the Allies, declared war on Germany soon after, but Germany conquered Poland and quickly defeated most of Europe. In 1941, the United States entered the war on the side of the Allies.

Ⓐ Finally, in May 1945, Germany surrendered. Near the end of the war, Allied troops were stunned to find Nazi concentration camps where six million Jews and other victims had been murdered. This mass slaughter was called the Holocaust.

Step 1 Read the passage. Look for reliable information about the subject.

SUBJECT Germany in World War II

IMPORTANT INFORMATION ABOUT THE SUBJECT Ⓐ Germany, Italy, and Japan formed an alliance. Germany's invasion of Poland in 1939 started World War II. Germany soon defeated most of Europe. In 1945, Germany surrendered and Allied troops discovered evidence of the Holocaust.

Step 2 Form your own opinion.

Decide the subject's importance or wider meaning. What do you believe about the subject?

MY OPINION *Hitler and the Nazi Party's actions in the 1930s and 1940s dramatically violated human rights.*

Step 3 Find facts to support your opinion.

If you cannot find facts to support your opinion, you must revise it.

FACTS THAT SUPPORT MY OPINION *Hitler was a powerful leader of the Nazis. They murdered six million Jews and other victims during the Holocaust.*

TIP To distinguish facts from opinions, ask yourself, "Can this sentence be proven?" If evidence can prove a statement, it is a fact. If the sentence cannot be proven because it expresses a thought or emotion, it is an opinion.

APPLY THE SKILL

Turn to *Europe Geography & History,* Section 2.2, "Classical Greece." Read the "Greek Achievements" passage. Provide an opinion about what you have read and list facts to support your opinion. Use a chart like the one at right to organize your ideas.

OPINION	EVIDENCE
Greece's golden age was a period of extraordinary achievements.	

Identify Problems and Solutions

Throughout history, people have faced problems and learned how to solve them. When you **identify problems,** you find the difficulties people faced. When you **identify solutions,** you learn how people tried to fix their problems. To identify problems and solutions, follow the steps shown at right.

Step 1 Read the text and determine what problem or problems people faced.

Step 2 Determine what caused the problems. There may be multiple causes.

Step 3 Identify the solutions people used to resolve, or fix, the problems.

Step 4 Determine if the solutions succeeded. Ask yourself, "Was the problem solved?"

GUIDED MODEL

Impact of the Revolution

The Industrial Revolution had a tremendous impact on how people worked and lived. Cities grew rapidly because people migrated to them for factory jobs. Standards of living rose, and a prosperous middle class grew.

(A) However, factory workers often faced harsh conditions. (B) Laborers worked as many as 14 hours a day. Child labor was common. Children as young as ten years old worked in factories and mines.

(B) Many workers lived in small, crowded houses in neighborhoods where open sewers were common. Diseases spread quickly in these buildings. (C) Over time, the workers' quality of life improved as public health acts were passed to provide sewage systems and larger buildings.

Step 1 Read the passage and determine what problem people faced.

PROBLEM (A) Factory workers faced harsh conditions during the Industrial Revolution.

Step 2 Determine what caused the problem.

CAUSES (B) long work hours, child labor, crowded lodgings, open sewers

Step 3 Identify the solutions people used to resolve the problem.

SOLUTIONS (C) Larger buildings for workers to live in were built, and sewage systems for improved sanitation were installed.

Step 4 Determine if the solutions succeeded.

SUCCESS? Yes. Laws regulated working and living conditions, which are now much cleaner and safer. Sewers reduced disease. Children are not permitted to work in factories.

TIP A chart like the one below is a useful tool for recording information about a problem and solution. Organize your chart to have separate rows for a description of the problem, its cause, the solutions, and the success of the solution.

APPLY THE SKILL

Turn to *Human & Physical Geography,* Section 2.5, "Habitat Preservation." Read the "Natural Habitats" passage and the "Habitat Loss and Restoration" passage. Identify the problems and solutions using a chart like the one at right.

What was the problem?	
What were the causes?	
What was the solution?	
Did the solution work?	

Analyze Data

Information, or data, can be collected in charts, databases, graphs, diagrams, models, and maps. Regardless of its format, you can **analyze data** to draw conclusions, make comparisons, identify trends, and improve your understanding of the information. To analyze data, follow the steps at right.

Step 1 Identify the data source, which could be a chart, graph, database, model, diagram, or map. Read its title to determine what information it contains.

Step 2 Read any headings or subheadings to see how the information is organized.

Step 3 Study the data to answer questions, draw conclusions, make comparisons, and identify trends.

GUIDED MODEL

A ECONOMIC INDICATORS OF SELECTED COUNTRIES*

B COUNTRY	GDP PER CAPITA (U.S. $)	LITERACY RATE (percent)
B Afghanistan	366	28.0
Brazil	8,536	90.0
China	3,422	93.3
Ethiopia	321	35.9
Germany	44,525	99.0
Mexico	10,249	92.8
United States	47,210	99.0

Source: The World Bank and the United Nations

* **GDP per capita figures are for 2008, while literacy rate is for 2007**

TIP Data can be represented in ways other than charts and graphs. When analyzing data from a model, diagram, or map, examine all callouts, symbols, illustrations, and text boxes. Important data may be found where you least expect it.

Step 1 Identify the data source and determine what information it contains.

A *This data source is a chart. Its title is "Economic Indicators of Selected Countries." That tells me it includes economic data on various countries.*

Step 2 Read any headings or subheadings to see how the information is organized.

In a chart, columns go up and down, and rows go from left to right. Other data sources, such as graphs, models, and diagrams, organize information in different ways.

This chart gives two types of economic indicator data recorded for each country. B *The columns record the economic indicator data.* B *The rows list information for each country.*

Step 3 Study the data to answer questions, draw conclusions, make comparisons, and identify trends.

Using the data in this chart, I can compare the data in the literacy rate column to the data in the GDP per capita column. I can draw the conclusion that a high literacy rate is tied to a high GDP per capita.

APPLY THE SKILL

Turn to *Human & Physical Geography,* Section 1.2, "Earth's Complex Structure." Examine the "Tectonic Plate Movements" diagram. Then analyze the data in the diagram to answer the following questions.

1. What type of data does this diagram provide?
2. What type of plates are shown?
3. What are the four types of plate movements?
4. What is subduction?
5. Compare divergence and convergence.
6. Based on your analysis of the data in the diagram, what conclusions can you draw about the movement of Earth's plates?

Distinguish Fact and Opinion

It is important to **distinguish fact from opinion** to separate someone's personal beliefs from concepts or events that are known to be true. To distinguish facts from opinions, follow the steps shown at right.

Step 1 Read the passage and identify facts: information that can be proven to be true.

Step 2 Identify opinions: claims or feelings about a topic or statements of personal beliefs.

Step 3 Ask yourself, "Can this statement be checked to see if it is true?" If the answer is yes, you should decide how or where you can verify it.

GUIDED MODEL

Changes in Eastern Europe

Ⓐ After communism fell in Yugoslavia in 1991, civil war broke out among its ethnic groups. The war ended in 1995. Over time, Yugoslavia was broken up into several new democratic countries.

Ukraine has also had setbacks. In 2004, the Ukrainian people staged the Orange Revolution and peacefully removed their prime minister, Viktor Yanukovych. Ⓑ Many believed that he was corrupt and being controlled by Russia. However, their new leader disappointed the Ukrainians. Ⓐ In 2010, they re-elected Yanukovych.

TIP When you identify personal beliefs in a historical text, you should also check for bias or a writer's prejudiced point of view. Look for words, phrases, or statements that reflect the positive or negative opinions of a particular group, social class, or political party.

Step 1 Read the passage and identify information that can be proven to be true.

FACTS Ⓐ Communism in Yugoslavia fell in 1991. Civil war broke out among ethnic groups. The war ended in 1995. Yugoslavia was broken into several new countries. In 2004, the Ukrainian people staged the Orange Revolution and peacefully removed their prime minister, Viktor Yanukovych. In 2010, Ukrainians re-elected Yanukovych.

Step 2 Identify opinions.

Look for personal beliefs, judgments, and feelings. Key words such as "could be" and "believed" are clues that a statement is an opinion.

OPINIONS Ⓑ Many believed that Viktor Yanukovych was corrupt and being controlled by Russia.

Step 3 Ask yourself, "Can this statement be checked to see if it is true?" If the answer is yes, decide how and where to verify it.

STATEMENT After communism fell in Yugoslavia in 1991, civil war broke out among its ethnic groups.

WHERE TO CHECK encyclopedias, history books, reliable online sources

FACT OR OPINION fact

APPLY THE SKILL

Turn to *Europe Geography & History*, Section 1.4, "Protecting the Mediterranean." Read the "Marine Reserves" passage. Use a chart like the one at right to record statements of fact and opinion and decide where you could verify it. After you have checked the statement, record whether it is a fact or opinion.

Statement: Scandola is a marine reserve that prohibits people from fishing, swimming, or anchoring their boats.
Where to check:
Fact or opinion:

Evaluate

When you read an informational text, you must **evaluate**, or assess what you have read. Sometimes you evaluate a passage to determine if its claims are believable. You may also evaluate to understand someone's actions. To evaluate something that took place, follow the steps shown at right.

Step 1 Identify the action or event you want to evaluate.

Step 2 Gather evidence about the positive impact of the action or event.

Step 3 Gather evidence about the negative impact of the action or event.

Step 4 Decide if the evidence is adequate.

Step 5 Form your evaluation of the action.

GUIDED MODEL

(A) Eastern European countries have had mixed results since changing to a market economy. (B) Poland has had the greatest success. It has a fast-growing economy and exports goods throughout Europe. (C) Other countries have been slower to establish new businesses and become competitive. They have also experienced rises in prices and unemployment.

(B) The leaders of many eastern European countries wish to integrate with the rest of Europe. They want to join the European Union and NATO, a military alliance of democratic states in Europe and North America.

TIP Making a list of the positive and negative outcomes of a decision, event, or action can help you evaluate. Read the passage and list the positives and negatives. Then review your list and form your evaluation.

Step 1 Identify the action or event.

(A) the impact of Eastern Europe's adoption of a market economy

Step 2 Gather evidence about the positive impact of the action or event.

(B) Poland's economy has grown quickly. Poland now exports goods throughout Europe. It has caused many eastern European leaders to reach out to the rest of Europe.

Step 3 Gather evidence about the negative impact of the action or event.

(C) Some countries have been slow to establish new businesses and become competitive. They have experienced rises in prices and unemployment.

Step 4 Decide if the evidence is adequate.

There is evidence to support both the positive and negative effects of the action. The evidence seems factual.

Step 5 Form your evaluation of the action.

Changing to a market economy has had negative effects on some eastern European countries, However, it has been successful in others and has encouraged Eastern Europe to deepen connections with the rest of the world.

APPLY THE SKILL

Turn to *Europe Geography & History,* Section 2.5, "Middle Ages and Christianity." Read "The Growth of Towns" passage. Record evidence from the passage and determine whether it is positive or negative using a chart like the one at right. Then write a brief evaluation of the impact of the growth of towns on Europe.

EVIDENCE	POSITIVE/NEGATIVE
Trade and business developed.	

Synthesize

When you read, you take in information, details, clues, and concepts. When you **synthesize**, you combine all of that data to form an overall understanding of what you have read. To synthesize, follow the steps shown at right.

Step 1 Look for solid, factual evidence.

Step 2 Look for explanations that connect facts.

Step 3 Think about what you have experienced or already know about the topic.

Step 4 Use evidence, explanations, and your prior knowledge to form a general understanding of what you have read.

GUIDED MODEL

Immigrants in Australia

Ⓐ Like Europe, Australia has an aging population. Ⓑ The Australian government believes that immigration can help change that trend. As a result, each year the government identifies gaps in the country's workforce. It then determines the number of skilled immigrants that can come to Australia. Between 2008 and 2009, more than 110,000 immigrants came on this skilled migration program. Most of Australia's immigrants came from Great Britain, India, and China.

TIP To synthesize, you must be able to determine what information is the most important. Once you have identified the most important facts, organize them, find explanations for them, and fit them in with what you already know. Synthesis occurs as you extract the most important information from a passage and give it personal meaning.

Step 1 Look for factual evidence.

Identifying the facts will help you base your synthesis on reliable evidence.

FACT Ⓐ Like Europe, Australia has an aging population.

Step 2 Look for explanations that connect facts.

In this passage, the facts have a problem-solution connection.

EXPLANATION Ⓑ Australia believes that immigration can help change its aging population trend.

Step 3 Think about what you already know.

I know that immigration made the United States diverse. I assume that the cultural diversity of Australia has grown due to that country's skilled migration program.

Step 4 Use evidence, explanations, and your prior knowledge to form a general understanding of what you have read.

Because of its aging population, Australia has a government-regulated immigration policy to draw workers from other countries. Through this program, Australia is both solving the problem of its aging population and increasing its cultural diversity.

APPLY THE SKILL

Turn to *Europe Geography & History*, Section 2.3, "The Republic of Rome." Read "The Roman Way" passage. Use a chart like the one at right to organize the evidence. Then write a brief synthesis statement about the importance of Roman values in the development of Rome.

Evidence: Romans applied the values of self-control, working hard, doing one's duty, and pledging loyalty to Rome.
Supporting explanation:
Synthesis:

Analyze Primary and Secondary Sources

Primary sources are materials written or provided by people who have had personal experience with an event. **Secondary sources** are materials written by people who did not witness or experience an event directly. To analyze primary and secondary sources, follow the steps at right.

Step 1 Identify whether the material is a primary or secondary source.

Step 2 Determine the quality and credibility of the primary or secondary source.

Step 3 Determine the main idea of the secondary source material.

Step 4 Identify the author of the primary source and his or her main point.

GUIDED MODEL

Mandela's Inspiration

Ⓐ Enlightenment thinkers asserted that Ⓑ people have natural rights, such as life, liberty, and property. They inspired Nelson Mandela in his struggle to end apartheid in South Africa. For his efforts, Mandela received the 1993 Nobel Peace Prize. The following is from his speech.

Ⓒ *The value of our shared reward will and must be measured by the joyful peace which will triumph.*

Thus shall we live, because we will have created a society which recognizes that all people are . . . entitled . . . to life, liberty, prosperity, human rights, and good governance.

Ⓓ—Nelson Mandela, 1993

TIP When you analyze a primary or secondary source, separate facts from opinions. These opinions may reflect the author's bias and the beliefs and ideas of his or her time.

Step 1 Identify whether the material is a primary or secondary source.

Most textbook passages are secondary sources. An observation from an expert or eyewitness is likely to be a primary source. The secondary source material in this textbook passage is marked with Ⓐ.

Step 2 Determine the main idea of the secondary source material.

MAIN IDEA Ⓑ People have natural rights that all humans should enjoy.

Step 3 Determine the quality and credibility of the primary or secondary source.

The primary source in this passage is marked with Ⓒ. Ask yourself, "Is the author a reliable source of information? *Nelson Mandela is a political activist and Nobel Peace Prize winner. He is a credible source. The secondary source is my textbook, which is also a reliable source.*

Step 4 Identify the author of the primary source and his or her main point.

AUTHOR Ⓓ Nelson Mandela

MAIN IDEA Ⓑ People have natural rights that all humans should enjoy.

APPLY THE SKILL

Turn to *Europe Geography & History,* Section 1.4, "Protecting the Mediterranean." Read the "Under the Sea" passage. Determine what parts of the passage are primary and secondary sources. Then use a chart like the one at right to analyze the passage.

Quality of the source:
Main idea of secondary source material:
Author of primary source material:
Information provided by primary source:

Analyze Visuals

Visuals, such as charts, graphs, maps, photos, artwork, and diagrams, illustrate ideas within a text. When you **analyze visuals**, you determine what information is being presented and how it relates to other information about that topic. To analyze visuals, follow the steps shown at right.

Step 1 Study the visual and determine what information it provides.

Step 2 Determine how the information in the visual relates to other information provided in a text.

Step 3 Analyze the information presented in the visual to enhance your understanding.

GUIDED MODEL

Solstices and Equinoxes

The moment at which summer and winter start is called a B **solstice**. June 20 or 21 is the summer solstice in the Northern Hemisphere. On December 21 or 22, the Northern Hemisphere has its winter solstice.

The beginning of spring and autumn is called an B **equinox**. In the Northern Hemisphere, the spring equinox occurs around March 21, and the autumn equinox occurs around September 23.

A EARTH'S FOUR SEASONS: NORTHERN HEMISPHERE

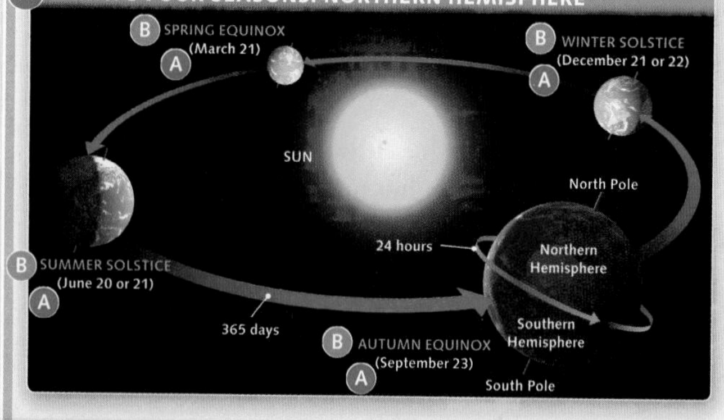

B SPRING EQUINOX A (March 21)
B WINTER SOLSTICE A (December 21 or 22)
SUN
North Pole
24 hours
Northern Hemisphere
B SUMMER SOLSTICE A (June 20 or 21)
365 days
B AUTUMN EQUINOX A (September 23)
Southern Hemisphere
South Pole

Step 1 Study the visual and determine what information it provides.

Examine any titles, labels, or captions for clues.

A *This diagram is titled "Earth's Four Seasons: Northern Hemisphere." The dates on the visual tell me the position of Earth in relationship to the sun throughout the year.*

Step 2 Determine how the information in the visual relates to other information provided in a text.

B *The summer and winter solstices and spring and autumn equinoxes are described in the text. The visual illustrates the information from the text.*

Step 3 Analyze the information presented in the visual to enhance your understanding.

Ask yourself: What information does the visual show? Why was this visual used? Does the visual represent information accurately? What does the visual show that is not explained in the text?

TIP Look for patterns and connections between items and information in a visual and the surrounding text. Study colors and symbols to understand what they represent.

APPLY THE SKILL

Turn to *Human & Physical Geography,* Section 1.3, "Earth's Landforms." Analyze the visual and answer the following questions.

1. What type of information does this visual provide?
2. Which text passage does this visual enhance?
3. Use the information in the related text passage and the visual to explain what the continental shelf is and where it is located.
4. What does the visual show you about continental rise that is not explained in the passage?

Interpret Physical Maps

Physical Maps provide information about the earth's physical features such as lakes, rivers, and mountains. You can learn about elevation, or relief, and absolute and relative location by studying physical maps. To read a physical map, follow the steps at right.

Step 1 Read the title of the map.

Step 2 Use the map legend.

Step 3 Use the map scale to measure distance.

Step 4 Use the compass rose or directional pointer to determine direction.

Step 5 Use the latitude and longitude gridlines to determine the region's location on Earth.

GUIDED MODEL

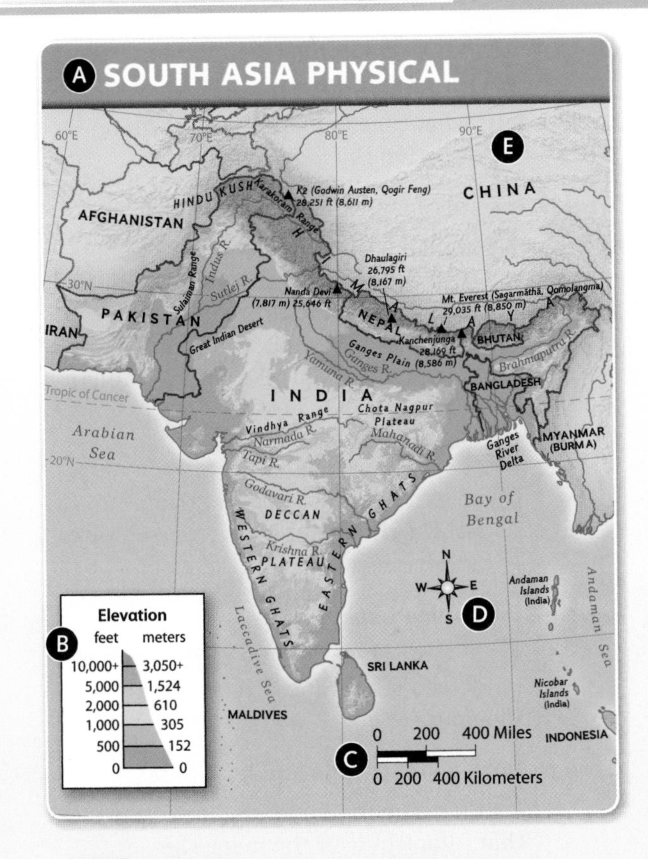

TIP Making a chart is a good way to record important information from a physical map. Use a chart to record the map's title, legend information, scale information, latitude and longitude, and location.

Step 1 Read the title of the map.

A Read the title of the map to find out what type of map it is and what kind of information the map presents.

Step 2 Use the map legend.

B A map legend explains the symbols used on the map. On a physical map, the legend usually provides information about physical features, such as mountains. The map also has a color-coded elevation scale to show how far above sea level each area is.

Step 3 Use the map scale.

C Use the map scale to help you determine the distances between points on the map.

Step 4 Determine direction.

D Use the compass rose or directional pointer to help you determine direction on the map.

Step 5 Determine latitude and longitude.

E Examine the numbered gridlines on the map. The horizontal lines represent latitude. The vertical lines represent longitude.

APPLY THE SKILL

Turn to *Europe Geography & History*, Section 1.1, "Physical Geography." Interpret the map to answer the following questions.

1. What is the map about?
2. What type of map is this and what region does it represent?
3. What is the highest elevation in the country of Romania?
4. What is the approximate distance between the easternmost point in France and the westernmost point in Italy, in both miles and kilometers?

Interpret Political Maps

Political maps provide information about human-made features such as cities, capitals, and borders between countries. Unlike physical maps, political maps do not focus on the physical features of a country or region. To read a political map, follow the steps shown at right.

Step 1 Read the title of the map.

Step 2 Use the map legend.

Step 3 Use the map scale to measure distance.

Step 4 Use the compass rose or directional pointer to determine direction.

Step 5 Use the latitude and longitude gridlines to determine the region's location on Earth.

GUIDED MODEL

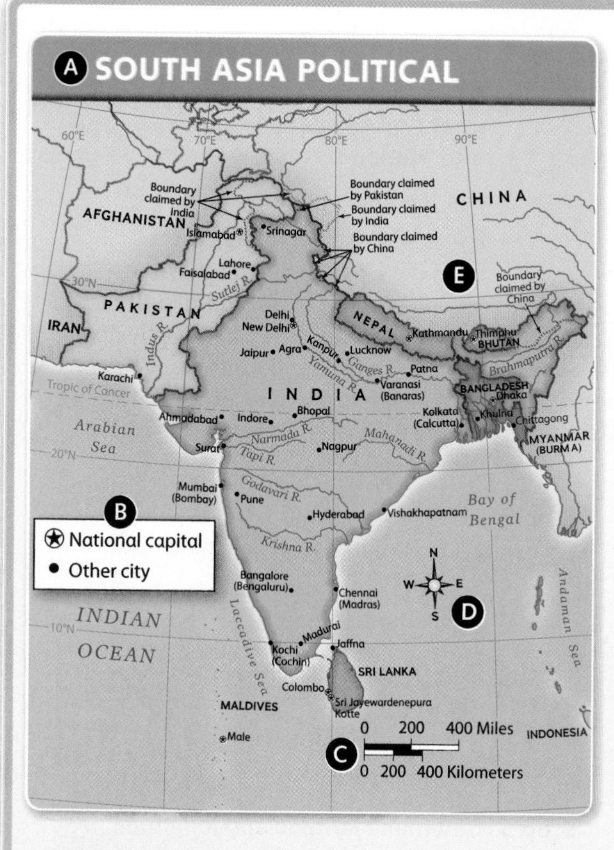

TIP On many political maps, countries or states are shown in different colors. This makes it easy to distinguish their borders. Cities are often designated by a dot of varying sizes based on the city's population.

Step 1 Read the title of the map.

A Read the title of the map to find out what type of map it is and what information the map represents.

Step 2 Use the map legend.

B A map legend explains the symbols used on the map. Symbols for capital cities and major cities are usually found in the legend of a political map. On this map, New Delhi, India's capital, is representated by a star.

Step 3 Read the map scale.

C Use the map scale to help you determine the distances between cities or countries on the map.

Step 4 Determine direction.

D Use the compass rose or directional pointer to determine direction on the map.

Step 5 Determine latitude and longitude.

E Examine the numbered gridlines that intersect over the map. The horizontal lines represent latitude, and the vertical lines represent longitude. Latitude and longitude lines can help you establish the location of the countries or major cities shown.

APPLY THE SKILL

Turn to the chapter introduction for *Europe Geography & History*. Locate the Europe Political map next to the Preview the Chapter page. Interpret the map and answer the following questions.

1. What type of map is this and what region does it represent?
2. What is the capital city of Spain?
3. What is the approximate distance between the easternmost and westernmost coasts of Iceland?
4. Where is Berlin located in relation to Warsaw?

Create Sketch Maps

In social studies textbooks, maps are a common visual aid. However, sometimes it is helpful to draw your own map to help you understand a place better or to visualize features described in the text. To **create a sketch map** of a country, follow the steps shown at right.

Step 1 Determine which map you are going to draw and give it a title.

Step 2 Sketch the outline of the location.

Step 3 Add important political and physical features to your map.

Step 4 Add a compass rose to your map.

Step 5 If appropriate, sketch and label surrounding countries or regions.

GUIDED MODEL

Step 1 Determine which map you are going to draw and give it a title.

A Study existing maps and text describing that geographical location.

Step 2 Sketch the outline of the location.

Ask yourself, "What shape is this location? What borders it?" Remember that sketch maps do not have to be perfect. Everyone's sketch map will look different.

Step 3 Add important political and physical features to your map.

B This student chose to draw and label the Ebro River in Spain.

Step 4 Add a compass rose to your map. C

Step 5 If appropriate, sketch and label surrounding countries or regions.

D On this sketch map, the country of Portugal has been identified.

TIP Sketch your map on graph paper so that you can draw your location and its surrounding area to scale. Also, use a pencil when you sketch so that you can easily erase any mistakes.

APPLY THE SKILL

Examine the Europe Political map next to the Preview the Chapter page in *Europe Geography & History*. Then examine the Europe Physical map and read the passage "A Peninsula of Peninsulas" in Section 1.1, "Physical Geography." Use a chart like the one at right to help you create a sketch map of Ireland.

Map title:
Capital:
Major cities:
Surrounding countries:

Create Charts and Graphs

To organize information, it is useful to **create charts** and **graphs.** Charts simplify and summarize information. Graphs present numerical information. To create charts and graphs, follow the steps shown at right.

Step 1 Determine whether you should use a chart or graph to represent your data.

Step 2 Give your chart or graph a title to tell what kind of information it shows.

Step 3 Create your chart or graph using appropriate labels for the data.

GUIDED MODEL

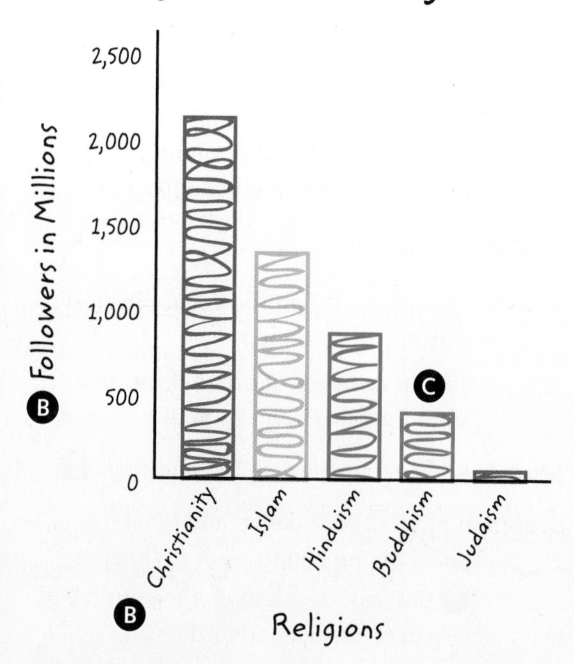

Step 1 Determine whether you should use a chart or graph to represent your data.

I want to represent the number of people who practice the major world religions: Christianity, Islam, Hinduism, Buddhism, and Judaism. The data is numerical, so I will create a graph. I will use a bar graph to compare the data and show which religion has the most followers and which religion has the least.

Step 2 Give your chart or graph a title to tell what kind of information it shows.

Ⓐ *I will call the bar graph "Major World Religions."*

Step 3 Create your chart or graph using appropriate labels for the data.

Ⓑ *My horizontal axis will represent the five major world religions. My vertical axis will represent the number of followers in millions. I will label the axes accordingly.*

Ⓒ *I will record my data on the bar graph using a different color to represent each world religion.*

TIP Different visuals are used to represent different types of data. Line graphs are useful to compare changes over time. Bar graphs compare quantities. Pie graphs show percentages of a whole. Charts can be structured in many different ways, but they always simplify and organize information.

APPLY THE SKILL

Turn to *Europe Geography & History,* and locate the Chapter Review. Use the "Miles of Railway Track in Selected European Countries" chart to create a line graph. To simplify the graph, include only the four countries with the highest totals. A line graph representing 1840 is shown at right.

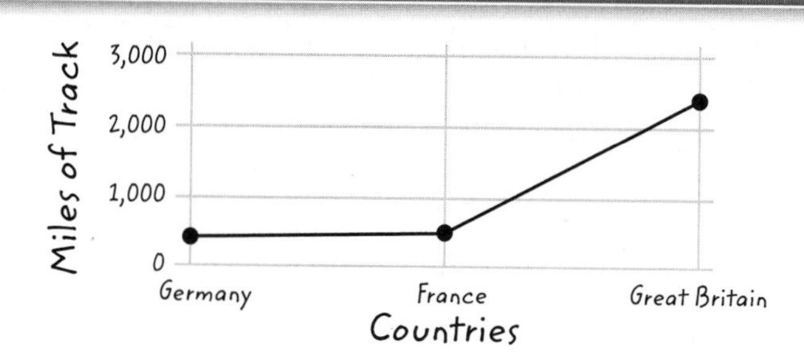

Interpret Charts

A chart is a way to represent information or data visually. In a chart, information is organized, simplified, and summarized. To **interpret charts,** follow the steps shown at right.

Step 1 Read the title of the chart to find out what type of information the chart presents.

Step 2 Check the source of the data in the chart for reliability.

Step 3 Read any headings, subtitles, or labels to understand how the chart is organized.

Step 4 Examine the data in the chart.

Step 5 Summarize the information.

GUIDED MODEL

A POPULATION OF MEDITERRANEAN CITIES (IN MILLIONS)

C City	1960	2015 (projected)
C Athens, Greece	2.2	3.1
Barcelona, Spain	1.9	2.73
Istanbul, Turkey	1.74	11.72
Marseille, France	0.8	1.36
Rome, Italy	2.33	2.65

B **Source:** UN, 2002

TIP When you interpret a chart, compare the data and draw conclusions from the information. For example, in the chart above, you might conclude that Istanbul is the fastest-growing city on the Mediterranean. You might also conclude that, as a result, the city may face challenges in housing and employing its large population.

Step 1 Read the title of the chart to find out what type of information it represents.

TITLE A Population of Mediterranean Cities (in millions)

Step 2 Check the source of the data in the chart for reliability.

SOURCE B The UN, or United Nations, is a known and reliable source.

Step 3 Read any headings, subtitles, or labels to understand how the chart is organized.

C This chart is organized in columns and rows. The rows feature major Mediterranean cities, and the columns give data for 1960 and 2015.

Step 4 Examine the data in the chart.

The data tells how certain cities are predicted to grow over a period of 55 years. Its purpose is to report past data and project future data.

Step 5 Summarize the information.

The data in this chart predicts that Istanbul, Turkey, will have the greatest growth and Rome, Italy, the least.

APPLY THE SKILL

Turn to *Human & Physical Geography*, Section 1.5, "Waters of the Earth." Interpret the "Longest World Rivers" chart to answer the following questions.

1. What information does the chart present?
2. How is the chart organized?
3. Where is the Amazon River located?
4. What is the length of the Mississippi-Missouri River?
5. What is the longest river in the world?
6. Of the two Asian rivers on the chart, which is longer?

Interpret Graphs

A graph is another way to represent information or data in picture form rather than written text. In a graph, data can be represented using numbers, symbols, or pictures. Common graphs include pie graphs, line graphs, and bar graphs. To **interpret graphs,** follow the steps shown at right.

Step 1 Read the title of the graph and identify what type of graph it is.

Step 2 Check the source of the data in the graph for reliability.

Step 3 Read the labels in the graph.

Step 4 Examine the data in the graph and look for patterns.

Step 5 Summarize the information in the graph.

GUIDED MODEL

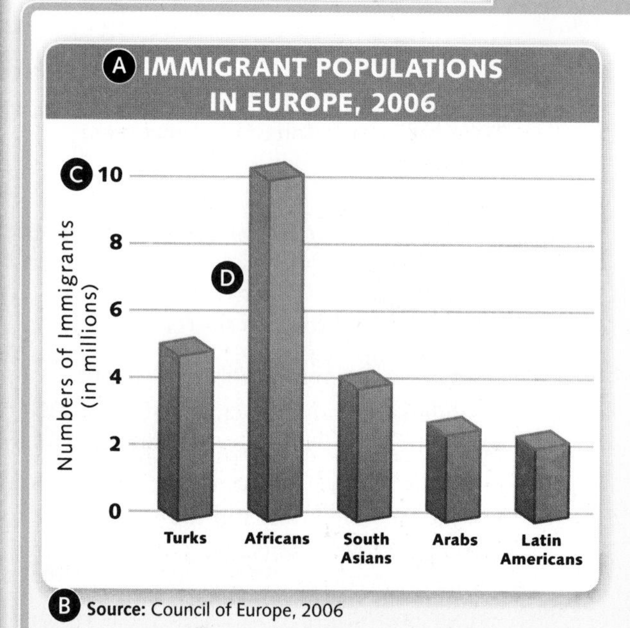

B **Source:** Council of Europe, 2006

TIP A pie graph is used to compare parts of a whole, and each "slice" represents a percentage. Remember that the percentages represented by the slices in a pie graph always add up to 100.

Step 1 Read the title of the graph and identify what type of graph it is.

TITLE A Immigrant Populations in Europe, 2006; This is a bar graph, which is often used to compare quantities.

Step 2 Check the source of the data in the graph for reliability.

SOURCE B Council of Europe, which is a reliable source on Europe.

Step 3 Read the labels in the graph.

C The vertical axis is labeled "Numbers of Immigrants." Its scale goes to ten million. The horizontal axis is labeled with five immigrant groups.

Step 4 Examine the data and look for patterns.

D The graph compares the numbers of immigrants to Europe from various regions in a single year.

Step 5 Summarize the information in the graph.

The graph shows that Africans made up the largest group of immigrants in Europe in 2006. The smallest group came from Latin America.

APPLY THE SKILL

Turn to "Compare Across Regions: World Languages" at the end of the Europe unit. Study the Number of Languages Spoken by Continent graph and use it to answer the following questions.

1. What do the colors of the bars represent?
2. What do the words written on the bars represent?
3. On which continent are the most languages spoken?
4. On which continent are the fewest languages spoken?

Create and Interpret Time Lines

Another visual way to organize, represent, and review information is by creating a time line. When you **create a time line,** you record events and dates chronologically along an axis from left to right or from top to bottom. To create a time line, follow the steps at right.

Step 1 Decide what your time line will show and give your time line a title.

Step 2 Determine time line dates and place them in chronological order.

Step 3 Plot major events on the time line in the appropriate locations.

Step 4 Note the patterns that emerge on your time line.

GUIDED MODEL

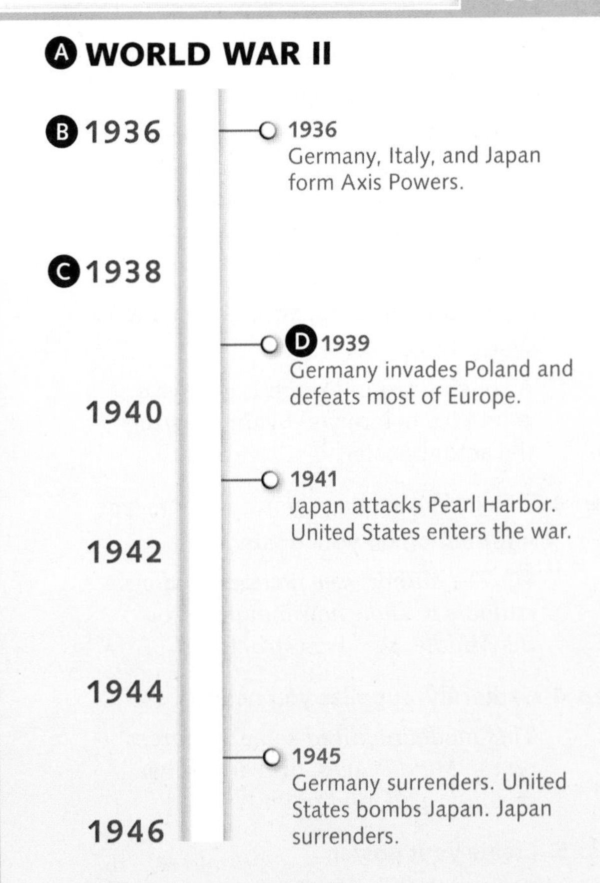

TIP When you record B.C. dates on a time line, be careful to place them in the correct order. The higher the number, the further back in time it occurred. For example, 850 B.C. took place 350 years before 500 B.C. The year A.D. 1 took place immediately after 1 B.C. No zero year exists.

Step 1 Decide what your time line will show and give your time line a title.

Time lines often feature events that are thematically related.

I will create a time line to record the major events that led up to and took place during World War II.

TITLE Ⓐ World War II

Step 2 Determine time line dates and place them in chronological order.

Determine your start and end dates. Make sure the interval dates in between are regular.

Ⓑ *The events occur from 1936 to 1945. My start date will be 1936, and my end date will be 1946.* Ⓒ *I will put an interval date every two years.*

Step 3 Plot major events on the time line in the appropriate locations.

Ⓓ *For an event that occurs on a date that isn't labeled, I will draw a line from the time line and label the date.*

Step 4 Note the patterns that emerge.

On my time line, I see that Germany was powerful in the late 1930s and that the United States played an important role in the war.

APPLY THE SKILL

Turn to *Europe Geography & History*, Section 2.2, "Classical Greece." Study the time line and use it to answer the following questions.

1. What title might you give the time line?
2. What is the time span of the time line?
3. What are the intervals between major dates on the time line?
4. When did the Greeks defeat Persia?
5. What patterns do you see on the time line?

Build Models

You can **build models** to show information about a place, an event, or a concept in a visual way. Models can be any visual representation of information, but the most common examples are posters, diagrams, mobiles, and dioramas. To create a poster, follow the steps shown at right.

Step 1 Determine what your poster will show and research the topic.

Step 2 Brainstorm ideas for your poster and sketch them.

Step 3 Think of visual ways you can represent information on your poster.

Step 4 Gather the supplies you need.

Step 5 Create your poster.

GUIDED MODEL

TIP Dioramas are models that show a three-dimensional place, event, or scene, such as a battle or extreme type of weather. A diorama is built inside a small box, such as a shoe box. The inside of the box is painted with a background, and the scene is constructed using craft supplies and common household objects.

Step 1 Determine what your poster will show and research the topic.

This poster shows the structure of a manor during the Middle Ages. To create this model, the author researched the Middle Ages and the feudal system.

Step 2 Brainstorm ideas for your poster and sketch them.

A rough draft or sketch is always a good idea before you begin drawing the actual poster.

Step 3 Think of visual ways you can represent information on your poster.

Ⓐ *The author uses pictures and callouts to show how a manor from the Middle Ages was structured.*

Step 4 Gather the supplies you need.

This model required some resources on the Middle Ages, drawing paper, a pencil or pen, and paint.

Step 5 Create your poster.

This model was drawn from a bird's eye view, which shows how the parts of the manor relate to each other.

APPLY THE SKILL

Turn to *Human & Physical Geography,* Section 2.4, "Natural Resources." Read the "Earth's Resources" and "Categories of Resources" passages. Create a poster or other type of model to represent how Earth's natural resources are categorized and show examples of each type. A sample poster is shown at right.

Earth's Natural Resources

Coal
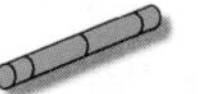
Sugarcane

Wheat

Copper

Another type of model is a diagram, which can help observers visualize something that may not be visible in real life. Diagrams are also often used to explain how things work or to show how something is divided into parts. To build a diagram, follow these steps.

Step 1 Determine what your diagram will show and research the topic.

Step 2 Brainstorm ideas for your diagram and sketch them.

Step 3 Think of visual ways you can represent information on your diagram.

Step 4 Gather the supplies you will need.

Step 5 Create your diagram.

GUIDED MODEL

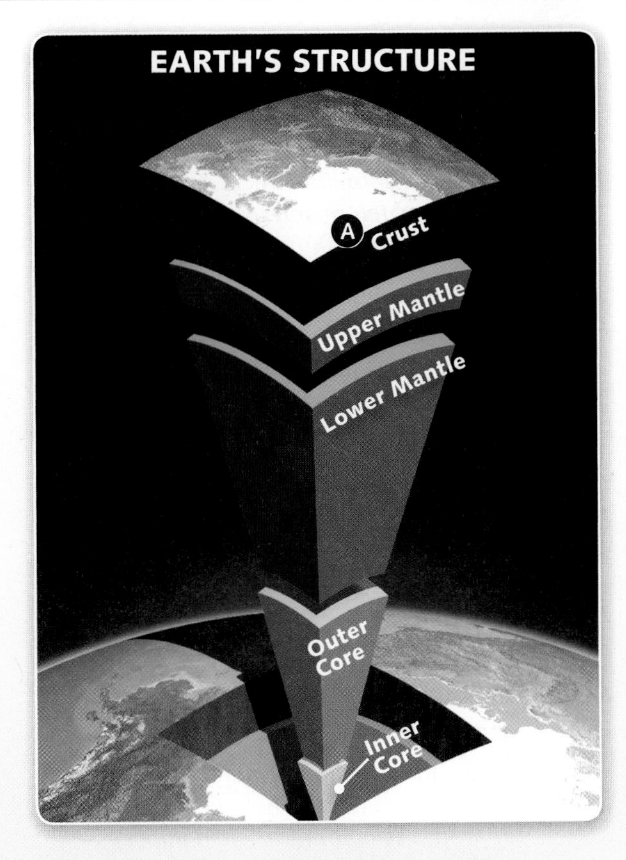

TIP A mobile is a three-dimensional model. Mobiles are especially useful for showing how objects relate spatially to each other, such as how the planets circle the sun. Mobiles can be created by hanging objects from a simple wire hanger using yarn or thread.

Step 1 Determine what your diagram will show and research the topic.

This model shows the structure of Earth's layers. It is helpful because we cannot normally see them.

Step 2 Brainstorm ideas for your diagram and sketch them.

It is a good idea to determine and make a rough sketch of the approximate thicknesses of each layer ahead of time.

Step 3 Think of visual ways you can represent information on your diagram.

A *This diagram shows the layers of the earth, using a different color to represent each layer. The names of the layers are labeled.*

Step 4 Gather the supplies you will need.

This diagram required paints, markers, or colored pencils, as well as paper and a pen or pencil.

Step 5 Create your diagram.

This model clearly shows Earth's five layers, their approximate thicknesses, and their correct order.

APPLY THE SKILL

Turn to *Europe Geography & History,* Section 1.2, "A Long Coastline." Read the "Exploration and Settlement" passage. Draw a diagram like the one at right to represent the structure of a European fishing village. Use visual elements rather than text to represent information.

Create Databases

A database is a collection of information, or data, organized in a chart format so it can easily be viewed, used, and updated. You can use a computer program to **create databases** that allow you to search through the data to find only what you need. To create a database, follow these steps.

Step 1 Determine the information your database will contain and give it a title.

Step 2 Enter the column headings and/or row headings for your database.

Step 3 Enter the data in the appropriate columns and rows.

GUIDED MODEL

Ⓐ FAMOUS EUROPEAN EXPLORERS

Ⓑ Explorer's Name	Explorer's Country	Date of Exploration	Place Explored
Ⓒ Bartolomeu Dias	Portugal	1488	Southern tip of Africa
Vasco de Gama	Portugal	1498	Southern tip of Africa; also reached India and opened up trade with Asia
Christopher Columbus	Italy (but he was working for Spain)	1492	Islands in the Caribbean; continents of North and South America
Jacques Cartier	France	1530s	Northern part of North America
Sir Francis Drake	England	1577	Sailed around the world

TIP You might want to use index cards to write and organize the information you want to use in your database. Once you have gathered all of your information, you can put it in a written chart format or use database software to organize it.

Step 1 Determine the information your database will contain and give it a title.

Ⓐ *This database includes data from a passage on European exploration. It records data about which explorers went to which regions. An appropriate title is "Famous European Explorers."*

Step 2 Enter the column headings and/or row headings for your database.

Remember, columns go up and down, and rows go from left to right. Database headings most commonly appear across the top or down the left side of the database, or in both places.

Ⓑ *I have four categories of data, so I will use four columns, one per category. I will label each column with a heading that describes the data below it.*

Step 3 Enter the data in the appropriate columns and rows.

Ⓒ *I will extract information from the text I have read and insert it in the appropriate column in the database. I will use a separate row to provide information on each explorer.*

APPLY THE SKILL

Turn to *Human & Physical Geography*, Section 2.2, "World Climate Regions." Read about the five climate regions featured in this section. Then create a database like the one at right to record the names of each climate region, a description of its weather and plant life, and a list of places that have that climate.

WORLD CLIMATE REGIONS

CLIMATE REGION	WEATHER	PLANT LIFE	LOCATIONS

Create Graphic Organizers

While you read, you can **create graphic organizers** such as charts, diagrams, and time lines to take notes and explore how the information is related. To create a graphic organizer, follow the steps shown at right.

Step 1 Determine your needs.

Step 2 Determine what type of graphic organizer you need.

Step 3 Draw your graphic organizer.

Step 4 Use information from the passage, plus any relevant maps or other features, to fill in your graphic organizer.

GUIDED MODEL

A PROBLEM
Earthquakes can cause severe damage and endanger people's lives.

↓

B CONTRIBUTING FACTORS
In some areas where earthquakes occur, buildings cannot withstand the intense shaking.

↓

C SOLUTION
Scientists are working hard to predict earthquakes. Engineers try to design buildings that keep people safe and minimize damage.

TIP You can use a cause-and-effect chain to help you understand the results of a particular event. Note that the chain can contain a single cause and multiple effects, or it can have many causes and only one effect.

Step 1 Determine your needs.

First, think about the type of information you are recording. Then determine what you will be doing with the information. Are you gathering facts, comparing two things, or recording a series of events?

Step 2 Determine what type of graphic organizer you need.

A chart can be used to record facts. A Venn diagram helps compare and contrast two things. A time line records a series of events over time. For this example, a problem-solution chart was created to identify a problem and understand how people are trying to solve it.

Step 2 Draw your graphic organizer.

Step 4 Use information from the passage, plus any relevant maps or other features to fill in your graphic organizer.

This student identified:

A the problem, in this case, the dangers posed by earthquakes;

B contributing factors, the situations that add to the problem;

C a possible solution to the problem.

APPLY THE SKILL

Turn to *Europe Geography & History*, Section 3.2, "The Industrial Revolution." Read "The Revolution Begins" passage. Then create a main idea and details graphic organizer like the one at right and use it to record important information from the text.

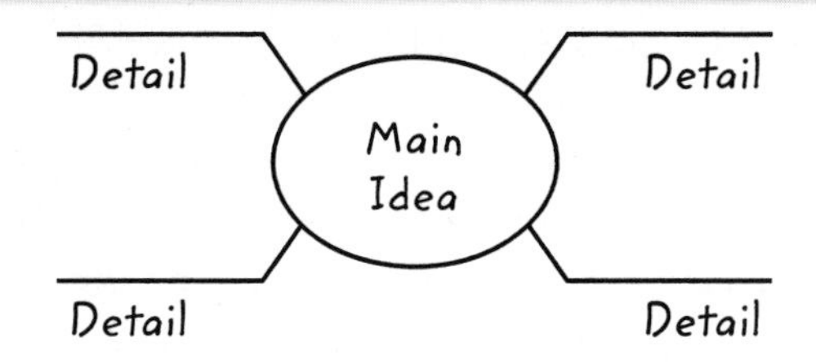

Conduct Internet Research

When you **conduct Internet research**, you search and access source material from all over the world via the Internet. As you probably know, the World Wide Web (or "www") is part of the Internet that allows you access to online information and data. To conduct Internet research, follow the steps shown at right.

Step 1 Choose and access a search engine.

Step 2 Type key words or phrases into the search field.

Step 3 Examine the search results.

Step 4 Visit the suggested Web sites and perform more specific searches for information.

Step 5 Use other reliable sources to verify any information you gather.

GUIDED MODEL

A elephant habitat | Search

B

ELEPHANTS- Habitat & Distribution
ELEPHANTS- Habitat & Distribution...Discover animal, environmental, and zoological career facts as you explore in-depth topic coverage via SeaWorld, ...
www.seaworld.org/.../elephants- habitat & distribution.htm -

Elephant Facts - Defenders of Wildlife
Get the facts on **elephant**. **Elephants** are the largest land-dwelling mammals on earth. ... of **elephants' habitat** will become significantly hotter and drier, ...
www.defenders.org › Wildlife and Habitat -

African **Elephants**, African **Elephant** Pic ...
African **elephants** are the largest of Earth's **land** mammals. Their enormous ears help them to keep cool in the hot African climate. ...
animals.nationalgeographic.com/animals/.../african-**elephant**/ -

Elephants Habitat | Animal Habitats
A quick look into the **elephant's habitat**. In studying an **elephant's habitat**, the first thing one encounters is the fact that there are several species of ...
www.animalhabitats.org/**elephants_habitat**/**elephants_habitat**.htm -

Step 1 Choose and access a search engine.

Step 2 Type key words into the search field.

Be as specific as possible to narrow your search and yield better results.

A *I typed "elephant habitat" into the search field and clicked "search."*

Step 3 Examine the search results.

Remember that Web addresses that end with ".edu," ".gov," and ".org" are often more reliable than ".com" sites. Review the addresses and summaries of each Web site and decide which ones to use.

B *This Web site's domain is ".org", which is usually a reliable source, and the information is relevant to my topic, so I will go to this Web site.*

Step 4 Visit the suggested Web sites and perform more specific searches for information.

Once I'm on the site, I will search "elephant habitat preservation" to find more specific information.

Step 5 Use other reliable sources to verify any information you gather.

I found the same information in an online encyclopedia and on the National Geographic Web site. This tells me the information is reliable.

TIP When you use the Internet to do research for a report or project, print the source material you find online. This will provide you with a record of the Web sites you have used and allow you to review the source materials as you write your report without having to go back online.

APPLY THE SKILL

Think of a topic from this book you would like to research. Brainstorm key words to use on a search engine. Then, with your teacher's permission, use a Web browser to conduct Internet research on your topic. If possible, print copies of the information you find. Write a brief paragraph explaining your search process and results.

Ancient Greece | Search

Evaluate Internet Sources

It is important to use only the most reliable and credible information as a resource. To do so, you must **evaluate Internet sources** to be sure they are accurate and reputable. To evaluate Internet sources, follow the steps shown at right.

Step 1 Examine the Web site's Internet address.

Step 2 Identify the Web site's author.

Step 3 Identify when the Web site was created or last updated.

Step 4 Verify the information from the Web site using other reliable sources.

Step 5 Evaluate the Internet source.

GUIDED MODEL

Step 1 Examine the Web site's Internet address.

A *This Web site's address is* www.cia.gov. *A government agency created the site, which makes it reliable.*

Step 2 Identify the Web site's author.

Sites that clearly name the author are more reliable than anonymous Web sites.

B *The author of this Web site is the Central Intelligence Agency, or CIA. It is expert at collecting data.*

Step 3 Identify when the Web site was created or last updated.

C *This Web site is updated every week, so the information is current.*

Step 4 Verify the information from the Web site using other reliable sources.

I found the information in an encyclopedia and on the National Geographic Web site.

Step 5 Evaluate the Internet source.

I believe this is a reliable source because it is created by the government, updated frequently, and has the same information as other reliable sources.

TIP Another way to evaluate an Internet source is to determine its intended audience. Ask yourself to whom the author is writing, and study the writing style, vocabulary, and tone. Determine if the author's objective appears to be to inform, to explain, or to persuade, and whether he or she provides sufficient evidence.

APPLY THE SKILL

With your teacher's permission, visit the following Web address: www.nationalgeographic.com. Use the steps above to evaluate this Internet site as a reliable source of information. Write a brief paragraph explaining your evaluation of the site.

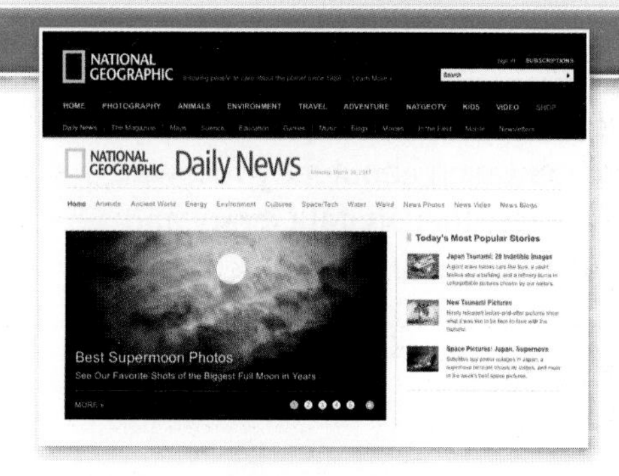

Create Multimedia Presentations

You can **create multimedia presentations** on any topic using media such as photographs, video clips, and audio recordings. To create a multimedia presentation, follow the steps shown at right.

Step 1 Determine the topic you will be presenting and the types of media you will use.

Step 2 Research your topic.

Step 3 Put your presentation together.

GUIDED MODEL

Unlike many species of crab, the one shown here can swim as well as walk.

TIP The **Magazine Maker** allows you to search for images and information on a wide variety of topics and create your own magazine by importing pictures and text. Use this program as you create your multimedia presentation.

Step 1 Determine the topic you will be presenting and the types of media you will use.

Certain types of media enhance topics more effectively than others. Once you have determined your topic, decide which media work best with it. Would your topic be enhanced by images, or would audio clips or music be more relevant? Would a video clip add meaning to your presentation? What about a map or graph? You decide.

A This student used a photo from the **Digital Library** and wrote her own caption for it.

Step 2 Research your topic.

Use reliable library and online sources to research your topic. Write a brief script that provides information on your topic and ties your media together.

Step 3 Put your presentation together.

Combine your informational written script with the media you have selected. Make sure everything flows together well. You might use presentation software to display your information. Rehearse your presentation a number of times to identify and correct any problems.

APPLY THE SKILL

Select a topic from this text and create a multimedia presentation about it by following the steps above. Use the **Digital Library** and **Magazine Maker** to create your presentation.

HANDBOOK
ECONOMICS & GOVERNMENT

PART I ECONOMICS

agriculture *n., the development of plants and animals to provide food.* Agricultural products include crops such as wheat, corn, and barley. Agriculture also includes animals that have been domesticated, or tamed. These animals are often called livestock, and they include cattle, sheep, pigs, and horses.

business cycle *n., a period during which a country's economic activities increase and then decrease in a relatively predictable pattern.* A business cycle has four phases. During expansion, businesses do well. At the peak of the cycle, economic activity begins to slow. During a contraction, economic activity continues to decrease. A contraction is also called a recession. The trough is the lowest level of economic activity. Then business starts to improve, and a new business cycle begins.

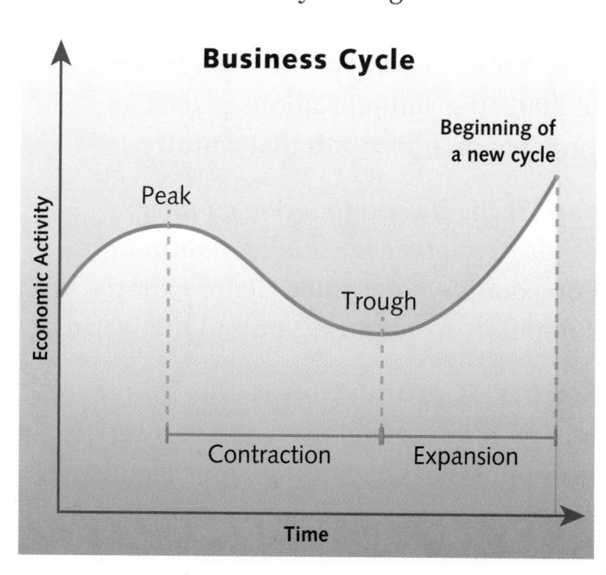

capitalism *n., an economic system in which private individuals or groups own the resources and produce goods for a profit.* In capitalism, private individuals or groups decide to produce goods or offer services. They offer those goods and services for sale in markets, which are places where people buy and sell. Businesses raise capital, or money, to create new products and hire workers. Capitalism is also called free enterprise because people are free to start businesses.

Bookstore, Los Angeles, California

command economy *n., an economic system in which the government controls a country's economic activities.* The government owns and controls the factories, farms, and stores in a country.

communism *n., an economic and political system in which the government owns and controls economic activities.* Communism is a type of command economy. In a Communist system, the government owns and operates factories, farms, and other types of economic activities. For example, in the steel industry, government officials decide what kind of steel and how much steel will be produced. North Korea, Vietnam, and Cuba have Communist systems. Communist economies have been much less efficient than free enterprise economies in producing goods and services. See **Part II Government** for the definition of communism as a political system.

corporation *n., a company in which people own shares, or parts of the company.* A corporation sells stock, or shares of the company, to raise money to create products and services. Shareholders often receive a dividend, which is a share of the profits. Multinational corporations are corporations that operate in several countries. People purchase shares of a corporation on a stock exchange.

depression *n., a deep and long-lasting contraction of economic activities.* During a depression, business activity falls dramatically. Businesses hire few workers. The unemployment rate, or percent of people without work, rises. In the United States, the Great Depression lasted from 1929 until the early 1940s. A recession is also a contraction in business. It is less severe and shorter than a depression.

A church in New York City distributes food in a breadline during the Great Depression.

developed nation *n., a country with highly-developed industries, a high standard of living, and private ownership of most businesses.* Developed nations have industries that make products like automobiles and computers. Less developed countries have fewer industries and rely on agriculture. They also have lower standards of living. The United States is a developed nation. Many countries in Africa, Asia, and Latin America are less developed countries.

economy *n., a country's system for producing and exchanging goods and services.* A country's economic development is based on its level of economic activity. If a country has a great deal of industry, it has a high level of economic development.

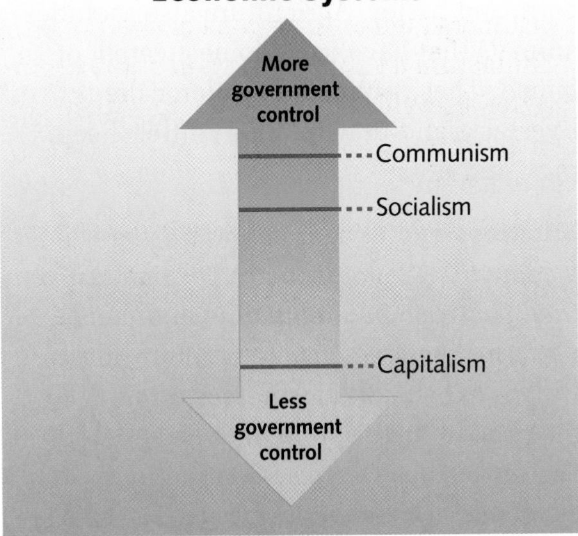

embargo *n., a government ban on trade with another country.* Countries set embargoes to show their disapproval of another country's activities. For example, after North Korea tested a nuclear weapon in 2006, the United Nations placed an embargo on trade with that country.

export *n., a good that one country sends to another for sale or distribution.* For example, the United States exports computers to countries around the world.

factors of production *n., the things that go into producing a good or a service.* Economists have identified four factors of production: land, labor, capital, and entrepreneurs. Land includes all natural resources, such as oil and silver. Labor is the work that people do. Capital is the machinery and other tools that are used to create a good or service. Entrepreneurs are people who start businesses.

free enterprise *n., an economic system in which businesses are privately owned, people buy and sell goods in free markets, and individuals freely make decisions whether to buy or sell.* The system is also called capitalism. The United States has a free enterprise system.

In a free enterprise system, producers are motivated by self-interest. To meet consumers' demand, producers decide to manufacture a good, such as an automobile, or offer a service. By doing so, they hope to make a profit. Consumers are also motivated by self-interest. They try to buy the best goods and services they can at the lowest price offered in a market.

Government makes sure that businesses compete fairly. It also ensures that food, medicines, and other products are safe. It provides services, such as defense, that are important to a country. The government also builds infrastructure, such as transportation facilities and roads.

Free Enterprise System

Individuals
- Create demand
- Offer labor

Supplies goods

Demand goods

Product Market
- Sells goods and services
- Offers labor

Producers
- Create goods
- Hire Labor

Send goods to market

Government
- Collects taxes
- Offers services
- Regulates economy

gross domestic product (GDP) *n., the total value of all the goods and services that a country produces in a specific time period, such as a year.* The GDP is an important measure of an economy's strength. Economists measure the GDP by adding together four kinds of goods and services. One is the goods and services that consumers buy. Another is the machines and other items that companies buy for their businesses. A third is goods and services that government buys. Fourth is the goods and services that a country exports to other countries.

GDP per capita is a country's GDP divided by the country's population. It shows how much the country produces per person.

Gross Domestic Product

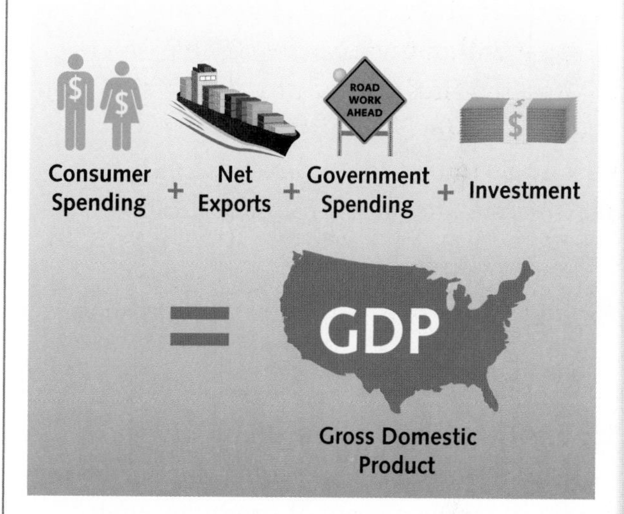

import *n., a good that one country receives from another for sale or distribution.* For example, the United States imports many cars and trucks from automobile manufacturers in Japan. Japan in turn, imports oil from Saudi Arabia.

industry *n., a group of businesses that produce a similar product or service.* For example, the film industry produces feature films. Common industries in the United States include construction, computers, pharmaceuticals, and electronics. In many industries, businesses take a raw material and turn it into a finished product. For example, in the clothing industry, businesses turn cotton, wool, and other materials into clothes.

inflation *n., an increase in the price of goods and services in a country.* In any given year, the average price of goods and services may go up. This is called an increase in the price level. The rate of increase is called the inflation rate. In 1980, for example, the United States had an inflation rate of about 14 percent. This means that the average prices that year were about 14 percent higher than in 1979.

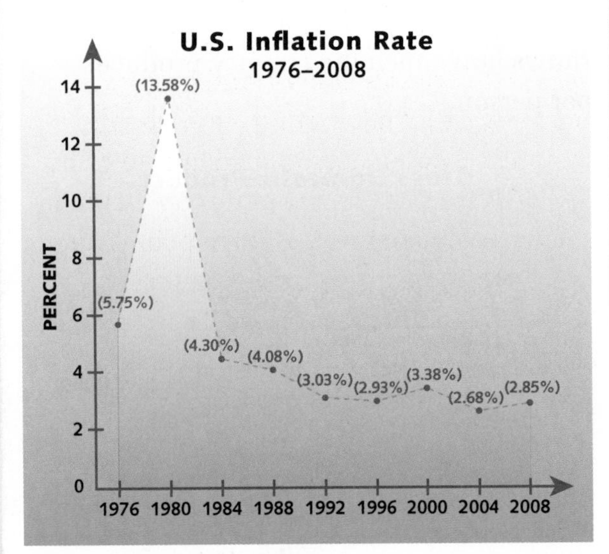

Source: U.S. Bureau of Labor Statistics

manufacturing *n., the production of physical products to be sold.* Manufacturing creates a wide variety of products, such as automobiles, steamships, airplanes, computers, and furniture. The term refers to the creation of items by machine and by hand. Light manufacturing means the creation of relatively small things, such as the circuits in computers. Heavy manufacturing means the creation of large objects, such as diesel engines for railroads. In 2010, manufacturing employed about 17 million workers in the United States and Canada.

market economy *n., an economic system in which people and businesses choose freely to buy and sell in markets.* A market is a place where people buy and sell goods and services. In a market economy, individuals make choices to buy and sell. For example, an individual might decide to earn a living by making and selling T-shirts. After making them, the person would try to interest stores in selling the T-shirts. A store is a common type of market. The government establishes rules, such as the rule that stores should be safe and clean.

mechanization *n., the use of machines instead of humans or animals to perform tasks.* For example, in 1764, the Englishman James Hargreaves invented the spinning jenny. It was a machine that could spin yarn from wool. The machine allowed operators to create yarn much faster than they could by hand.

Spinning jenny

monopoly *n., the situation in which one company controls the production or selling of a product or service.* For example, suppose that a person owns the only grocery store in an isolated town. Because the store has no competition, the owner can set high prices on food. If someone else opens another grocery store in town, competition results. Because the stores are competing, they will have to lower prices and improve quality.

national debt *n., the total amount of money that the federal government owes.* If a government spends more money than it receives in taxes, it uses deficit spending. It must borrow money from individuals, companies, or other governments. As a government uses deficit spending, its national debt grows.

The national debt of the United States has grown rapidly since 1980. Critics say that the government should raise taxes, cut spending, or do a combination of the two to balance its budget.

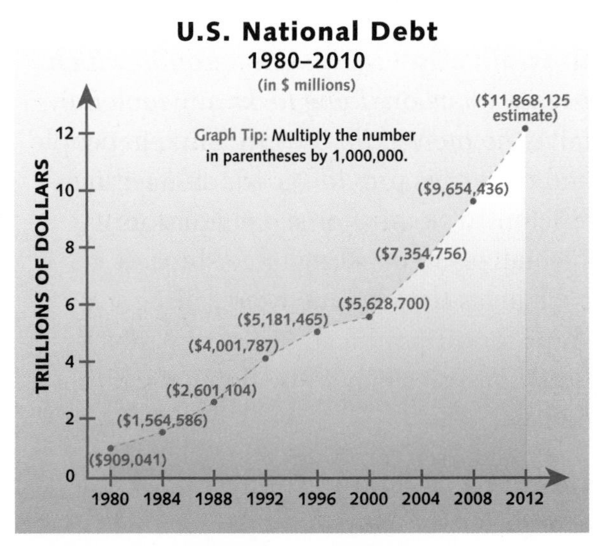

Source: U.S. Office of Management and Budget

natural resources *n., resources such as oil, coal, timber, and water that exist naturally in a place.* Having ample natural resources can help a country to develop its economy. For example, the United States has numerous natural resources, including timber and fresh water. On the other hand, Japan has become wealthy even though it has few energy resources like oil. It has developed its wealth by inventing new technologies and building quality products.

Oil pump

opportunity cost *n., the opportunity that a person gives up when he or she chooses to buy one item instead of another.* For example, suppose that Benita must decide between buying a new flat-screen television or taking a vacation. If she chooses the television, her opportunity cost is the vacation she did not take.

poverty *n., the lack of enough money to buy necessary things like food, clothing, and shelter.* Poverty is a worldwide problem. It causes disease and hunger. Poor people often do not have adequate shelter and suffer greatly during cold or hot weather. Economists say that more than one billion people in the world are poor. The U.S. government measures the percentage of its population that is poor. The graph below shows how the rate has changed over the past several years.

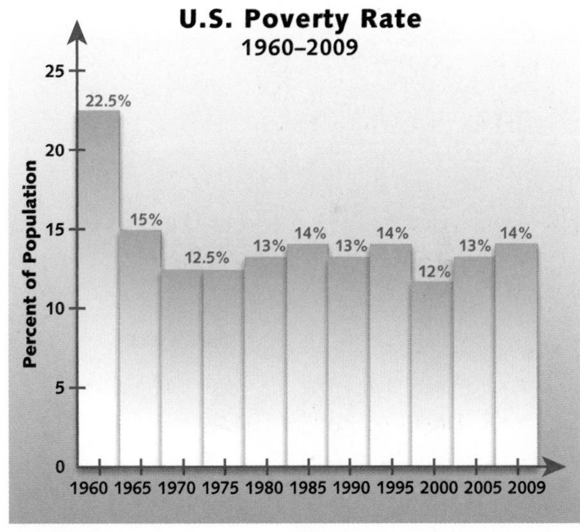

Source: U.S. Census Bureaus

raw materials *n., the materials that are used in manufacturing a final product.* Raw materials often come from natural resources, such as wood, oil, and iron ore. For example, the basic raw material in creating a table is wood. Cotton and wool are raw materials used in clothing.

retail goods *n., goods that are sold directly to consumers.* When you go to a store and purchase a DVD, you are buying a retail good. Today, merchants sell retail goods in a variety of ways. They sell them in stores, through vending machines, over the telephone, or on the Internet.

service industries *n., companies, nonprofit organizations, and government organizations that provide services rather than products.* Services include many activities, such as health care, education, financial operations, retail selling, and legal advice. In the United States, more than 75 percent of people work in service industries. Less developed countries, on the other hand, have fewer people working in service industries.

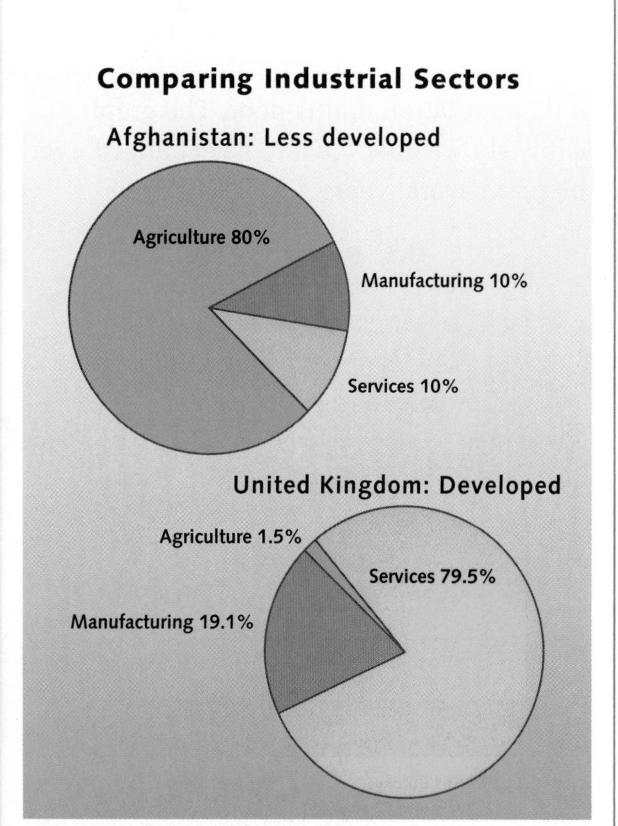

Source: www.NationMaster.com

socialism *n., an economic system in which the government owns and operates most businesses.* Socialism first developed in Europe in the early 1800s. Socialists wanted to eliminate poverty and improve working conditions. During the 1840s, a German socialist named Karl Marx said that the government should own all businesses. His ideas gave rise to communism.

Today, Sweden and some other European countries are often called democratic socialist countries. This means the government owns fewer businesses than in Communist countries, but it does provide many services, such as health care. In addition, taxes are high. Many economists claim that socialism is less efficient than free enterprise.

specialization *n., the situation in which people focus on doing tasks at which they have the most skill.* By specializing, people and countries provide goods and services efficiently. South Korea, for example, specializes in building large ships. It exchanges the revenue from selling the ships for oil produced in Saudi Arabia. Both countries benefit by doing what they do best.

standard of living *n., the level of economic well-being that people in a country have.* In a country with a high standard of living, people have the income to purchase such items as automobiles. In a country with a low standard of living, most of the population struggles to have enough food, clothing, and shelter.

stock market or stock exchange *n., a market where people buy and sell stocks and bonds.* Companies are able to raise money in one of two ways. They can sell stocks, which are shares of ownership in the company, or they can issue bonds. A bond is a written agreement to borrow money, and the company pays the money back with interest.

People may want to sell stocks or bonds to raise money or make a profit. The sale of stocks and bonds takes place on a stock exchange or a stock market. Brokers handle both the buying and selling of stocks and bonds. The largest stock market in the United States is the New York Stock Exchange on Wall Street in Manhattan.

strike *n., a situation in which workers stop working in order to win higher wages or benefits.* Strikes are most often called by unions. A union is an organization that represents workers in a certain industry. If a union's workers believe that wages are not high enough, they may go on strike. The strike brings the company's operations to a halt. A wildcat strike occurs when workers go on strike without the union's support.

Auto workers on strike, Naperville, Illinois, 2007

supply and demand *n., the economic forces that decide the price and amount of a product or service.* The supply of a good or service is the amount that a business is willing and able to offer for sale. If the company can get a higher price, it will create more of the product or service. The demand is the amount of a good or service that consumers are willing and able to buy at a given price. Usually, the demand for a good or service goes down as the price rises. The price at which supply equals demand is called the equilibrium price.

The graph shows how supply and demand work. The gray line stands for demand. The green line stands for supply. The lines meet at equilibrium, when supply equals demand. At prices above equilibrium, demand for the product goes down. At prices below equilibrium, consumer demand rises.

Supply and Demand

Price
Demand
Equilibrium
Supply
Quantity

tariff *n., a tax that a country places on goods imported from another country.* Countries use tariffs to protect their own companies from competition with other countries. For example, to protect auto manufacturers, a country might place tariffs on cars from other countries. The country's own manufacturers would thrive. However, consumers might pay more for their cars because there is less price competition.

trade *n., the exchange of services and goods.* Trade occurs when people cannot make things themselves but can get them from other people. For example, Chile grows asparagus. It exports the vegetable to countries around the world. With the money it earns, it imports automobiles from the United States, oil from Saudi Arabia, and so forth. This is an example of international trade. The trade within a country is called domestic trade.

unemployment rate *n., the percentage of people in a society who cannot find work.* Unemployment occurs when a person is willing and able to work but cannot find a job. The unemployment rate is the percentage of people who are looking for jobs but cannot find them. For example, if the unemployment rate is 8 percent, 8 out of 100 people could work, but do not have a job and are not able to find one. During a recession or depression, the unemployment rate rises. Unemployed people sometimes receive government aid.

wholesale goods *n., goods that producers sell to other business firms, such as stores.* Wholesalers buy goods in large quantities. They then sell those goods to retailers, who sell them to customers. For example, a factory in China might produce thousands of toys for children. A wholesaler buys the toys from the factory and ships the toys to retail stores around the world. Then the retailer sells them to you, the consumer. Without wholesalers, the modern economy would not work efficiently.

PART II GOVERNMENT

citizenship *n., membership in a state or a nation, with full rights and responsibilities.* A citizen of a country owes loyalty to that country. He or she is expected to perform certain duties, such as obeying the laws and paying taxes. In return, a citizen has certain rights, such as the right to vote. Civic participation is the way in which citizens participate in their government and society. Citizens let their representatives know what they think about important issues and participate in other ways.

Responsibilities of U.S. Citizens

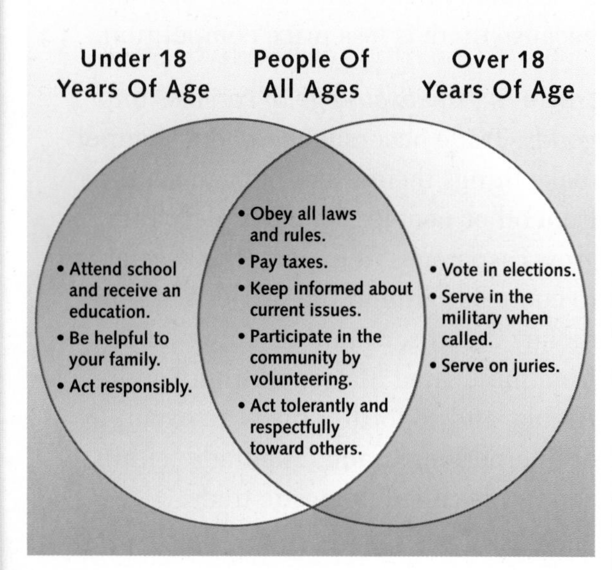

communism *n., an economic and political system in which the government owns and controls economic activities.* A communist government exerts great power over its people. This type of government is called totalitarian because it totally controls its people's lives. In the Soviet Union (1917–1991), the Communist Party outlawed all other parties, and the population had no freedom of speech, religion, or other rights.

constitution *n., a statement that explains the basic principles and rules of an organization.* The constitution of a government explains how leaders are selected, how laws are passed, and how laws are interpreted and enforced. The U.S. Constitution was ratified, or agreed to, by the states in 1789. In 1791, the states approved of the first 10 amendments, which are known as the Bill of Rights. These amendments guarantee rights, such as the right to speak freely. The U.S. Constitution can be changed, but doing so requires significant majorities to propose amendments, and an even greater majority of states to ratifiy the change.

democracy *n., a form of government in which the citizens of a nation or state hold the power to pass laws and select leaders.* The United States is a representative democracy in which people elect fellow citizens to serve in a legislature and pass laws. They also elect their leaders, such as the President of the United States. Representative democracies are also known as democratic republics or federal republics (see definition on this page). Democracy had its origins in Greek city-states.

dictatorship *n., a form of government in which a ruler holds total power.* A dictator usually comes to power through force. One of the most brutal dictators was Adolf Hitler of Nazi Germany, who ruled from 1933 to 1945. He created the policy that led to the Holocaust, or the murder of 6 million Jews and other people during World War II.

Adolf Hitler

executive branch *n., the branch of the federal government that implements and enforces laws.* In the United States, the president is the leader of the executive branch. He or she is the commander-in-chief of the armed forces. The president signs bills from Congress into law. The executive branch then makes sure that the laws are applied. To help enforce the laws, the president has a cabinet, and each member of the cabinet is responsible for a specific area of the federal government.

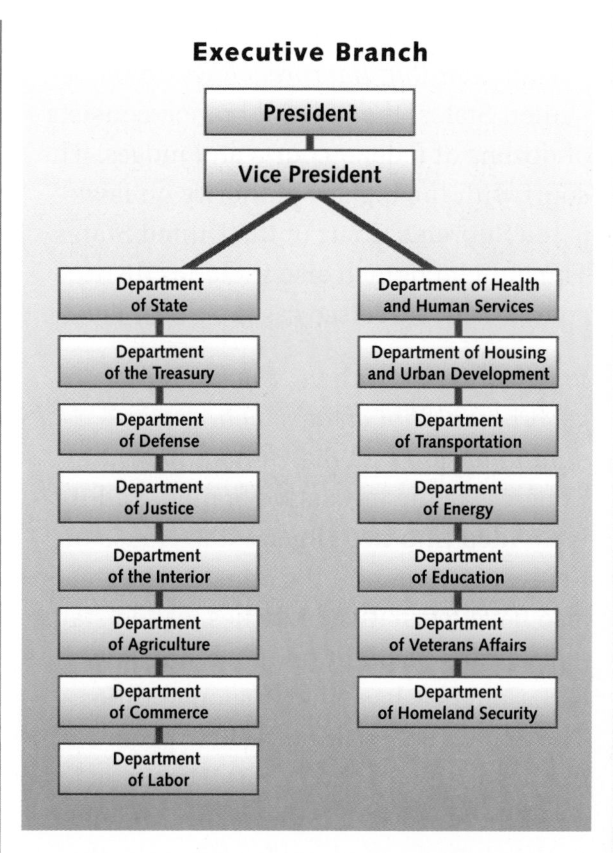

fascism *n., a form of government that has a dictator and controls political, economic, and other activities.* The two most well-known examples of fascism in the 20th century were Italy under Benito Mussolini (1922–1943) and Germany under Adolf Hitler (1933–1945). Mussolini came to power in Italy in 1922. He promised to solve the economic crisis that Italy faced after World War I. He soon became the country's dictator. Fascism in Italy ended in 1943 when the country surrendered during World War II.

federal republic *n., a form of government in which the population elects representatives to pass and carry out laws.* Many Western countries, including the United States and Canada, are federal republics. In federalism, a national, or federal, government shares power with states or provinces. The federal government has certain powers, such as national defense. The states or provinces control other activities, such as education.

judicial branch *n., the branch of government that interprets laws.* In the United States, the judicial branch consists of dozens of federal courts and judges. The court with the highest authority on laws is the Supreme Court of the United States. The judicial branch also includes district courts, which oversee cases at local levels.

legislative branch *n., the branch of government that creates, changes, or eliminates laws.* In the United States, Congress is the legislative branch, and it is divided into two Houses: the House of Representatives and the Senate. The states also have legislative branches, called state legislatures. In addition to passing laws, the U.S. Congress has other responsibilities. The Senate, for example, must approve of treaties with other countries. Democracies around the world have legislative branches. In the United Kingdom, for example, the Parliament passes laws.

Parliament, United Kingdom

limited government; unlimited government *n., a limited government places limits on governmental powers. An unlimited government does not place limits on governmental powers.* The United States has a limited government. The Constitution places limits on its powers. For example, the First Amendment states that the government does not have the power to limit people's free speech.

In unlimited government, the power of the government is not limited by a constitution or other document. Absolute monarchies and dictatorships are unlimited governments. From 1917 until 1991, the Soviet Union had an unlimited government.

monarchy *n., a form of government in which a king or a queen rules a country.* A monarch inherits the throne, or position, from his or her parent. Until the 1700s, monarchy was a common form of government. Britain, France, and Russia all had monarchs. In Russia, kings were known as czars. Then, in 1789, the French Revolution limited the powers of the French king, King Louis XVI. In 1793, the revolutionary government executed Louis and formed a republic. Through the 1800s and early 1900s, governments around the world became more democratic. They either abolished the monarchies completely or limited the powers of kings and queens.

oligarchy *n., a kind of government in which a small group of people holds power.* In an oligarchy, the rulers usually are either the military or the wealthy classes. Oligarchies usually rule for their own interests, not for the interests of people in lower classes. Myanmar (also known as Burma) is an oligarchy. It is ruled by a small group of military officers.

totalitarian *adj., a form of government in which a dictator or small group holds total control over the lives of the people in a country.* In a totalitarian country, the party in power is usually intolerant of other viewpoints and ideas. For example, in the 1930s, the Nazi Party rose to power in Germany by promising to bring back prosperity and return Germany to glory. The Nazi Party shut down all other parties, stating they were obstacles to these goals.

From the beginning of human history, people have asked questions about what life means. What will happen after I die? What is the right way to act? Religion helps people find answers to questions like those.

World Religions OVERVIEW

A religion is an organized system of beliefs and practices. Thousands of religions exist in the world. Most teach that one or more gods, or supreme powers, exist. To help people relate to the divine power, most religions teach a set of beliefs and a moral or ethical code. These codes of conduct also teach people how to treat each other. Another way that religion affects human relationships is in the building of community. Groups of people who share beliefs often gather together to worship and celebrate important events. For a map showing the distribution of the world's religions, see Chapter 2.

WORLD POPULATION'S RELIGIOUS AFFILIATIONS

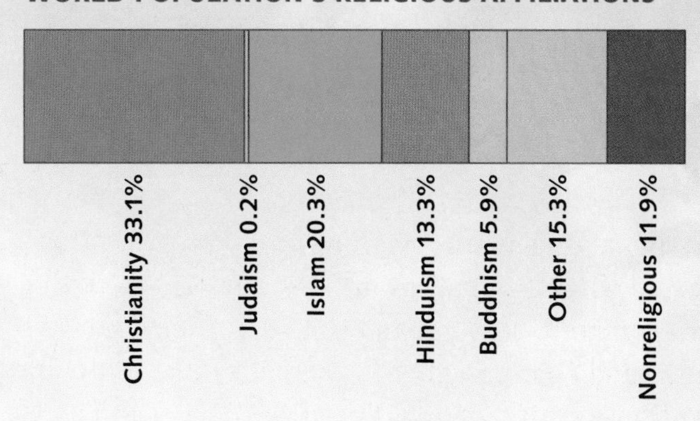

Source: Encyclopedia Brittanica

CHRISTIANITY

Historical Origins

Christianity is based on the life and teachings of Jesus of Nazareth, also called Jesus Christ by Christians. Christians believe he was the son of God who died to save humanity from sin. He was a Jew who lived in the first century near Jerusalem, which was then part of the Roman Empire. Roman rulers put him to death fearing he might lead a revolt. The life of Jesus is recorded in the New Testament of the Christian Bible, which also contains stories of his followers and letters outlining Christian beliefs.

Detail of Jesus in Leonardo DaVinci's *The Last Supper*

Central Beliefs

Most Christians believe there is only one God, who exists in three forms: Father, Son, and Holy Spirit. The New Testament teaches that after Jesus was executed, he rose from the dead and ascended to heaven. Christians believe they gain salvation by believing in Jesus and following his teachings. Christians gather at places of worship called churches. Their religious leaders are called either priests or ministers.

Spread of Christianity

Jesus' followers, called disciples, carried their faith around the Mediterranean world. In the A.D. 300s, Christianity became the official religion of the Roman Empire. Later, during the period of colonization, Europeans spread Christianity around the globe. It is now the largest and most widespread religion.

Christians wave palm fronds in Managua, Nicaragua, as they celebrate Palm Sunday, the first day of Holy Week.

ISLAM

Historical Origins

Islam teaches that in the year 610, an Arab trader named Muhammad was visited by the angel Gabriel. The angel told Muhammad that he was God's messenger. Muslims believe that through a series of these visits, Muhammad received the words of the Qur'an, or sacred book. According to Islam, Muhammad was the last prophet that Allah, the Muslim name for God, sent to humanity. Muslims believe he is a direct descendant of Abraham, who is also the founder of Judaism.

A 17th-century Turkish ceramic tile from a mosque

Central Beliefs

Muslims, or followers of Islam, believe there is only one God, the same God worshipped by Jews and Christians. The word *Islam* means surrender, and the goal of Islam is to surrender to the will of Allah. Muslims do this by practicing the Five Pillars of Islam. These are professing faith, praying five times a day, giving to charity, fasting, and making a journey to Mecca to reenact Abraham's dedication. Muslims worship in mosques.

Spread of Islam

In the centuries after Muhammad's death, Muslims spread their religion by conquest. Islamic rulers took control of Southwest Asia, Central Asia, North Africa, and parts of India and Spain. Today Islam continues to spread around the world through migration and conversion. It is the world's second largest religion.

Muslims engage in prayer at a mosque in Delhi, India, during Ramadan, a holy month of fasting.

JUDAISM

Historical Origins

Judaism, the religion of the Jewish people, dates back more than 4,000 years. Its founder was Abraham, who lived in Mesopotamia. According to the Hebrew Bible, God told Abraham to move to Canaan in present-day Israel and Lebanon. God made an agreement with Abraham to bless his descendants. They later became known as Hebrews or Israelites. The Hebrew Bible contains books of law, history, and prophecy. Another important work is the Talmud, a collection of scholarly writings.

Moses, a descendant of Abraham, is rescued from the Nile.

Central Beliefs

Judaism was the first major religion to teach monotheism, or the belief in one god. Jews believe God is the creator of the whole universe and has given them special responsibilities. Jews are to live holy lives by treating others well and pursuing justice. Today, Jews worship in synagogues, and their leaders are called rabbis.

Spread of Judaism

For centuries, Judaism was practiced primarily in what is present-day Israel. Several times in history, empires conquered the region and drove many Jews from the area. The last major event occurred in A.D. 135 when Rome punished Jewish rebels attempting to regain independence. As Jews spread out around the world, Judaism spread with them.

A worshipper holds up a Torah scroll during a Passover blessing at the Western Wall in Jerusalem, Israel.

BUDDHISM

Historical Origins

Buddhism is based on the teachings of Siddhartha Gautama, known as the Buddha or "enlightened one." He was born a prince in India in the 400s or 500s B.C. Siddhartha was protected from seeing sickness, death, poverty, or old age until he was 29. However, after he learned about suffering, he left his palace to lead a religious life. While he was meditating several years later, Buddhists believe he received enlightenment about the meaning of life.

A 14th-century painting of Buddha

Central Beliefs

Buddhists believe that a law of cause and effect called karma controls the universe. Buddhists teach that suffering occurs because people desire what they do not have. A person who gives up desire and other negative emotions will achieve a state called nirvana, or the end of suffering. The basic beliefs of Buddhism are summarized in the Four Noble Truths. The actions that help people achieve nirvana are called the Eightfold Path.

FOUR NOBLE TRUTHS

- Suffering is a part of life.
- Selfishness is the cause of suffering.
- It is possible to move beyond suffering.
- There is a path that leads to the end of suffering.

Spread of Buddhism

During its first century, Buddhism spread across northern India. Over time, missionaries and travelers carried it to the Himalayas, Central Asia, and China. China spread Buddhism to Japan and Korea. In the 1800s, immigrants introduced Buddhism to the United States. In the late 20th century, the religion gained popularity in the United States and other Western countries.

Buddhist monks in Siem Reap, Cambodia, celebrate the birthday of Buddha.

HINDUISM

Historical Origins

Hinduism, one of the world's oldest religions, originated in India in about 1500 B.C. Scholars believe that it developed from the beliefs of a group of Indo-European people who spoke Sanskrit. The sacred writings of Hindus include the Vedas, which are poems and hymns, and the Puranas, which are sacred stories. Other Hindu texts such as the *Mahabharata*, of which the Bhagavad Gita is a part, teach Hindu beliefs in the form of epic poems.

Scene from the Bhagavad Gita, a sacred Hindu text

Central Beliefs

In general, Hindus believe in one eternal force called Brahman. This divine spirit takes the form of many gods and goddesses. The most important deities are Brahma, the creator; Vishnu, the preserver; and Shiva, the destroyer. Hindus believe that souls are constantly being reborn. Karma, the negative or positive effect of one's actions, determines if the soul moves to a higher or lower state of being. Hindu religious practice includes worship, study, and rituals such as bathing in the Ganges River.

Spread of Hinduism

Hinduism spread from India through parts of Southeast Asia, but now is practiced by few people in that region. In general, Hinduism has remained mainly a religion of the Indian people. Nearly 80 percent of Indians are Hindus. Indian immigrants have brought Hinduism to the United States.

Women light candles in observance of Diwali, the Hindu Festival of Lights, in Jaipur, India.

SIKHISM

Historical Origins

Followers of Sikhism are called Sikhs, which means "learner." A teacher named Guru Nanak established the religion in India in the late 1400s. After his death, a line of nine other teachers, or gurus, followed him. Sikhs believe that all ten gurus were inspired by a single spirit. They also believe that after the tenth guru died, this spirit inhabited the Sikhs' sacred scripture, the Guru Granth Sahib or Adi Granth.

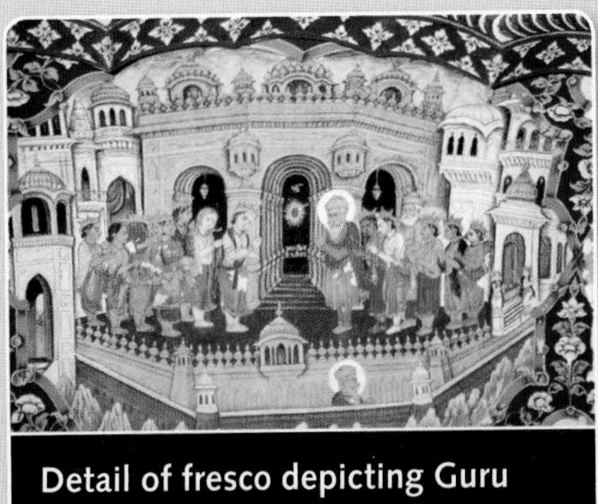

Detail of fresco depicting Guru Nanak Sikh

Central Beliefs

Sikhs believe in one god who does not take physical form. Like Hindus, they believe in reincarnation—the rebirth of the soul. The goal of Sikhism is to form a close, loving relationship with God. Sikh practices include prayer several times a day, worship, and meditation. Sikhs do not use tobacco or alcohol, and they often follow a strict dress code, which includes never cutting their hair.

Spread of Sikhism

Sikhism is practiced by nearly 25 million people, most of whom live in the Punjab region of northwest India. However, immigrants have introduced the religion to Western countries, including the United States, Canada, and the United Kingdom.

A Sikh pilgrim visits the Golden Temple—Sikhism's holiest site—in Amritsar, India.

CONFUCIANISM

Historical Origins

Confucianism is an ethical system and philosophy based on the teachings of a Chinese public official and teacher named Kongfuzi. He is called Confucius in Western countries. He lived from 551 to 479 B.C. His goals were to revive traditional values and establish education as a way to improve society. The sayings and writings of Confucius were collected in a work called the *Analects*.

A 12th-century Chinese painting portrays Confucian filial piety.

Central Beliefs

One of the most important things in Confucianism is the concept of filial piety, in which children obey and honor their parents. Confucius applied this idea to other areas. For example, subjects should obey their rulers. Confucius taught that education, right relationships, and moral behavior would create an orderly society.

Spread of Confucianism

Confucianism became a way of life in China. Because China had a major influence on East Asia and Southeast Asia, Confucianism spread throughout these regions. It continues to be a strong cultural force in China and other Asian countries such as South Korea, Japan, and Singapore.

Celebrants participate in a festival in Qufu City, Shandong Province, China, the birthplace of Confucius.

ONGOING ASSESSMENT

RESEARCH LAB

GeoJournal

Express Ideas Through Speech Get together in a group and prepare a panel discussion in which you analyze what the major world religions have in common.

Step 1 Appoint each person in your group to become an expert on a different religion. Each person should review the pages in the text that cover their assigned religion.

Step 2 Each student should summarize the religion they studied. As a group, identify characteristics that all or most of the religions and ethical systems have in common.

Step 3 Hold a panel discussion before the class and present your conclusions. After you have finished, allow time for questions.

A

abolition *n.*, the movement to end slavery, before and during the Civil War in the United States

absolute location *n.*, the exact point where a place is located, identified by latitude and longitude coordinates

abstract *adj.*, an artistic style that stresses form and color over realism

acknowledge *v.*, to recognize

Acropolis *n.*, a rocky hill in Athens, Greece, that served as a fortress for the ancient city's most important buildings

adapt *v.*, to adjust or modify to fit

aging population *n.*, a demographic trend that occurs as the average age of a population rises

Alamo *n.*, a fort in Texas where 200 Texans lost a battle during the U.S.-Mexican War

Alexander the Great *n.*, the conqueror who extended the Macedonian Empire and spread Greek culture throughout Eurasia from 334–323 B.C.

alliance *n.*, a partnership between countries

Alps *n.*, a European mountain chain

amendment *n.*, a formal change to a law

ancestry *n.*, the family one is descended from, going back in time; heritage

Amazon River Basin *n.*, located in South America, the largest river basin on Earth

Andes Mountains *n.*, mountain range extending 5,500 miles along the western side of South America

Angel Falls *n.*, the highest waterfall in the world, located in Venezuela

annexation *n.*, the adding of territory to a country

apartheid (uh PAHRT hyt) *n.*, the legal separation of the races; a system that denied black South Africans their rights

aqueduct *n.*, a transport system for carrying water long distances, sometimes raised on a bridge

aquifer *n.*, a layer of rock beneath Earth that contains water

Aral Sea *n.*, a saltwater lake in Central Asia that has been greatly reduced in size due to diversion for irrigation of the rivers that flow into it

archipelago (ahr kuh PEH lug goh) *n.*, a chain of islands

arid *adj.*, very dry, having almost no rainfall

aristocrat *n.*, a member of the upper class

artifact *n.*, an object made by humans from a past culture

Aryans (AIR ee uhnz) *n.*, nomads who migrated from Central Asia into the Indus Valley around 2000 B.C.

assimilate *v.*, to be absorbed into a society's culture

Atacama Desert *n.*, a desert located on the western side of the Andes Mountains

Atahualpa (AH tah WAHL pah) *n.*, the last emperor of the Incas before they were conquered by the Spanish in 1533

Augustus *n.*, the first emperor of Rome in 27 B.C.

autonomy *n.*, a country or people's self-governance

Aymara (eye MAHR uh) *n.*, an indigenous culture who lives in the Andes Mountains of Peru and Bolivia

Aztec *n.*, a people who settled in the area of modern Mexico City, A.D. 1325–1525

B

barbarian *n.*, a soldier or warrior considered to be culturally less developed than those being fought; the term originally comes from the German tribes who invaded the Roman Empire in A.D. 235

Baroque period *n.*, a period from 1600–1750 in which music had complicated patterns and themes

bay *n.*, a body of water surrounded on three sides by land

benefit *v.*, to be useful to

Berlin Wall *n.*, wall that divided communist East Berlin from democratic West Berlin; it was torn down in 1989

Bill of Rights *n.*, the first ten amendments to the U.S. Constitution

biodiversity *n.*, the variety of species in an ecosystem

biofuel *n.*, an alternative fuel that is a mixture of ethanol and gasoline

Black Sea *n.*, an inland sea bounded by Europe, Russia, Georgia, and Turkey

Bolshevik *n.*, a political party in Russia led by Lenin that overthrew the czar in 1917

butte *n.*, a hill or mountain with steep sides and a flat top

C

Caesar, Julius *n.*, a general who became the ruler of Rome, 46–44 B.C.

Calypso *n.*, a type of folk music that started in Trinidad

canal *n.*, a human-made waterway through land for boats and ships

Candomblé *n.*, a Brazilian religion that combines African spiritualism and Catholicism

canopy *n.*, a roof over a rain forest created by treetops

capital *n.*, a country's wealth and infrastructure

Caribbean Sea *n.*, a tropical sea in the Western Hemisphere bounded by Mexico, Central America, the Greater Antilles, and the Lesser Antilles

cartographer *n.*, a mapmaker

cash crop *n.*, a farm crop sold for profit

Caspian Sea *n.*, the largest enclosed body of water on Earth, bounded by Russia, Kazakhstan, Turkmenistan, Azerbaijan, and Iran

categorize *v.*, to group information; to classify

Catherine the Great *n.*, the empress of Russia from 1762 to 1796

cenote (se NO tay) *n.*, an underground pool of water

Chernobyl *n.*, a city in the Ukraine in which a nuclear reactor exploded in 1986

Christianity *n.*, a religion based on the life and teachings of Jesus of Nazareth

citizen *n.*, a person living within a territory who has rights and responsibilities granted by the government

city-state *n.*, an independent state made up of a city and the territories depending on it

civil war *n.*, a war between opposing groups of citizens in the same country

civilization *n.*, a society with a highly developed culture, politics, and technology

Classical period *n.*, a period from 1750 to 1900 in which music followed the standard rules of form and complexity, such as sonatas or symphonies

climate *n.*, the average condition of the atmosphere of an area over a long period of time, including temperature, precipitation, and seasonal changes

climograph *n.*, a graph showing a region's climate through average precipitation and temperature

coastal plains *n.*, the lowlands next to the seacoast

Cold War *n.*, a long period of political tension without fighting between the United States and the Soviet Union, roughly from 1948 to 1991

collective farm *n.*, in the Soviet Union, a large farm where workers grew food to be distributed to the entire population.

colonize *v.*, to build settlements and develop trade in lands that a country controls

colony *n.*, an area controlled by a distant country

Columbian Exchange *n.*, the exchange of plants, animals, and disease between the Americas, Europe, and Africa that began in the 1500s

commercial agriculture *n.*, the business of producing crops to sell

Common Market *n.*, the European Economic Community formed in 1957

commonwealth *n.*, a nation that governs itself but is part of a larger country

communal *adj.*, shared

communism *n.*, a system of government in which a single political party controls the government and the economy

concentration camp *n.*, an area where Jews and others were held during World War II and murdered by the Nazis

condensation *n.*, the process of water vapor turning into liquid droplets due to cooling in the hydrologic cycle

conquistador *n.*, a Spanish soldier and leader that explored the newly discovered territories of North America, South America, Central America, and the Caribbean beginning in the 1500s, primarily in search of gold and silver

Constantine *n.*, Roman Emperor, 306–337 B.C., who made Christianity the official religion of the empire

constitution *n.*, a document that organizes a government and states its powers

consumer *n.*, a person who buys goods

contaminated *adj.*, infected; unfit for use because of the presence of unsafe elements

contiguous *adj.*, connected in one block

continent *n.*, a large landmass on Earth's surface; Earth has seven continents

continental drift *n.*, the slow movement of continents on tectonic plates

continental shelf *n.*, the edge of a continent that extends under water into the sea

convert *v.*, to persuade someone to change religious beliefs

cordillera *n.*, a system of several parallel mountain ranges

Cortés, Hernán *n.*, the Spanish conquistador who defeated the Aztecs in 1525

cosmopolitan *adj.*, bringing together many different cultures and influences

Counter-Reformation *n.*, a movement within the Roman Catholic Church to reform its own practices

coup (KOO) *n.*, a sudden, illegal takeover of government by force

Creole *n.*, a blended language of European and non-European languages

critical *adj.*, extremely important, needed for survival

Crusades *n.*, military expeditions of the Roman Catholic Church to take back holy lands in the Middle East from Muslim control, 1096–1291

cuisine *n.*, the food and cooking traditions common to a certain region

culture *n.*, a group's way of life, including types of food, shelter, clothing, language, behavior, and ideas

culture region *n.*, an area that is unified by common cultural traits

currency *n.*, a form of money

current *n.*, the continuous movement of air or water in the same direction

cyclone *n.*, a storm with rotating winds, called a typhoon in the Eastern Hemisphere; a hurricane in other parts of the world

czar *n.*, term used for an emperor in Russia

D

dam *n.*, a barrier that controls the flow of water

Danube River *n.*, a river that starts in Germany and ends at the Black Sea

Declaration of Independence *n.*, the document declaring U.S. independence from the British Empire, adopted July 4, 1776

deforestation *n.*, the practice of cutting down forests for crops or urban use

democracy *n.*, a form of government by the people, in which citizens often elect representatives to govern them

democratization *n.*, the process of becoming a democracy

demographics *n.*, the characteristics of a human population, such as age, income, and education

descendants *n.*, future generations; relatives of a past family member

dialect *n.*, a regional variation of a main language

dictator *n.*, a ruler with complete control

diplomacy *n.*, discussion between groups or countries to resolve disputes or disagreements

displace *v.*, to force a people to leave their homes

distinct *adj.*, easily recognized

distort *v.*, to change the usual shape or appearance; twist out of shape

diversify *v.*, to give variety to

diversity *n.*, variety

drought *n.*, a long period with little or no precipitation

Dry Pampas *n.*, dry grasslands in western Argentina

due process *n.*, in the United States, rules that authorities must follow in dealing with citizens

dynamic *adj.*, continuously changing

E

earthquake *n.*, a shaking of Earth's crust generally caused by the collision or sliding of tectonic plates

Eastern Hemisphere *n.*, the half of Earth east of the prime meridian

economic sector *n.*, a subdivision or smaller part of an economy, such as industry and agriculture

economy *n.*, a system in which people produce, sell, and buy things

ecosystem *n.*, a community of living organisms and their natural environment, or habitat

ecotourism *n.*, a way of visiting natural areas that conserves the natural resources of the region

El Niño (ehl NEEN yoh) *n.*, a reversal of usual wind and ocean currents

elevation *n.*, the height of a physical feature above sea level

eliminate *v.*, to get rid of

Emancipation Proclamation *n.*, an 1863 document that freed all slaves living in Confederate held territory during the U.S. Civil War

empire *n.*, a group of peoples or states ruled by a strong single ruler

Enlightenment *n.*, a social movement of the 1700s that worked for the education and rights of the individual

enlist *v.*, to volunteer for military service

entrepreneur *n.*, a person who starts up a new business

entrepreneurship *n.*, the characteristics of creativity and risk existing in a person or society

epic poem *n.*, a long poem that tells of a hero's adventures

epidemic *n.*, an outbreak of a disease affecting a great number of the population in a particular community

equator *n.*, an imaginary circle around Earth that is the same distance from the North and South Poles and divides Earth in half; the center or 0° line of latitude

equinox *n.*, when day and night are of equal length; occurs twice each year on March 21 or 22 and September 21 or 22

erosion *n.*, the process by which rocks and soil slowly break apart and are worn away

erratic *adj.*, inconsistent or irregular

essential *adj.*, necessary

ethanol *n.*, a liquid alcohol removed from sugarcane or corn that can be used as a fuel alone or blended with gasoline

euro *n.*, the common currency of the European Union

European Union (EU) *n.*, an economic organization composed of 27 European member countries (2011)

eurozone *n.*, the countries that have adopted the euro as money

evaporation *n.*, the process of water turning into vapor and rising into the atmosphere due to the sun's heating; part of the hydrologic cycle

excavate *v.*, to carefully uncover or dig up

exchange *v.*, to convert money into another currency

exile *n.*, a state of absence from one's home country

exile *v.*, to force to leave a country

expand *v.*, to make larger

exploit *v.*, to take advantage of; to use selfishly or unfairly for one's own profit

export revenue *n.*, money or income received for goods sold to another country

export *v.*, to send to another country for aid or profit

extinction *n.*, the dying out of a species or type of living thing

F

factory system *n.*, a way of working in which each person works on only one part of a product

fallout *n.*, radioactive particles from a nuclear explosion that fall through the atmosphere

federal system *n.*, a system of government with a strong central government and local government units

fertile *adj.*, able to produce a great deal of fruit, crops, or offspring

fertilizer *n.*, a substance added to soil to enrich it

feudal system *n.*, a social structure during the Middle Ages consisting of a king, lords, vassals, and serfs

fiber optics *n.*, glass fibers used to send digital code quickly across great distances

fjord (fee ORD) *n.*, a deep, narrow bay

food security *n.*, easy access to enough food

foremost *adj.*, first in rank or importance, leading

fortify *v.*, to strengthen

free enterprise economy *n.*, a system in which privately-owned businesses create goods and services; also called a market economy or capitalism

fuel cell *n.*, a small unit that, like a battery, combines chemicals to make energy

fuse *v.*, to blend

G

Gadsden Purchase *n.*, a sale in 1853 from Mexico to the United States that established the current U.S. southwestern border

gaucho *n.*, a cowboy of South America

gauge *n.*, the measurement of the width of railroad tracks

Genghis Khan (JEHNG-gihs KAHN) *n.*, the Mongol ruler who established an empire in Central Asia in the early 1200s

genre *n.*, a form of literature such as poem, play, or novel

geoglyph *n.*, a large geometric design or animal shape drawn on the ground

Geographic Information Systems (GIS) *n.*, computer-based devices that show data about specific locations

geographic pattern *n.*, a similarity among places

Gettysburg Address *n.*, the speech Abraham Lincoln made in 1863 honoring soldiers who died at the Battle of Gettysburg during the U.S. Civil War

glacier *n.*, a large mass of ice and packed snow

glasnost (GLAHS-nuhst) *n.*, the Soviet Union's policy of openness that encouraged people to speak openly about government, introduced by Mikhail Gorbachev in the 1980s

global *adj.*, worldwide

Global Positioning System (GPS) *n.*, a satellite system based in space that finds absolute location and time anywhere on Earth

global warming *n.*, the increase in Earth's average temperature since the mid-20th century

globalization *n.*, the development of a world economy based on free trade and the use of foreign labor

globe *n.*, a three-dimensional, or spherical, model of Earth

golden age *n.*, a period of great wealth, culture, and democracy in Greece

Gorbachev, Mikhail (mih KYL GAWR buh chawf) *n.*, a leader in the Soviet Union 1985–1991; President from 1990 until it was dissolved in 1991

government *n.*, an organization that keeps order, sets rules, and provides services for a society

Grand Canyon *n.*, rock formation in southwestern United States that has been cut deeply by the Colorado River over millions of years

grasslands *n.*, wide open areas used for grazing and crops

Great Depression *n.*, worldwide economic downturn in the 1930s, marked by poverty and high unemployment

Great Lakes *n.*, five large freshwater lakes between Canada and the United States

Great Plains *n.*, a flat area of land east of the Rocky Mountains

Great Recession *n.*, a downturn in the economies of the United States, Canada, and other countries, beginning in 2007

greenhouse gas *n.*, a gas that traps the sun's heat over Earth

gross domestic product (GDP) *n.*, the total value of all goods and services produced in a country in a given year

Guaraní (GWAH rah NEE) *n.*, an indigenous people who live in the lowlands on the Paraguay and Paraná rivers in South America

guillotine (GHEE uh teen) *n.*, a machine used to execute people during the French Revolution

Gutenberg, Johannes *n.*, a German printer who invented the printing press in 1450

H

habitat *n.*, the natural environment of a living plant or animal

half-life *n.*, the time needed for half the atoms in a radioactive substance to decay

harbor *n.*, a place where ships can land protected from the open sea

hemisphere *n.*, one-half of Earth

heritage *n.*, a tradition passed down from ancestors

Hermitage Museum *n.*, a museum of art and culture in St. Petersburg, Russia

Hidalgo, Miguel *n.*, a Catholic priest who led a revolt in Mexico in 1810

hieroglyphics (HY ruh GLIHF ihks) *n.*, an ancient system of writing that uses pictures and symbols

Hitler, Adolf *n.*, German head of state from 1933 to 1945

Holocaust *n.*, the mass slaughter by the Nazis of six million Jews and others during World War II

hotspot *n.*, an unusually hot part of Earth's mantle

Human Development Index (HDI) *n.*, a set of data used by geographers to compare quality of life in different countries including health, education, and standard of living

human rights *n.*, the political, economic, and cultural rights that all people should have

hunter-gatherer *n.*, a person who hunts animals and gathers plants and fruits for food

hurricane *n.*, a strong storm with swirling winds and heavy rainfall

hybrid *n.*, a vehicle that can use either electricity or gas to run

hydroelectric power *n.*, a source of energy that uses flowing water to produce electricity

I

immigrant *n.*, a person who takes up permanent residence in another country

immigrate *v.*, to move to a new country or region

impact *n.*, an effect that produces change

Impressionism *n.*, an artistic style in which artists used light and color in short strokes to capture a moment in time

incorporate *v.*, to include; to combine with something already formed

indigenous *adj.*, native to the area in which something is found

indulgence *n.*, a fee paid to the church to relax the penalty for a sin during the Middle Ages

Industrial Revolution *n.*, a period in the 1700s and 1800s in which workers in factories began to use machines and power tools for large-scale industry

industrialization *n.*, the shift to large-scale production using machines

infrastructure *n.*, the basic systems of a society such as roads, bridges, sewers, and electricity

Institutional Revolutionary Party (PRI) *n.*, a political party that controlled Mexico's government from 1929 to 2000

Internet *n.*, a communications network

intersection *n.*, a place where people meet or paths cross

invader *n.*, an enemy who enters a country by force

Iron Curtain *n.*, an imaginary boundary that separated Communist and non-Communist countries in Europe during the Cold War

isolated *adj.*, cut off from others

isthmus *n.*, a narrow strip connecting two large land areas

K

Kievan Rus *n.*, the state established by the Varangian Russes in 882 that became part of modern Russia

kimono *n.*, traditional Japanese women's clothing

kinship *n.*, blood or family relationship

Kremlin *n.*, a historic complex of palaces, armories, and churches in Moscow and the seat of Russian government

L

land reform *n.*, the breaking up of large estates to give land to the poor

landmass *n.*, a very large area of land

language family *n.*, a group of related languages

latitude *n.*, an imaginary line around Earth that runs east to west, showing location relative to the equator

Lenin, V.I. *n.*, the Bolshevik leader who overthrew the czar in the Russian Revolution of 1917 and led the new government

liberate *v.*, to set someone or something free

Llanos *n.*, grasslands in northern South America

Locke, John *n.*, an English philosopher of the late 1600s who helped inspire the American Revolution and the Enlightenment

locks *n.*, devices in canals used to raise or lower ships between the waterways being connected

longitude *n.*, an imaginary line running north to south from the North Pole to the South Pole that shows location relative to the prime meridian

Louisiana Purchase *n.*, the land Thomas Jefferson bought in 1803, doubling the size of the United States

L'Ouverture, Toussaint (too SAN loh ver CHOOR) *n.*, a former slave who led Haiti's successful revolt for independence from France

lowland *n.*, a low-lying area

Luther, Martin *n.*, a German monk whose actions in 1517 led to the Reformation to address corruption in the Roman Catholic Church

M

Machu Picchu (MAH choo PEE choo) *n.*, a complex city built on a mountain by the Incas in the 1400s

malnutrition *n.*, the lack of enough food or nourishment

Manifest Destiny *n.*, the idea that the United States had the right to expand its territory to the Pacific Ocean

manufacturing *n.*, the use of machines to make raw materials into usable products

map *n.*, a two-dimensional, flat representation of Earth

marine life *n.*, the plants and animals living in the ocean

marine reserve *n.*, an ocean area set aside to protect ocean life from humans

marketing *n.*, advertising and promotion for a product or business

mass media *n.*, communication from a single source with the potential to reach large audiences

Maya *n.*, a civilization that lived in the Yucatán and northern Central America from 100 B.C. to A.D. 900

Mediterranean climate *n.*, a climate that has hot, dry summers and mild, rainy winters

megacity *n.*, a large city with more than 10 million people

Mesa Central *n.*, in Mexico, the southern area of the Mexican Plateau

mestizo *n.*, a person who has mixed European and Native American ancestry

methane *n.*, a colorless, odorless natural gas released from carbon

Mexican Cession *n.*, the land ranging from Texas to California given by Mexico to the United States in the 1848 Treaty of Guadalupe Hidalgo

Mexican Plateau *n.*, a flat area of land that lies between the two ranges of the Sierra Madre Mountains in Mexico

Middle Ages *n.*, a period in Western Europe after the fall of the Roman Empire, from about 500 to 1500

Middle Passage *n.*, the months-long trip across the Atlantic Ocean in which enslaved Africans were brought to European colonies in the Americas

migrate *v.*, to move from one place to another

missionary *n.*, a person sent by a religious organization to convert others to that religion

mobile *adj.*, movable

modify *v.*, to change, or make less extreme

Mongol Empire *n.*, an empire established in Central Asia in the early 1200s by Genghis Khan

monopoly *n.*, the complete control of the market for a service or product

monotheistic *adj.*, related to a religious belief in one god

monotheistic religion *n.*, a system of belief in one god or deity

Montezuma *n.*, an Aztec ruler who was killed by the Spanish conquistador Hernán Cortés

mouth *n.*, the place where a river empties into a sea

multi-party democracy *n.*, a political system in which elections include candidates from more than one party

multitudes *n.*, large numbers

Mundurukú (moon doo ROO koo) *n.*, an indigenous people of Brazil

mural *n.*, a large painting on a wall

N

Napoleon *n.*, Napoleon Bonaparte, the leader of France who conquered other European countries to build an empire, 1804–1815

National Action Party (PAN) *n.*, a political party in Mexico that won election in 2000

nationalism *n.*, a strong sense of loyalty to one's country

nationalize *v.*, to place a private industry under government control

natural rights *n.*, rights such as life, liberty, and property that people possess at birth

naturalization *n.*, the process that someone born in another country follows to become a citizen

navigable *adj.*, wide or deep enough to be traveled easily by boats and ships

navigation *n.*, the science of finding position and planning routes

Nazi Germany *n.*, Germany as led by the Nazi party from 1933–1945

neutrality *n.*, the refusal to take sides or become involved

nomad *n.*, a person who moves from place to place

nonrenewable fossil fuel *n.*, a source of energy such as oil, natural gas, or coal that is in limited supply

nonrenewable resource *n.*, a source of energy that is limited and cannot be replaced, such as oil

North American Free Trade Agreement (NAFTA) *n.*, a 1994 agreement that made trade and investment easier among Canada, Mexico, and the United States

North Atlantic Drift *n.*, a warm ocean current that warms the waters around the northwest part of Russia

North Pole *n.*, the northernmost point on Earth, opposite the South Pole, where all lines of longitude meet

Northern European Plain *n.*, a vast lowland that stretches from France to Russia

Northern Hemisphere *n.*, the half of Earth north of the equator

novel *n.*, a long work of fiction with complex characters and plot

Olmec *n.*, an organized society who lived along the southern coast of the Gulf of Mexico in 1000 B.C.

opera *n.*, a performance that tells a story through words and music

oppose *v.*, to object to

Orange Revolution *n.*, the Ukraine's peaceful removal of its prime minister in 2004

Pampas *n.*, the grassy plain in Argentina

Panama Canal Zone *n.*, the area in which the Panama Canal was built

parliamentary democracy *n.*, a government system in which the chief executive is the prime minister, chosen by the party with the most seats in Parliament

patrician *n.*, a wealthy landowner in ancient Rome

Pearl Harbor *n.*, a U.S. naval base in Hawaii that the Japanese bombed in 1941 and that brought about U.S. entry into World War II

peat *n.*, the material from very old decayed plants that burns like coal

peninsula *n.*, a body of land surrounded on three sides by water

perestroika (pehr ih STORY kuh) *n.*, reform in the structure of the economy introduced by Mikhail Gorbachev in the Soviet Union in 1985

permafrost *n.*, permanently frozen ground

perspective *n.* an artistic way of showing objects as they appear to people in terms of relative distance or depth, as if in three dimensions

pesticide *n.*, a chemical that kills harmful insects and weeds

Peter the Great *n.*, Peter Romanov, who ruled Russia as czar from 1682 to 1725

petroleum *n.*, raw material used to produce oil

philosopher *n.*, a person who examines questions about the universe and searches for the truth

pioneer *n.*, a settler of new land

pipeline *n.*, a series of connected pipes used to transport liquids or gases

Pizarro, Francisco *n.*, a Spanish conquistador who in 1533 overthrew the Incan emperor and founded the city of Lima, Peru

plain *n.*, a level area on Earth's surface

plantation *n.*, a large farm that grows crops for profit

plateau *n.*, a plain high above sea level that usually has a cliff on all sides

plebeian *n.*, a farmer or lower class person in ancient Rome

poacher *n.*, a person who hunts or fishes illegally

polder *n.*, land in the Netherlands that has been reclaimed from the sea for farming

policy *n.*, the official guidelines and procedures of an organization or government

polytheistic religion *n.*, a system of belief in many deities or gods

port *n.*, a harbor for ships where goods are exchanged

Port-au-Prince *n.*, the capital city of Haiti

precipitation *n.*, the process of water falling to Earth as rain, snow, or hail

predominant *adj.*, main, most common, superior to others

prehistoric *adj.*, before written history

prime meridian *n.*, the line of 0° longitude that runs from the North to the South Pole, and passes through Greenwich, England

privatization *n.*, the process of government-owned businesses becoming privately owned

profitable *adj.*, making money, financially successful

projection *n.*, a way of showing Earth's curved surface on a flat map

promote *v.*, to encourage

propaganda *n.*, information made to influence people's opinions or advance an organization or party's ideas

proportional representation *n.*, a system in which a political party gets the same percentage of seats as its percentage of votes

prosperous *adj.*, economically strong

protest *v.*, to object to

province *n.*, a smaller part of a larger country, especially in Canada

Q

Quechua (KEHCH wah) *n.*, a people who live in the Andes Mountains of Peru, Ecuador, and Bolivia

R

radical *n.*, a person who wants an extreme change or holds an extreme political position

radioactive *adj.*, giving off energy caused by the breakdown of atoms

rain forest *n.*, a forest with warm temperatures, high humidity, and thick vegetation that receives more than 100 inches of rain per year

rain shadow *n.*, a dry region on one side of a mountain range

rainshadow effect *n.*, process in which moist air rises up a mountain range and then cools and falls as precipitation, leaving the other side of the range mostly dry

raw materials *n.*, unfinished or natural materials such as minerals, oil, or coal used to make finished products

rebellion *n.*, a revolt or resistance to authority

recession *n.*, a slowdown in economic growth

Reconstruction *n.*, the effort to rebuild and reunite the United States following the Civil War

reform *n.*, change aimed at correcting a problem

Reformation *n.*, the movement in the 1500s to reform Christianity

region *n.*, a group of places with common traits

regulate *v.*, to control

reign *n.*, the period of rule for a king, queen, emperor, or empress

Reign of Terror *n.*, a movement in France led by Maximilien Robespierre in which 40,000 people were beheaded in 1793–94

relative location *n.*, the position of a place in relation to other places

relief *n.*, the change in elevation from one place to another

religious tolerance *n.*, the acceptance of different religions to be practiced at the same time, without prejudice

remittance *n.*, money sent to a person in another place

remote *adj.*, hard to reach, isolated

Renaissance *n.*, meaning "rebirth," a period in the 1300–1500s in which culture and the arts flourished

renewable resource *n.*, a raw material or energy source that replaces itself over time

reparation *n.*, after a war, money paid as punishment, usually by the aggressors in the conflict

republic *n.*, a form of government in which officials are elected by the people to govern

reserve *n.*, land set aside for a special purpose such as farming, preserving habitats, or housing specific groups of people; a future supply (of oil)

restore *v.*, to bring back

revenue *n.*, income

revolution *n.*, an action to overthrow a government by citizens or colonists

Rhine River *n.*, a river that starts in Switzerland and ends in the North Sea

Ring of Fire *n.*, an area along the rim of the Pacific Ocean where tectonic plates meet, causing many active volcanoes and earthquakes

Rio de Janeiro *n.*, a city in Brazil that will host the 2016 Olympic Games

Romantic Period *n.*, an artistic period in the early 1800s when artists painted landscapes and natural scenes to convey emotions

roots *n.*, cultural origins

Russian Revolution *n.*, the revolution in Russia in 1917 that overthrew the czar and put the Bolsheviks in power

Russification *n.*, the policy of putting Russians in charge of Soviet republics during the 1970s and 1980s

ruthless *adj.*, cruel

S

Santa Anna *n.*, the Mexican president and general who won the battle of the Alamo but lost the U.S.-Mexican War

São Paolo *n.*, the largest city in Brazil

saturate *v.*, to soak thoroughly

scale *n.*, the part of a map that indicates how big an area of Earth is shown

scarcity *n.*, a shortage of something

scorched earth policy *n.*, in 1812, practice in which Russian troops, as they retreated from Napoleon's army, burned crops and resources that could supply the enemy

secede *v.*, to formally withdraw

secular *adj.*, worldly, not connected to a religion

seismic *adj.*, having to do with earthquake activity or movement

semiarid *adj.*, somewhat dry, with very little rainfall

serf *n.*, from the 1500s to the 1800s, a poor Russian or European peasant farmer who rented land from a landlord and had few rights

Siberia *n.*, a huge region in central and eastern Russia

significant *adj.*, important

Silk Roads *n.*, ancient trade routes that connected Southwest and Central Asia with China

slash-and-burn *n.*, a farming method of clearing land by cutting down and burning forest and vegetation

Slavs *n.*, people who came from around the Black Sea or Poland and settled in the Ukraine and western Russia around A.D. 800

slum *n.*, an area in a city that is crowded, with poor housing and bad living conditions

smartphone *n.*, a handheld device that combines communication and software applications

socialism *n.*, a system of government in which the government controls economic resources

solstice *n.*, the point at which the sun is farthest north or farthest south of the Equator; the beginning of summer and winter

South Pole *n.*, the southernmost point on Earth, opposite the North Pole, where all lines of longitude meet

Southern Hemisphere *n.*, the half of Earth south of the equator

sovereignty *n.*, a country's control over its own affairs

Soviet Union *n.*, the Union of Soviet Socialist Republics, a country made up of Russia and other Eurasian states, from 1922 to 1991

soybean *n.*, a type of bean grown for food and industrial products

spatial thinking *n.*, a way of thinking about space on Earth's surface, including where places are located and why they are there

standard of living *n.*, the level of goods, services, and material comforts of people in a country

staple *n.*, a basic part of people's diets

state *n.*, a defined territory with its own government

steel *n.*, a strong metal made from iron combined with other metals

steppe *n.*, a very large plain of dry grassland

strike *n.*, a work stoppage by employees who refuse to work

subsistence farmers *n.*, farmers who grow food for their families to eat, not to sell

subsistence farming *n.*, farming to grow only enough food for families to eat, not to sell

surplus *n.*, extra

suspension bridge *n.*, a bridge that is hung from two or more cables

sustainable *adj.*, capable of being continued without damaging the environment or using up resources permanently

symbol *n.*, an object or idea that can be used to represent another object or idea

T

taiga (TY guh) *n.*, the large forest area that stretches through northern Russia, Canada, and other northern countries

Taino (TY noh) *n.*, the native people of the Caribbean

tariff *n.*, a tax on imports and exports

tax *n.*, a fee paid to a government for public services

tectonic plate *n.*, a section of Earth's crust that floats on Earth's mantle

temperate *adj.*, mild, in terms of climate

terrace *n.*, a flat surface that is built into a hillside

terraced *adj.*, flat fields cut into slopes or mountainsides

terrain *n.*, the physical features of the land

terrorism *n.*, a type of warfare using violence to achieve political results, typically carried out by individuals or small groups

textile *adj.*, related to cloth or clothing

theme *n.*, topic

tolerance *n.*, acceptance of others' beliefs

topography *n.*, physical features of land

tornado *n.*, a storm with powerful winds that follows an unpredictable path

tourism *n.*, the travel business or industry

Trail of Tears *n.*, the route the Cherokees took during their forced migration from the southeast United States to Oklahoma in the 1830s

transcontinental *adj.*, across an entire continent

transform *v.*, to remake or change

transition *n.*, a change from one activity or stage to another

transpiration *n.*, a process by which plants and trees release water vapor into the air

treaty *n.*, an agreement between two or more countries

Treaty of Guadalupe Hidalgo *n.*, the agreement in 1848 in which Mexico gave up the area from Texas to California to the United States

Treaty of Tordesillas (tor duh SEE uhs) *n.*, a treaty in 1494 that divided South American land between the Spanish and Portuguese

Treaty of Versailles *n.*, a peace treaty that ended World War I in 1919

trench *n.*, a long ditch that protects soldiers from enemy gunfire

triangular trade *n.*, trade among three continents: the Americas, Europe, and Africa

tributary *n.*, a small river that drains into a larger river

tribute *n.*, fees paid to another ruler or country for protection or as a token of submission

troubadour *n.*, a singer during the Middle Ages who performed songs about knights and love

tsunami (soo NAH mee) *n.*, a large, powerful ocean wave

tundra *n.*, flat treeless land found in arctic and subarctic regions

Tupinambá (too pee NAAM baa) *n.*, a people who lived near the mouth of the Amazon River and along the Atlantic Coast around 3000 B.C.

tyranny *n.*, a harsh government

U

United Nations (UN) *n.*, an organization of countries formed in 1945 to keep peace among countries and protect human rights

Universal Declaration of Human Rights *n.*, an agreement approved by the United Nations that defines the rights that people all over the world should have

uplands *n.*, hills, mountains, and plateaus

Ural Mountains *n.*, a mountain range that separates the Northern European Plain from the West Siberian Plain in Russia

utilize *v.*, to make practical use of

V

vegetation *n.*, plant life

venue *n.*, the location for an organized event

veto *v.*, to reject a decision made by another government body

viceroy *n.*, a governor of Spain's colonies in the Americas who represented the Spanish king and queen

volcano *n.*, a mountain that erupts in an explosion of molten rock, gases, and ash

vulnerable *adj.*, open, able to be hurt by outside forces

W

waterway *n.*, a navigable route for traveling and transport

weapon of mass destruction (WMD) *n.*, a weapon that causes great harm to large numbers of people

weather *n.*, the condition of the atmosphere at a particular time, including temperature, precipitation, and humidity for a particular day or week

Western Hemisphere *n.*, the half of Earth west of the prime meridian

Wet Pampas *n.*, humid grasslands in eastern Argentina

wind turbine *n.*, an engine powered by wind to generate electricity

Y

Yanomami (yaa noh MAA mee) *n.*, an indigenous people who still live in the Amazon River Basin as hunter-gatherers

yurt *n.*, a traditional felt tent of Central Asia

A

abolition [abolicionismo] *s.*, movimiento para terminar con la esclavitud, antes y después de la Guerra Civil en los Estados Unidos

absolute location [ubicación absoluta] *s.*, punto exacto donde está ubicado un lugar, identificado por medio de las coordenadas de latitud y longitud

abstract [abstracto] *adj.*, estilo artístico que enfatiza la forma y el color por sobre el realismo

acknowledge [reconocer] *v.*, admitir, agradecer

Acropolis [Acrópolis] *s.*, colina rocosa ubicada en Atenas, Grecia, que servía de fortaleza para los edificios más importantes de la ciudad antigua

adapt [adaptar] *v.*, ajustar o modificar para que sea apropiado

aging population [envejecimiento de la población] *s.*, tendencia demográfica que ocurre cuando aumenta la edad media de una población

Alamo [El Álamo] *s.*, fuerte ubicado en Texas donde 200 texanos perdieron una batalla durante la guerra entre Estados Unidos y México

Alexander the Great [Alejandro Magno] *s.*, conquistador que extendió el Imperio Macedónico y difundió la cultura griega por Eurasia desde 334–323 A.C.

alliance [alianza] *s.*, sociedad entre países

Alps [Alpes] *s.*, cadena montañosa europea

al-Qaeda [Al Qaeda] *s.*, grupo terrorista con sede en el suroeste de Asia

Amazon River Basin [cuenca del río Amazonas] *s.*, la cuenca fluvial más grande de la Tierra, ubicada en América del Sur

amendment [enmienda] *s.*, cambio formal que se hace a una ley

ancestry [ascendencia] *s.*, familia de la cual se desciende, remontándose al pasado; herencia

Andes Mountains [cordillera de los Andes] *s.*, cordillera que se extiende 5,500 millas a lo largo del costado occidental de América del Sur

Angel Falls [Salto Ángel] *s.*, el salto de agua más alto del mundo, ubicado en Venezuela

annexation [anexión] *s.*, acción de añadir territorio a un país

apartheid [apartheid] *s.*, separación legal de las razas; sistema que privaba de sus derechos a los sudafricanos negros

aqueduct [acueducto] *s.*, sistema de transporte para llevar agua a grandes distancias, a veces elevado sobre un puente

aquifer [acuífero] *s.*, capa de roca subterránea que contiene agua

Aral Sea [mar Aral] *s.*, lago de agua salada ubicado en el centro de Asia, cuyo tamaño se ha reducido mucho como consecuencia del desvío para irrigación de los ríos que desembocan en él

archipelago [archipiélago] *s.*, cadena de islas

arid [árido] *adj.*, muy seco, que casi no recibe lluvia

aristocrat [aristócrata] *s.*, miembro de la clase alta

artifact [artefacto] *s.*, objeto hecho por seres humanos de una cultura pasada

Aryans [arios] *s.*, nómades que migraron desde el centro de Asia al valle del Indo alrededor del año 2000 A.C.

assimilate [asimilarse] *v.*, incorporarse en la cultura de una sociedad

Atacama Desert [desierto de Atacama] *s.*, desierto ubicado en el lado occidental de la cordillera de los Andes

Atahualpa [Atahualpa] *s.*, el último emperador de los incas antes de ser conquistados por los españoles en 1533

Augustus [Augusto] *s.*, primer emperador de Roma, en 27 A.C.

autonomy [autonomía] *s.*, auto-gobierno de un país o pueblo

Aymara [aimara] *s.*, cultura indígena que habita en la cordillera de los Andes de Perú y Bolivia

Aztec [aztecas] *s.*, pueblo que se estableció en el área de la actual ciudad de México, 1325–1525 D.C.

B

barbarian [bárbaro] *s.*, soldado o guerrero considerado culturalmente menos desarrollado que aquellos contra los que luchaba; el término proviene originalmente de las tribus germanas que invadieron el Imperio Romano en el año 235 D.C.

Baroque period [Barroco] *s.*, período entre 1600 y 1750 en el cual la música presentaba esquemas y temas complicados

bay [bahía] *s.*, masa de agua rodeada por tierra en tres de sus lados

benefit [beneficiar] *v.*, hacer bien

Berlin Wall [Muro de Berlín] *n.*, muro que dividía Berlín Este communista de Berlín Oeste demucrático

Bill of Rights [Declaración de Derechos] *s.*, primeras diez enmiendas a la Constitución estadounidense

biodiversity [biodiversidad] *s.*, variedad de especies que viven en un ecosistema

biofuel [biocombustible] *s.*, combustible alternativo que es una mezcla de etanol y gasolina

Black Sea [mar Negro] *s.*, mar interior que limita con Europa, Rusia, Georgia y Turquía

Bolshevik [bolchevique] *s.*, partido político de Rusia liderado por Lenin que derrocó al zar en 1917

butte [cerro testigo] *s.*, colina o montaña con laderas escarpadas y cima plana

C

Caesar, Julius [César, Julio] *s.*, general que se convirtió en gobernante de Roma, 46–44 A.C.

Calypso [calipso] *s.*, tipo de música folclórica que se inició en Trinidad

canal [canal] *s.*, vía fluvial construida por el hombre a través de la tierra para botes y barcos

Candomblé [candomblé] *s.*, religión brasileña que combina el espiritualismo africano y el catolicismo

canopy [enramada] *s.*, techo creado sobre un bosque tropical por las copas de los árboles

capital [capital] *s.*, riqueza e infraestructura de un país

Caribbean Sea [mar Caribe] *s.*, mar tropical ubicado en el Hemisferio Occidental, que limita con México, América Central, las Antillas Mayores y las Antillas Menores

cartographer [cartógrafo] *s.*, persona que hace mapas

cash crop [cultivo comercial] *s.*, cultivo que se vende para obtener una ganancia

Caspian Sea [mar Caspio] *s.*, la masa de agua endorreica más grande de la Tierra, que limita con Rusia, Kazajistán, Turkmenistán, Azerbaiyán e Irán

categorize [categorizar] *v.*, agrupar información; clasificar

Catherine the Great [Catalina la Grande] *s.*, emperadora de Rusia desde 1762 hasta 1796

cenote [cenote] *s.*, un depósito de agua subterránea

Chernobyl [Chernóbil] *s.*, ciudad ubicada en Ucrania, en la cual un reactor nuclear explotó en 1986

Christianity [cristianismo] *s.*, religión basada en la vida y en las enseñanzas de Jesús de Nazaret

citizen [ciudadano] *s.*, persona que vive dentro de un territorio y tiene derechos y responsabilidades garantizadas por el gobierno

city-state [ciudad estado] *s.*, estado independiente compuesto por una ciudad y los territorios que dependen de ella

civil war [guerra civil] *s.*, guerra entre grupos opuestos de ciudadanos del mismo país

civilization [civilización] *s.*, sociedad con cultura, política y tecnología altamente desarrolladas

Classical period [Clasicismo] *s.*, período comprendido entre 1750 y 1900 en el cual la música seguía las reglas establecidas para la forma y la complejidad, como en el caso de las sonatas o las sinfonías

climate [clima] *s.*, el promedio de las condiciones de la atmósfera de un área durante un largo período de tiempo, incluidas la temperatura, la precipitación y los cambios estacionales

climograph [gráfica del clima] *s.*, gráfica que muestra el clima de una región a través de la precipitación y la temperatura medias

coastal plains [llanuras costeras] *s.*, tierras bajas ubicadas junto a la orilla del mar

Cold War [Guerra Fría] *s.*, largo período de tensión política sin lucha armada entre los Estados Unidos y la Unión Soviética, desde aproximadamente 1948 hasta 1991

collective farm [granja colectiva] *s.*, en la Unión Soviética, una granja grande donde los trabajadores cultivaban alimentos para distribuirlos a toda la población

colonize [colonizar] *v.*, construir asentamientos y desarrollar el comercio en tierras que controla un país

colony [colonia] *s.*, área controlada por un país lejano

Columbian Exchange [Intercambio Colombino] *s.*, intercambio de plantas, animales y enfermedades entre las Américas, Europa y África que comenzó en el siglo XVI

commercial agriculture [agricultura comercial] *s.*, negocio de producir cultivos para vender

Common Market [Mercado Común] *s.*, Comunidad Económica Europea formada en 1957

commonwealth [mancomunidad] *s.*, nación que se gobierna a sí misma pero es parte de un país mayor

communal [comunal] *adj.*, compartido

communism [comunismo] *s.*, sistema de gobierno en el cual un único partido político controla el gobierno y la economía

concentration camp [campo de concentración] *s.*, área donde los judíos y otras personas fueron detenidos durante la Segunda Guerra Mundial y asesinados por los nazis

condensation [condensación] *s.*, proceso por el cual el vapor de agua se convierte en gotitas de líquido debido al enfriamiento durante el ciclo hidrológico

conquistador [conquistador] *s.*, soldado y líder español que exploraba los territorios recientemente descubiertos de América del Norte, América del Sur, América Central y el Caribe a partir del siglo XVI, fundamentalmente en busca de oro y plata

Constantine [Constantino] *s.*, emperador romano, 306–337 A.C., que hizo del cristianismo la religión oficial del imperio

constitution [constitución] *s.*, documento que organiza un gobierno y enuncia sus poderes

consumer [consumidor] *s.*, persona que compra bienes

contaminated [contaminado] *adj.*, infectado; inapropiado para el uso debido a la presencia de elementos peligrosos

contiguous [contiguo] *adj.*, conectado en un bloque

continent [continente] *s.*, gran masa de tierra sobre la superficie de la Tierra; la Tierra tiene siete continentes

continental drift [deriva continental] *s.*, movimiento lento de los continentes sobre las placas tectónicas

continental shelf [plataforma continental] *s.*, borde de un continente que se extiende bajo el agua y se adentra en el mar

convert [convertir] *v.*, persuadir a alguien para que cambie sus creencias religiosas

cordillera [cordillera] *s.*, sistema de varias cadenas de montañas paralelas

Cortés, Hernán [Cortés, Hernán] *s.*, conquistador español que derrotó a los aztecas en 1525

cosmopolitan [cosmopolita] *adj.*, que reúne muchas culturas e influencias diferentes

Counter-Reformation [Contrarreforma] *s.*, movimiento dentro de la Iglesia Católica Romana para reformar sus propias prácticas

coup [golpe de estado] *s.*, toma repentina e ilegal del gobierno por medio de la fuerza

Creole [criollo] *s.*, lengua que mezcla elementos de idiomas europeos y no europeos

critical [crítico] *adj.*, extremadamente importante, necesario para sobrevivir

Crusades [Cruzadas] *s.*, expediciones militares de la Iglesia Católica Romana para recuperar las tierras santas ubicadas en Medio Oriente que estaban bajo control musulmán, 1096–1291

cuisine [cocina] *s.*, alimentos y tradiciones culinarias comunes a cierta región

culture [cultura] *s.*, modo de vida de un grupo, que incluye el tipo de alimentación, vivienda, vestido, idioma, comportamiento e ideas

culture region [región cultural] *s.*, área que está unificada por rasgos culturales comunes

currency [moneda] *s.*, forma de dinero

current [corriente] *s.*, movimiento continuo de aire o agua que fluye en la misma dirección

cyclone [ciclón] *s.*, tormenta con vientos giratorios, llamada tifón en el Hemisferio Oriental; denominada huracán en otras partes del mundo

czar [zar] *s.*, término utilizado en Rusia para designar al emperador

D

dam [presa] *s.*, barrera que controla el flujo de agua

Danube River [río Danubio] *s.*, río que nace en Alemania y desemboca en el mar Negro

Declaration of Independence [Declaración de Independencia] *s.*, documento que declara que los Estados Unidos son independientes del Imperio Británico, adoptado el 4 de julio de 1776

deforestation [deforestación] *s.*, práctica de talar bosques para despejar la tierra y utilizarla para cultivos o uso urbano

democracy [democracia] *s.*, forma de gobierno del pueblo, en la cual los ciudadanos suelen elegir representantes para que los gobiernen

democratization [democratización] *s.*, proceso de convertirse en una democracia

demographics [demografía] *s.*, características de una población humana, tales como la edad, el ingreso y la educación

descendants [descendientes] *s.*, generaciones futuras; parientes de un miembro de la familia que vivió en el pasado

dialect [dialecto] *s.*, variante regional de un idioma principal

dictator [dictador] *s.*, gobernante con control total

diplomacy [diplomacia] *s.*, discusión entre grupos o países para resolver disputas o desacuerdos

displace [desplazar] *v.*, obligar a un pueblo a dejar su hogar

distinct [definido] *adj.*, fácilmente reconocible

distort [distorsionar] *v.*, modificar la forma o apariencia usual; deformar

diversify [diversificar] *v.*, dar variedad a algo

diversity [diversidad] *s.*, variedad

drought [sequía] *s.*, largo período con escasa o ninguna precipitación

Dry Pampas [Pampa seca] *s.*, región árida ubicada en el oeste de Argentina

due process [debido proceso] *s.*, en los Estados Unidos, reglas que las autoridades deben seguir al tratar con los ciudadanos

dynamic [dinámico] *adj.*, que cambia continuamente

E

earthquake [terremoto] *s.*, sacudida de la corteza de la Tierra generalmente causada por el choque o el deslizamiento de las placas tectónicas

Eastern Hemisphere [Hemisferio Oriental] *s.*, la mitad de la Tierra ubicada al este del primer meridiano

economic sector [sector económico] *s.*, subdivisión o parte más pequeña de una economía, tal como la industria y la agricultura

economy [economía] *s.*, sistema en el cual las personas producen, venden y compran cosas

ecosystem [ecosistema] *s.*, comunidad de organismos vivos y su medio ambiente o hábitat natural

ecoturism [ecoturismo] *s.*, manera de visitar las áreas naturales que conserva los recursos naturales de la región

El Niño [El Niño] *s.*, inversión del viento y las corrientes oceánicas usuales

elevation [elevación] *s.*, altura de un accidente geográfico sobre el nivel del mar

eliminate [eliminar] *v.*, deshacerse de

Emancipation Proclamation [Proclama de Emancipación] *s.*, documento de 1863 que liberó a todos los esclavos que vivían en el territorio en poder de los Confederados durante la Guerra Civil estadounidense

empire [imperio] *s.*, grupo de pueblos o estados gobernados por un único gobernante fuerte

Enduring Voices Project [Proyecto Voces Perdurables] *s.*, proyecto de National Geographic para estudiar y conservar las lenguas que están en riesgo de desaparición

Enlightenment [Ilustración] *s.*, movimiento social del siglo XVIII que trabajó a favor de la educación y de los derechos del individuo

enlist [alistarse] *v.*, presentarse como voluntario para el servicio militar

entrepreneur [empresario] *s.*, persona que comienza un negocio nuevo

entrepreneurship [espíritu empresarial] *s.*, características de creatividad y riesgo que existen en una persona o sociedad

epic poem [poema épico] *s.*, poema largo que relata las aventuras de un héroe

epidemic [epidemia] *s.*, brote de una enfermedad que afecta a una gran parte de la población de una comunidad en particular

equator [ecuador] *s.*, círculo imaginario alrededor de la Tierra que está a la misma distancia del Polo Norte y del Polo Sur y divide la Tierra por la mitad; la línea central o de 0° de latitud

equinox [equinoccio] *s.*, momento en que el día y la noche tienen la misma duración; ocurre dos veces al año, el 21 o 22 de marzo y el 21 o 22 de septiembre

erosion [erosión] *s.*, proceso por el cual las rocas y el suelo se rompen lentamente y se desgastan

erratic [errático] *adj.*, inconsistente o irregular

essential [esencial] *adj.*, necesario

establish [establecer] *v.*, instituir

ethanol [etanol] *s.*, alcohol líquido extraído de la caña de azúcar o del maíz que se puede usar como combustible, solo o mezclado con gasolina

euro [euro] *s.*, moneda común de la Unión Europea

European Union (EU) [Unión Europea (UE)] *s.*, organización económica compuesta por 27 países miembros europeos (2011)

eurozone [eurozona] *s.*, países que han adoptado el euro como moneda

evaporation [evaporación] *s.*, proceso por el cual el agua se convierte en vapor y sube a la atmósfera debido al calentamiento del Sol; una parte del ciclo hidrológico

excavate [excavar] *v.*, quitar la tierra cuidadosamente o hacer un hoyo

exchange [cambiar] *v.*, convertir el dinero en otra moneda

exile [exilar] *v.*, obligar a alguien a dejar un país

exile [exilio] *s.*, situación en la que se está ausente del propio país natal

expand [ampliar] *v.*, hacer más grande

exploit [explotar] *v.*, aprovecharse de algo; usar de manera egoísta o injusta para beneficio propio

export [exportar] *v.*, enviar a otro país para obtener ayuda o un beneficio

export revenue [ingresos por exportaciones] *s.*, dinero o ingresos recibidos por los bienes vendidos a otro país

extinction [extinción] *s.*, desaparición de una especie o de un tipo de ser vivo

F

factory system [sistema fabril] *s.*, modo de trabajar en el cual cada persona trabaja en solo una parte de un producto

fallout [lluvia radiactiva] *s.*, partículas radiactivas provenientes de una explosión nuclear que caen a través de la atmósfera

federal system [sistema federal] *s.*, sistema de gobierno con un gobierno central fuerte y unidades gubernamentales locales

fertile [fértil] *adj.*, capaz de producir una gran abundancia de frutos, cultivos o crías

fertilizer [fertilizante] *s.*, sustancia añadida al suelo para enriquecerlo

feudal system [sistema feudal] *s.*, durante la Edad Media, estructura social que consistía en un rey, señores, vasallos y siervos

fiber optics [fibra óptica] *s.*, fibras de vidrio utilizadas para enviar un código digital rápidamente a través de grandes distancias

fjord [fiordo] *s.*, bahía angosta y profunda ubicada

food security [seguridad alimentaria] *s.*, fácil acceso a alimentos suficientes

foremost [principal] *adj.*, primero en rango o importancia, que lidera

fortify [fortificar] *v.*, fortalecer

fossil [fósil] *s.*, restos conservados de plantas y animales antiguos

fragmented country [país fragmentado] *s.*, país que está físicamente dividido en partes separadas, tales como una cadena de islas, y/o política o culturalmente dividido

free enterprise economy [economía de libre empresa] *s.*, sistema en el cual las empresas de propiedad privada producen bienes y servicios, también llamada economía de mercado o capitalismo

fuel cell [celda de combustible] *s.*, unidad pequeña que, al igual que una pila o batería, combina sustancias químicas para producir energía

fuse [fusionar] *v.*, mezclar

G

Gadsden Purchase [Compra de Gadsden] *s.*, venta de territorio realizada en 1853 por México a los Estados Unidos que estableció la frontera suroccidental estadounidense actual

gaucho [gaucho] *s.*, vaquero de América del Sur

gauge [trocha] *s.*, medida del ancho de las vías del ferrocarril

generation [generación] *s.*, grupo de individuos que nacen y viven aproximadamente en la misma época

Genghis Khan [Gengis Kan] *s.*, gobernante mongol que estableció un imperio en el centro de Asia, a principios del siglo XIII

genre [género] *s.*, forma literaria, como un poema, una obra de teatro o una novela

geoglyph [geoglifo] *s.*, figura geométrica grande o forma de animal dibujada sobre el suelo

Geographic Information Systems (GIS) [Sistemas de Información Geográfica (SIG)] *s.*, aparatos computarizados que presentan datos sobre lugares específicos

geographic pattern [patrón geográfico] *s.*, similitud entre lugares

Gettysburg Address [Discurso de Gettysburg] *s.*, discurso que Abraham Lincoln pronunció en 1863 en honor a los soldados que murieron en la batalla de Gettysburg durante la Guerra Civil estadounidense

Giza [Giza] *s.*, ciudad ubicada junto al Nilo, donde se construyeron diez pirámides en el antiguo Egipto

glacier [glaciar] *s.*, masa grande de hielo y nieve acumulada

glasnost [glásnost] *s.*, política de la Unión Soviética de apertura que animó al pueblo a hablar abiertamente acerca del gobierno, introducida por Mijaíl Gorbachov en la década de 1980

global [global] *adj.*, mundial

Global Positioning System (GPS) [Sistema de Posicionamiento Global (GPS)] *s.*, sistema de satélites con base en el espacio que encuentra la ubicación absoluta y la hora de cualquier lugar de la Tierra

global warming [calentamiento global] *s.*, aumento de la temperatura media de la Tierra desde mediados del siglo XX

globalization [globalización] *s.*, desarrollo de una economía mundial basada en el libre comercio y el uso de mano de obra extranjera

globe [globo terráqueo] *s.*, modelo tridimensional, o esférico, de la Tierra

golden age [edad dorada] *s.*, período de gran riqueza, cultura y democracia en Grecia

Gorbachev, Mikhail [Gorbachov, Mijaíl] *s.*, líder de la Unión Soviética, 1985–1991; presidente desde 1990 hasta que fue disuelta en 1991

government [gobierno] *s.*, organización que mantiene el orden, establece reglas y proporciona servicios para una sociedad

Grand Canyon [Gran Cañón] *s.*, formación rocosa ubicada en el suroeste de los Estados Unidos que el río Colorado ha excavado profundamente durante millones de años

grasslands [praderas] *s.*, áreas abiertas y amplias apropiadas para el pastoreo y los cultivos

Great Depression [Gran Depresión] *s.*, descenso económico mundial ocurrido en la década de 1930, marcado por la pobreza y una alta tasa de desempleo

Great Lakes [Grandes Lagos] *s.*, cinco lagos grandes de agua dulce ubicados entre Canadá y los Estados Unidos

Great Leap Forward [Gran Salto Adelante] *s.*, plan de Mao Zedong para hacer que la economía de China creciera más rápidamente, 1958–1961

Great Plains [Grandes Llanuras] *s.*, área de tierra baja y llana ubicada al este de las montañas Rocosas

Great Recession [Gran Recesión] *s.*, descenso de las economías de los Estados Unidos, Canadá y otros países, que comenzó en 2007

greenhouse gas [gas invernadero] *s.*, gas que atrapa el calor del Sol sobre la Tierra

gross domestic product (GDP) [producto interno bruto (PIB)] *s.*, valor total de todos los bienes y servicios producidos en un país en un año dado

Guaraní [guaraníes] *s.*, pueblo indígena que vive en los valles de los ríos Paraguay y Paraná, en América del Sur

guillotine [guillotina] *s.*, máquina utilizada para ejecutar a las personas durante la Revolución Francesa

Gutenberg, Johannes [Gutenberg, Johannes] *s.*, impresor alemán que inventó la imprenta en 1450

H

habitat [hábitat] *s.*, medio ambiente natural de una planta o animal vivo

half-life [vida media] *s.*, tiempo necesario para que la mitad de los átomos de una sustancia radiactiva se desintegren

harbor [embarcadero] *s.*, lugar donde los barcos pueden atracar protegidos del mar abierto

hemisphere [hemisferio] *s.*, una mitad de la Tierra

heritage [herencia] *s.*, tradición que se transmite de los ancestros a los descendientes

Hermitage Museum [museo Hermitage] *s.*, museo de arte y cultura ubicado en San Petersburgo, Rusia

Hidalgo, Miguel [Hidalgo, Miguel] *s.*, sacerdote católico que lideró una revuelta en México en 1810

hieroglyphics [jeroglíficos] *s.*, sistema antiguo de escritura que emplea imágenes y símbolos

Hitler, Adolf [Hitler, Adolf] *s.*, jefe de estado alemán desde 1933 hasta 1945

hotspot [punto caliente] *s.*, parte inusualmente caliente del manto de la Tierra

Human Development Index (HDI) [Índice de Desarrollo Humano (IDH)] *s.*, conjunto de datos utilizados por los geógrafos para comparar la calidad de vida en diferentes países, incluyendo salud, educación y estándar de vida

human rights [derechos humanos] *s.*, derechos políticos, económicos y culturales que todas las personas deben tener

hunter-gatherer [cazador-recolector] *s.*, persona que caza animales y recolecta plantas y frutos para alimentarse

hurricane [huracán] *s.*, tormenta fuerte con vientos giratorios y lluvia intensa

hybrid [híbrido] *s.*, vehículo que puede funcionar a electricidad o gasolina

hydroelectric power [energía hidroeléctrica] *s.*, fuente de energía que emplea agua en movimiento para producir electricidad

I

immigrate [inmigrar] *v.*, mudarse a otro país o región

impact [impacto] *s.*, efecto que produce un cambio

Impressionism [Impresionismo] *s.*, estilo artístico en el cual los artistas usaban la luz y el color en pinceladas cortas para capturar un momento de tiempo

incorporate [incorporar] *v.*, incluir; combinar con algo ya formado

indigenous [indígena] *adj.*, originario del área donde algo se encuentra

indulgence [indulgencia] *s.*, durante la Edad Media, tarifa pagada a la iglesia para disminuir el castigo por un pecado

Industrial Revolution [Revolución Industrial] *s.*, período de los siglos XVIII y XIX durante el cual los trabajadores de las fábricas comenzaron a usar máquinas y herramientas eléctricas para producir a gran escala

industrialization [industrialización] *s.*, paso a una producción a gran escala mediante el uso de máquinas

infrastructure [infraestructura] *s.*, sistemas básicos de una sociedad, tales como las carreteras, los puentes, las cloacas y el tendido eléctrico

Institutional Revolutionary Party (PRI) [Partido Revolucionario Institucional (PRI)] *s.*, partido político que controló el gobierno de México desde 1929 hasta 2000

interior [interior] *s.*, tierra que está lejos de la costa del mar

Internet *n.*, una red de comunicación

intersection [encrucijada] *s.*, lugar donde se encuentran las personas o se cruzan los caminos

invader [invasor] *s.*, enemigo que entra a un país por la fuerza

Iron Curtain [Cortina de Hierro] *s.*, en Europa, frontera imaginaria que separaba los países comunistas de los no comunistas durante la Guerra Fría

isolated [aislado] *adj.*, separado de los demás

isthmus [istmo] *s.*, franja estrecha de tierra que une dos áreas grandes de tierra

K

Kievan Rus [Rus de Kiev] *s.*, estado fundado por los rusos Varegos en 882, que pasó a formar parte de la Rusia moderna

kimono [kimono] *s.*, vestimenta femenina tradicional del Japón

kinship [parentesco] *s.*, vínculo por consanguinidad o relación familiar

Kremlin [Kremlin] *s.*, complejo histórico de palacios, arsenales e iglesias ubicado en Moscú, sede del gobierno ruso

L

L'Ouverture, Toussaint [L'Ouverture, Toussaint] *s.*, ex esclavo que lideró la exitosa revuelta haitiana para lograr independizarse de Francia

land reform [reforma agraria] *s.*, división de las fincas de gran extensión para dar tierras a los pobres

landmass [masa de tierra] *s.*, área muy extensa de tierra

language family [familia de lenguas] *s.*, grupo de lenguas relacionadas

latitude [latitud] *s.*, línea imaginaria que se extiende de este a oeste alrededor de la Tierra y que indica la ubicación en relación con el ecuador

launch [lanzamiento] *v.*, para empezar

Lenin, V.I. [Lenin, V.I.] *s.*, líder bolchevique que destituyó al zar en la Revolución Rusa de 1917 y tomó el mando del gobierno nuevo

liberate [liberar] *v.*, conceder la libertad a algo o a alguien

Llanos [llanos] *s.*, praderas del norte de Sudamérica

Locke, John [Locke, John] *s.*, filósofo inglés de fines del siglo XVII que contribuyó a inspirar la Revolución Norteamericana y la Ilustración

locks [esclusas] *s.*, compartimentos que se usan en los canales para subir o bajar los barcos entre vías fluviales conectadas

longitude [longitud] *s.*, línea imaginaria que corre del Polo Norte al Polo Sur y que indica la ubicación en relación con el primer meridiano

Lost Decade [Década Perdida] *s.*, en Japón, la década de 1990, en la que disminuyó la producción porque las empresas estaban fuertemente endeudadas

Louisiana Purchase [Compra de la Luisiana] *s.*, tierras compradas por Thomas Jefferson en 1803 que duplicaron el tamaño de los Estados Unidos

lowland [tierras bajas] *s.*, área de poca altura

Luther, Martin [Lutero, Martín] *s.*, monje alemán cuyos actos condujeron, en 1517, a la Reforma para enfrentar la corrupción de la Iglesia Católica

M

Machu Picchu [Machu Picchu] *s.*, ciudad compleja construida por los Incas sobre una montaña en el siglo XV

malnutrition [desnutrición] *s.*, insuficiencia de alimentos o nutrientes

Manifest Destiny [Destino Manifiesto] *s.*, idea de que Estados Unidos tiene el derecho de expandir su territorio hacia el océano Pacífico

manufacturing [manufactura] *s.*, uso de máquinas para convertir las materias primas en productos útiles

map [mapa] *s.*, representación plana, bidimensional, de la Tierra

marine life [vida marina] *s.*, plantas y animales que viven en el océano

marine reserve [reserva marina] *s.*, área del océano resguardada para proteger a los animales marinos de los seres humanos

marketing [marketing] *s.*, publicidad y promoción de un producto o negocio

mass media [medios masivos de comunicación] *s.*, comunicación proveniente de una única fuente, con el potencial de llegar a grandes audiencias

Maya [maya] *s.*, civilización que vivió en el Yucatán y en la parte norte de Centroamérica de 100 A.C. a 900 D.C.

medicinal plant [planta medicinal] *s.*, planta que se usa para tratar enfermedades

meditation [meditación] *s.*, práctica de usar la concentración para calmar los pensamientos y controlarlos

Mediterranean climate [clima mediterráneo] *s.*, clima de veranos calurosos y secos e inviernos templados y lluviosos

megacity [megalópolis] *s.*, ciudad grande, que tiene más de 10 millones de habitantes

Mesa Central [Mesa Central] *s.*, en México, la parte sur de la meseta mexicana

mestizo [mestizo] *s.*, persona que tiene una mezcla de ancestros europeos y americanos nativos

methane [metano] *s.*, gas natural incoloro que se libera a partir del carbono

Mexican Cession [Cesión Mexicana] *s.*, tierras que abarcan desde Texas hasta California, entregadas por México a los Estados Unidos en 1848, de acuerdo con el tratado de Guadalupe Hidalgo

Mexican Plateau [Meseta Mexicana] *s.*, área de tierras llanas que se encuentran entre las dos cordilleras montañosas de Sierra Madre, en México

Middle Ages [Edad Media] *s.*, período en Europa occidental posterior a la caída del Imperio Romano, desde aproximadamente 500 a 1500

Middle Passage [Pasaje del Medio] *s.*, viaje para cruzar el océano Atlántico, que tomaba meses, en el cual los africanos esclavizados eran llevados a las colonias europeas en América

migrate [emigrar] *v.*, trasladarse de un lugar a otro

missionary [misionero] *s.*, persona enviada por una iglesia a convertir a otras personas a esa religión

mobile [móvil] *adj.*, que se puede mover

modify [modificar] *v.*, cambiar o hacer menos extremo

Mongol Empire [Imperio Mongol] *s.*, imperio establecido en Asia Central por Gengis Kan a comienzos del siglo XIII

monopoly [monopolio] *s.*, control total del mercado para un servicio o producto

monotheistic [monoteísta] *adj.*, relativo a la creencia religiosa en un solo dios

monotheistic religion [religión monoteísta] *s.*, sistema de creencias basadas en un solo dios o deidad

monsoon [monzón] *s.*, viento estacional que trae lluvias intensas durante parte del año

Montezuma [Moctezuma] *s.*, líder azteca asesinado por el conquistador español Hernán Cortés

mouth [desembocadura] *s.*, lugar donde un río desemboca en el mar

multi-party democracy [democracia multipartidaria] *s.*, sistema político en el cual las elecciones incluyen candidatos de más de un partido

multitudes [multitudes] *s.*, grandes cantidades

Mundurukú [mundurukú] *s.*, pueblo indígena de Brasil

mural [mural] *s.*, pintura de gran tamaño realizada sobre una pared

N

Nairobi [Nairobi] *s.*, capital de Kenia

Napoleon [Napoleón] *s.*, emperador de Francia que conquistó otros países europeos y formó un imperio, 1804–1815

National Action Party (PAN) [Partido Acción Nacional (PAN)] *s.*, partido político mexicano que ganó las elecciones del año 2000

nationalism [nacionalismo] *s.*, profundo sentimiento de lealtad al propio país

nationalize [nacionalizar] *v.*, dar al gobierno el control de una empresa privada

natural rights [derechos naturales] *s.*, derechos como la vida, la libertad y la propiedad, que las personas poseen desde su nacimiento

naturalization [naturalización] *s.*, proceso que permite que una persona nacida en otro país se convierta en ciudadano

navigable [navegable] *adj.*, suficientemente ancho o profundo para que los barcos o botes puedan navegar sin inconvenientes

navigation [navegación] *s.*, ciencia de averiguar la posición y planear rutas marítimas

Nazi Germany [Alemania nazi] *s.*, Alemania bajo el régimen del partido nazi, de 1933 a 1945

neutrality [neutralidad] *s.*, negativa a tomar partido o a involucrarse

nomad [nómada] *s.*, persona que se desplaza de un lugar a otro

nonrenewable fossil fuel [combustible fósil no renovable] *s.*, fuente de energía, como el petróleo, el gas natural o el carbón, cuya provisión es limitada

nonrenewable resource [recurso no renovable] *s.*, fuente de energía que es limitada, y no se puede reemplazar, como el petróleo

North American Free Trade Agreement (NAFTA) [Tratado de Libre Comercio de América del Norte (NAFTA)] *s.*, acuerdo firmado en 1994 que facilitó el comercio y la inversión entre Canadá, México y los Estados Unidos

North Atlantic Drift [Corriente del Atlántico Norte] *s.*, corriente marina cálida que calienta las aguas que bañan la parte noroeste de Rusia

North Pole [Polo Norte] *s.*, punto ubicado en el extremo norte de la Tierra, opuesto al Polo Sur, donde convergen todas las líneas de longitud

Northern European Plain [Llanura del Norte de Europa] *s.*, vastas tierras bajas que se extienden desde Francia hasta Rusia

Northern Hemisphere [Hemisferio Norte] *s.*, la mitad de la Tierra que se encuentra al norte del ecuador

novel [novela] *s.*, extensa obra de ficción, con trama y personajes complejos

O

Olmec [olmecas] *s.*, sociedad organizada que vivió junto a la costa sur del golfo de México en 1000 A.C.

opera [ópera] *s.*, representación que cuenta una historia mediante música y palabras

oppose [oponerse] *v.*, objetar

Orange Revolution [Revolución Naranja] *s.*, destitución pacífica del primer ministro de Ucrania en 2004

P

Pampas [Pampa] *s.*, llanuras cubiertas de pasto de la Argentina

Panama Canal Zone [Zona del Canal de Panamá] *s.*, área en la que se construyó el canal de Panamá

parliamentary democracy [democracia parlamentaria] *s.*, sistema de gobierno en el cual el poder ejecutivo está presidido por el primer ministro, que es elegido por el partido que posee la mayoría de los escaños del Parlamento

patrician [patricio] *s.*, rico terrateniente de la antigua Roma

Pearl Harbor [Pearl Harbor] *s.*, base naval de los EE.UU. ubicada en Hawái, bombardeada por los japoneses en 1941, lo cual provocó el ingreso de EE.UU. en la Segunda Guerra Mundial

peat [turba] *s.*, material que se forma a partir de la descomposición de plantas muy antiguas y que arde como el carbón

peninsula [península] *s.*, masa de tierra rodeada por agua en tres de sus lados

perestroika [perestroika] *s.*, reformas en la estructura económica de la Unión Soviética introducidas por Mijaíl Gorbachov en 1985

permafrost [permafrost] *s.*, suelo que está permanentemente congelado

perspective [perspectiva] *s. modo artístico de mostrar los objetos de la manera en que son vistos por las personas*, en términos de distancia o profundidad relativa, como si estuvieran en tres dimensiones

pesticide [pesticida] *s.*, sustancia química que mata insectos y malezas nocivas

Peter the Great [Pedro el Grande] *s.*, Pedro Romanov, zar que gobernó Rusia desde 1682 hasta 1725

petroleum [petróleo] *s.*, materia prima que se usa para producir combustibles

philosopher [filósofo] *s.*, persona que examina las preguntas sobre el universo y busca la verdad

pioneer [pionero] *s.*, colono de tierras nuevas

pipeline [tubería] *s.*, serie de tubos o caños conectados para transportar líquidos o gases

Pizarro, Francisco [Pizarro, Francisco] *s.*, conquistador español que en 1533 derrocó al emperador y fundó la ciudad de Lima, Perú

plain [llanura] *s.*, área plana de la superficie terrestre

plantation [plantación] *s.*, granja de gran tamaño que produce cultivos para obtener ganancias

plateau [meseta] *s.*, llanura ubicada a gran altura sobre el nivel del mar que a menudo tiene un precipicio en todos sus lados

plebeian [plebeyo] *s.*, agricultor o persona de clase baja de la antigua Roma

poacher [cazador o pescador furtivo] *s.*, persona que caza o pesca de manera ilegal

poaching [cazar o pescar furtivamente] *s.*, caza o pesca ilegal

polder [pólder] *s.*, tierras de los Países Bajos ganadas al mar que se destinan a la agricultura

policy [política] *s.*, pautas y procedimientos oficiales de una organización o gobierno

polytheistic religion [religión politeísta] *s.*, sistema de creencias basadas en varios dioses o deidades

port [puerto] *s.*, embarcadero para barcos donde se intercambian mercancías

Port-au-Prince [Puerto Príncipe] *s.*, capital de Haití

precipitation [precipitación] *s.*, proceso que hace caer agua sobre la Tierra, en forma de lluvia, nieve o granizo

predominant [predominante] *adj.*, principal, más común, superior a los demás

prehistoric [prehistórico] *adj.*, anterior a la historia escrita

prime meridian [primer meridiano] *s.*, línea de longitud de 0° que se extiende desde Polo Norte al Polo Sur y que pasa por Greenwich, Inglaterra

privatization [privatización] *s.*, proceso por el que las empresas que eran propiedad del gobierno pasan a manos privadas

profitable [rentable] *adj.*, que hace ganar dinero, financieramente exitoso

projection [proyección] *s.*, modo de mostrar la superficie curva de la Tierra sobre un mapa plano

promote [fomentar] *v.*, animar, estimular

propaganda [propaganda] *s.*, información que se difunde para influir sobre la opinión de las personas o para promover las ideas de un partido u organización

proportional representation [representación proporcional] *s.*, sistema en el cual un partido político consigue un porcentaje de escaños igual al porcentaje de votos que obtuvo

prosperous [próspero] *adj.*, económicamente fuerte

protest [protestar] *v.*, objetar

province [provincia] *s.*, parte más pequeña en que se divide un país, especialmente Canadá

Q

Quechua [quechuas] *s.*, pueblo que vive en la cordillera de los Andes de Perú, Ecuador y Bolivia

R

radical [radical] *s.*, persona que busca un cambio extremo o sostiene una posición política extrema

radioactive [radiactivo] *adj.*, que emite energía producida por la ruptura de un átomo

rain forest [bosque tropical] *s.*, bosque de temperatura cálida, humedad elevada y vegetación espesa que recibe más de 100 pulgadas de lluvia al año

rain shadow [sombra orográfica] *s.*, región seca ubicada sobre uno de los lados de una cordillera

rainshadow effect [efecto de la sombra orográfica] *s.*, proceso en el cual el aire húmedo asciende por una ladera de la cordillera y luego se enfría y cae en forma de precipitación, dejando el otro lado de la cordillera mayormente seco

raw materials [materias primas] *s.*, materiales naturales o sin terminar, como minerales, petróleo o carbón, que se usan para elaborar productos terminados

rebellion [rebelión] *s.*, revuelta o resistencia a la autoridad

recession [recesión] *s.*, desaceleración del crecimiento económico

Reconstruction [Reconstrucción] *s.*, esfuerzo realizado para reconstruir y unificar a los Estados Unidos después de la guerra civil

reform [reforma] *s.*, cambio que apunta a corregir un problema

Reformation [Reforma] *s.*, movimiento que surgió en el siglo XVI para reformar el cristianismo

region [región] *s.*, conjunto de sitios con características comunes

regulate [regular] *v.*, controlar

reign [reinado] *s.*, período de mando de un rey, reina, emperador o emperadora

Reign of Terror [El Terror] *s.*, movimiento francés liderado por Maximilien Robespierre, en el cual fueron decapitadas 40,000 personas, durante el período 1793–94

reincarnation [reencarnación] *n.*, el nacimiento de un alma en otra vida

relative location [ubicación relativa] *s.*, la posición de un lugar en relación con otros

relief [relieve] *s.*, cambio en la elevación de un lugar a otro

religious tolerance [tolerancia religiosa] *s.*, aceptación de distintas religiones para que sean profesadas al mismo tiempo, sin prejuicios

remittance [remesa] *s.*, dinero enviado a una persona que se encuentra en otro lugar

remote [remoto] *adj.*, difícil de llegar, aislado

Renaissance [Renacimiento] *s.*, período que se desarrolló entre los siglos XIV y XVI, donde florecieron el arte y la cultura

renewable resource [recurso renovable] *s.*, materia prima o fuente de energía que se reemplaza a sí misma con el paso del tiempo

reparation [reparación] *s.*, dinero que, después de una guerra, pagan como castigo normalmente los agresores que iniciaron el conflicto

republic [república] *s.*, forma de gobierno en la cual las personas eligen funcionarios para que gobiernen

reserve [reserva] *s.*, tierras destinadas a propósitos especiales, como la agricultura, la preservación de los hábitats o para ser usadas como vivienda por determinados grupos de personas; futuro suministro (de petróleo)

resistance [resistencia] *n.*, oposición

restore [restaurar] *v.*, recuperar

revenue [rentas] *s.*, ingresos

Revolution [revolución] *s.*, acción de ciudadanos o colonos cuyo objetivo es derrocar un gobierno

Rhine River [río Rin] *s.*, río que nace en Suiza y desemboca en el mar del Norte

Ring of Fire [Anillo de Fuego] *s.*, área que se extiende a lo largo de las riberas del océano Pacífico, donde chocan las placas tectónicas, lo que genera terremotos y una gran actividad volcánica

Rio de Janeiro [Río de Janeiro] *s.*, ciudad de Brasil que albergará los juegos olímpicos de 2016

river basin [cuenca de un río] *s.*, área baja por la que fluye un río

Romantic Period [Romanticismo] *s.*, período artístico de comienzos del siglo XIX, en el cual los artistas pintaban paisajes y escenas de la naturaleza para transmitir emociones

roots [raíces] *s.*, orígenes culturales

Russian Revolution [Revolución Rusa] *s.*, revolución que tuvo lugar en Rusia en 1917, en la cual el zar fue depuesto y los bolcheviques tomaron el poder

Russification [rusificación] *s.*, política de designar ciudadanos rusos a cargo de las repúblicas soviéticas durante las décadas de 1970 y 1980

ruthless [despiadado] *adj.*, cruel

Santa Anna [Santa Anna] *s.*, presidente y general mexicano, que venció en la batalla de El Álamo, pero perdió la guerra entre los Estados Unidos y México

São Paolo [San Pablo] *s.*, la ciudad más grande de Brasil

saturate [saturar] *v.*, remojar completamente

scale [escala] *s.*, parte de un mapa que indica el tamaño en el que se muestra un área de la Tierra

scarcity [escasez] *s.*, falta o carencia de algo

scorched earth policy [táctica de tierra quemada] *s.*, práctica llevada adelante por las tropas rusas en 1812, quienes, a medida que retrocedían ante el avance del ejército de Napoleón, quemaban los cultivos y todos los recursos que pudieran servir al enemigo para abastecerse

secede [separarse] *v.*, retirarse formalmente, dejar de ser parte

secular [secular] *adj.*, terrenal, sin vínculo con una religión

seismic [sísmico] *adj.*, relacionado con la actividad o movimiento producidos por los terremotos

semiarid [semiárido] *adj.*, algo seco, con muy poca lluvia

serf [siervo] *s.*, campesino ruso o europeo, pobre y con pocos derechos, que alquilaba tierras a un terrateniente entre los siglos XVI y XIX

Siberia [Siberia] *s.*, enorme región del centro y este de Rusia

significant [significativo] *adj.*, importante

Silk Roads [rutas de la seda] *s.*, antiguas rutas comerciales que unían el sudoeste y el centro de Asia con China

slash-and-burn [tala y quema] *s.*, método agrícola que consiste en despejar las tierras talando y quemando el bosque y la vegetación

Slavs [eslavos] *s.*, pueblo originario de los alrededores del mar Negro o Polonia, que se instaló en Ucrania y el oeste de Rusia alrededor del año 800 D.C.

slum [barriada] *s.*, área densamente poblada de una ciudad, con viviendas precarias y malas condiciones de vida

smartphone [teléfono inteligente] *s.*, dispositivo portátil que combina la comunicación con aplicaciones de software

socialism [socialismo] *s.*, sistema de gobierno en el que el gobierno controla los recursos económicos

solstice [solsticio] *s.*, punto en que el Sol se encuentra a la distancia máxima, al sur o al norte, del ecuador; inicio del invierno y del verano

South Pole [Polo Sur] *s.*, punto más austral de la Tierra, opuesto al Polo Norte, donde convergen todas las líneas de longitud

Southern Hemisphere [Hemisferio Sur] *s.*, la mitad de la Tierra que se encuentra al sur del ecuador

sovereignty [soberanía] *s.*, control de un país sobre sus propios asuntos

Soviet Union [Unión Soviética] *s.*, Unión de las Repúblicas Socialistas Soviéticas, país formado por Rusia y otros estados euroasiáticos, que existió desde 1922 hasta 1991

soybean [soja] *s.*, tipo de frijol que se cultiva como alimento y para elaborar productos industriales

spatial thinking [pensamiento espacial] *s.*, manera de pensar en el espacio que está sobre la superficie de la Tierra, incluyendo la ubicación de los distintos lugares y por qué se encuentran allí

standard of living [nivel de vida] *s.*, nivel de acceso de los habitantes de un país a bienes, servicios y comodidades materiales

staple [alimento básico] *s.*, constituyente básico de la dieta de las personas

state [estado] *s.*, territorio determinado que posee un gobierno propio

steel [acero] *s.*, metal de gran dureza elaborado a partir del hierro combinado con otros metales

steppe [estepa] *s.*, llanura muy extensa de praderas secas

strike [huelga] *s.*, interrupción del trabajo por parte de empleados que se niegan a trabajar

subsistence farmers [agricultores de subsistencia] *s.*, agricultores que producen cultivos para alimentar a sus familias, no para vender

subsistence farming [agricultura de subsistencia] *s.*, agricultura que solamente produce cultivos para que las familias se alimenten, no para vender

suspension bridge [puente colgante] *n.*, colgados de un puente que se cuelga de dos o más cables

sustainable [sustentable] *adj.*, capaz de ser continuado sin dañar el medio ambiente o sin agotar los recursos de manera permanente

symbol [símbolo] *s.*, objeto o idea que se puede usar para representar otro objeto o idea

T

taiga [taiga] *s.*, extensa área de bosques que se extiende a través del norte de Rusia, Canadá y otros países del norte

Taino [taínos] *s.*, pueblo nativo del Caribe

tariff [arancel] *s.*, impuesto sobre las importaciones y exportaciones

tax [impuesto] *s.*, suma que se paga al gobierno para contar con servicios públicos

tectonic plate [placa tectónica] *s.*, sección de la corteza terrestre que flota sobre el manto terrestre

temperate [templado] *adj.*, benigno, en términos del clima

terrace [terraza] *s.*, superficie llana construida sobre la ladera de un monte

terraced [abancalado] *adj.*, campo llano excavado en la pendiente o ladera de una montaña

terrain [terreno] *s.*, características físicas de la tierra

terrorism [terrorismo] *s.*, tipo de guerra que se vale de la violencia para obtener resultados políticos; es normalmente empleada por grupos reducidos o individuos

textile [textil] *adj.*, relativo a la tela o ropa

theme [tema] *s.*, tópico

tolerance [tolerancia] *s.*, aceptación de las creencias de los demás

topography [topografía] *s.*, características físicas de la tierra

tornado [tornado] *s.*, tormenta de vientos muy fuertes que sigue una trayectoria impredecible

tourism [turismo] *s.*, industria o negocio de los viajes

Trail of Tears [Sendero de Lágrimas] *s.*, ruta que siguieron los cheroquis durante su forzada emigración desde el sudeste de los Estados Unidos hasta Oklahoma, en la década de 1830

transcontinental [transcontinental] *adj.*, que atraviesa todo un continente

transform [transformar] *v.*, rehacer o cambiar

transition [transición] *s.*, cambio de una actividad o etapa a otra

transpiration [transpiración] *s.*, proceso mediante el cual las plantas y los árboles liberan vapor de agua en el aire

trans-Saharan [transahariano] *adj.*, que atraviesa el desierto del Sahara

Trans-Siberian Railroad [ferrocarril transiberiano] *n.*, el ferrocarril de servicio continuo más largo del mundo, que une Moscú con el este de Rusia, atravesando Siberia

treaty [tratado] *s.*, acuerdo entre dos o más países

Treaty of Guadalupe Hidalgo [Tratado de Guadalupe Hidalgo] *s.*, acuerdo firmado en 1848, por el cual México cedió a los Estados Unidos el área que se extiende desde Texas hasta California

Treaty of Tordesillas [Tratado de Tordesillas] *s.*, tratado firmado entre españoles y portugueses en 1494, a través del cual se dividieron la posesión de las tierras de Sudamérica

Treaty of Versailles [Tratado de Versalles] *s.*, tratado de paz que puso fin, en 1919, a la Primera Guerra Mundial

trench [trinchera] *s.*, zanja extensa que protege a los soldados del fuego enemigo

triangular trade [comercio triangular] *s.*, comercio entre tres continentes: América, Europa y África

tributary [tributario] *s.*, río pequeño que fluye hacia un río más grande

tribute [tributo] *s.*, sumas pagadas a otro país o gobernante a cambio de protección o como muestra de sumisión

troubadour [trovador] *s.*, cantante de la Edad Media que interpretaba canciones sobre caballeros y el amor

tsunami [tsunami] *s.*, ola enorme y muy potente que se forma en el océano

tundra [tundra] *s.*, tierras llanas y sin árboles que se encuentran en las regiones árticas y subárticas

Tupinambá [tupinambás] *s.*, pueblo que vivió cerca de la desembocadura del río Amazonas y a lo largo de la costa atlántica hacia el año 3000 A.C.

tyranny [tiranía] *s.*, gobierno duro o severo

U

United Nations (UN) [Naciones Unidas (ONU)] *s.*, organización de países formada en 1945, con el objetivo de mantener la paz entre los países y proteger los derechos humanos

Universal Declaration of Human Rights [Declaración Universal de los Derechos Humanos] *s.*, acuerdo aprobado por las Naciones Unidas que define los derechos que deben tener todas las personas del mundo

uplands [tierras altas] *s.*, colinas, montañas y mesetas

Ural Mountains [montes Urales] *s.*, cordillera que separa la Llanura del Norte de Europa de la Llanura de Siberia Occidental, en Rusia

utilize [utilizar] *v.*, hacer uso práctico de algo

V

vegetation [vegetación] *s.*, formas de vida vegetal

venue [sede] *s.*, ubicación donde tiene lugar un suceso programado

veto [veto] *v.*, rechazar una decisión tomada por otro órgano de gobierno

viceroy [virrey] *s.*, gobernador de las colonias españolas en América, en representación del rey y la reina de España

volcano [volcán] *s.*, montaña que, al entrar en erupción, explota y lanza roca derretida, gases y ceniza

vulnerable [vulnerable] *adj.*, abierto, que puede ser lastimado por fuerzas externas

W

waterway [vía fluvial] *s.*, ruta navegable que se usa para los viajes y el transporte

weapon of mass destruction (WMD) [arma de destrucción masiva] *s.*, arma que produce un daño inmenso a grandes cantidades de personas

weather [tiempo atmosférico] *s.*, condiciones de la atmósfera en un momento determinado, incluyendo la temperatura, precipitación y humedad de un día o una semana determinada

Western Hemisphere [Hemisferio Occidental] *s.*, la mitad de la Tierra que se encuentra al oeste del primer meridiano

Wet Pampas [Pampa húmeda] *s.*, región húmeda en el este de la Argentina

wind turbine [aerogenerador] *s.*, motor propulsado por el viento para generar electricidad

Y

Yanomami [yanomamis] *s.*, pueblo indígena cuyos habitantes son cazadores recolectores y continúan viviendo en la cuenca del río Amazonas

yurt [yurta] *s.*, carpa tradicional de fieltro de Asia Central

A

B

D

F

G

H

I

M

N

O

P

Q

R

S

T

SKILLS INDEX

Text Acknowledgments

210: Excerpts from *El Libertador: Writings of Simón Bolivar* by Simón Bolívar, edited by David Bushness, translated by Fred Fornoff. Copyright © 2003 by Oxford University Press. Reprinted by permission of Oxford University Press. All rights reserved.

National Geographic School Publishing

National Geographic School Publishing gratefully acknowledges the contributions of the following National Geographic Explorers to our program and to our planet:

Greg Anderson, National Geographic Fellow
Katey Walter Anthony, 2009 National Geographic Emerging Explorer
Ken Banks, 2010 National Geographic Emerging Explorer
Katy Croff Bell, 2006 National Geographic Emerging Explorer
Christina Conlee, National Geographic Grantee
Alexandra Cousteau, 2008 National Geographic Emerging Explorer
Thomas Taha Rassam (TH) Culhane, 2009 National Geographic Emerging Explorer
Jenny Daltry, 2005 National Geographic Emerging Explorer
Wade Davis, National Geographic Explorer-in-Residence
Sylvia Earle, National Geographic Explorer-in-Residence
Grace Gobbo, 2010 National Geographic Emerging Explorer
Beverly Goodman, 2009 National Geographic Emerging Explorer
David Harrison, National Geographic Fellow
Kristofer Helgen, 2009 National Geographic Emerging Explorer
Fredrik Hiebert, National Geographic Fellow
Zeb Hogan, National Geographic Fellow
Shafqat Hussain, 2009 National Geographic Emerging Explorer
Beverly and Dereck Joubert, National Geographic Explorers-in-Residence
Albert Lin, 2010 National Geographic Emerging Explorer
Elizabeth Kapu'uwailani Lindsey, National Geographic Fellow
Sam Meacham, National Geographic Grantee
Kakenya Ntaiya, 2010 National Geographic Emerging Explorer
Johan Reinhard, National Geographic Explorer-in-Residence
Enric Sala, National Geographic Explorer-in-Residence
Kira Salak, 2005 National Geographic Emerging Explorer
Katsufumi Sato, 2009 National Geographic Emerging Explorer
Cid Simoes and Paola Segura, 2008 National Geographic Emerging Explorers
Beth Shapiro, 2010 National Geographic Emerging Explorer
José Urteaga, 2010 National Geographic Emerging Explorer
Spencer Wells, National Geographic Explorer-in-Residence

Photographic Credits

vi (left column, top to bottom) ©Gemma Atwal, ©Ken Banks, ©Rebecca Hale/National Geographic Stock. (right column, top to bottom) ©Christina Conlee, ©Tyrone Turner/National Geographic Stock. (b) ©Kip Evans Photography. vii (left column, top to bottom) ©Rebecca Hale/National Geographic Stock, ©National Geographic Society, Explorer Programs and Strategic Initiatives. (right column, top to bottom) ©Mark Thiessen/National Geographic Stock, ©Rebecca Hale/National Geographic Stock, ©Victor Sanchez de Fuentes. (b) ©Mauricio Ramos. viii (left column, top to bottom) ©Beth Shapiro, ©Victor Sanchez de Fuentes. (right column, top to bottom) ©Rachel Etherington, ©David Evans/National Geographic Society. (b) ©Paul Hoekman/ViaNica.com (c) ©Brian Wallace. 8 (bkg) ©Richard Barnes/National Geographic Stock (bl) ©Mitchell Funk/Photographer's Choice/Getty Images (cl) ©NASA Goddard Space Flight Center. (tl) ©David Evans/National Geographic Society. 10 ©Stephen Alvarez/National Geographic Stock. 11 ©Sunpix Travel/Alamy. 12 (bc) ©Mike Theiss/National Geographic Stock (cl) ©Blakeley/Alamy (cr) ©Michael S. Yamashita/National Geographic Stock (tl) ©John Wark/Wark Photography, Inc. 13 ©Mark Remaley /Precision Aerial Photo. 15 (b) ©Michael S. Yamashita/National Geographic Stock (t) ©Ma Wenxiao/Sinopictures/Photolibrary. 16 ©Susan Byrd/National Geographic My Shot/National Geographic Stock. 20 ©Stephen Alvarez/National Geographic Stock. 23 ©PictureLake/Alamy. 26 ©Brooks Kraft/Corbis. 28 (bc) ©James Forte/National Geographic Stock (bl) ©Kenneth Garrett/National Geographic Stock (br) ©Michael Poliza/National Geographic Stock (bkg) ©National Geographic Maps. 29 (bcl) ©N.C. Wyeth/National Geographic Stock (bcr) ©Justin Guarlglia/National Geographic Stock (bl) ©Kenneth Garrett/National Geographic Stock (br) ©Abraham Nowitz/National Geographic Stock 33 (b) ©David Trood/Getty Images (tr) ©Peter Carsten/National Geographic Stock. 34 ©Andrew Hasson/Alamy. 38 ©George H.H. Huey/Corbis. 41 (tc) ©Images & Volcans/Photo Researchers, Inc. (tr) ©Chris Cheadle/Getty Images. 44 ©Bill Hatcher/National Geographic Stock. 46 (bc) ©Michael Doolittle/Alamy (cl) ©Daniel Dempster Photography/Alamy (tl) ©Tom Bean/Alamy. 47 (tl) ©Frank Krahmer/Corbis (tr) ©imagebroker/Alamy. 50 ©Panoramic Images/Getty Images. 52 ©Michael Nichols/National Geographic Stock. 53 ©Russ Bishop/Alamy. 54 ©Kip Evans Photography. 55 ©NASA Goddard Space Flight Center. 56 (bc) ©Gordon Wiltsie/National Geographic Stock (bl) ©Ralph Lee Hopkins/National Geographic Stock (bkg) ©George Grall/National Geographic Stock. (br) ©Paul Nicklen/National Geographic Stock. 57 (bcl) ©Norbert Rosing/National Geographic Stock (bcr) ©William Albert Allard/National Geographic Stock (bl) ©Stuart Franklin/National Geographic Stock (br) ©Priit Vesilind/National Geographic Stock. 58 (b) ©David R. Frazier Photolibrary, Inc./Alamy (cl) ©Olivier Asselin/Alamy. 59 ©Greg Elms/Lonely Planet Images. 63 ©John Stanmeyer/National Geographic Stock. 64 ©Nic Bothma/epa/Corbis. 65 ©Alain Nogues/Corbis. 67 ©Michael Dunning/Photographer's Choice/Getty Images. 72 ©Romeo Gacad/AFP/Getty Images. 74 (b) ©Tetra Images/Corbis (bkg) ©John Burcham, National Geographic Stock (c) ©Walter

Meayers Edwards, National Geographic Stock (t) ©Beth Shapiro. 77 ©Petra Engle/National Geographic Stock. 78 (b) ©Phil Schermeister/National Geographic Stock (bkg) ©Menno Boermans/Aurora Photos/Corbis (t) ©SeBuKi/ Alamy. 80 ©Mike Grandmaison/Corbis. 82 ©Bill Hatcher/National Geographic Stock. 84 ©Daniel H. Bailey/Corbis. 86 (bkg) ©Jean-Pierre Lescourret/Corbis (cr) ©Mauricio Ramos. 88 ©Jon Arnold Images Ltd/ Alamy. 89 ©American School Private Collection/ Peter Newark American Pictures/The Bridgeman Art Library Nationality. 90 ©Thomas Sbampato/Photolibrary. ©91 ©The Granger Collection. 92 (l) ©The Granger Collection (r) ©photostock1/Alamy. 93 ©Visions LLC/Photolibrary. 95 ©William Manning/Corbis. 96 ©Bettmann/Corbis. 98 (l) ©Corbis (r) ©American School Private Collection/ Courtesy of Swann Auction Galleries/ The Bridgeman Art Library. 99 ©Corbis. 100 ©Bettmann/Corbis. 101 ©Lynn Johnson/National Geographic Stock. 102 ©The Art Archive/Museo Ciudad Mexico/Gianni Dagli Orti. 103 (l) ©Kenneth Garrett/National Geographic Stock (r) ©David R. Frazier Photolibrary, Inc./Alamy. 104 ©The Stapleton Collection/The Bridgeman Art Library. 105 ©The Stapleton Collection/The Bridgeman Art Library. 106 ©The Granger Collection. ©107 (b) ©The Granger Collection (t) ©Look and Learn Magazine Ltd/The Bridgeman Art Library. 108 (l) ©North Wind Picture Archives/Alamy (r) ©Randy Faris/Corbis. 109 (l) ©Corbis (r) ©Charles & Josette Lenars/Corbis. 110 ©North Wind Picture Archives/Alamy. 111 ©Hulton Archive/Getty Images. 114 ©Joe McNally/National Geographic Stock. 116 ©Jennifer Shaffer/National Geographic School Publishing. 117 ©Mike Theiss/National Geographic Society Image Sales. 118 ©Jennifer Shaffer/National Geographic School Publishing/Art Institute of Chicago. 120 ©Car Culture/Corbis. 123 (c) ©James Forte/National Geographic Stock (tr) ©Michael Dunning/Photographer's Choice/Getty Images. 124 ©Marjorie Kamys Cotera/ Daemmrich Photography/The Image Works. 125 (b) ©Robb Kottmyer (t) ©Roger Meno. 126 ©Tono Labra/ Photolibrary. 127 ©Keith Dannemiller/Alamy. 128 ©STR/ Reuters/Corbis. 130 ©Alfredo Guerrero/epa/Corbis. 131 ©Blaine Harrington III/Alamy. 136 ©Corbis Premium RF/ Alamy. 138 (b) ©Georgios Kollidas/Alamy (bkg) ©Raul Touzon, National Geographic Stock (c) ©Stephen Alvarez, National Geographic Stock. 141 ©Konrad Wothe/ Minden Pictures. 142 (b) ©Jon Arnold Images Ltd/Alamy (bkg) ©Menno Boermans/Aurora Photos/ Corbis (t) © Danny Lehman/Corbis. 144 ©Dr. Richard Roscoe/Visuals Unlimited, Inc. 147 © Stuart Westmorland/Corbis. 148 (b) ©Paul Hoekman (t) ©Bryan Wallace. 150 (bc) ©Roy Toft/National Geographic Stock (bl) ©Michael Nichols/National Geographic Stock (bkg) ©Paul Sutherland/National Geographic Stock (br) ©Steve Winter/National Geographic Society Image Sales. 151 (bcl) ©Michael Melford/National Geographic Stock (bcr) ©Bobby Haas/National Geographic Stock (bl) ©Christian Ziegler/National Geographic Stock (br) ©Roy Toft/ National Geographic Stock. 152 ©Steve Winter/National Geographic Stock. 155 ©The Bridgeman Art Library. 156 (br) ©Georgios Kollidas/Alamy (r) ©Hemis /Alamy. 157 ©Stuwdamdorp/Alamy. 158 (bl) ©Creativ Studio Heinemann/Westend61/Corbis (c) ©INTERFOTO /Alamy. 159 (br) ©Reuters/Corbis (tl) ©Bettmann/Corbis (tl) ©Creativ Studio Heinemann/Westend61/Corbis. 162 ©Walter Bibikow/JAI/Corbis. 164 ©Martin Gray/National Geographic Stock. 165 ©Danita Delimont/Alamy. 166 ©Nico Tondini/Photolibrary. 167 ©Rick Gerharter/Lonely Planet Images. 168 ©National Geographic Stock. 169 ©Frans Lanting/Corbis. 170 ©Photolibrary. 172 ©Yuan Man/Xinhua Press/Corbis. 173 ©Logan Abassi/UN Handout/Corbis. 174 ©JS Callahan/tropicalpix/Alamy. 175 ©Michael Dunning/Photographer's Choice/Getty Images. 176 ©Lonely Planet Images /Alamy. 178 (bkg) ©Christian Heeb/Aurora Photos (c) ©Roy Toft/National Geographic Stock (r) ©Micahel & Patricia Fogden/ Minden Pictures/National Geographic Stock. 179 ©Arterra Picture Library/Alamy. 183 (l) ©Jacques Marais/ Getty Images (r) ©Danny Lehman/Corbis. 184 ©Christian Zeigler/National Geographic Stock. 186 (b) ©Frans Lanting/Corbis (bkg) ©Rod Smith/National Geographic My Shot/National Geographic Stock (cl) ©David R. Frazier Photolibrary, Inc./Alamy (tl) ©Photograph by Victor Sanchez de Fuentes. 189 ©Nick Gordon/Oxford Scientific (OSF)/Photolibrary. 190 (bkg) ©Menno Boermans/Aurora Photos/Corbis (bl) ©Colin Monteath/ Minden Pictures/National Geographic Stock (br) ©John Eastcott and Yva Momatiuk/National Geographic Stock. 192 ©Ivan Kashinsk/National Geographic Stock. 196 (bkg) ©Melissa Farlow/National Geographic Stock (br) ©Aldo Sessa/Tango Stock/Getty Images. 199 (br) ©Michael Nichols/National Geographic Stock (tr) ©Michael Dunning/Photographer's Choice/Getty Images. 200 ©Ethan Welty/ Aurora Photos/Alamy. 201 ©Gnter Wamser/F1online digitale Bildagentur GmbH/Alamy. 202 ©Christina Conlee. 203 (b) ©Christina Conlee (t) ©Robert Clark/National Geographic Stock. 206 (bc) ©Maria Stenzel/National Geographic Stock (c) ©Peruvian School/Museo Arqueologia, Lima, Peru/Boltin Picture Library/The Bridgeman Art Library International (tr) ©McConnell, James Edwin/Private Collection/Look and Learn/The Bridgeman Art Library International. 207 ©Cro Magnon/Alamy. 208 (bc) ©The Art Archive/ Bibliothque des Arts Dcoratifs Paris/Gianni Dagli Orti (tr) ©The Art Archive/Bibliothque des Arts Dcoratifs Paris/Gianni Dagli Orti. 209 ©The Art Archive/ Kharbine-Tapabor. 211 ©Luis Marden/National Geographic Stock. 215 ©Florian Kopp/imagebroker/ Alamy. 216 ©Paolo Aguilar/epa/Corbis. 219 ©Corey Wise/Lonely Planet Images/Getty Images. 220 ©Mike Theiss/National Geographic Stock. 221 ©Richard Nowitz/National Geographic Stock. 223 ©Ivan Alvarado/ Reuters/Corbis. 224 ©Jeremy Hoare/Alamy. 225 (b) ©James P. Blair/National Geographic Stock (c) ©Nicolas Misculin/Reuters (t) ©Kit Houghton/Corbis. 226 ©Jennifer Shaffer/National Geographic School Publisingg. 228 ©Keren Su/Corbis. 229 (t) ©Imaginechina/Corbis. 231 ©Robert Clark/National Geographic Stock. 232 ©Sebastiao Moreira/epa/Corbis. 234 (b) ©Mike Theiss/National Geographic Stock (bl) ©Charles Dharapak/Pool/Reuters. 240 ©Pete McBride/ National Geographic Stock. 242 (b) ©Bob Krist, National Geographic Stock (bkg) ©Richard List, Corbis (c) ©Anne Keiser, National Geographic Stock. 245 ©Atlantide Phototravel/Corbis. 246 (b) ©Yann Arthus-Bertrand/ Corbis (bkg) ©Menno Boermans/Aurora Photos/Corbis (t) ©Douglas Pearson/Corbis. 248 ©Owi-Diasign/ Photolibrary. 250 ©National Geographic Stock. 252 ©Octavio Aburto. 254 (bc) ©Anne Keiser/National Geographic Stock (bl) ©Agnieszka Pruszek/National Geographic My Shot/National Geographic Stock (bkg) ©Jim Richardson/National Geographic Stock (br) ©Steve Raymer/National Geographic Stock. 255 (bcl) ©Greg Dale/National Geographic Stock (bcr) ©James P. Blair/ National Geographic Stock (bl) ©Richard Nowitz/ National Geographic Stock (br) ©James L. Stanfield/ National Geographic Stock. 256 ©Panoramic Images/ National Geographic Stock. 258 (l) ©Richard Nowitz/ National Geographic Stock (r) ©The Art Gallery Collection/Alamy. 259 ©PoodlesRock/Corbis. 260 ©Jean-Pierre Lescourret/Corbis. 262 (l) ©Hoberman

Collection/Corbis (r) ©The Bridgeman Art Library International. 263 ©North Wind Picture Archives/Alamy. 266 (bl) ©The Bridgeman Art Library (br) ©Underwood & Underwood/Corbis. 267 (bl) ©Doug Taylor/Alamy (br) ©The Bridgeman Art Library (t) ©The Bridgeman Art Library. 268 ©Richard Schlect/National Geographic Stock. 270 ©Paul Thompson/Corbis. 272 (l) ©The Bridgeman Art Library (r) ©The Gallery Collection/ Corbis. 273 (b) ©Peter Horree/Alamy (t) ©The Bridgeman Art Library. 274 ©The Bridgeman Art Library. 275 ©SCANFOTO/X00729/Reuters/Corbis. 276 (l) ©Stefano Bianchetti/Corbis (r) ©Clynt Garnham/Alamy. 277 (l) ©Michael Nicholson/Corbis (r) ©Michael Nicholson/ Corbis. 279 ©DC Premiumstock/Alamy. 282 ©Rudy Sulgan/Corbis. 284 ©MARKA /Alamy. ©286 (l) ©Leonardo da Vinci (1452-1519) Louvre, Paris, France/ Giraudon/ The Bridgeman Art Library (r) Claude Monet (1840-1926) Musee Marmottan, Paris, France/ Giraudon/ The Bridgeman Art Library Nationality. 287 ©Arnaud Chicurel/Hemis/Corbis. 288 ©The Gallery Collection/ Corbis. 289 ©Columbia/The Kobal Collection. 290 ©Jon Arnold/JAI/Corbis. 291 ©Sergiy Koshevarov/ StockPhotoPro. 292 ©Photolibrary. 294 ©Paul Seheult/ Eye Ubiquitous/Corbis. 295 ©Perutskyi Petro/ Shutterstock Photos. 296 ©Gregory Wrona/Alamy. 299 ©Michael Dunning/Photographer's Choice/Getty Images. 304 ©Grand Tour/Corbis. 306 (b) ©Gerd Ludwig/Corbis (bkg) ©Photolibrary (c) ©Gordon Wiltsie, National Geographic Stock (t) ©Rebecca Hale, National Geographic Stock. © 309 ©Klaus Nigge/National Geographic Stock. 310 (bkg) ©Menno Boermans/Aurora Photos/Corbis (br) ©Bruno Morandi/Robert Harding World Imagery/Corbis (tl) ©Maxim Toporskiy/Alamy. 312 ©Denis Sinyakov/Reuters/Corbis. 314 ©Cary Wolinsky/ National Geographic Stock. 316 ©National Geographic Stock. 318 ©Gerd Ludwig/National Geographic Stock. 319 (tl) ©U.S. Geological Survey (tr) ©NASA. 320 (l) ©Sisse Brimberg/National Geographic Society (r) ©James L. Stanfield/National Geographic Society. 321 (l) ©Massimo Pizzotti/Getty (r) ©Dallas and John Heaton/ Photolibrary. 322 ©Richard Klune/Corbis. 323 ©imagebroker/Alamy. 324 ©The Bridgeman Art Library (r) ©Cary Wolinsky/National Geographic Society. 325 (l) ©The Bridgeman Art Library. 326 ©North Wind Picture Archives/Alamy. 328 (l) ©Bettmann/Corbis (r) ©The Art Archive. 329 (b) ©Bettmann/Corbis (t) ©Thomas Johnson/Sygma/Corbis. 332 ©Paul Harris/JAI/Corbis. 334 ©Arne Hodalic/Corbis. 335 (tc) ©Michael Runkel/Robert Harding World Imagery/Corbis (tr) ©Maria Stenzel/ National Geographic Stock (tr) ©Sean Sprague/ Photolibrary. 336 ©Olaf Meinhardt/VISUM /Fotofinder. 339 ©Kristel Richard/Grand Tour/Corbis. 340 ©Shepard Sherbell/CORBIS SABA. 342 ©Imagesource/Photolibrary. 344 ©Oleg Nikishin/Stringer/Getty Images. 347 (c) ©iStockphoto ©Michael Dunning/Photographer's Choice/Getty Images. 352 ©iStockphoto. 354 (b) ©Ingo Arndt/Minden Pictures/National Geographic Stock (bkg) ©David Alan Harvey/National Geographic Stock (c) ©Mitsuaki Iwago/Minden Pictures/National Geographic Stock (t) ©Kakenya Ntaiya. 357 ©Top-Pics TBK/Alamy. 358 (bkg) ©Menno Boermans/Aurora Photos/Corbis (bl) ©tbkmedia/Alamy (tr) ©Michael Poliza/National Geographic Stock. 360 ©Michael Nichols/National Geographic Stock. 362 ©Philippe Bourseiller/Getty Images. 364 ©Ian Nichols/National Geographic Stock. 366 ©Mike Hutchings/Reuters. 368 (bc) ©Beverly Joubert/National Geographic Stock (bl) ©Beverly Joubert/ National Geographic Stock. 370 ©Gerald Hoberman/ Hoberman Collection UK/Photolibrary. 372 (bl) ©The Trustees of the British Museum/Art Resource (br) ©ADB Travel/dbimages/Alamy. 373 ©HIP/Art Resource. 376 ©Private Collection/Look and Learn/The Bridgeman Art Library International. 377 (bl) ©Mary Evans Picture Library/The Image Works (br) ©Bruce Dale/National Geographic Stock. 378 (bc) ©James L. Stanfield/National Geographic Stock (bl) ©Tim Laman/National Geographic Stock (bkg) ©George Steinmetz/National Geographic Stock (br) ©Annie Griffiths/National Geographic Stock. 379 (bcl) ©Roy Toft/National Geographic Stock (bcr) ©Tino Soriano/National Geographic Stock (bl) ©Jodi Cobb/National Geographic Stock (br) ©Ed Kashi/ National Geographic Stock. 382 ©Ralph Lee Hopkins/ National Geographic Stock. 384 ©Vanessa Burger/Images of Africa Photobank Alamy. 386 ©Paul Gilham - FIFA/ FIFA via Getty Images. 387 ©David Alan Harvey/ National Geographic Stock. 388 (bc) ©Nigel Pavitt/John Warburton-Lee Photography/Alamy (bl) ©Sean Sprague/ Still Pictures/Photolibrary. 389 (cl) ©Michael Nichols/ National Geographic Stock (tr) ©Suzi Eszterhas/Minden Pictures/National Geographic Stock. 390 ©Jane Goodall Institute. 391 (bkg) ©Gerry Ellis/ Minden Pictures/ National Geographic Stock (br) ©Wade Davis/Ryan Hill. 392 ©Finbarr O'Reilly/Reuters. 394 ©George Steinmetz/ Corbis. 396 (b) ©Pascal Maitre /National Geographic Stock (tr) ©Joerg Boethling/Alamy. 398 ©Louise Gubb/ Corbis. 399 ©Michael Dunning/Photographer's Choice/ Getty Images. 400 ©Frederic Courbet/Still Pictures/ Photolibrary. 402 ©Ulrich Doering/Alamy. 403 ©Trinity Mirror/Mirrorpix/Alamy. 406 (bl) ©Chris Stenger/FN/ Minden Pictures/National Geographic Stock (br) ©Walker, Lewis W./National Geographic Stock. 407 (bkg) ©Clement Philippe/Arterra Picture Library/Alamy (bkg) ©Tim Fitzharris/Minden Pictures/National Geographic Stock (cf) ©Mattias Klum /National Geographic Stock (l) ©Tom Vezo/Minden Pictures/National Geographic Stock (rbkg) ©Ted Wood/Aurora Photos (rf) ©Thomas Lehne/ Alamy. 408 ©Anup Shah/Corbis. R39 ©Fresco J Linga/My Shot/National Geographic Stock. R40 ©George Grall/ National Geographic Stock. R41 ©SERDAR/Alamy. R42 ©Bettmann/Corbis. R44 ©Hulton Archive/Getty Images. R45 ©Svabo/Alamy. R47 ©Scott Olson/Getty Images. R49 ©Lordprice Collection/Alamy. R50 ©Parbul TV via Reuters TV/Reuters/Corbis. R51 ©Markus Altmann/ Corbis. R52 (b) ©Mario Lopez/epa/Corbis (t) ©Alinari Archives/Corbis. R53 (b) ©Anindito Mukherjee/epa/ Corbis (t) ©Werner Forman/Art Resource. R54 (b) ©Ronen Zvulun/Reuters/Corbis (t) ©The Art Archive/Museo del Prado Madrid. R55 (b) ©Mak Remissa/epa/Corbis (t) ©Rubin Museum of Art/Art Resource. R56 (b) ©Joe McNally/National Geographic Stock (t) ©Biju/Alamy. R57 (b) ©Robert Harding World Imagery/Corbis (t) ©Art Directors & TRIP/Alamy. R58 (b) ©Christian Kober/ Photolibrary (t) ©National Palace Museum Taiwan/The Art Archive.

Map Credits

Mapping Specialists, LTD., Madison, WI.
National Geographic Maps, National Geographic Society

Illustrator Credits

Precision Graphics

NATIONAL GEOGRAPHIC
World Cultures and Geography
GEO

Southeast Asia North America

Russia & the Eurasian Republics

Southeast Asia

Central America & the Caribbean

Africa

Australia, the Pacific Realm & Antarctica

South Asia

Europe

Southeast Asia

Central America & the Caribbean

East Asia

North America

Southwest Asia & North Africa

AFRICA Southeast Asia

AUSTRALIA, THE PACIFIC REALM & ANTARCTICA

East Asia

EUROPE

SOUTHWEST ASIA & NORTH AFRICA

SUB-SAHARAN AFRICA

SOUTH AMERICA

South Asia

Central America & the Caribbean

East Asia

NORTH AMERICA

SOUTHWEST ASIA & NORTH AFRICA

Southeast Asia

North America

East Asia

Russia & the Eurasian Republics

Europe

Southeast Asia

Central America & the Caribbean

Subsaharan Africa

Australia, the Pacific Realm & Antarctica

South Asia

Southeast Asia

Central America & the Caribbean

East Asia

North America

Southwest Asia & North Africa

North America

HEAD

AFRICA Southeast Asi

Australia,
the Pacific Realm
& Antarctica

East
Asia

South America

Southwest Asia & North Africa

Europe

Europe

Southwest Asia
& North Africa

Southeast As

Central Amer
& the Carib

Sub-Saharan
Africa

South
America

Austra
the Pacific
& Ant

Russia
& the Eurasian
Republics

East
Asia

Australia,
the Pacific Realm
& Antarctica

South A

Southeast

Central Amer
& the Carib

East
Asia

North
Ameri

South
Asia

Southwest As
& North Afric

North Am

utheast Asia

North America

ast sia

South America

Southwest Asia & North Africa

Europe

Russia & the Eurasian Republics

pe

sia

ca

Southeast Asia

Central America & the Caribbean

aharan

rica

Australia, the Pacific Realm & Antarctica

Europe

South Asia

South Asia

Southeast Asia

Central America & the Caribbean

the Pacific Realm & Antarctica

East Asia

North America

Sub-Saharan Africa

Southwest Asia & North Africa

South America

North America